Circle

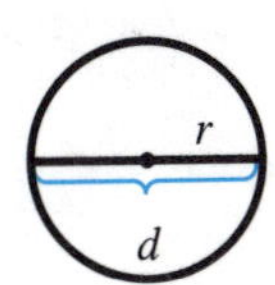

r is the radius $\quad$ d is the diameter

$d = 2r$

Area $A = \pi r^2$

Circumference $C = 2\pi r$ or $C = \pi d$

lid

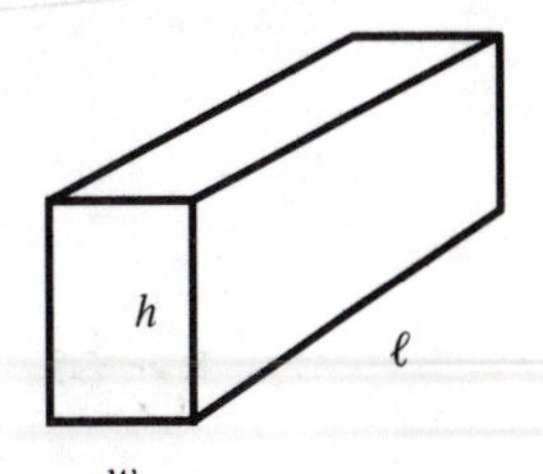

Volume $V = w\ell h$

Sphere

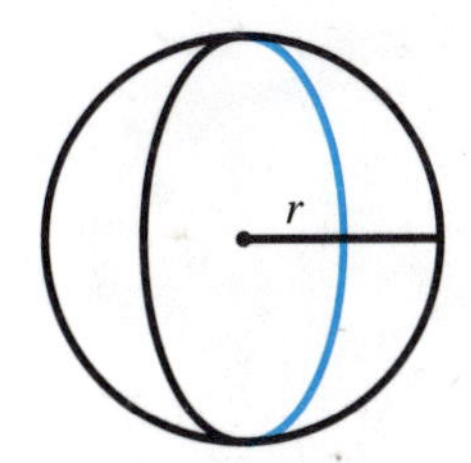

Volume $V = \dfrac{4}{3}\pi r^3$

Right Circular Cylinder

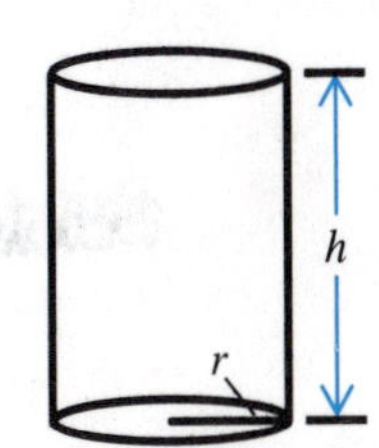

Volume $V = \pi r^2 h$

Surface Area $S = 2\pi r^2 + 2\pi rh$

Right Circular Cone

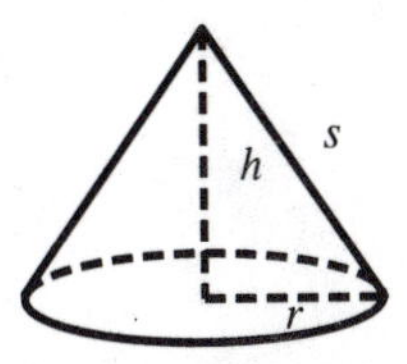

Volume $V = \dfrac{1}{3}\pi r^2 h$

Surface Area $A = \pi r^2 + \pi rs$

Right Pyramid

Volume $V = \dfrac{1}{3} Bh$

where B is the area of the base

ELEMENTARY

Algebra

Second Edition

Larry R. Mugridge *Kutztown University*

ELEMENTARY Algebra

Second Edition

SAUNDERS COLLEGE PUBLISHING

HARCOURT BRACE COLLEGE PUBLISHERS

Fort Worth Philadelphia San Diego New York Orlando Austin San Antonio Toronto Montreal London Sydney Tokyo

Text Typeface: 10/12 Times Roman
Compositor: York Graphic Services, Inc.
Acquisitions Editor: Deirdre A. Lynch
Developmental Editor: Ellen Newman
Managing Editor: Carol Field
Project Editor: York Production Services
Manager of Art and Design: Carol Bleistine
Art and Design Coordinator: Christine Schueler
Text Designer: Tracy Baldwin
Cover Designer: Lawrence R. Didona
Text Artwork: York Graphic Services, Inc.
Director of EDP: Tim Frelick
Production Manager: Carol Florence
Marketing Manager: Monica Wilson

Cover Credit: W. Cody/Westlight

Printed in the United States of America

ELEMENTARY ALGEBRA, second edition

ISBN: 0-03-072991-2

Library of Congress Catalog Card Number: 93-48371

4567 048 987654321

This book is dedicated to my children,
Lyn and Jason

Preface

This book is intended for students who have no background in elementary algebra, but who have had some experience with arithmetic. Each concept is introduced by example before the definition is given, and the student is encouraged to become an active participant. A complete pedagogical system that is designed to motivate the student and make mathematics more accessible includes the following features:

Key Features

Chapter Overviews Each chapter begins with an Overview to introduce the student to the material and demonstrate its relevance to the real world.

Objectives Each section contains Objectives to help the student focus on skills to be learned in the chapter.

Learning Advantages These hints occur in Chapters 1 through 7 to give students additional help; they immediately follow the objectives of each section. For more detailed explanations on studying algebra, I recommend the pamphlet *How to Study Mathematics* by James Margenau and Michael Sentlowitz (The National Council of Teachers of Mathematics, Inc., Reston, Virginia, 1977, from which the Learning Advantages are adapted).

Examples with Solutions Approximately 700 examples with complete, worked-out solutions are provided; explanations are highlighted in blue print. Some of the examples contain steps that can be done mentally once the student has achieved a certain level of proficiency (although performing these steps mentally may not be possible at first);

a blue dashed box calls the student's attention to this portion of the equation. A yellow triangle indicates the end of the solution.

Common Error Boxes

Wherever appropriate, students are shown errors that are commonly made. The correct approach is then illustrated and explained.

Strategies

After developing a mathematical technique, such as factoring a trinomial, a summary of the technique is given as a strategy. This will provide the student with a reference when doing problems from the exercise set.

Graphing Calculator Crossovers

Many of the examples are followed by problems that are specifically designed to be solved with a graphing calculator. These problems are similar to the examples and illustrate the role of the graphing calculator in solving problems. Graphing Calculator Crossovers are set off by a symbol for easy reference, and may be used at the instructor's discretion.

Warming Up

Preceding each set of exercises, these true-false questions can be used as a classroom activity.

Exercises

Graded by level of difficulty, the exercises are carefully constructed so that there is odd-even pairing. In addition, each exercise problem is related to an example from the section. The numbers used in the examples and exercises have been carefully selected so that the student becomes confident in using fractions and decimals in addition to the integers. The numbers, however, do not become so complicated as to ''turn off'' the student or teacher. All problems have been thoroughly reviewed for accuracy.

Some of the exercises are specifically designed to be solved with either a calculator or graphing calculator. They may be assigned at the instructor's discretion and are marked with a symbol for identification. There are also problems in each section called ''SAY IT IN WORDS'' that require students to express mathematics in written form.

Beginning with Chapter 2, each exercise set contains **Review Exercises** to help the student understand and remember concepts studied in earlier sections. The **Enrichment Exercises** at the end of each section are challenging problems that the more capable students will be able to solve. The Answer Section in the back of the book contains answers to the odd-numbered exercises and the Review Exercises and all of the Enrichment Exercises.

Applications

Each chapter begins with a real-world photograph and application posing a problem that is solved later in the chapter. Whenever appropriate, word problems have been included throughout the text so that the student constantly practices translating word phrases and statements into mathematical equations. The word problems are realistic and are culled from applications in geometry, physics, business, economics, history, and psychology. The structure of problem solving is emphasized throughout the text, beginning with the simple conversion of word phrases into mathematical expressions. Many concepts introduced in early chapters are repeated using different applications.

Team Project To emphasize cooperative learning skills, each chapter includes a research project that encourages students to work together.

Chapter Summary and Review Each chapter concludes with a comprehensive review which includes definitions and strategies learned in the chapter. All terms are keyed to the section number for easy reference. Marginal examples are included to further illustrate the concepts reviewed.

Review Exercise Set These exercises are included at the end of the chapter and represent all types of problems to ensure that the student has attained a level of proficiency and is comfortable proceeding to the next chapter. All exercises are keyed to sections so the student can refer to the text for assistance. Answers to all of these exercises can be found`in the back of the book.

Chapter Test Each chapter concludes with a chapter test. The answers to all problems begin on page A.1.

Pedagogical Use of Color This text uses color in the figures and to highlight the various pedagogical features throughout the text. Multiple colors are useful, for example, to distinguish between two lines that are graphed simultaneously, or to highlight important statements. The complete color system is described in more detail on page xvii.

Changes in the Second Edition

A number of changes and improvements have been made in preparing the new edition of this text. Many of these changes are in response to comments and suggestions offered by users and reviewers of the manuscript. The following represent the major changes in the second edition:

1. Approximately 30% of the exercises have been changed, for a total of almost 1500 new problems. We have also added problems in each section that require students to express mathematics in written form; this section is called "Say It In Words." All of the exercises have been thoroughly reviewed for accuracy.
2. Each exercise set is now preceded by Warming Up, a group of true-false questions that can be used as a classroom activity.
3. New Graphing Calculator Crossovers in appropriate chapters contain explanation and examples specifically designed to be solved with a graphing calculator to enhance the graphing of functions and conic sections. They are optional and may be assigned at the discretion of the instructor. The Calculator

Crossovers from the first edition have been moved to the end-of-section exercises, which also contain exercises for the graphing calculator; each type of calculator exercise is marked with its own symbol for easy reference.

4. Mental Math symbols have been deleted. A blue dashed box now indicates the portion of the equation that can be done mentally once the student has achieved a certain level of proficiency.
5. Each chapter now begins with a "Connection," a real-world application and photograph posing a problem that is solved later in the chapter.
6. A team project has been added to each chapter to emphasize cooperative learning skills.
7. Some sections have been streamlined and combined (e.g., addition and multiplication properties of equalities in Sections 2.2 and 2.3; solving inequalities in Sections 2.7 and 2.8; and multiplication and division properties of exponents in Sections 4.1 and 4.2).
8. Coverage of Linear Equations and Their Graphs has been moved from Chapter 6 to Chapter 3 to give students earlier exposure to real-world applications involving linear equations in two variables.
9. Treatment of greatest common factor and factoring by grouping has been expanded and moved to the beginning of Chapter 5.

An Overview of the Book

The main thrust of this book is to enable students to develop algebraic skills to be used to solve word problems. This is the primary aim of each chapter; we start with developing a skill and then use it to solve word problems. In addition, many geometric problems are included. A chapter-by-chapter overview follows:

Chapter 1: Operations and Variables

Since the foundation of algebra is arithmetic, it is reviewed comprehensively in this chapter, starting with addition, subtraction, multiplication, and division of fractions. Next, we review basic symbols and order of operations. The set of real numbers is examined along with the subsets of natural, whole, rational, and irrational numbers. Then, we review the four basic operations on real numbers: addition, subtraction, multiplication, and division. In Section 1.6, we study variables and variable expressions. We then introduce the important concept of writing verbal statements as algebraic expressions. In the final section we deal with the properties of real numbers.

Chapter 2: Linear Equations and Inequalities

In this chapter, algebraic expressions are used to construct linear equations and inequalities. In Section 2.2, we investigate properties of equality that will be used to solve linear equations, which constitutes Section 2.3. The skill of translating word problems to mathematics and applying these skills to word problems that require solving linear equations is developed. In Section 2.6 we solve linear inequalities, and, in the final section, we solve word problems using formulas from business (simple interest, cost-revenue-profit) as well as from geometry and trigonometry (area and volume formulas of regions and solids, and the sum of the three angles of a triangle being 180°).

Chapter 3: Linear Equations and Their Graphs

In this chapter, we start the investigation of the connection between analytic geometry and its applications by developing linear equations and their graphs. In Section 3.2 we find the slope of a line given two points on the line and the slope of a line from the graph. The difference between zero slope and undefined slope is explained, and the geometric meaning of positive and negative slope is discussed. In the next section, we find the equation of a line given (a) the slope and y-intercept, (b) the slope and a point on the line, or (c) two points on the line. In Section 3.4 graphing linear inequalities and using linear inequalities in applications are illustrated. Many applications of linear equations are included, such as linear depreciation, linear cost-revenue-profit, and linear supply and demand equations. In the last section we discuss relations and functions, establishing the vertical line test to determine when a relation is also a function. The domain and range of a function are defined, and the $f(x)$ notation is introduced.

Chapter 4: Exponents and Polynomials

In this chapter we study properties of exponents and simplify expressions using $a^m a^n = a^{m+n}$ and $(a^m)^n = a^{mn}$. We use these rules to introduce scientific notation. The remainder of the chapter deals with the algebra of polynomials, beginning with multiplying and dividing monomials. In the remaining sections we introduce polynomials, discussing the types of polynomials, the degree of a polynomial, and how to simplify polynomials. The remainder of the chapter covers the addition, subtraction, multiplication, and division of polynomials.

Chapter 5: Factoring

Continuing our study of polynomials, we develop techniques to factor them in this chapter. Coverage includes factoring by grouping, factoring of monomials, general trinomials, special polynomials such as the difference of two squares, perfect square trinomials, and the sum or difference of two cubes. We then state a strategy for general factoring. These techniques are then used to solve quadratic equations that are factorable, which, in turn, leads to applications of quadratic equations. The applications include using the Pythagorean Theorem, solving geometric problems, and understanding applications to business.

Chapter 6: Rational Expressions

In the first section, we define rational expressions, evaluate them, and determine where a rational expression is undefined. Next, we multiply and divide rational expressions, putting the answer in lowest terms. In Section 6.3 we add and subtract rational expressions. Complex fractions are discussed in the next section, where we develop methods to simplify complex fractions into rational expressions. We then solve equations containing rational expressions and introduce formulas that involve rational expressions such as the total resistance formula. Ratio and proportion are covered, showing how proportions are used to solve problems involving similar triangles. The final topic of the chapter is applications of rational expressions in which direct and inverse variation, along with various applications of these concepts, are developed.

Chapter 7: Systems of Equations

In this chapter, we continue to investigate the connection between analytic geometry and its applications. We show that with a system of two equations in two variables we are able to study more complex applications. We start our study of systems of equations by finding solutions by graphing, particularly in the following two cases: (a) when the system has no solutions and (b) when a system has infinitely many solutions. In the next two sections we use two algebraic methods for solving a system: the elimination method and the substitution method. In the last two sections, the technique of graphing linear inequalities is developed and then used to solve word problems using systems of linear inequalities. Many of the problems show the initial steps to solve linear programming problems in two variables by the graphing method.

Chapter 8: Roots and Radicals

We return to the concept of exponents and properties such as $a^m a^n = a^{m+n}$, $(a^m)^n = a^{mn}$, and $a^{-n} = 1/a^n$, where m and n are integers. In this chapter, we define numbers raised to rational powers, connecting this concept with taking the nth root of a number. In Section 8.2, we simplify radical expressions, and in the next two sections we study the algebra of radical expressions. In Section 8.5, radical equations and word problems using radical equations are solved; the connection is then made between taking roots and rational exponents. This chapter concludes by defining a complex number in terms of i and covering the algebra of complex numbers.

Chapter 9: Quadratic Equations

One of the main goals of this book is developing techniques for solving equations. In the last chapter we return to the problem of solving the general quadratic equation $ax^2 + bx + c = 0$, where the trinomial may not be factorable. We begin with a discussion of using the square root method for solving quadratic equations and then develop the completing of the square method. The quadratic formula is developed in Section 9.3 and then used to find any complex solutions to quadratic equations.

Ancillary Package

The following supplements are available to accompany this text:*

*For the instructor, we have an Instructor's Manual with solutions to all the exercises; Prepared Tests with six tests for each chapter, as well as midterm and final examinations and a Diagnostic test; a Computerized Test Bank for the IBM and Macintosh computers; and a Printed Test Bank containing tests generated from the Computerized Test Bank.

Student Solutions Manual and Study Guide (Linda Holden, Indiana University)

Contains step-by-step solutions to one fourth of the problems in the exercise sets (every other odd-numbered problem) in addition to providing the student with a short summary of the important concepts in each chapter.

State Requirements

Particular attention has been paid to the testing requirements for various states (e.g., TASP, ELM, CLAST, etc.) Please see pages xix–xxii for additional information on how this text meets the requirements for your state.

Videotapes

A complete set of videotapes (18 hours), free to adopters, has been created and scripted by the author. Keyed to the text, the videotapes explain, using computer graphics, examples with corresponding practice problems. The student can participate by stopping the tape to work the practice problems on his or her own and then checking the solutions by continuing the tape. The tapes are available in VHS format and include a list of topics and the amount of time spent on each section.

MathCue Tutorial Software (George W. Bergeman, Northern Virginia Community College)

Available for both Macintosh and IBM, this software allows students to test their skills and to pinpoint and correct weak areas. MathCue presents worked examples and displays annotated, step-by-step solutions to problems answered incorrectly. Students may also choose to see the solutions to problems answered correctly and to view a partial solution if they need help to begin solving a problem. The software includes two useful review capabilities: Missed Problems Review and Disk Review. As the student works, MathCue keeps track of problems that are answered incorrectly. When the work on a topic is completed, the students are given the option of reviewing all problems answered incorrectly. In addition, the Disk Review feature provides a quick and efficient review of all the topics on the disk.

MathCue Solution Finder Software

This expert system approach allows students to enter problems, receive help, and check their answers as if they were working with a tutor. Software tracks students' progress and refers them to sections of the text for help. Available for IBM and Macintosh.

Acknowledgments

This text was prepared with the assistance of many professors who reviewed the manuscript throughout the course of its development. I wish to acknowledge the following people and express my appreciation for their suggestions, criticisms, and encouragement:

Ronald E. Bailey, Pennsylvania State University, York Campus
Helen Burrier, Kirkwood Community College

Don Chandler, El Paso Community College
Carol DeVille, Louisiana Tech University
Arthur P. Dull, Diablo Valley College
Robert Eicken, Illinois Central College
Mark Gidney, Lees McRae College
Jack Gill, Miami-Dade Community College
Debra Hall, Indiana University at Fort Wayne
Linda Holden, Indiana University at Bloomington
Nancy Hyde, Broward Community College
Bill Jordan, Seminole Community College
Lloyd Koontz, Eastern Illinois University
J. Stanley Laughlin, Idaho State University
Beverly Mick, Cuesta College
Len Mrachek, Hennepin Technical Center
Catherine B. Pace, Louisiana Tech University
Thomas Ribley, Valencia Community College
G. Thomas Riggle, Cuyahoga Community College
John Snyder, Sinclair Community College
Lora Stewart, Consumnes River College
Kathleen Conway Stiehl, University of Missouri
June Oliverio-Wallace, Montgomery County Community College
Kenneth J. Word, Central Texas College

I would also like to thank the following reviewers who were invaluable in ensuring the accuracy of the text: **June Oliverio-Wallace,** Montgomery County Community College, and **Debra Hall,** Indiana University at Fort Wayne. I would also like to thank **Regina Brunner** for her contribution to the text, and **Melissa Handy** and **Robert Gemin** for their excellent photographs.

Special thanks to the following people for their work on the various ancillary items that accompany *Elementary Algebra,* second edition:

George W. Bergeman, Northern Virginia Community College (MathCue Tutorial Software and MathCue Solution Finder)

Roger Willig, Montgomery County Community College (Prepared Tests portion of the Instructor's Testing Manual)

Mary Chabot, Mt. San Antonio College (Computerized Test Bank—software and printout in the Instructor's Testing Manual)

Linda Holden, Indiana University at Bloomington (Student Solutions Manual and Study Guide)

Debra Hall, Indiana University at Fort Wayne (Instructor's Resource Manual)

I want to thank the dedicated staff at Saunders College Publishing for their tireless energy: Deirdre Lynch, Mathematics Editor, with her insightful analysis of the situation; Ellen Newman, Developmental Editor, who spent long hours, that frequently shortened the evening, on the book; and Susan Bogle, Project Editor with York Production Services, who guided the book to a successful conclusion with her expertise in converting the manuscript to a polished full-color book.

Larry R. Mugridge
Kutztown University

Pedagogical Use of Color

The various colors in the text figures are used to improve clarity and understanding. Many figures with three-dimensional representations are shown in various colors to make them as realistic as possible. Color is used in those graphs where different lines are being plotted simultaneously and need to be distinguished.

In addition to the use of color in the figures, the pedagogical system in the text has been enhanced with color as well. We have used the following colors to distinguish the various pedagogical features:

PROPERTY

STRATEGY

GRAPHING CALCULATOR CROSSOVER

DEFINITION

COMMON ERROR

RULE

Table No. Table Title

Col. Head	*Col. Head*

How Text Meets State Requirements

TASP Skills

The following table lists the Texas TASP skills and their location in *Elementary Algebra.*

Description of TASP Skill	***Chapter, Section***
Use number concepts and computation skills	1.1, 1.2, 1.3, 1.4, 1.5
Solve word problems involving integers, fractions, or decimals (including percents, ratios, and proportions)	1.1, 1.2, 1.3, 1.4, 1.5
Interpret information from a graph, table, or chart	3.1, 3.2, 3.3, 3.4
Graph numbers or number relationships	1.3
Solve one- and two-variable equations	2.2, 2.3, 3.3
Solve word problems involving one and two variables	2.4, 2.5, 2.7, 3.4, 5.7, 6.7
Understand operations with algebraic expressions	1.6, 2.1, 2.2, 2.7, 4.1, 4.3, 4.4, 4.5, 4.6, 4.7
Solve problems involving quadratic equations	5.7
Solve problems involving geometric figures	6.6
Apply reasoning skills	1.7, 2.4, 2.5, 2.7, 3.4, 5.7, 6.7 All Enrichment Exercises

ELM Skills

The following table lists the California ELM skills and the location of those skills in *Elementary Algebra.*

Description of ELM Skill	*Chapter, Section*
Whole numbers and their operations	1.2–1.5
Fractions and their operations	1.1, 1.3
Exponentiation and square roots	1.2, 8.1
Applications (percents and word problems)	2.4, 2.5
Simplification of a polynomial by grouping—one and two variables	4.3, 4.4
Evaluation of a polynomial—one and two variables	4.3
Addition and subtraction of polynomials	4.4
Multiplication of a monomial with a polynomial	4.5
Multiplication of two binomials	4.5, 4.6
Squaring a binomial	4.6
Division of polynomials with a monomial divisor—no remainder	4.7
Division of polynomials with a linear binomial divisor—no remainder	4.7
Factoring polynomials by finding common factors	5.1
Factoring a trinomial	5.2, 5.3
Factoring a difference of squares	5.4
Simplification of a rational expression by cancellation of common factors—one and two variables	6.1
Evaluation of a rational expression	6.1
Addition and subtraction of rational expressions	6.3
Multiplication and division of rational expressions	6.2
Simplification of a compound rational expression	6.4
Definition of exponentiation with positive exponents	4.1
Laws of exponents with positive exponents	4.1
Simplification of an expression with positive exponents	4.1
Definition of exponentiation with integral exponents	4.1

Description of ELM Skill	***Chapter, Section***
Laws of exponents with integral exponents	4.1
Simplification of an expression with integral exponents	4.1
Scientific notation	4.2
Definition of radical sign	8.1
Simplification of products under a single radical	8.2
Addition and subtraction of radical expressions	8.3
Multiplication of radical expressions	8.4
Solution of a simple radical equation	8.5
Solution of a linear equation in one unknown with numerical coefficients	2.3
Solution of a linear equation in one unknown with literal coefficients	2.7
Solution of a simple equation in one unknown which is reducible to a linear equation	2.3
Solution of a linear inequality in one unknown with numerical coefficients	2.6
Solution of two linear equations in two unknowns with numerical coefficients—by substitution	7.3
Solution of two linear equations in two unknowns with numerical coefficients—by elimination	7.2
Solution of a quadratic equation from factored form	5.6
Solution of a quadratic equation by factoring	5.6
Graphing a point on the number line	1.3
Graphing linear inequalities in one unknown	2.6
Graphing a point in the coordinate plane	3.1
Graphing a simple linear equation: $y = mx$, $y = b$, $x = b$	3.1
Reading data from a graph	3.4
Measurement formulas for perimeter and area of triangles, squares, rectangles, and parallelograms	2.7
Measurement formulas for circumference and area of circles	2.7
The Pythagorean Theorem	5.7

CLAST Skills

The following table lists the Florida CLAST skills and their location in *Elementary Algebra.*

Skill Number	*Description of CLAST Skill*	*Chapter, Section*
I.A.1a	Add, subtract rational numbers	1.1, 1.4
I.A.1b	Multiply, divide rational numbers	1.1, 1.5
I.B.2a	Calculate distance	1.1
I.B.2b	Calculate areas	2.7
I.C.1a	Add, subtract real numbers	1.4
I.C.1b	Multiply, divide real numbers	1.5
I.C.2	Order of operations	1.4, 1.5
I.C.3	Scientific notation	4.2
I.C.4	Solve linear equations and inequalities	2.3, 2.6
I.C.5	Use given formulas	2.7
I.C.6	Find function values	3.5
I.C.7	Factor quadratic expressions	5.1–5.5
I.C.8	Find roots of quadratic equations	9.1, 9.2, 9.3
II.A.1	Meaning of exponents	4.1, 8.6
II.A.3	Equivalent forms of rationals	1.1
II.A.4	Order relation	1.2, 1.3
II.C.1	Properties of operations	1.4, 1.5
II.C.2	Checking equations or inequalities	2.3, 2.6
II.C.3	Proportion and variation	6.6
II.C.4	Regions of coordinate plane	3.1
IV.B.1	Solve perimeter, area, volume problems	2.7
IV.B.2	Pythagorean property	5.7
IV.C.1	Solve real-world problems	2.4, 2.5, 2.7, 3.4, 5.7, 6.6, 6.7
IV.C.2	Solve problems involving structure and logic of algebra	2.1, 2.2

To the Student

Why must I need to know this? When will I ever use it? You may have asked these questions about algebra and mathematics at some time in your high school or college career.

Everyone agrees that if your major is in a science such as physics or chemistry, then you must take mathematics courses. If, on the other hand, your major is in an area such as journalism or history, then is mathematics important?

A student majoring in journalism needs to understand math in order to deal with misleading statistics, to gain insight into business trends, and to evaluate economic plans. No matter what your future job may be, you will advance faster than those who lack mathematical skills.

You will obtain some fundamental mathematical skills by successfully completing this course. The key to success is to practice daily. Read each assigned section carefully, and complete the homework assignment. The even-numbered problems are paired with the previous odd-numbered ones, so there is opportunity for repetitive practice. Repetition is the key to acquiring skills, so it is very important to get in the habit of doing the homework on a regular basis.

In this book, you will find that mathematics plays a role in many different areas. Furthermore, knowledge of mathematics will help you think logically and to reason through problems in all areas of your life.

This book has been written with you in mind. As you advance through this course, you will find many things to help guide you. For example, the **Learning Advantages** at the beginning of each section in the first seven chapters will help you study and prepare for tests. You will find clear examples of each new concept. As you begin this course, remember to relax—learning mathematics can be a rewarding experience.

Contents

Operations and Variables

CHAPTER 1

CONNECTIONS

The gestation period of the American Elk ranges from 270 to 285 days. See Exercise 73 in Section 1.4.

Overview

The foundation of algebra is arithmetic. In order to perform *algebraic* operations, a thorough understanding of performing *arithmetic* operations is necessary. Therefore, in this first chapter, we review arithmetic starting with a review of adding, subtracting, multiplying, and dividing fractions. Next, we review basic symbols and order of operations. The set of real numbers is examined in Section 1.3, along with its various subsets. Sections 1.4 and 1.5 deal with the four basic operations on real numbers: addition, subtraction, multiplication, and division. In Section 1.6, we study variables and variable expressions. At this time, we introduce the important concept of writing verbal statements as algebraic expressions. Converting verbal statements into algebraic expressions is the basis for using algebra to solve word problems, and this will be practiced throughout the text. We next study the structure of the set of real numbers together with the properties of the four basic operations of addition, subtraction, multiplication, and division. The final section deals with the properties of real numbers.

1.1 Fractions—A Review

OBJECTIVES

- *To write fractions in lowest terms*
- *To multiply and divide fractions*
- *To add and subtract fractions with common denominators*
- *To add and subtract fractions with unlike denominators*

LEARNING ADVANTAGE

Before starting on each homework assignment, first study the notes your instructor gave you and then read the appropriate section(s) from the book. If you have trouble on a particular exercise problem, find the example that most closely pertains to it. Remember that the problems in the exercise set are paired so that each even-numbered problem matches the previous odd-numbered problem.

In our daily living we encounter the **whole numbers**

$$0, 1, 2, 3, 4, \ldots$$

as well as **fractions,** such as $\frac{1}{2}$, $\frac{4}{5}$, and $\frac{3}{9}$. In this section we review some basic concepts involving fractions. In particular, we will review the techniques used to add, subtract, multiply, and divide fractions. These four operations will be discussed in greater detail in later sections. The purpose here is to strengthen your ability to work with fractions, since these same techniques will be used in algebra. In the fraction $\frac{a}{b}$, the number $\boldsymbol{a}$ is called the **numerator** and the number $\boldsymbol{b}$ is the **denominator.** The fraction is in **lowest terms** if a and b have no common factors. For example, the fraction $\frac{2}{3}$ is in lowest terms, while $\frac{4}{6}$ is not in lowest terms, since the numerator and the denominator have a **common factor** of 2.

To write a fraction in lowest terms, we make use of the following statement.

RULE

The value of a fraction remains the same, when the numerator and denominator are divided by the same nonzero number.

$$\frac{a \cdot c}{b \cdot c} = \frac{a}{b}, \qquad c \neq 0$$

Example 1 Write each fraction in lowest terms.

(a) $\frac{6}{21}$ **(b)** $\frac{60}{105}$

Solution **(a)** We start by writing both numerator and denominator as products.

$$\frac{6}{21} = \frac{2 \cdot \overset{1}{\cancel{3}}}{7 \cdot \underset{1}{\cancel{3}}} \qquad \text{Divide the numerator and the denominator by 3.}$$

$$= \frac{2}{7}$$

Notice that slashes were used to indicate that we divided both numerator and denominator by the common factor 3.

Dividing the numerator and denominator by the common factor 3 can be written in a more compact form:

$$\frac{6}{21} = \frac{\overset{2}{\cancel{6}}}{\underset{7}{\cancel{21}}} = \frac{2}{7}$$

This shorter process, however, does not show directly that 3 is a common factor.

(b) Since the last digit of the numerator is 0 and the last digit of the denominator is 5, they have a common factor of 5. Therefore,

$$\frac{60}{105} = \frac{12 \cdot \cancel{5}}{21 \cdot \cancel{5}}$$

Divide the numerator and the denominator by 5.

$$= \frac{12}{21}$$

Although $\frac{12}{21}$ is a partially reduced form of $\frac{60}{105}$, it is not the final answer, because 12 and 21 have a common factor of 3.

$$\frac{12}{21} = \frac{4 \cdot \cancel{3}}{7 \cdot \cancel{3}}$$

Divide the numerator and the denominator by 3.

$$= \frac{4}{7}$$

Therefore,

$$\frac{60}{105} = \frac{4}{7}$$

The shorthand method to reduce this fraction looks like this:

$$\frac{60}{105} = \frac{\overset{\overset{4}{\cancel{12}}}{\cancel{60}}}{\underset{\underset{7}{\cancel{21}}}{\cancel{105}}} = \frac{4}{7}$$

COMMON ERROR

When reducing a fraction to lowest terms, be sure that the numerator and denominator are written as products and not as sums. Compare the following two sequences of steps.

Sequence 1: $\frac{12}{15} = \frac{4 \cdot 3}{5 \cdot 3} = \frac{4 \cdot \cancel{3}}{5 \cdot \cancel{3}} = \frac{4}{5}$ Correct.

Sequence 2: $\frac{9}{10} = \frac{1 + 8}{2 + 8} = \frac{1 + \cancel{8}}{2 + \cancel{8}} = \frac{1}{2}$ Incorrect.

Sequence 1 is correct since 3 is a common factor. Sequence 2 is incorrect, since 8 is not a common *factor*.

We next consider combining fractions using the operations of addition, subtraction, multiplication, and division. Since the methods for multiplying and dividing fractions are simpler than for addition and subtraction, we start with multiplication. To multiply two fractions, we use the following definition.

DEFINITION

Multiplication of Two Fractions

Given two fractions $\frac{a}{b}$ and $\frac{c}{d}$, then

$$\frac{a}{b} \cdot \frac{c}{d} = \frac{a \cdot c}{b \cdot d}$$

That is, the answer from multiplying two fractions is a new fraction obtained by multiplying the numerators and then multiplying the denominators. The answer to a multiplication of numbers is called a **product.** For example, the product of 3 multiplied by 4 is 12; that is, $3 \cdot 4 = 12$. The numbers 3 and 4 are called **factors.**

Example 2 Find the product and write the answer in lowest terms.

(a) $\frac{3}{4} \cdot \frac{5}{7}$ **(b)** $\frac{2}{3} \cdot \frac{1}{5} \cdot \frac{4}{5}$ **(c)** $\frac{4}{5} \cdot \frac{10}{3}$

Solution **(a)** $\frac{3}{4} \cdot \frac{5}{7} = \frac{3 \cdot 5}{4 \cdot 7} = \frac{15}{28}$ Multiply numerators together and multiply denominators together.

(b) To find the product of three (or more) fractions, simply multiply all numerators and then all denominators.

$$\frac{2}{3} \cdot \frac{1}{5} \cdot \frac{4}{5} = \frac{2 \cdot 1 \cdot 4}{3 \cdot 5 \cdot 5} = \frac{8}{75}$$

(c) Write the numerator and denominator as products, then look for common factors.

$$\frac{4}{5} \cdot \frac{10}{3} = \frac{4 \cdot 10}{5 \cdot 3}$$

$$= \frac{4 \cdot 2 \cdot \overset{1}{\cancel{5}}}{\underset{1}{\cancel{5}} \cdot 3}$$ Divide the numerator and denominator by 5.

$$= \frac{4 \cdot 2 \cdot 1}{1 \cdot 3}$$

$$= \frac{8}{3}$$

Consider again the multiplication problem $\frac{4}{5} \cdot \frac{10}{3}$ of Example 2(c). Since any factor of a numerator ends up as a factor of the numerator in the product and any factor of a denominator ends up as a factor in the denominator of the product, we can "divide out" common factors *before* multiplying the two fractions. Therefore, Example 2(c) can be done in the following way.

$$\frac{4}{5} \cdot \frac{10}{3} = \frac{4}{\cancel{5}_{1}} \cdot \frac{\cancel{10}^{2}}{3} \quad \text{Divide out 5.}$$

$$= \frac{4 \cdot 2}{1 \cdot 3}$$

$$= \frac{8}{3}$$

Example 3 Multiply.

(a) $3 \cdot \frac{4}{5}$ **(b)** $\frac{25}{45} \cdot 12$

Solution **(a)** First write 3 as $\frac{3}{1}$, then multiply.

$$3 \cdot \frac{4}{5} = \frac{3}{1} \cdot \frac{4}{5}$$

$$= \frac{3 \cdot 4}{1 \cdot 5}$$

$$= \frac{12}{5}$$

(b) $\frac{25}{45} \cdot 12 = \frac{25}{45} \cdot \frac{12}{1}$ $\quad 12 = \frac{12}{1}$

$$= \frac{\cancel{25}^{5}}{\cancel{45}_{\cancel{9}_{3}}} \cdot \frac{\cancel{12}^{4}}{1} \quad \text{Divide 25 and 45 by 5, then divide 9 and 12 by 3.}$$

$$= \frac{5 \cdot 4}{3 \cdot 1}$$

$$= \frac{20}{3}$$

In Section 1.7, we will study properties of real numbers. One particular property that we preview here is the concept of *reciprocals.* Two numbers are **reciprocals** of each other if their product is one. For example, $\frac{2}{5}$ and $\frac{5}{2}$ are reciprocals since

$$\frac{2}{5} \cdot \frac{5}{2} = 1$$

In general, the two fractions $\frac{a}{b}$ and $\frac{b}{a}$, where a and b are both unequal to zero, are reciprocals, since

$$\frac{a}{b} \cdot \frac{b}{a} = 1$$

The reciprocal is used to divide two fractions.

RULE

Division of Two Fractions

For two fractions $\frac{a}{b}$ and $\frac{c}{d}$, where b, c, and d are nonzero,

$$\frac{a}{b} \div \frac{c}{d} = \frac{a}{b} \cdot \frac{d}{c}$$

That is, to divide two fractions, multiply the first fraction by the reciprocal of the second fraction. We say "Invert the second fraction and multiply." Inverting a fraction means writing its reciprocal. The answer to a division problem is called the **quotient.**

Example 4 Find the quotient and write the answer in lowest terms.

(a) $\frac{4}{3} \div \frac{5}{2}$ **(b)** $\frac{2}{3} \div \frac{4}{7}$

Solution **(a)** $\frac{4}{3} \div \frac{5}{2} = \frac{4}{3} \cdot \frac{2}{5}$ Invert and multiply.

$$= \frac{4 \cdot 2}{3 \cdot 5}$$

$$= \frac{8}{15}$$

(b) $\frac{2}{3} \div \frac{4}{7} = \frac{2}{3} \cdot \frac{7}{4}$ Invert and multiply.

$= \frac{\overset{1}{\cancel{2}}}{3} \cdot \frac{7}{\underset{2}{\cancel{4}}}$ 2 is a common factor.

$= \frac{1 \cdot 7}{3 \cdot 2}$

$= \frac{7}{6}$ ◀

Example 5 Divide:

(a) $\frac{2}{3} \div 4$ **(b)** $3 \div \frac{6}{5}$

Solution **(a)** First write 4 as $\frac{4}{1}$.

$\frac{2}{3} \div 4 = \frac{2}{3} \div \frac{4}{1}$

$= \frac{\overset{1}{\cancel{2}}}{3} \cdot \frac{1}{\underset{2}{\cancel{4}}}$ Invert and multiply; divide out 2.

$= \frac{1}{6}$

(b) Writing 3 as $\frac{3}{1}$,

$3 \div \frac{6}{5} = \frac{3}{1} \div \frac{6}{5}$

$= \frac{\overset{1}{\cancel{3}}}{1} \cdot \frac{5}{\underset{2}{\cancel{6}}}$ Invert and multiply; divide out 3.

$= \frac{5}{2}$ ◀

When adding or subtracting fractions, recall that the fractions must have the same denominator.

RULE

Addition and Subtraction of Fractions

To add or subtract fractions, the denominators must be the same. Combine the numerators, keeping the common denominator.

$$\frac{a}{b} + \frac{c}{b} = \frac{a+c}{b} \qquad \frac{a}{b} - \frac{c}{b} = \frac{a-c}{b}$$

Example 6 Add or subtract as indicated. Write the answer in lowest terms.

(a) $\frac{3}{5} + \frac{6}{5}$ **(b)** $\frac{7}{2} - \frac{5}{2}$

Solution **(a)** $\frac{3}{5} + \frac{6}{5} = \frac{3+6}{5}$ Add the numerators.

$$= \frac{9}{5}$$

(b) $\frac{7}{2} - \frac{5}{2} = \frac{7-5}{2} = \frac{2}{2} = 1$

To add or subtract fractions with *unlike denominators,* first rewrite each fraction so that the fractions have a common denominator. To do this, we use the following property.

$$\frac{a}{b} = \frac{a \cdot c}{b \cdot c}, \qquad c \neq 0$$

That is, the value of a fraction remains the same when both numerator and denominator are multiplied by the same nonzero number, c. We use this property to change the denominator of a fraction while keeping its value.

Example 7 Add or subtract as indicated. Write the answer in lowest terms.

(a) $\frac{3}{8} + \frac{1}{2}$ **(b)** $\frac{7}{12} - \frac{1}{3}$ **(c)** $\frac{2}{3} + \frac{1}{2}$

Solution **(a)** A common denominator of the two fractions is 8. Therefore, multiply the numerator and denominator of $\frac{1}{2}$ by 4.

$$\frac{3}{8} + \frac{1}{2} = \frac{3}{8} + \frac{1 \cdot 4}{2 \cdot 4}$$

$$= \frac{3}{8} + \frac{4}{8}$$

$$= \frac{3 + 4}{8} \qquad \text{Add numerators.}$$

$$= \frac{7}{8}$$

(b) A common denominator of the two fractions is 12, so we start by multiplying the numerator and denominator of $\frac{1}{3}$ by 4.

$$\frac{7}{12} - \frac{1}{3} = \frac{7}{12} - \frac{1 \cdot 4}{3 \cdot 4}$$

$$= \frac{7}{12} - \frac{4}{12}$$

$$= \frac{7 - 4}{12} \qquad \text{Subtract numerators.}$$

$$= \frac{3}{12}$$

$$= \frac{1}{4} \qquad \text{Reduce to lowest terms.}$$

(c) A common denominator of $\frac{2}{3}$ and $\frac{1}{2}$ is 6. Therefore,

$$\frac{2}{3} + \frac{1}{2} = \frac{2 \cdot 2}{3 \cdot 2} + \frac{1 \cdot 3}{2 \cdot 3}$$

$$= \frac{4}{6} + \frac{3}{6}$$

$$= \frac{4 + 3}{6} \qquad \text{Add numerators.}$$

$$= \frac{7}{6}$$

Example 8 Add or subtract as indicated.

(a) $4 + \frac{2}{3}$ **(b)** $1 - \frac{3}{5}$

Solution **(a)** First, write 4 as $\frac{4}{1}$. Now, a common denominator of $\frac{4}{1}$ and $\frac{2}{3}$ is 3. Therefore, write $\frac{4}{1}$ as $\frac{4 \cdot 3}{1 \cdot 3}$ or $\frac{12}{3}$.

$$4 + \frac{2}{3} = \frac{4}{1} + \frac{2}{3} \qquad 4 = \frac{4}{1}.$$

$$= \frac{4 \cdot 3}{1 \cdot 3} + \frac{2}{3}$$ Multiply numerator and denominator by 3.

$$= \frac{12}{3} + \frac{2}{3}$$

$$= \frac{12 + 2}{3}$$ Add numerators.

$$= \frac{14}{3}$$

(b) First write 1 as $\frac{1}{1}$. The common denominator of $\frac{1}{1}$ and $\frac{3}{5}$ is 5. Therefore, multiply the numerator and denominator of $\frac{1}{1}$ by 5. Here are the details:

$$1 - \frac{3}{5} = \frac{1}{1} - \frac{3}{5}$$ Replace 1 by $\frac{1}{1}$.

$$= \frac{1 \cdot 5}{1 \cdot 5} - \frac{3}{5}$$ Multiply numerator and denominator by 5.

$$= \frac{5}{5} - \frac{3}{5}$$

$$= \frac{5 - 3}{5}$$ Subtract numerators.

$$= \frac{2}{5}$$

Example 9 Add: $4\frac{1}{2} + 1\frac{3}{4}$.

Solution We first change each mixed number into an improper fraction.

$$4\frac{1}{2} = 4 + \frac{1}{2} = \frac{4}{1} + \frac{1}{2} = \frac{8}{2} + \frac{1}{2} = \frac{8 + 1}{2} = \frac{9}{2}$$

$$1\frac{3}{4} = 1 + \frac{3}{4} = \frac{4}{4} + \frac{3}{4} = \frac{4 + 3}{4} = \frac{7}{4}$$

Next, we add the two fractions.

$$4\frac{1}{2} + 1\frac{3}{4} = \frac{9}{2} + \frac{7}{4}$$

$$= \frac{18}{4} + \frac{7}{4}$$ Replace $\frac{9}{2}$ by $\frac{9 \cdot 2}{2 \cdot 2}$, which is $\frac{18}{4}$.

$$= \frac{25}{4} \text{ or } 6\frac{1}{4}$$

Example 10 A 68-gallon tank is $\frac{3}{4}$ full of water. How much water is in the tank?

Solution Since the 68-gallon tank is $\frac{3}{4}$ full, the amount of water in the tank is $\frac{3}{4}$ of 68; that is, $\frac{3}{4}$ times 68. Therefore,

$$\text{Amount of water} = \frac{3}{4} \cdot 68$$

$$= \frac{3}{\cancel{4}_{1}} \cdot \cancel{68}^{17} \quad \text{Reduce to lowest terms.}$$

$$= 51$$

A 68-gallon tank is $\frac{3}{4}$ full of water

There are 51 gallons of water in the tank.

Decimals and Percents

A **decimal** is a number written using a decimal point. For example,

$$0.3 = \frac{3}{10}, \qquad 0.02 = \frac{2}{100}, \qquad 0.61 = \frac{61}{100}$$

Decimals are used in working with percents. The word **percent** means "per hundred." For example, 42% (read forty-two percent) means 42 per hundred. That is,

$$42\% = \frac{42}{100} = 0.42$$

See Appendix A for more details.

WARMING UP

Answer true or false.

1. $\frac{2}{4} = \frac{1}{2}$ **2.** $\frac{6}{8} = \frac{3}{4}$ **3.** $\frac{1}{4} + \frac{1}{4} = \frac{1}{2}$ **4.** $\frac{2}{3} \cdot \frac{1}{2} = \frac{1}{2}$

5. $\frac{3}{2} - \frac{1}{2} = 2$ **6.** $\frac{1}{4} \div \frac{1}{3} = \frac{1}{12}$ **7.** $\frac{1}{2} + \frac{1}{3} = \frac{5}{6}$ **8.** $1\frac{1}{2} + 1\frac{1}{2} = 3$

EXERCISE SET 1.1

For Exercises 1–10, write each fraction in lowest terms.

1. $\frac{4}{6}$ **2.** $\frac{4}{8}$ **3.** $\frac{5}{15}$ **4.** $\frac{6}{9}$

5. $\frac{6}{3}$ 6. $\frac{10}{5}$ 7. $\frac{12}{10}$ 8. $\frac{15}{6}$

9. $\frac{12}{6}$ 10. $\frac{15}{3}$

For Exercises 11–30, find the product and write the answer in lowest terms.

11. $\frac{2}{3} \cdot \frac{1}{3}$ 12. $\frac{1}{4} \cdot \frac{1}{2}$ 13. $\frac{5}{4} \cdot \frac{1}{2} \cdot \frac{3}{2}$

14. $\frac{4}{3} \cdot \frac{1}{5} \cdot \frac{2}{3}$ 15. $\frac{3}{5} \cdot \frac{3}{2} \cdot \frac{1}{2}$ 16. $\frac{1}{4} \cdot \frac{1}{3} \cdot \frac{5}{2}$

17. $\frac{2}{5} \cdot \frac{10}{3}$ 18. $\frac{6}{7} \cdot \frac{14}{5}$ 19. $\frac{3}{7} \cdot \frac{14}{9}$

20. $\frac{12}{5} \cdot \frac{10}{9}$ 21. $\left(\frac{4}{12}\right)\left(\frac{9}{2}\right)$ 22. $\left(\frac{3}{4}\right)\left(\frac{16}{6}\right)$

23. $\frac{8}{15} \cdot \frac{5}{6} \cdot \frac{9}{8}$ 24. $\frac{4}{9} \cdot \frac{3}{16} \cdot \frac{8}{5}$ 25. $2 \cdot \frac{2}{3}$

26. $5 \cdot \frac{1}{6}$ 27. $3 \cdot \frac{2}{9}$ 28. $4 \cdot \frac{5}{8}$

29. $\frac{3}{2} \cdot 4$ 30. $\frac{4}{5} \cdot 15$

For Exercises 31–46, find the quotient and write the answer in lowest terms.

31. $\frac{1}{2} \div \frac{2}{3}$ 32. $\frac{3}{4} \div \frac{1}{3}$ 33. $\frac{2}{3} \div \frac{4}{3}$ 34. $\frac{5}{3} \div \frac{5}{2}$

35. $\frac{4}{7} \div \frac{3}{14}$ 36. $\frac{3}{16} \div \frac{9}{8}$ 37. $\frac{4}{6} \div \frac{1}{3}$ 38. $\frac{5}{3} \div \frac{5}{9}$

39. $\frac{2}{3} \div \frac{8}{1}$ 40. $\frac{4}{7} \div 4$ 41. $\frac{3}{2} \div 3$ 42. $\frac{2}{5} \div 4$

43. $1 \div \frac{1}{3}$ 44. $2 \div \frac{4}{7}$ 45. $\frac{5}{6} \div 10$ 46. $\frac{8}{9} \div 2$

For Exercises 47–90, add or subtract as indicated. Write the answer in lowest terms.

47. $\frac{2}{5} + \frac{1}{5}$ 48. $\frac{1}{7} + \frac{4}{7}$ 49. $\frac{3}{4} + \frac{3}{4}$ 50. $\frac{3}{10} + \frac{2}{10}$

51. $\frac{3}{2} + \frac{1}{2}$ 52. $\frac{3}{8} + \frac{5}{8}$ 53. $\frac{2}{3} - \frac{1}{3}$ 54. $\frac{4}{7} - \frac{2}{7}$

55. $\frac{5}{6} - \frac{1}{6}$ 56. $\frac{5}{4} - \frac{3}{4}$ 57. $\frac{9}{10} - \frac{3}{10}$ 58. $\frac{7}{8} - \frac{5}{8}$

59. $\frac{2}{3} + \frac{1}{6}$ **60.** $\frac{1}{2} + \frac{1}{4}$ **61.** $\frac{3}{8} + \frac{3}{4}$ **62.** $\frac{5}{12} + \frac{2}{3}$

63. $\frac{3}{4} - \frac{1}{2}$ **64.** $\frac{5}{6} - \frac{2}{3}$ **65.** $\frac{3}{2} - \frac{5}{6}$ **66.** $\frac{5}{3} - \frac{5}{12}$

67. $\frac{1}{6} - \frac{1}{18}$ **68.** $\frac{11}{12} - \frac{3}{4}$ **69.** $\frac{1}{3} + \frac{3}{2}$ **70.** $\frac{1}{5} + \frac{2}{3}$

71. $\frac{3}{4} + \frac{2}{3}$ **72.** $\frac{1}{4} + \frac{1}{5}$ **73.** $\frac{2}{3} - \frac{1}{2}$ **74.** $\frac{5}{4} - \frac{3}{5}$

75. $\frac{3}{2} - \frac{2}{5}$ **76.** $\frac{5}{7} - \frac{1}{2}$ **77.** $1 + \frac{1}{4}$ **78.** $1 + \frac{2}{3}$

79. $2 - \frac{1}{2}$ **80.** $3 - \frac{2}{5}$ **81.** $\frac{8}{3} - 1$ **82.** $\frac{7}{2} - 3$

83. $1\frac{1}{2} + 2\frac{1}{2}$ **84.** $3\frac{1}{4} + 2\frac{1}{4}$ **85.** $5\frac{1}{2} + 1\frac{1}{4}$ **86.** $6\frac{2}{3} + 2\frac{1}{6}$

87. $2\frac{3}{4} + 1\frac{3}{4}$ **88.** $3\frac{5}{6} + 2\frac{1}{3}$ **89.** $4\frac{2}{3} + 4\frac{1}{2}$ **90.** $2\frac{3}{4} + 1\frac{2}{3}$

91. A rectangle has two sides that are each $1\frac{1}{3}$ feet long and two other sides that are each $2\frac{1}{2}$ feet long. Find the total distance around the rectangle.

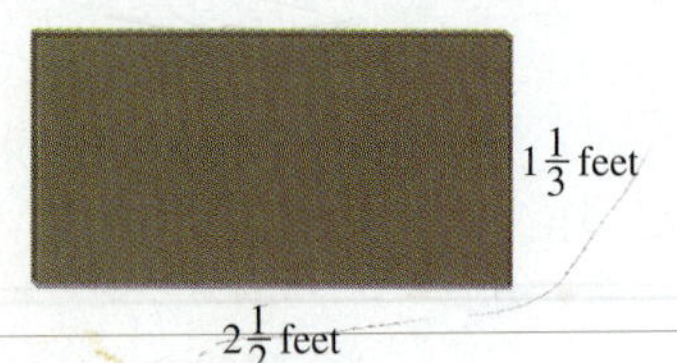

92. Kathy mowed $\frac{1}{3}$ of the lawn and Henry mowed $\frac{1}{4}$ of the lawn. How much of the lawn is mowed?

93. If a pizza is cut into eight pieces and you eat three pieces, how much of the pizza is left?

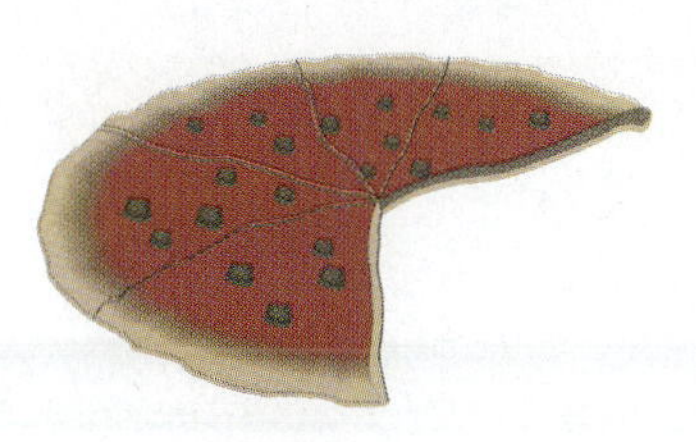

94. Joe drank $\frac{1}{3}$ of a pitcher of cola and Frank drank $\frac{1}{4}$ of the pitcher. How much cola is left?

$\frac{1}{3} + \frac{1}{4}$ of the cola has been drunk

95. A recipe for oatmeal cookies calls for $1\frac{2}{3}$ teaspoons of cinnamon. If Jay plans to triple the recipe, how much cinnamon should he use?

96. A heating oil company is contracted to deliver $3\frac{2}{5}$ tanks of oil each month for four months. How many tanks of oil will be delivered during the four months?

97. Tom and Linda are studying elementary algebra together. One of the homework problems is to find the sum:

$$\frac{8}{12} + \frac{12}{18}$$

Tom says that a common denominator is 36, therefore,

$$\frac{8}{12} + \frac{12}{18} = \frac{8 \cdot 3}{12 \cdot 3} + \frac{12 \cdot 2}{18 \cdot 2} = \frac{24}{36} + \frac{24}{36}$$

$$= \frac{48}{36} = \frac{48}{36} = \frac{4}{3}$$

Linda says that his solution is correct, but that she has a shorter way to do the problem. What is Linda's way?

98. What is wrong with the following sequence of steps?

$$\frac{7}{21} = \frac{7}{7 \cdot 3} = \frac{7}{7 \cdot 3} = \frac{0}{3} = 0$$

SAY IT IN WORDS

99. Explain how you would reduce a fraction to lowest terms.

100. What is your method for dividing two fractions?

ENRICHMENT EXERCISES

Simplify.

1. $\frac{3}{8} + \frac{1}{12}$

2. $\frac{4}{13} - \frac{5}{26}$

3. $\frac{2}{3} \cdot \frac{1}{8} + \frac{5}{6} \cdot \frac{3}{4}$

4. $1 - \left(\frac{1}{6} + \frac{1}{3}\right)$

For Exercises 5 and 6, what fraction of the pie graph is represented by the yellow region?

5.

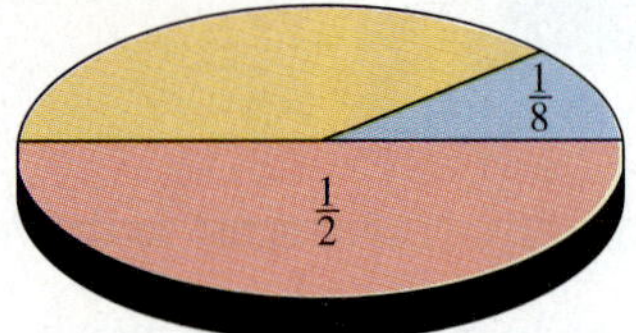

6.

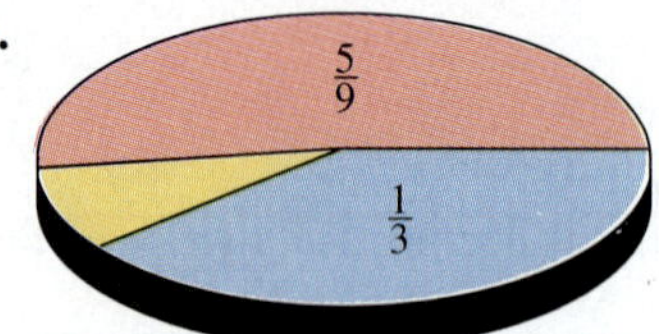

7. A taxicab company charges 80 cents for the first $\frac{1}{4}$ mile and 50 cents for each additional $\frac{1}{4}$ mile. How much is the cab fare for a 3-mile trip?

Answers to Enrichment Exercises begin on page A.1.

1.2 Basic Symbols and Order of Operations

OBJECTIVES

- *To review basic symbols*
- *To convert verbal statements into symbols*
- *To simplify numerical expressions*

LEARNING ADVANTAGE

An important factor in successfully completing an algebra course is to ***regularly attend class.*** *Furthermore, be an active member of the class. Do not hesitate to ask questions, even ones you think might be too elementary. Your teacher is understanding, and there are probably other students who have the very same question.*

In this section we review some of the basic symbols of arithmetic. These same symbols will be used in the study of algebra. A **numerical expression** is a group of symbols used to represent a number. For example,

$$2 - 1, \qquad 5 + 7, \qquad \text{and} \qquad (8 - 2) - 3$$

each are numerical expressions. The **value** of a numerical expression is the number that it represents. For example, $11 + 3$ represents the number 14.

We may *connect* numerical expressions to make mathematical sentences. For example, $2 + 2 = 4$ is a sentence formed by using the equality symbol to connect the two expressions $2 + 2$ and 4. Besides equality, there are other symbols used to make mathematical sentences.

The following is a list of common symbols. We let the letters a and b represent any two numerical expressions.

Symbol	Meaning	Examples
$a = b$	a is equal to b	$2 + 3 = 5$, $\frac{3}{6} = \frac{1}{2}$
$a \neq b$	a is not equal to b	$2 + 1 \neq 4$, $3 + 5 \neq 6 - 2$
$a < b$	a is less than b	$4 < 7$
$a \leq b$	a is less than or equal to b	$9 \leq 9$, $4 \leq 7$
$a > b$	a is greater than b	$23 > 19$
$a \geq b$	a is greater than or equal to b	$9 \geq 9$, $7 \geq 4$

Inequality statements such as $3 < 4$ and $4 > 3$ both have the same meaning. For instance, $3 < 4$ means "3 is less than 4," which is the same as saying "4 is greater than 3," that is, $4 > 3$.

You are familiar with the four basic operations on numbers—addition, subtraction, multiplication, and division.

Operation	Symbols	In Words
Addition	$a + b$	The *sum* of a and b
Subtraction	$a - b$	The *difference* of a and b
Multiplication	$a \cdot b$, ab, $(a)(b)$, $a(b)$, $(a)b$	The *product* of a and b
Division	$a \div b$, $\frac{a}{b}$, a/b, $b\overline{)a}$	The *quotient* of a and b

Example 1

	Symbols	In Words
(a)	$2 + 5$	The sum of 2 and 5.
(b)	$6 > 4$	Six is greater than 4.
(c)	$3 \leq 10$	Three is less than or equal to 10.
(d)	$3 - 2 = 1$	The difference of 3 and 2 is 1.

When a number is repeatedly multiplied, we can use the *exponent notation* to express the multiplication. For example, $6 \cdot 6 = 6^2$. This is read "six squared." Six is called the **base** and two is called the **exponent.**

Example 2 Find the value of each expression.

(a) $5^2 = 5 \cdot 5 = 25$ — 5^2 is read "five squared"

(b) $7^3 = 7 \cdot 7 \cdot 7 = 343$ — 7^3 is read "seven cubed."

(c) $3^4 = 3 \cdot 3 \cdot 3 \cdot 3 = 81$ — 3^4 is read "three to the fourth power."

(d) $16^1 = 16$ — 16 raised to the first power is 16.

Replacing a numerical expression by the number that it represents is called **simplifying** the expression. For example,

$$14 - (2 + 5) = 14 - 7 = 7$$

$$\frac{2 + 18}{4} = \frac{20}{4} = 5$$

When simplifying an expression, it is important to obey rules that govern the *order of operations.* For example, the expression $7 + 4 \cdot 9$ might be simplified in two different ways. Only one of them is correct.

$$7 + 4 \cdot 9 = 11 \cdot 9 = 99 \quad \text{Incorrect.}$$

$$7 + 4 \cdot 9 = 7 + 36 = 43 \quad \text{Correct.}$$

Since a numerical expression should simplify to a single number, the following order of operations has been established.

STRATEGY

Order of Operations Convention

1. Simplify all powers.
2. Perform all multiplications and divisions from left to right.
3. Perform all additions and subtractions from left to right.

Example 3 Simplify each numerical expression.

(a) $14 - 5 + 8$

(b) $\frac{15}{3} + 2 \cdot 4$

(c) $3^2 - 2^3 + 5 \cdot 3$

(d) $4 \cdot 10 + 5^2 - 8 \div 2$

Solution (a) $14 - 5 + 8 = 9 + 8$ — Perform the subtraction.

$= 17$

(b) We perform the division and multiplication first, then add the two resulting numbers.

$$\frac{15}{3} + 2 \cdot 4 = 5 + 8$$

$$= 13$$

(c) $3^2 - 2^3 + 5 \cdot 3 = 9 - 8 + 15$ Simplify the powers and perform the multiplication.

$= 16$ Perform the subtraction and addition.

(d) $4 \cdot 10 + 5^2 - 8 \div 2 = 40 + 25 - 4$ Perform the multiplication, simplify the power, and perform the division.

$= 61$ Perform the addition and subtraction.

CALCULATOR CROSSOVER

Does your calculator follow the order of operations convention? Try the following experiment. Use your calculator to find the value of $12 + 4 \div 2$ by entering the numbers and operations as they appear from left to right.

If your calculator displays the correct answer 14, then it does obey the rule that division is performed before addition. Your calculator has an algebraic operating system.

If your calculator displays the incorrect answer 8, then it performs the operations in the order in which they occurred. A way to obtain the correct answer is to divide 4 by 2 first, then add 12.

In more complicated numerical expressions, grouping symbols such as parentheses are used to change the order of operations. For example,

$$4 \cdot 7 - 5 \quad \text{means} \quad 28 - 5 \quad \text{or} \quad 23$$

but

$$4(7 - 5) \quad \text{means} \quad 4(2) \quad \text{or} \quad 8$$

In the second expression, the grouping of the difference of 7 and 5 means that the subtraction of 5 from 7 is performed *before* multiplying by 4.

There are three basic grouping symbols: parentheses, ( ), brackets, [ ], and braces, { }.

RULE

Simplifying an Expression Containing Grouping Symbols

Use the order of operations convention within each grouping starting with the innermost and working outward until a final answer is reached.

Example 4 Simplify each numerical expression.

(a) $(3 + 2)(8 - 4)$ **(b)** $4 + 3[5(6 - 2) - 14]$

Solution **(a)** We first simplify within each pair of parentheses.

$$(3 + 2)(8 - 4) = (5)(4)$$
$$= 20$$

(b) We start first with the innermost grouping and work outward.

$$4 + 3[5(6 - 2) - 14] = 4 + 3[5(4) - 14]$$ Proceed inside the brackets and perform the multiplication before subtraction.

$$= 4 + 3[20 - 14]$$ Continue working within the brackets.

$$= 4 + 3[6]$$ Do the multiplication before addition.

$$= 4 + 18$$
$$= 22$$

Example 5 Write the following expressions in words.

(a) $2 + 1$ **(b)** $2(4 + 1)$ **(c)** $(9 + 6) \div 3$

Solution **(a)** This expression is the sum of two and one.

(b) Notice that in this expression, $4 + 1$ is grouped using parentheses indicating that 4 and 1 are added *before* multiplying by 2. We say, "two times the sum of 4 and one."

(c) Since the quantity $9 + 6$ is grouped, we say "the sum of nine and six, the result divided by three."

In an expression such as $\frac{2 + 18}{1 + 3}$, the fraction bar is a symbol that indicates that the numerator and denominator are each considered as a single expression. Therefore,

$$\frac{2 + 18}{1 + 3} \quad \text{means} \quad \frac{20}{4} \quad \text{or} \quad 5$$

Example 6 Simplify the following expressions.

(a) $\frac{30 - 2 \cdot 3}{3 \cdot 4}$ **(b)** $\frac{2^3 - 2^2 + 11}{(2 - 1)^2 + 2}$ **(c)** $\frac{(15 - 7) \cdot 7}{(1 + 1) \cdot 4}$

Solution **(a)** To simplify this fraction, we evaluate the numerator and the denominator, and then divide.

$$\frac{30 - 2 \cdot 3}{3 \cdot 4} = \frac{30 - 6}{12}$$
$$= \frac{24}{12}$$
$$= 2$$

(b) We again evaluate the numerator and the denominator and then divide the resulting two numbers.

$$\frac{2^3 - 2^2 + 11}{(2-1)^2 + 2} = \frac{8 - 4 + 11}{1^2 + 2}$$
$$= \frac{15}{3}$$
$$= 5$$

(c) $$\frac{(15-7)\cdot 7}{(1+1)\cdot 4} = \frac{\overset{\overset{1}{\cancel{4}}}{\cancel{8}}\cdot 7}{\underset{1}{\cancel{2}}\cdot \underset{1}{\cancel{4}}}$$ Divide numerator and denominator by 2, then by 4.

$$= \frac{1\cdot 7}{1\cdot 1}$$
$$= 7$$

WARMING UP

Answer true or false.

1. $6^2 = 36$

2. $10^1 = 1$

3. $2^3 = 8$

4. $3^2 + 1^2 = 16$

5. $7^1 = 0$

6. $5^3 = 15$

7. $1^1 = 1$

8. $(4-2)(3) = 6$

9. $2(5+8) = 18$

10. $(2+3)^2 = 13$

EXERCISE SET 1.2

For Exercises 1–8, write in words the meaning of the symbols.

1. $2 + 3$

2. $4 - 1$

3. $6 < 7$

4. $4 \neq 3$

5. $9 \div 3$

6. $9 > 5$

7. $5 - 2 = 3$

8. $4 \div 1 = 4$

For Exercises 9–16, write in symbols the meaning of the words.

9. The difference of 7 and 4.

10. The difference of 9 and 8.

11. The quotient of 8 and 4.

12. The quotient of 12 and 3.

13. 5 is greater than 4.

14. 4 is less than 5.

15. The quotient of 15 and 5 is 3.

16. The difference of 9 and 5 is 4.

For Exercises 17–24, find the value of each expression.

17. 4^3 **18.** 3^1 **19.** 3^2 **20.** 2^4

21. 2^1 **22.** 6^3 **23.** 50^2 **24.** 100^2

For Exercises 25–40, simplify each numerical expression.

25. $5 + 2 + 3$

26. $7 + 1 + 4$

27. $6 - 2 - 1$

28. $7 - 1 - 4$

29. $4 \cdot 2 + 3 \cdot 1$

30. $3 \cdot 2 + 2 \cdot 6$

31. $4^2 + 1^2$

32. $3^2 + 2^2$

33. $5^2 - 4^2 + 1^3$

34. $6^2 + 2^3 - 3^2$

35. $(4 - 1)^2 + (3 + 2)^2$

36. $(5 - 3)^2 + (1 + 2)^2$

37. $8 \cdot 2 - 9 \div 3$

38. $15 \div 3 + 2 \cdot 3$

39. $3[10 - 2(5 - 1)]$

40. $2[(6 - 1) \cdot 3 - 14]$

For Exercises 41–44, write in symbols the meaning of the words.

41. Three times the sum of 5 and 2.

42. Four times the difference of 8 and 3.

43. The difference of 7 and 4, the result squared.

44. The sum of 3 and 1, the result divided by 2.

For Exercises 45–52, simplify.

45. $\dfrac{3 - 1}{5 + 1}$ **46.** $\dfrac{4 - 2}{8 - 4}$ **47.** $\dfrac{4 + 5}{4 - 1}$

48. $\dfrac{8 - 2}{1 + 5}$ **49.** $\dfrac{2^2 + 3^2}{5 + 8}$ **50.** $\dfrac{5^2 - 10}{4^2 + 4}$

51. $\dfrac{(3 - 1)^2 \cdot 6}{2(6 + 2)}$ **52.** $\dfrac{(7 - 3) \cdot 2^2}{(3 - 1)^3 \cdot 4}$

For Exercises 53–60, solve.

The value of a number raised to a power can be found by using a calculator. Use the $\boxed{y^x}$ key to find the value of each expression. For example, to find 3^6 using a calculator, perform the following sequence of steps:

Enter	*Press*	*Enter*	*Press*	*Display*
3		6	$\boxed{=}$	729

53. 7^5 **54.** 11^3 **55.** 64^4 **56.** 21^5

57. 3^{14} **58.** 2^{25} **59.** 4^{10} **60.** 289^3

SAY IT IN WORDS

61. Is $(2 + 3)^2 = 2^2 + 3^2$? To answer this question, first find $(2 + 3)^2$, then $2^2 + 3^2$.

62. Is $(1 + 2)^3 = 1^3 + 2^3$? Why?

63. Is the quotient of 4 and 2 the same as the quotient of 2 and 4? Explain

64. Is the product of 3 and 5 the same as the product of 5 and 3? Explain.

65. Explain your method for simplifying a numerical expression.

66. Why do we need the order of operation rules?

ENRICHMENT EXERCISES

The memory capacity of a personal computer is usually measured in kilobytes and is frequently a power of 2. For example, some computers have 64K (kilobytes) of memory. Notice that $64 = 2^6$.

Express the following memory capacities as powers of 2.

1. 128K **2.** 256K **3.** 512K

4. Is $\left(\frac{1}{2}\right)^2$ greater than $\frac{1}{2}$? Explain.

Answers to Enrichment Exercises begin on page A.1.

1.3 The Real Numbers

OBJECTIVES

- *To graph real numbers on a number line*
- *To use numbers to describe real-life situations*
- *To find the absolute value of a number*
- *To compare numbers*

LEARNING ADVANTAGE

When reading this book, be sure to concentrate on what is being expressed. In mathematics, the English language is used in a very precise manner. For example, consider the two phrases:

"4 less than 5" *and* *"4 is less than 5"*

Although they look similar, mathematically they each have two entirely different concepts:

"4 less than 5" *means* $5 - 4$

"4 is less than 5" *means* $4 < 5$

A **set** is a collection of objects, called **elements.** We use capital letters like A, B, and C to represent sets and we list the elements of a set by enclosing them within a pair of braces. For example, $A = \{a, b, c, d\}$ is the set of the first four letters on the alphabet and $B = \{1, 2, 3, 4, 5, 6\}$ is the set of the first six counting numbers.

If a is a member of the set A, we write $a \in A$. If a is not a member of A, we write $a \notin A$. For example, $5 \in \{1, 3, 5, 7\}$ while $6 \notin \{1, 3, 5, 7\}$. The **empty set** is the set with no elements and is denoted by $\varnothing$ or $\{ \ \}$.

Set A is a **subset** of set B if every member of A is also a member of B and we write $A \subseteq B$. For example, $\{2, 4, 6\}$ is a subset of $\{1, 2, 3, 4, 5, 6\}$, that is,

$$\{2, 4, 6\} \subseteq \{1, 2, 3, 4, 5, 6\}$$

In this section we talk about the set of real numbers and some of its subsets. To visualize real numbers and to observe how they relate to each other, we represent them as points on a line, called a **number line.** To construct a number line:

1. Draw a straight line, then choose a convenient point and label it zero. This point is called the **origin.** The origin separates the line into two parts, the negative part and the positive part, as shown in the figure below.
2. We now make a **choice of scale** by marking off equally spaced distances in both directions from the origin. Label the marks to the right by the **positive integers** 1, 2, 3, . . . and the marks to the left by the **negative integers** −1, −2, −3, . . . , where the dots mean "and so on."

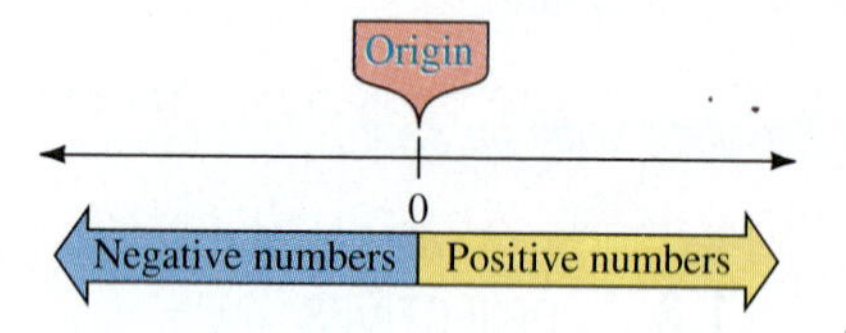

Notice that the negative numbers lie to the left of zero and the positive numbers lie to the right of zero. The number zero itself is neither positive nor negative.

The **set of real numbers** can be classified as *counting* or *natural numbers, whole numbers, integers, rational numbers,* and *irrational numbers.* Each of these are *subsets* of the real numbers. The counting numbers are the numbers 1, 2, 3, and so on. Using the set notation of enclosing the members of the set within braces, we write the set of **counting numbers** as

$$\{1, 2, 3, \ldots\}$$

The set of **whole numbers** is given by

$$\{0, 1, 2, 3, \ldots\}$$

Notice that the set of whole numbers consists of the counting numbers and the number 0.

The positive integers (counting numbers), the negative integers, and zero comprise the set of **integers.** The set of integers can be split into two parts, the **even integers** and the **odd integers.**

The set of integers: $\{\ldots, -3, -2, -1, 0, 1, 2, 3, \ldots\}$
The set of even integers: $\{\ldots, -4, -2, 0, 2, 4, \ldots\}$
The set of odd integers: $\{\ldots, -5, -3, -1, 1, 3, 5, \ldots\}$

There are many numbers in addition to the integers. Numbers such as

$$\frac{1}{2}, \qquad -\frac{3}{7}, \qquad \text{and} \qquad 2.25$$

are examples of *rational numbers.* A **rational number** is any number that can be written as a ratio $\frac{a}{b}$, where a and b are integers.

Every rational number may be expressed as a decimal. The decimal may terminate, as in $\frac{1}{4} = 0.25$, or it may not terminate, as in $\frac{23}{11} = 2.090909\ldots$ Notice that in the decimal expansion for $\frac{23}{11}$, the block consisting of the two digits 0 and 9 repeats without end. Conversely, any decimal that either terminates after a certain number of digits or has a repeating block of digits is a rational number.

An **irrational number** is a number that cannot be written as a ratio of integers. Examples of irrational numbers are π, $\sqrt{2}$, and $-\sqrt{3}$. The set of **real numbers** consists of the rational numbers together with the irrational numbers.

Example 1 List all numbers from the set

$$\left\{-9, -6, -\frac{3}{2}, 0, \frac{2}{6}, 0.7, 1, \sqrt{2}, \sqrt{3}, 2, \frac{12}{3}\right\}$$

that are

(a) Counting numbers.
(b) Whole numbers.
(c) Negative integers.
(d) Odd integers.
(e) Integers.
(f) Noninteger rational numbers.
(g) Irrational numbers.
(h) Real numbers.

Solution **(a)** The counting numbers are 1, 2, and $\frac{12}{3} = 4$.

(b) The whole numbers are 0, 1, 2, and $\frac{12}{3}$.

(c) The negative integers are -9 and -6.

(d) There are two odd integers: -9 and 1.

(e) The integers are -9, -6, 0, 1, 2, and $\frac{12}{3}$.

(f) The noninteger rational numbers are $-\frac{3}{2}$, $\frac{2}{6}$, and $0.7 = \frac{7}{10}$.

(g) There are two irrational numbers in the set: $\sqrt{2}$ and $\sqrt{3}$.
(h) Every number in the set is a real number.

The relationships among the types of numbers we have introduced can be shown in a family tree of numbers.

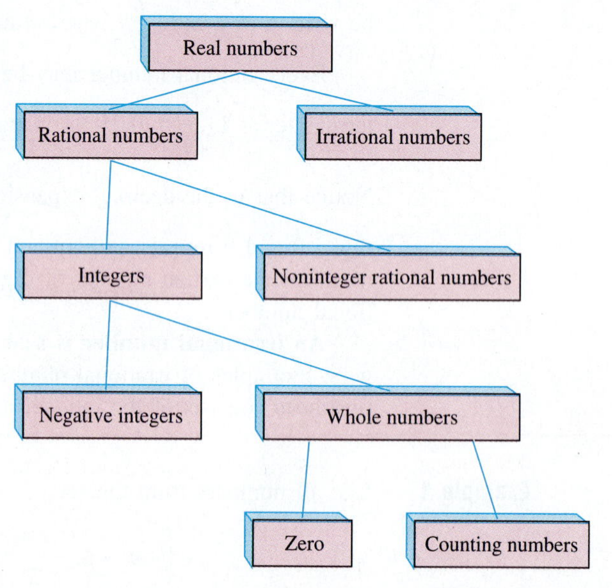

To **graph** a real number means to locate it on a number line. Every real number, rational or irrational, has a position on the number line, with positive numbers lying to the right of zero and negative numbers lying to the left of zero.

Example 2 Draw a number line, then graph the numbers from the set

$$\left\{-2, 0, \frac{1}{2}, \pi\right\}$$

Solution A number line is drawn in the figure below with the given numbers graphed as indicated. We use 3.14 to approximate the irrational number π, and, therefore, it lies between the integers 3 and 4, but close to 3.

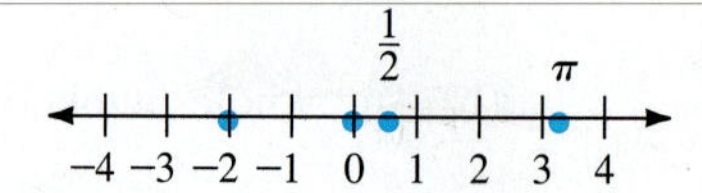

The positive and negative numbers indicate opposite directions. On the number line, positive numbers lie to the right of zero and negative numbers lie to the left of zero. Therefore, the **sign of a number** is used to measure direction. For example, a *profit* of $100 is indicated by the number +100 or simply 100. A *loss* of $100 can be indicated by −100. As another example, if the number 12 indicates 12 miles north, then −12 means 12 miles south.

Example 3 Represent each quantity as either a positive number or a negative number.

(a) Fred loses \$250 in a poker game. −250

(b) The fullback gained 23 yards on the play. 23

(c) The company's earnings were down 14%. −14%.

Solution **(a)** -250 **(b)** 23 **(c)** -14

In addition to the direction from the origin on the number line, another important concept is the distance that a number is located from the origin. We call this distance the *absolute value* of the number.

DEFINITION

The **absolute value** of a number is the distance that it is located from the origin. If x is a number, the absolute value of x is written as $|x|$.

For example, consider the number 4 as shown on the following number line.

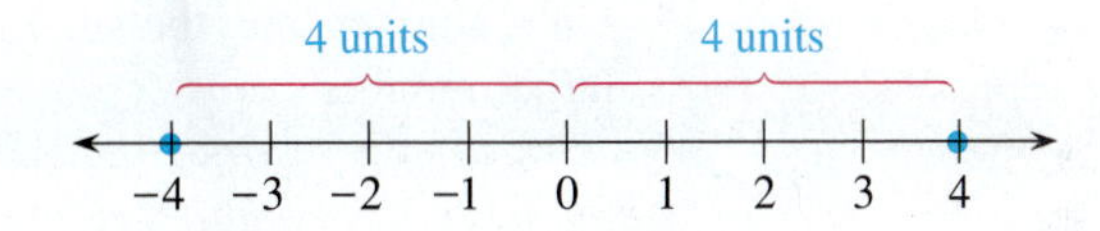

The number 4 is 4 units from the origin, therefore the absolute value of 4 is 4. We write $|4| = 4$. Similarly, the absolute value of -4 is also 4, since -4 is also 4 units from the origin. That is, $|-4| = 4$.

Since the absolute value of a number is the distance from the origin, the absolute value is either positive or zero, but never negative.

Example 4 Find the absolute value of each number.

(a) 3 **(b)** 0 **(c)** -12

(d) $-\dfrac{4}{3}$ **(e)** -10.3 **(f)** $-\sqrt{2}$

Solution

(a) $|3| = 3$

(b) $|0| = 0$ The number zero is 0 units from zero.

(c) $|-12| = 12$ The number -12 is 12 units from zero.

(d) $\left|-\dfrac{4}{3}\right| = \dfrac{4}{3}$

(e) $|-10.3| = 10.3$

(f) $|-\sqrt{2}| = \sqrt{2}$

The number line gives a visual way to compare two numbers with respect to size. For example, $3 < 4$ since 3 lies to the left of 4 on the number line as shown in the figure below.

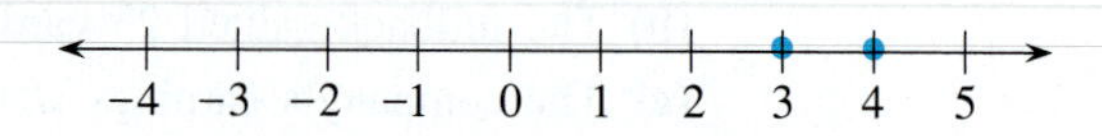

3 is less than 4

Similarly, $-1 > -2$, since -1 lies to the right of -2 on the number line shown below.

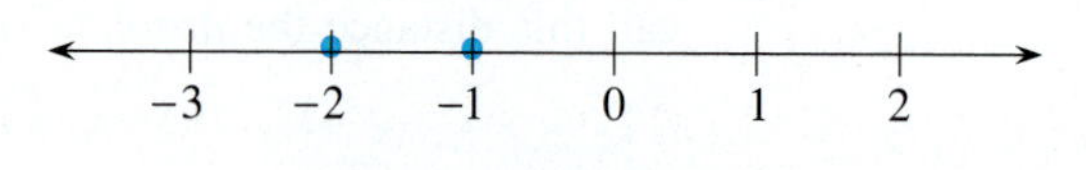

-1 is greater than -2

We can use these examples to illustrate the rule for comparing two numbers.

RULE

Comparing Two Numbers

On a number line, the smaller of two numbers lies to the left of the larger number.

Example 5 Place $<$, $>$, or $=$ between the two numbers to make the statement true.

(a) $-1 \quad 0$ **(b)** $\frac{3}{4} \quad \frac{1}{2}$ **(c)** $|-3| \quad |3|$

Solution For (a) and (b), we graph the points on a number line.

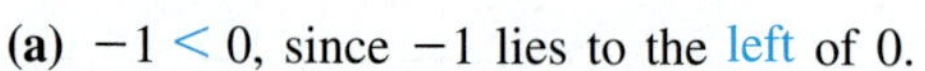

(a) $-1 < 0$, since -1 lies to the left of 0.

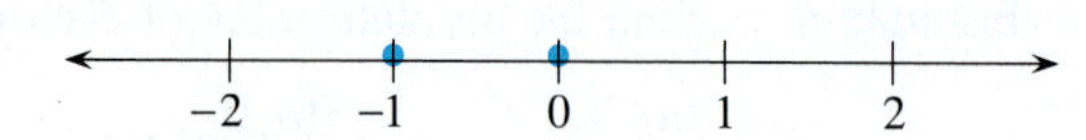

(b) $\frac{3}{4} > \frac{1}{2}$, since $\frac{3}{4}$ lies to the right of $\frac{1}{2}$.

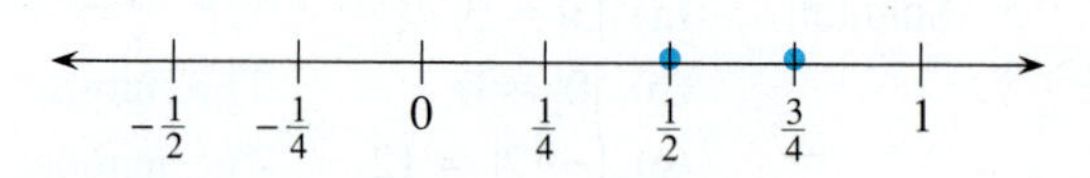

(c) Since $|-3| = 3$ and $|3| = 3$, $|-3| = |3|$. ◀

We are now ready to compare numbers without using a number line. When comparing two fractions, convert each one using a common denominator.

Example 6 Write the set of numbers described, then graph the numbers on a number line.

(a) The set of whole numbers less than 5.

(b) The set of negative integers greater than −4.

(c) The set of integers between −2 and 4.

Solution **(a)** The set described is $\{0, 1, 2, 3, 4\}$ and these numbers are graphed on the following number line.

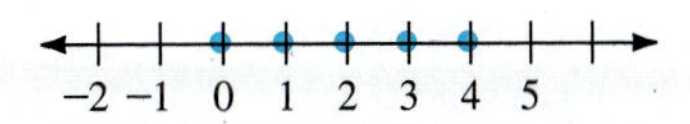

(b) This set is $\{-3, -2, -1\}$ and is graphed below.

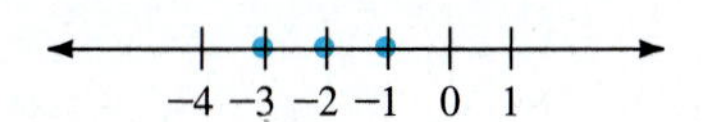

(c) The set of integers between −2 and 4 does not include −2 or 4. The set is given by $\{-1, 0, 1, 2, 3\}$ and is shown below.

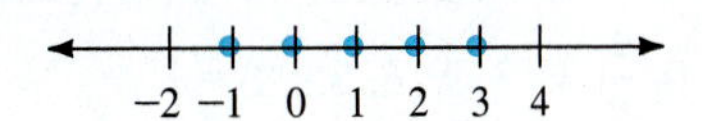

Two decimals such as 3.45 and 3.46 can be compared by comparing the corresponding digits *after* the decimal point. The two numbers are the same until the hundredth's place. In this place, the first number has a 5 and the second number has a 6. Therefore, $3.45 < 3.46$.

Example 7 List three consecutive even integers with the middle one being 6.

Solution The even integer immediately before 6 is 4, and the even integer immediately after 6 is 8. See the figure.

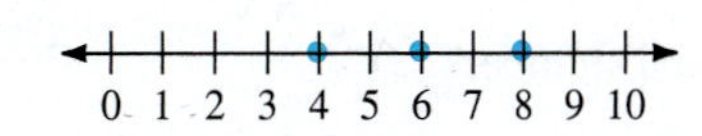

The three consecutive even integers are 4, 6, and 8.

WARMING UP

Answer true or false.

1. $3 \in \{1, 2, 3\}$

2. $\varnothing = \{0\}$

3. $\{a, b, c\} \subseteq \{a, b\}$

4. $\{0, 1, 2, 3\}$ is a subset of the set of whole numbers.

5. 12 is a counting number.

6. −3 is an integer.

7. $\sqrt{3}$ is a rational number.

8. Zero is an irrational number.

9. $|-7| = 7$

10. $|-2 + 3| = 5$

EXERCISE SET 1.3

1. Draw a number line, then graph the numbers from the set

$$\{-1, 0, 0.5, 1, \sqrt{3}\}$$

The irrational number $\sqrt{3}$ is approximately equal to 1.7.

2. Draw a number line, then graph the numbers from the set

$$\left\{-\frac{3}{2}, \frac{1}{2}, \sqrt{5}, 3\right\}$$

The irrational number $\sqrt{5}$ is approximately equal to 2.2.

3. List all numbers from the set

$$\left\{-\frac{10}{5}, -\frac{1}{2}, 0, \frac{1}{2}, \sqrt{3}, \pi, 6, 7.8, 8\frac{1}{2}\right\}$$

that are

(a) Counting numbers.

(b) Whole numbers.

(c) Negative integers.

(d) Integers.

(e) Noninteger rational numbers.

(f) Rational numbers.

(g) Irrational numbers.

(h) Real numbers.

4. List all numbers from the set

$$\left\{-7, -2, -\sqrt{2}, 0, 0.5, \sqrt{5}, 4, 5\frac{1}{2}, \frac{12}{2}\right\}$$

that are

(a) Counting numbers.

(b) Whole numbers.

(c) Negative integers.

(d) Integers.

(e) Noninteger rational numbers.

(f) Rational numbers.

(g) Irrational numbers.

(h) Real numbers.

For Exercises 5–10, represent each quantity as either a positive number or a negative number.

5. Jamie deposited \$45 in his savings account.

6. In football, the quarterback was tackled for a loss of 18 yards.

7. The Tennis Boutique had a loss of \$10,000 last year.

8. During the night, the temperature dropped 15 degrees.

9. The population of Smallville, Pa., decreased by 234 people last year.

10. The basketball team won by three points.

11. If 100 means 100 miles east, what does -100 mean?

12. If $-2{,}000$ means 2,000 feet below sea level, what does 2,000 mean?

13. Find the absolute value of each number.
(a) 12 **(b)** -42
(c) -0.1 **(d)** 0
(e) -15 **(f)** $-\frac{2}{3}$
(g) $-\sqrt{7}$

14. Find the absolute value of each number.
(a) 5 **(b)** -4
(c) 0.2 **(d)** -0.6
(e) -1.4 **(f)** $\frac{1}{2}$
(g) $-\pi$

For Exercises 15–24, place either $<$ or $>$ between the two numbers to make a true statement.

15. $1 \square 5$

16. $-1 \square -5$

17. $\frac{1}{2} \square -\frac{1}{2}$

18. $0.33 \square 0.3$

19. $-\frac{1}{3} \square -\frac{1}{2}$

20. $-\frac{4}{3} \square -\frac{7}{6}$

21. $2 \square \frac{1}{2}$

22. $-2 \square -\frac{1}{2}$

23. $0.65 \square 0.66$

24. $-0.71 \square -0.72$

For Exercises 25–28, write the set of numbers described, then graph the numbers on a number line.

25. The set of whole numbers less than 4.

26. The set of counting numbers less than 6.

27. The set of negative integers greater than or equal to -4.

28. The set of integers between -3 and 3.

29. List three consecutive odd integers such that the smallest one is 5.

30. List three consecutive even integers such that the largest one is 12.

31. List three consecutive even integers such that the middle one is 0.

32. List three consecutive odd integers such that the middle one is -1.

For each pair of numbers, which is larger?

33. $\frac{43}{89}$ and 0.4831561

34. $\frac{65}{23}$ and $\frac{659}{233}$

35. $21^{0.45}$ and $5^{2.047}$

SAY IT IN WORDS

36. The thermometer is one example of an application of the number line. Give examples of two other real-world connections of the number line.

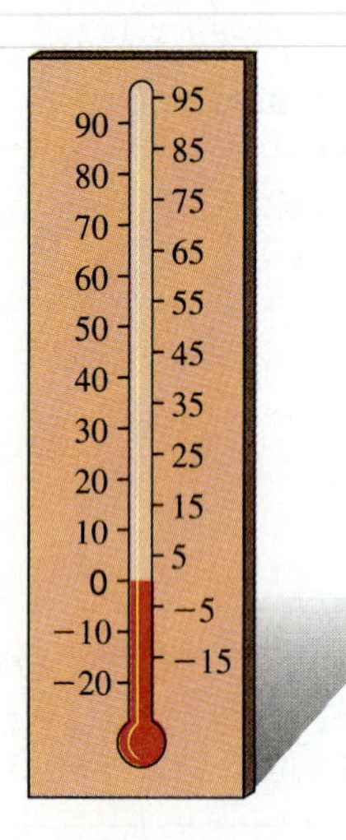

37. Explain your concept of the absolute value of a number.

ENRICHMENT EXERCISES

1. On a number line, graph the two numbers -1 and 5. By looking at the line, find the number between -1 and 5 that is twice as far from -1 as it is from 5.

2. If x is a positive number, what is $|x|$?

3. If x is a negative number, what is $|x|$?

4. Show that $|-x| = |x|$, where x is any number.

5. If a and b are numbers, show that $|a - b| = |b - a|$.

Answers to Enrichment Exercises begin on page A.1.

1.4 Addition and Subtraction

OBJECTIVES

- *To add numbers on a number line*
- *To find the opposite of a number*
- *To understand the rule for subtraction*
- *To translate word problems to addition and subtraction problems*

LEARNING ADVANTAGE *A major factor in successfully completing a mathematics course is to be consistent in your study habits. Spend time on this course each day. In fact, to achieve maximum benefit, divide your daily study time into two segments.*

From your previous experience with arithmetic, you know how to add two numbers. In this section, we will look at the geometric method of adding numbers by using the number line. For example, to evaluate $5 + 2$ on a number line, the 5 tells us to start at zero and draw an arrow 5 units in the positive direction (to the right). The tip of the arrow lands on the number 5 as shown below.

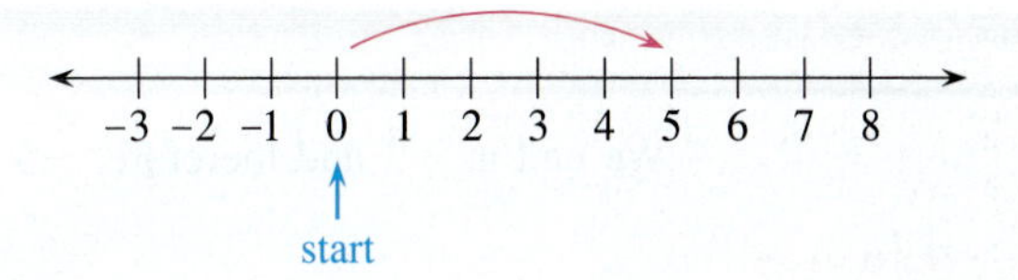

Next, the $+2$ tells us to start a second arrow from the tip of the first arrow and move 2 units in the positive direction. The arrow ends at the number 7 as shown in the figure below. This geometrically shows that $5 + 2 = 7$.

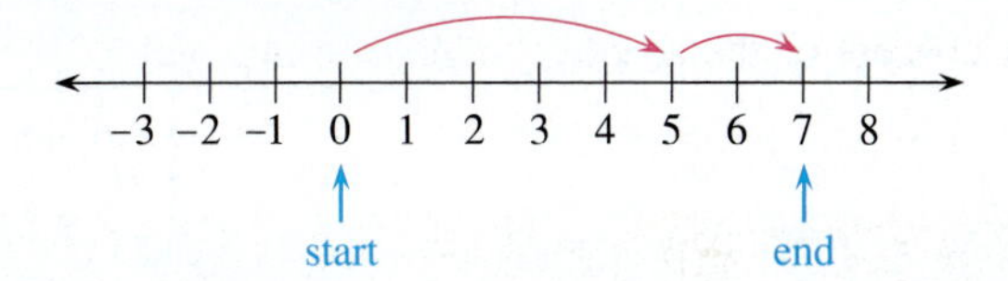

In summary, to add $5 + 2$ on the number line, start at the origin and move 5 units to the right (the *positive* direction) and then move 2 units again in the *positive* direction.

The following examples show how to use the number line to add $5 + (-2)$, $-5 + 2$, $-5 + (-2)$, and then other combinations of numbers.

Example 1 Use a number line to add $5 + (-2)$.

Solution On a number line draw an arrow from the origin 5 units to the right, the positive direction. Then draw an arrow 2 units to the left, the *negative* direction, as shown below. The tip of the second arrow lands on 3. Therefore, $5 + (-2) = 3$.

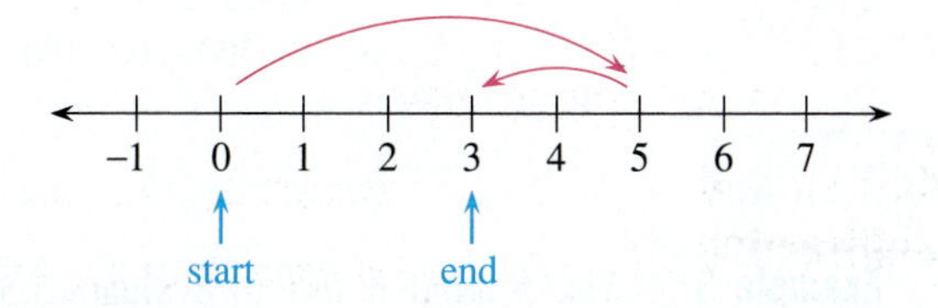

NOTE *If a number is positive, move to the right, the positive direction, on the number line. If a number is negative, move to the left, the negative direction.*

Example 2 Use a number line to add $-5 + 2$.

Solution Start at the origin and move 5 units in the negative direction. Then, move 2 units in the positive direction.

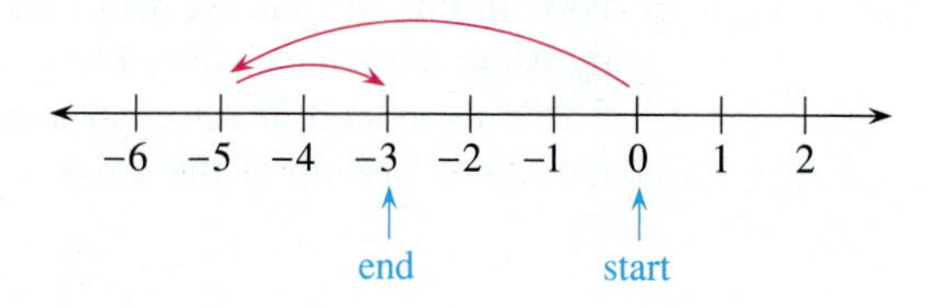

We end at -3 and therefore, $-5 + 2 = -3$.

Example 3 Use a number line to add $-5 + (-2)$.

Solution Start at the origin and move 5 units in the negative direction, then move 2 units in the negative direction.

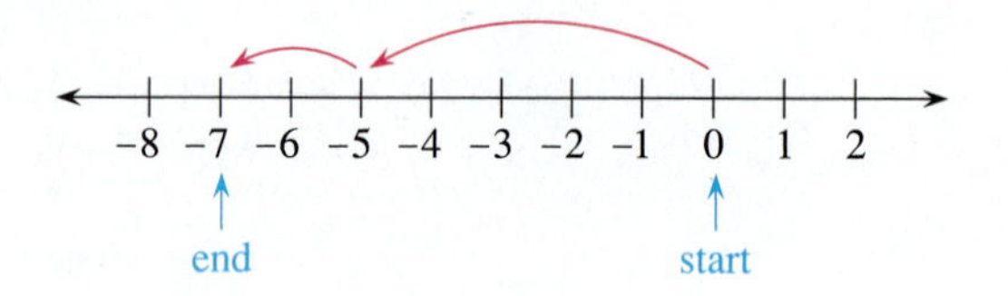

We end at -7, and so $-5 + (-2) = -7$.

Example 4 Use a number line to add $-200 + 700$.

Solution Since the numbers involved are in the hundreds, we draw a scale on the number line to conveniently fit the two numbers. The arrows and result are shown in the figure. Therefore, $-200 + 700 = 500$.

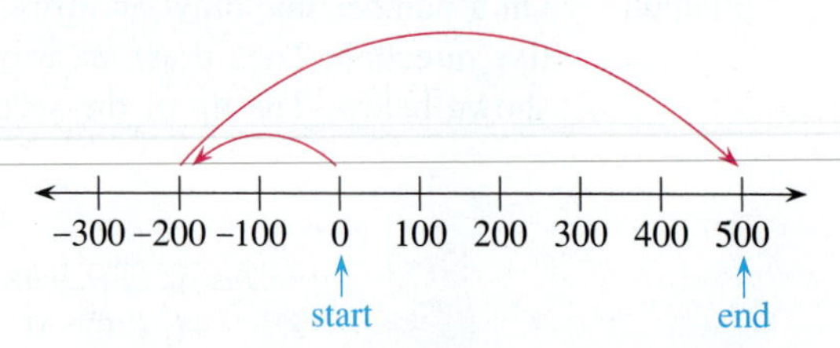

Example 5 Use a number line to evaluate $3.5 + (-3.5)$.

Solution First draw an arrow from the origin 3.5 units to the right. Then, draw the second arrow 3.5 units to the left as shown in the figure. Therefore, $3.5 + (-3.5) = 0$.

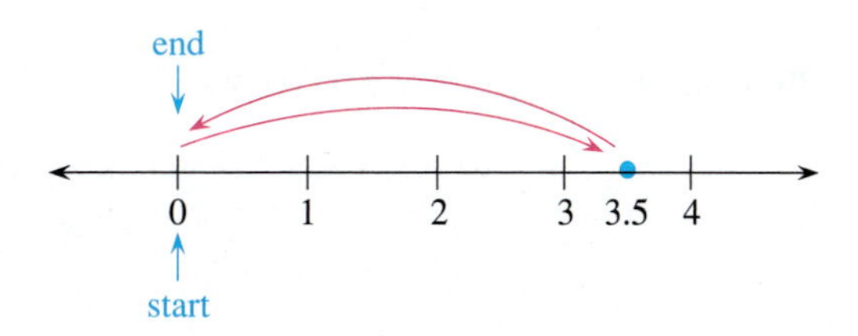

As illustrated in Example 5, the sum of a number a with $-a$ is zero, that is,

$$a + (-a) = 0$$

The number $-a$ is called the **opposite** or **additive inverse** of a. For example, since $3.5 + (-3.5) = 0$, the number -3.5 is the opposite of 3.5. Note also that $(-3.5) + 3.5 = 0$, so 3.5 is the opposite of -3.5. In general, we have

$$-(-a) = a$$

That is, the additive inverse of $-a$ is a.

Example 6

(a) The opposite of 34 is -34, since $34 + (-34) = 0$.

(b) The additive inverse of -6 is 6, since $-6 + 6 = 0$.

(c) Since $0 + 0 = 0$, the opposite of 0 is 0.

We now give the rule for adding two numbers, although we must keep in mind the mental picture of movement on the number line when adding numbers.

RULE

Adding Two Numbers

1. If the numbers are both positive or both negative, add their absolute values. The sum has the same sign as the two numbers.
2. If one number is positive and the other is negative, subtract the smaller absolute value from the larger absolute value. The result has the sign of the number with the larger absolute value.

This rule for adding two numbers should be memorized. It is more efficient to use than the graphing method for adding numbers.

Example 7

Use the rule for adding numbers to verify mentally that each answer is correct. Use a number line if you are not sure of the answer.

(a) $4 + 12 = 16$ **(b)** $-4 + 12 = 8$

(c) $4 + (-12) = -8$ **(d)** $-4 + (-12) = -16$

Example 8 Add $-5 + (-1) + 4$.

Solution We use the order of operations convention and add from left to right.

$$\begin{aligned} -5 + (-1) + 4 &= -6 + 4 \\ &= -2 \end{aligned}$$

Example 9 Simplify $[-3 + (-7)] + 3[12 + (-8)]$.

Solution First, we perform the operations within the brackets, multiply, and then add from left to right.

$$\begin{aligned} [-3 + (-7)] + 3[12 + (-8)] &= -10 + 3(4) \\ &= -10 + 12 \\ &= 2 \end{aligned}$$

The next example contains the absolute value of the sum of two numbers. Treat the absolute value symbols as you would parentheses. Do the operations within an absolute value to obtain a single number, then find the absolute value of this number.

Example 10 Simplify $|3 + (-5)| + (-6)$.

Solution

$$\begin{aligned} |3 + (-5)| + (-6) &= |-2| + (-6) \\ &= 2 + (-6) \\ &= -4 \end{aligned}$$

Example 11 The temperature this morning in Madison was 13 below zero. By noon, it had increased by 21 degrees. What is the temperature at noon?

Solution The temperature in the morning was -13 degrees and then increased by 21 degrees. Therefore, we find the sum of -13 and 21, $-13 + 21 = 8$. The temperature at noon is 8 degrees.

We now consider the operation of subtraction. A subtraction problem such as $5 - 2$ is called the **difference** of 5 and 2. Subtraction is related to addition as illustrated by the following examples.

$$\begin{aligned} 5 - 2 &= 3 &\quad \text{because} \quad 2 + 3 &= 5 \\ 4 - 6 &= -2 &\quad \text{because} \quad 6 + (-2) &= 4 \\ -1 - 3 &= -4 &\quad \text{because} \quad 3 + (-4) &= -1 \end{aligned}$$

Observe also the following relationship between subtraction and addition.

$$\begin{aligned} 5 - 2 &= 3 &\quad \text{and} \quad 5 + (-2) &= 3 \\ 4 - 6 &= -2 &\quad \text{and} \quad 4 + (-6) &= -2 \\ -1 - 3 &= -4 &\quad \text{and} \quad -1 + (-3) &= -4 \end{aligned}$$

From these relationships, we see that a subtraction problem can be converted to an addition problem. The general rule for subtraction is now stated.

RULE

Subtraction

For any two numbers a and b,

$$a - b = a + (-b)$$

That is, the difference of a and b is the same as the sum of a and the additive inverse of b.

For example, $9 - 5$ means $9 + (-5)$. Therefore,

$$\begin{aligned} 9 - 5 &= 9 + (-5) \\ &= 4 \end{aligned}$$

Example 12 Subtract:

(a) $12 - 5$ **(b)** $-14 - 23$ **(c)** $10 - (-3)$

Solution To find each difference, we first convert each problem to an addition problem by using the rule for subtraction.

(a) $$\begin{aligned} 12 - 5 &= 12 + (-5) \\ &= 7 \end{aligned}$$

(b) $$\begin{aligned} -14 - 23 &= -14 + (-23) \\ &= -37 \end{aligned}$$

(c) $$\begin{aligned} 10 - (-3) &= 10 + [-(-3)] \\ &= 10 + [3] && \text{Remember that } -(-3) = 3. \\ &= 13 \end{aligned}$$

◀

Example 13 Simplify $12 - (-4) - 15$.

Solution First change all subtractions to additions.

$$\begin{aligned} 12 - (-4) - 15 &= 12 + [-(-4)] + (-15) \\ &= 12 + 4 + (-15) && \text{Recall that } -(-a) = a. \\ &= 16 + (-15) && \text{Add left to right.} \\ &= 1 \end{aligned}$$

◀

Example 14 Simplify $3 - |7 - 12|$.

Solution

$$\begin{aligned} 3 - |7 - 12| &= 3 - |-5| && \text{Find the difference } 7 - 12. \\ &= 3 - 5 && |-5| = 5. \\ &= -2 \end{aligned}$$

◀

The next example contains multiplication and exponents along with addition and subtraction. Recall that by the order of operations convention, we simplify all powers and multiplication *before* addition and subtraction.

Example 15 Simplify $3^2 - 2^3 - 5 \cdot 2$.

Solution
$$\begin{aligned} 3^2 - 2^3 - 5 \cdot 2 &= 9 - 8 - 10 \\ &= 1 - 10 \\ &= -9 \end{aligned}$$

The next two examples in this section concern subtraction problems that are stated in words.

Example 16 Subtract 4 from -9.

Solution We first write the problem in terms of subtraction, then use the rule for subtraction to evaluate.

$$\begin{aligned} -9 - 4 &= -9 + (-4) \\ &= -13 \end{aligned}$$

Example 17 Find the difference of 14 and -6.

Solution Subtracting -6 from 14, we have

$$\begin{aligned} 14 - (-6) &= 14 + 6 \\ &= 20 \end{aligned}$$

As we have seen, the additive inverse plays an important role in subtraction. The concept of additive inverse can also be used in an algebraic definition of absolute value. Since the additive inverse of a negative number is positive, we can define the absolute value of a number x as follows:

$$|x| = \begin{cases} x, & \text{if } x \geq 0 \\ -x, & \text{if } x < 0 \end{cases}$$

In particular, the absolute value of a negative number is its additive inverse. For example, $|-7| = -(-7) = 7$.

TEAM PROJECT

(3 or 4 students)

TO DESIGN A NUMBER GAME

Project: To make a number game using the four operations of addition, subtraction, multiplication, and division.

Course of Action: Decide upon the nature of the game. Design a board for the game if it is a board game. If it is an action game, decide how it will be played.

As a group, write the rules of your team game. Play the game using problems from Exercise Set 1.4. Trade games with another team. Read the rules of their game. Play their game using problems from Exercise Set 1.4.

WARMING UP

Answer true or false.

1. $3 + (-2) = 1$

2. $-5 + (-1) = -6$

3. $-6 + 3 = 3$

4. $-1 - 4 = -5$

5. $|3 - 4| + (-1) = 0$

6. $\frac{3}{2} + \left(-\frac{5}{2}\right) = -\frac{1}{2}$

7. $40 + (-50) = -10$

8. $\frac{|2 + 3 - 7|}{2} = -1$

9. $2^3 - 5 = 1$

10. $|4 - 8 + 1| = 13$

EXERCISE SET 1.4

For Exercises 1–12, find the sum. You may want to use a number line to check your answers.

1. $3 + 2$

2. $5 + 7$

3. $-4 + 6$

4. $-5 + 8$

5. $9 + (-3)$

6. $4 + (-1)$

7. $-4 + (-2)$

8. $-1 + (-6)$

9. $-8 + (-2)$

10. $-4 + (-7)$

11. $-12 + (-12)$

12. $-14 + (-6)$

For Exercises 13–24, simplify each expression.

13. $6 + (-3) + (-2)$

14. $5 + (-9) + (-3)$

15. $4 + 3(6 - 4)$

16. $2(7 - 3) + (-3)$

17. $(-3)[4 + (-6)]$

18. $(7 - 5)3 + 2$

19. $4^2 + (5 - 4)8$

20. $(7 - 8)3 - 2^3$

21. $|4 + (-9)| - 5$

22. $-2 + |-3 + (-4)|$

23. $\frac{|7 - 10|}{2^2 + 2}$

24. $\frac{|1 - 3^2|}{|6 - 12|}$

For Exercises 25–32, find the additive inverse.

25. 4

26. 7

27. -6

28. $-\frac{1}{2}$

29. 0.4

30. -0.9

31. -0

32. 0

33. Find the sum of 1 and the additive inverse of 5.

34. Find the sum of -2 and the additive inverse of 3.

35. Two tennis players, Pete and Emmanuel, compete in a tournament. Records show that Pete's serve averages 92 mph, whereas, Emmanuel's serve averages 104 mph. How much faster than Pete's serve is Emmanuel's serve?

36. In 1993, July 1st was the 182nd day of the year. How many days were left in this year?

37. The world's record for the longest sandwich is 850 feet long. This is 85 feet longer than the previously held record. How long was the sandwich that held the record previously?

38. The final score of a basketball game was Hornets 113 and Celtics 94. What was the margin of victory?

For Exercises 39–54, simplify.

39. $8 - 3$

40. $7 - 6$

41. $-3 - 5$

42. $-4 - 2$

43. $12 - (3 - 5)$

44. $4 - (7 - 3)$

45. $-(3 - 8)$

46. $-(10 - 12)$

47. $-\frac{1}{2} - \frac{1}{2}$

48. $-\frac{3}{4} - \frac{1}{4}$

49. $|-12| + |10 - 13|$

50. $50 - |100 - 200|$

51. $|2 - 5| - |9 - 12|$

52. $|-2 - 4| + |3 - 5|$

53. $3 - [-4 - (5 - 10)]$

54. $16 - [12 - (6 - 14)]$

55. Subtract 5 from 8.

56. Subtract 9 from 2.

57. Subtract 3 from −4.

58. Subtract 6 from −8.

59. Subtract −5 from 1.

60. Subtract −10 from 3.

61. Subtract −2 from −14.

62. Subtract −7 from −8.

63. Find the difference of 6 and 2.

64. Find the difference of 12 and 8.

65. Find the difference of 4 and −1.

66. Find the difference of 10 and −2.

67. Find the difference of −1 and 5.

68. Find the difference of −6 and 3.

69. What number is $\frac{7}{2}$ less than $-\frac{9}{2}$?

70. What number is 0.4 less than 0.3?

71. The temperature at noon was $-1°$ and by 6 o'clock it dropped $12°$. What is the new temperature?

72. On one play the football team lost 15 yards, and on the next play the team gained 7 yards. What is the team's total gain or loss for the two plays?

73. The gestation period of the American Elk ranges from 270 to 285 days. What is the difference from the shortest to the longest length of the gestation period?

74. Water vapor in the atmosphere is present in concentrations from 0.1% to 1%. What is the difference in concentration of water vapor in the atmosphere from the lowest to the highest?

75. The distance from earth to the star Proxima is 1.31 parsecs. The distance from earth to the star Sirius is 2.67 parsecs. How much further from earth is Sirius than Proxima?

76. The span of the Golden Gate Bridge is 1280 meters. The span of the George Washington Bridge is 1067 meters. What is the difference in lengths of the two spans?

78. $3.26 + 7.95$

79. $8.07 + (-9.11)$

80. $-82.43 + (-71.98)$

81. $-0.057 + 0.0812$

82. $0.9017 - 1$

83. $1 - 0.762$

SAY IT IN WORDS

84. Explain your method for adding two numbers.

85. In your own words, how do you find the difference of two numbers?

ENRICHMENT EXERCISES

1. Which number in the set is 10,000 more than 999,999?

$$\{9{,}999{,}999,\ 1{,}099{,}999,\ 1{,}000{,}999,\ 1{,}009{,}999\}$$

For Exercises 2 and 3, find the missing number in the sequence.

2. 53, 43, 34, 26, _____, 13, 8, 4

3. 0.1, 0.09, 0.088, _____, 0.08766, 0.087655

4. During a football game, the home team has a first down on the 50-yard line. During the next three plays, the team gains 2 yards, loses 18 yards, and gains 7 yards. On what yard line is the ball located?

Answers to Enrichment Exercises begin on page A.1.

1.5 Multiplication and Division

OBJECTIVES

- *To learn the rules for multiplying and dividing signed numbers*
- *To translate word problems into multiplication and division problems*

LEARNING ADVANTAGE

When doing homework or taking a test, read the directions carefully. Furthermore, neatness really does count. Always use a pencil and never a pen when doing mathematics. When you make a mistake, erase rather than cross out. You may lose your train of thought if your work is messy.

The rules of multiplication depend upon the signs (positive or negative) of the numbers involved. For example, multiplying the two positive integers three and four can be thought of as repeated addition.

$$\begin{aligned} 3 \cdot 4 &= 4 + 4 + 4 \\ &= 12 \end{aligned}$$

Example 1 A box contains 18 cans of tennis balls. If each can contains three balls, how many tennis balls are in the box?

Solution If there are 18 cans containing three tennis balls each, then the total number of tennis balls can be obtained by adding $3 + 3 + 3 + \cdots + 3$, where there are 18 threes. However, repeated addition of the number three is the product $18 \cdot 3$. Therefore, there are $18 \cdot 3 = 54$ tennis balls in the box.

We next look at rules for multiplying two real numbers. The first rule is for multiplying two positive numbers.

RULE

Multiplying Two Positive Numbers

The product of two positive numbers is a positive number.

$$(\text{positive})(\text{positive}) = \text{positive}$$

Consider the following table of products.

First Factor		Second Factor		Product	
3	·	4	=	12	
					Subtract 3.
3	·	3	=	9	
					Subtract 3.
3	·	2	=	6	
					Subtract 3.
3	·	1	=	3	

Notice the pattern in this table. As we decrease the second factor by one each time, the product decreases by three each time. To continue this pattern,

$$3 \cdot 0 = 0$$

This statement is true in general.

PROPERTY

The Multiplication Property of Zero

For any number a, we have

$$a \cdot 0 = 0$$

Now, let us return to our table. So far the table looks like this:

$$\begin{aligned} 3 \cdot 4 &= 12 \\ 3 \cdot 3 &= 9 \\ 3 \cdot 2 &= 6 \\ 3 \cdot 1 &= 3 \\ 3 \cdot 0 &= 0 \end{aligned}$$

If we continue the pattern that whenever we decrease the second factor by one, the product decreases by three, then

$$\begin{aligned} 3 \cdot (-1) &= -3 \\ 3 \cdot (-2) &= -6 \\ 3 \cdot (-3) &= -9 \end{aligned}$$

From this, we can suggest the following rule.

RULE

The Product of Two Numbers of Different Signs

The product of a positive number and a negative number is negative.

$$(\text{positive})(\text{negative}) = \text{negative}$$

and

$$(\text{negative})(\text{positive}) = \text{negative}$$

In particular, if a and b are positive numbers,

$$a(-b) = -ab$$

and

$$(-a)b = -ab$$

Example 2 Find the products.

(a) $12 \cdot 3$ **(b)** $(0.4)(-6)$ **(c)** $\left(-\frac{2}{3}\right)\left(\frac{4}{7}\right)$

Solution

(a) $12 \cdot 3 = 36$

(b) $(0.4)(-6) = -(0.4)(6)$ $\quad a(-b) = -ab.$

$= -2.4$

(c) $\left(-\frac{2}{3}\right)\left(\frac{4}{7}\right) = -\left(\frac{2}{3}\right)\left(\frac{4}{7}\right)$ $\quad (-a)b = -ab.$

$= -\frac{2 \cdot 4}{3 \cdot 7}$ $\quad$ Multiply two fractions.

$= -\frac{8}{21}$

We next consider the product of two negative numbers. Let us observe a pattern in the following table.

First Factor		Second Factor		Product	
−3	·	4	=	−12	
					Add 3.
−3	·	3	=	−9	
					Add 3.
−3	·	2	=	−6	
					Add 3.
−3	·	1	=	−3	
					Add 3.
−3	·	0	=	0	

Notice the pattern that develops in the table. As we decrease the second factor by one each time, the product increases by three each time. Let us continue this pattern.

First Factor		Second Factor		Product	
-3	$\cdot$	0	$=$	0	
					Add 3
-3	$\cdot$	-1	$=$	3	
					Add 3
-3	$\cdot$	-2	$=$	6	
					Add 3
-3	$\cdot$	-3	$=$	9	
					Add 3
-3	$\cdot$	-4	$=$	12	

From this pattern we can suggest the following rule.

RULE

The Product of Two Negative Numbers

The product of two negative numbers is positive.

$$(\text{negative})(\text{negative}) = \text{positive}$$

In particular, if a and b are positive numbers,

$$(-a)(-b) = ab$$

Example 3 Simplify.

(a) $(-2)(-3.7)$ **(b)** $(-5)^2$ **(c)** -5^2

Solution **(a)** $(-2)(-3.7) = 2(3.7)$ $(-a)(-b) = ab.$

$= 7.4$

(b) $(-5)^2 = (-5)(-5) = 5 \cdot 5 = 25$

(c) $-5^2 = -(5 \cdot 5) = -25$

COMMON ERROR

Do not mistake $-a^2$ for $(-a)^2$. The term $-a^2$ means that you square a first, then use the minus sign. The term $(-a)^2$ means $(-a)(-a)$ which is a^2. For example, -3^2 means $-(3 \cdot 3)$ or -9, whereas, $(-3)^2$ means $(-3)(-3)$ or 9.

Another multiplication property that will be used in later chapters is that concerning -1.

PROPERTY

The Multiplication Property of the Number −1

For any number a,

$$(-1)a = -a$$

That is, negative one times a number a is the opposite of a.

Example 4

(a) $(-1)(5) = -5$ The opposite of 5 is -5.

(b) $(-1)\left(\frac{1}{2}\right) = -\frac{1}{2}$ The opposite of $\frac{1}{2}$ is $-\frac{1}{2}$.

(c) $(-1)(-6) = 6$ The opposite of -6 is 6.

(d) $(-1)(-1) = 1$ The opposite of -1 is 1.

Example 5 Fill in the blank.

(a) $-7 = (-1)$ ______ **(b)** $\frac{3}{11} = (-1)$ ______

Solution Use the multiplication property of -1: $-a = (-1)a$.

(a) The correct number is 7, since $(-1)7 = -7$.

(b) The answer is $-\frac{3}{11}$, since $(-1)\left(-\frac{3}{11}\right) = \frac{3}{11}$.

Just as subtraction was defined in terms of addition, division is defined in terms of multiplication. Recall from Section 1.1, two numbers whose product is 1 are called *reciprocals* of each other.

Example 6

(a) 2 and $\frac{1}{2}$ are reciprocals of each other, since $2\left(\frac{1}{2}\right) = 1$.

(b) $\frac{3}{4}$ and $\frac{4}{3}$ are reciprocals of each other, since $\left(\frac{3}{4}\right)\left(\frac{4}{3}\right) = 1$.

(c) 1 is its own reciprocal, since $1 \cdot 1 = 1$.

(d) 0 does not have a reciprocal, since 0 times any number is 0, *not* 1.

When we divide 12 by 4 to get 3, it is the same as 12 multiplied by $\frac{1}{4}$, $12 \div 4 = 12 \cdot \frac{1}{4}$. Therefore, 12 divided by 4 is the same as 12 times the reciprocal of 4. We can generalize this to any division problem.

RULE

Division

For all real numbers a and b, where b is not zero,

$$a \div b = \frac{a}{b} = a \cdot \frac{1}{b}$$

That is, a divided by b means a times the reciprocal of b.

We would not use the rule for division on problems such as $12 \div 4$. However, it is useful for problems as shown in the next example.

Example 7

(a) $\frac{3}{5} \div \frac{7}{4} = \left(\frac{3}{5}\right)\left(\frac{4}{7}\right)$ The reciprocal of $\frac{7}{4}$ is $\frac{4}{7}$.

$= \frac{3 \cdot 4}{5 \cdot 7}$ Multiply the two fractions.

$= \frac{12}{35}$

(b) $\frac{8}{5} \div 3 = \left(\frac{8}{5}\right)\left(\frac{1}{3}\right)$ The reciprocal of 3 is $\frac{1}{3}$.

$= \frac{8 \cdot 1}{5 \cdot 3}$

$= \frac{8}{15}$

(c) $0 \div 5 = 0 \cdot \left(\frac{1}{5}\right)$ The reciprocal of 5 is $\frac{1}{5}$.

$= 0$ $0 \cdot a = 0$.

In Example 7(c), notice that 0 divided by 5 is 0. In general, the number 0 has the following property.

RULE

For any nonzero number a,

$$\frac{0}{a} = 0 \div a = 0$$

That is,

zero divided by any nonzero number is zero.

Since zero does not have a reciprocal,

division by zero is not defined.

For example, $0 \div 3 = 0$, but $3 \div 0$ is not defined.

The rules for division of signed numbers are similar to the rules for multiplication of signed numbers.

RULE

Division of Signed Numbers

1. If a and b are both positive or both negative, then $\frac{a}{b}$ is positive.

$$\frac{\text{positive}}{\text{positive}} = \text{positive}$$

$$\frac{\text{negative}}{\text{negative}} = \text{positive}$$

2. If one number is positive and the other negative, then $\frac{a}{b}$ is negative.

$$\frac{\text{positive}}{\text{negative}} = \text{negative}$$

$$\frac{\text{negative}}{\text{positive}} = \text{negative}$$

Therefore, like signs give a positive quotient and unlike signs yield a negative quotient. These four possibilities are illustrated in the next example.

Example 8 **(a)** $\frac{21}{3} = 7$ **(b)** $\frac{-21}{-3} = 7$ **(c)** $\frac{21}{-3} = -7$ **(d)** $\frac{-21}{3} = -7$

The rules for division of signed numbers can be stated in the following form:

If a and b are numbers with b not equal to zero, then

$$\frac{-a}{b} = -\frac{a}{b} \quad \text{and} \quad \frac{a}{-b} = -\frac{a}{b}$$

For example,

$$\frac{-2}{3} = -\frac{2}{3} \quad \text{and} \quad \frac{2}{-3} = -\frac{2}{3}$$

Example 9 Perform the indicated operations.

(a) $\frac{4}{5} \div (-12)$ **(b)** $-6(-2)^2$ **(c)** $15 \div (-3)(-2)$

Solution **(a)**
$$\frac{4}{5} \div (-12) = \frac{4}{5} \cdot \frac{1}{-12}$$
Write division by -12 as multiplication by $1/-12$.
$$= \frac{4}{5}\left(-\frac{1}{12}\right)$$
$\frac{a}{-b} = -\frac{a}{b}$.
$$= -\frac{4}{5} \cdot \frac{1}{12}$$
$a(-b) = -ab$.
$$= -\frac{\overset{1}{\cancel{4}}}{5} \cdot \frac{1}{\underset{3}{\cancel{12}}}$$
Divide out 4.
$$= -\frac{1}{15}$$

(b) First square -2.
$$-6(-2)^2 = -6 \cdot 4 = -24$$

(c) Perform divisions and multiplications from left to right, so we start with $15 \div (-3)$.
$$15 \div (-3)(-2) = (-5)(-2) = 10$$

Example 10 Simplify: $\frac{2 \cdot 3 - 4 \cdot 5}{15 - 7}$.

Solution We start by simplifying numerator and denominator.
$$\frac{2 \cdot 3 - 4 \cdot 5}{15 - 7} = \frac{6 - 20}{8}$$
$$= \frac{-14}{8}$$
$$= -\frac{14}{8}$$
$\frac{-a}{b} = -\frac{a}{b}$.
$$= -\frac{7}{4}$$
Reduce to lowest terms by dividing numerator and denominator by 2.

Example 11 Place either = or ≠ between the two numerical expressions to make a true statement.

(a) $7 - 2(3) \stackrel{?}{=} 1$

(b) $(-2)^2 + 2(-2) + 1 \stackrel{?}{=} -(-2)$

Solution **(a)** We simplify the left side.

$$7 - 2(3) \stackrel{?}{=} 1$$
$$7 - 6 \stackrel{?}{=} 1$$
$$1 = 1$$

therefore, $7 - 2(3) = 1$.

(b) We simplify both sides.

$$(-2)^2 + 2(-2) + 1 \stackrel{?}{=} -(-2)$$
$$4 - 4 + 1 \stackrel{?}{=} 2$$
$$1 \neq 2$$

therefore, $(-2)^2 + 2(-2) + 1 \neq -(-2)$ ◀

WARMING UP

Answer true or false.

1. $7 \cdot 8 = 56$

2. $(-2)(8) = -16$

3. $(-3)(-4) = -12$

4. $(-1)(3) = -3$

5. $\left(\frac{2}{5}\right)\left(\frac{5}{6}\right) = \frac{2}{3}$

6. $12 \cdot \frac{5}{12} = 5$

7. $2 \div 2 = 1$

8. The reciprocal of 3 is -3.

9. $5 \div 0 = 0$

10. $\frac{-26}{-4} = \frac{13}{2}$

EXERCISE SET 1.5

For Exercises 1–28, find the product.

1. $5 \cdot 7$

2. $3 \cdot 13$

3. $12(-2)$

4. $8(-10)$

5. $(-4)(6)$

6. $(-7)(4)$

7. $(-1)(3)$

8. $(-1)(2.5)$

9. $(-0.1)(12)$

10. $(40)(-2.3)$

11. $(-1)(-17)$

12. $(-1)(-13)$

13. $(-2)(-3)$

14. $(-8)(-3)$

15. $(-1)\left(\frac{3}{4}\right)$

16. $(-1)\left(\frac{4}{9}\right)$

17. $(-1)\left(-\frac{7}{3}\right)$

18. $(-1)\left(-\frac{2}{15}\right)$

19. $\left(\frac{1}{2}\right)\left(\frac{3}{4}\right)$

20. $\left(\frac{2}{3}\right)\left(\frac{4}{9}\right)$

21. $\left(\frac{3}{5}\right)\left(\frac{5}{6}\right)$

22. $\left(\frac{1}{3}\right)\left(\frac{6}{7}\right)$

23. $\left(\frac{12}{3}\right)\left(\frac{4}{8}\right)$

24. $\left(\frac{22}{12}\right)\left(\frac{4}{11}\right)$

25. $7\left(\frac{3}{2}\right)$

26. $\left(\frac{4}{7}\right)3$

27. $\left(\frac{6}{5}\right)(10)$

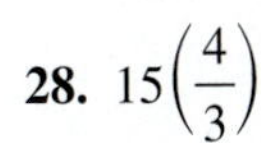

28. $15\left(\frac{4}{3}\right)$

For Exercises 29–38, find the reciprocal of each number.

29. 7

30. 40

31. -7

32. -6

33. $\frac{7}{8}$

34. $\frac{2}{7}$

35. $\frac{11}{2}$

36. $\frac{15}{4}$

37. $-\frac{5}{3}$

38. $-\frac{2}{5}$

For Exercises 39–58, find the quotient. If the division is not defined, write "not defined."

39. $\frac{3}{5} \div 2$

40. $\frac{6}{7} \div 5$

41. $\frac{1}{2} \div \frac{1}{3}$

42. $\frac{2}{5} \div \frac{1}{3}$

43. $0 \div 5$

44. $0 \div (-3)$

45. $\frac{4}{5} \div 2$

46. $\frac{9}{10} \div 3$

47. $9 \div 0$

48. $(-6) \div 0$

49. $30 \div (-5)$

50. $49 \div (-7)$

51. $\frac{-99}{3}$

52. $\frac{-24}{8}$

53. $\frac{-2}{-4}$

54. $\frac{-82}{-6}$

55. $\frac{4}{3} \div \left(-\frac{12}{7}\right)$

56. $\frac{8}{21} \div \left(-\frac{4}{7}\right)$

57. $-\frac{10}{3} \div \left(-\frac{2}{3}\right)$

58. $-\frac{35}{12} \div \left(-\frac{7}{6}\right)$

For Exercises 59–66, perform the indicated operations.

59. $\frac{3}{4} \div (-9)$

60. $8 \div \left(-\frac{4}{5}\right)$

61. $-5(-2)^2$

62. $-3(-1)^2$

63. $12 \div (-4)(-6)$

64. $10 \div (-5)(-7)$

65. $(-8) \div (-4)(-3)$

66. $(-15) \div (-5)(-5)$

For Exercises 67–70, simplify.

67. $\frac{|3 - 7| - 2^3}{(-2)(-3)}$

68. $\frac{4^2 - 3^2}{(4 - 3)^2}$

69. $|3^2 - 12| \div (-3)$

70. $16 \div |(-2)^3 - 10|$

For Exercises 71–78, replace $\stackrel{?}{=}$ with either $=$ or $\neq$ to make a true statement.

71. $4(-1) + 7 \stackrel{?}{=} 3$

72. $-2(-4) - 5 \stackrel{?}{=} -3$

73. $(-2)^2 + 3(-2) + 1 \stackrel{?}{=} 0$

74. $3(-1)^2 + (-1) - 2 \stackrel{?}{=} 0$

75. $\frac{2(4) + 1}{1 - 4} \stackrel{?}{=} -3$

76. $\frac{(-1)^2 + 3}{2^2 + (-1)} \stackrel{?}{=} \frac{2}{3}$

77. $\sqrt{4} + 1 \stackrel{?}{=} (-2)^2 - 1$

78. $|5 - 3^2| - 3 \stackrel{?}{=} 0$

79. Dan's Camera will print 24 slides for \$9.60. How much does each slide cost?

80. If five pounds of apples cost \$6, what is the cost per pound?

81. During one day, Ephraim drank five cups of coffee, two cups of tea, and five bottles of cola. If one cup of coffee, one cup of tea, and one bottle of cola contain 100 mg., 40 mg., and 60 mg. of caffeine, respectively, how many mg. of caffeine did he consume that day?

82. Fran bought 100 shares of Epic stock at \$18 per share and 50 shares of HH stock at \$28 per share. What was his total payment for the stock?

83. The success of a television show is based on a rating scale, where each rating point is equal to 921,000 households having watched the show. One weekly show had an average rating of 12.4 for the season. On average, how many households watched the show each week?

84. See Exercise 83. NBC projected a 16.9 rating for the coverage of the 1992 Summer Games from Barcelona. The actual rating was 18.8. How many more households watched the games than NBC had projected?

Many calculators have a button labeled 1/x to find reciprocals of numbers. For example, to find the reciprocal of 6.4, follow the sequence of steps:

Enter	*Press*	*Display*
6.4	1/x	0.15625

For Exercises 85–88, use a calculator to find the reciprocal of each number.

85. 2.5

86. 1.25

87. −0.004

88. $\frac{5}{6}$

SAY IT IN WORDS

89. Explain how you determine the sign of the product of two numbers.

90. State in your own words the rule for division of signed numbers.

ENRICHMENT EXERCISES

1. Bitsey plans to buy a television set. She has a choice of five brands, six picture sizes, and three sound systems. How many different choices (brand, picture size, and sound system) must she consider?

2. Find a pattern relating the two columns. Then, find a quick way to add the first 100 counting numbers.

Column 1	*Column 2*
$1 + 2$	$\frac{2 \cdot 3}{2}$
$1 + 2 + 3$	$\frac{3 \cdot 4}{2}$
$1 + 2 + 3 + 4$	$\frac{4 \cdot 5}{2}$

Answers to Enrichment Exercises begin on page A.1.

1.6 Variables

OBJECTIVES

- *To evaluate variable expressions and simplify*
- *To convert verbal statements to algebraic expressions*
- *To find solutions of equations*

LEARNING ADVANTAGE

An algebra book cannot be read like a novel. You must ***actively participate*** *by using pencil and paper. For example, if you are not sure of how a step was obtained, write the information and do the calculations yourself. If you find that you are losing concentration, put aside your work and take a break.*

The Pizza Place sells slices of pizza for 29 cents each. The following table gives the cost of buying one, two, three, or four slices of pizza.

Number of Slices	Cost
1	29 · 1¢ or \$.29
2	29 · 2¢ or \$.58
3	29 · 3¢ or \$.87
4	29 · 4¢ or \$1.16

Notice the pattern for the cost in terms of the number of slices. If we denote the number of slices by the letter x, then the cost of the x slices can be expressed as 29 times x or $29x$. We call x a *variable.*

A **variable** is a symbol that stands for a number or numbers. Variables are usually denoted by letters such as x, y, z, s, t, and so on. A variable may stand for a number or numbers that are unknown but are to be found, as in the equation

$$3x^2 - 2x + 1 = 0$$

Sometimes variables are used to describe a mathematical rule or property. For example, when we write $a - b = a + (-b)$, the variables a and b can be any real numbers.

A **constant** is a symbol whose value is fixed. For example, the symbols 1, 3, -19, and π are constants.

One of the primary goals in elementary algebra is to work with algebraic expressions.

DEFINITION

An **algebraic expression** contains constants and/or variables that are combined using the four operations of addition, subtraction, multiplication, and division. For example,

$$25y + 3, \qquad -4m^2, \qquad a(2a + 5), \qquad \frac{3x - y}{14}$$

are algebraic expressions.

When the variable is replaced by a constant, we can **evaluate** an algebraic expression to obtain a number. This number is called a **value of the expression.** For example, if x is replaced by 3, the algebraic expression $4x - 2$ becomes the numerical expression $4(3) - 2$, which simplifies to 10. Therefore, the value of $4x - 2$ is 10 when x is replaced by 3.

Example 1 Find the value of $x^2 + 3x - 1$ when x is replaced by 5.

Solution Wherever x appears in the expression, we replace it by 5,

$$\begin{aligned} x^2 + 3x - 1 &= (5)^2 + 3(5) - 1 \\ &= 25 + 15 - 1 \qquad \text{Simplify.} \\ &= 39 \end{aligned}$$

◀

Example 2 Evaluate $2u + 3v - 5$, if $u = 1$ and $v = 18$.

Solution We replace u with 1 and v with 18, and then simplify the result.

$$\begin{aligned} 2u + 3v - 5 &= 2(1) + 3(18) - 5 \\ &= 2 + 54 - 5 \\ &= 51 \end{aligned}$$

Example 3 Evaluate $3xz - 5yz - xy$, when $x = 4$, $y = 2$, and $z = 6$.

Solution

$$\begin{aligned} 3xz - 5yz - xy &= 3(4)(6) - 5(2)(6) - (4)(2) \qquad x = 4,\ y = 2,\ z = 6. \\ &= 72 - 60 - 8 \\ &= 4 \end{aligned}$$

Example 4 Describe in words the following algebraic expressions.

(a) $x + 5$ **(b)** $3(y - 2)$ **(c)** $(b + 9)^2$

Solution

(a) The sum of x and 5.

(b) Three times the difference of y and 2.

(c) The sum of b and 9, the result squared.

To use algebra in applications, we frequently convert word phrases into algebraic expressions. For example, suppose we are thinking of two numbers—one is seven more than the other. If we denote the smaller number by n, then the other number is given by $n + 7$.

Example 5 Write each word phrase as an algebraic expression.

(a) Six more than a number x.

(b) Eight less than a number m.

(c) Ten times a number z.

Solution

(a) Six more than a number x means $x + 6$.

(b) Eight less than a number m means $m - 8$.

(c) Ten times a number z is $10 \cdot z$ or simply $10z$.

COMMON ERROR

Notice that in part (b) of Example 5, the expression "eight less than a number" is *not* symbolized as $8 - m$. The correct answer is $m - 8$.

Example 6 Write each word phrase as an algebraic expression. Choose a letter for the variable.

(a) A number minus three.

(b) Five times a number.

(c) The difference of 52 and a number.

Solution Let x be the variable.

(a) A number x minus three is $x - 3$.

(b) Five times a number x is $5x$.

(c) The difference of 52 and a number x is $52 - x$.

Example 7 Express the number described in terms of the given variable.

(a) The cost of x pounds of bananas at 68 cents per pound.

(b) The age of Jason, if he is three years younger than Lyn. Lyn is n years old.

Solution **(a)** The cost of x pounds of bananas would be $68 \cdot x$ or $68x$ cents.

(b) Since Lyn is n years old and Jason is three years younger than Lyn, Jason is $n - 3$ years old.

An **equation** is formed by placing an equal sign between two numerical or variable expressions, called **sides** of the equation. For example,

$$7 + 5 = 12, \qquad 9 - 2x = 15, \qquad \text{and} \qquad a^2 - a = 4 - 3a$$

are equations.

An equation containing a variable becomes either a true or a false statement when the variable is replaced by a number. The number is called a **solution** of the equation if the equation becomes a true statement.

Example 8 Check if the given number is a solution of the equation.

(a) Equation: $2x + 1 = -1$
Given number: -1

(b) Equation: $2x^2 + 3x - 2 = 0$
Given number: 2

Solution **(a)** Replace x by -1 in the equation.

$$\begin{aligned} 2x + 1 &= -1 \\ 2(-1) + 1 &\stackrel{?}{=} -1 \\ -2 + 1 &\stackrel{?}{=} -1 \\ -1 &= -1 \end{aligned}$$

Since the equation becomes a true statement, -1 is a solution.

(b) Replace x by 2 in the equation

$$\begin{aligned} 2x^2 + 3x - 2 &= 0 \\ 2(2)^2 + 3(2) - 2 &\stackrel{?}{=} 0 \\ 2 \cdot 4 + 6 - 2 &\stackrel{?}{=} 0 \\ 8 + 6 - 2 &\stackrel{?}{=} 0 \\ 12 &\neq 0 \end{aligned}$$

Therefore 2 is not a solution.

Just as word phrases can be symbolized by algebraic expressions, entire sentences can be symbolized by equations. For example, the sentence "The sum of a number x and five is nine" can be symbolized by the equation $x + 5 = 9$.

The sum of a number x and 5 is nine.

$$x + 5 \qquad = \quad 9$$

Example 9 Write each sentence as an equation. Use x as the variable.

(a) A number plus seven is ten.

(b) Three times a number is 12.

(c) Eighteen divided by a number is twice the number.

Solution Keep in mind that the word "is" translates to "=."

(a) $x + 7 = 10$

(b) $3x = 12$

(c) $\dfrac{18}{x} = 2x$

Example 10 Write the following sentence as an equation using x as a variable: "The sum of the square of a number and six is five times the number." Then, determine if there are any solutions from the set

$$\{1, 2, 3, 4\}$$

Solution The equation that symbolizes the given sentence is

$$x^2 + 6 = 5x$$

Next, we replace x, in turn, by the constants in the set $\{1, 2, 3, 4\}$ to check for solutions.

$x = 1$:
$$\begin{aligned} x^2 + 6 &= 5x \\ (1)^2 + 6 &\stackrel{?}{=} 5(1) \\ 1 + 6 &\stackrel{?}{=} 5 \\ 7 &\neq 5 \end{aligned}$$

$x = 2$:
$$\begin{aligned} x^2 + 6 &= 5x \\ (2)^2 + 6 &\stackrel{?}{=} 5(2) \\ 4 + 6 &\stackrel{?}{=} 10 \\ 10 &= 10 \end{aligned}$$

$x = 3$:
$$\begin{aligned} x^2 + 6 &= 5x \\ (3)^2 + 6 &\stackrel{?}{=} 5(3) \\ 9 + 6 &\stackrel{?}{=} 15 \\ 15 &= 15 \end{aligned}$$

$x = 4$:
$$\begin{aligned} x^2 + 6 &= 5x \\ (4)^2 + 6 &\stackrel{?}{=} 5(4) \\ 16 + 6 &\stackrel{?}{=} 20 \\ 22 &\neq 20 \end{aligned}$$

Therefore, there are two numbers, 2 and 3, from the set that are solutions of the equation.

WARMING UP

Answer true or false.

1. When $x = 1$, the value of $2x - 1$ is 1.
2. When $x = 2$, the value of $x + 3$ is 6.
3. When $x = -3$, the value of $x^2 + 1$ is -8.
4. When $x = -5$, the value of $|x|$ is 5.
5. $3 + x$ means 3 times x.
6. $4(x + 1)$ means 4 times the sum of x and 1.
7. $x - 5$ means the difference of 5 and x.
8. If $y = -2$, then the value of $-|y|$ is -2.
9. $10x = 12$ means the product of 10 and x is 12.
10. $x - 3 = 5$ means the difference of x and 3 is 5.

EXERCISE SET 1.6

For Exercises 1–4, find the value of each algebraic expression when $x = 2$ and when $x = -3$.

1. $3x + 1$
2. $4x^2 + x + 1$
3. $-x^2 + 2x$
4. $\dfrac{6x + 6}{x + 1}$

For Exercises 5–8, find the value of each algebraic expression when $x = 3$ and when $x = -1$.

5. $x - 2$
6. $\dfrac{x^2 + 3x}{2}$
7. $2 - x^3 + 7x$
8. $\dfrac{1 - x}{x}$
9. Evaluate $3x + 5y - 2$ if $x = 5$ and $y = 2$.
10. Evaluate $(a + 2b)(a + 3b)$ for $a = 1$ and $b = 6$.
11. Evaluate $-6 + (3u - v)^3$ for $u = 2$ and $v = 4$.
12. Evaluate $m - 3n + 12$ for $m = 10$ and $n = 3$.
13. Evaluate $\dfrac{s + t}{3}$ for $s = -11$ and $t = 10$.
14. Evaluate $q^2 - (p + 1)^2 - 7$ if $p = -4$ and $q = 6$.
15. Evaluate each expression for $x = -4$.
    (a) $|x|$ **(b)** $|-x|$
    (c) $-|x|$ **(d)** $-|-x|$
16. Evaluate each expression for $y = -3$.
    (a) $|y|$ **(b)** $|-y|$
    (c) $-|y|$ **(d)** $-|-y|$
17. Describe in words the following algebraic expressions.
    (a) $2x$
    (b) $4(5 - m)$
    (c) $(a + 2)^3$
18. Describe in words the following algebraic expressions.
    (a) $3 - p$
    (b) $\dfrac{1}{2}(x - 4)$
    (c) $\dfrac{m + 1}{2}$
19. Write each word phrase as an algebraic expression.
    (a) The sum of a number x and 11.
    (b) Twelve less than a number a.
    (c) Twice a number y added to five.
    (d) Five less than three times a number b.
20. Write each word phrase as an algebraic expression.
    (a) A number x squared.
    (b) A number n subtracted from ten.
    (c) Nine more than four times a number k.
    (d) Six less than twice a number w.

21. Express the number described in terms of the given variable.
(a) The cost of x compact discs, if each disc costs seven dollars.
(b) The age of Lorrie, if she is twice as old as Tim. Tim is n years old.
(c) The amount that x dollars are worth in British pounds if one dollar is worth 1.56 British pounds.

22. Express the number described in terms of the given variable.
(a) The temperature at noon in Atlanta, if it is 21 degrees higher than yesterday's high of n degrees.
(b) The amount of money Jill has, if she has seven fewer dollars than Ted. Ted has m dollars.
(c) The cost of p boxes of cereal at $1.95 per box.

For Exercises 23–38, check whether the given number is a solution of the equation.

23. Equation: $4x - 3 = 5$
Given number: 2

24. Equation: $x + 4 = -1$
Given number: -5

25. Equation: $x^2 - x + 4 = 0$
Given number: 1

26. Equation: $x(2x - 1) = 10$
Given number: -2

27. Equation: $y^3 - y = 0$
Given number: -1

28. Equation: $(y - 1)^2 = 3y$
Given number: -4

29. Equation: $(x - 2)(2x + 1) = 0$
Given number: $-\frac{1}{2}$

30. Equation: $(3x + 2)(x - 4) = 0$
Given number: $-\frac{2}{3}$

31. Equation: $2(x + 1) = 3x$
Given number: 3

32. Equation: $x(1 - x) = x^2 - 7$
Given number: -3

33. Equation: $\sqrt{x} + 1 = 2$
Given number: 1

34. Equation: $2\sqrt{x} = 6$
Given number: 9

35. Equation: $|x| = -5$
Given number: -5

36. Equation: $|x| = 7$
Given number: -7

37. Equation: $\frac{x + 1}{2x - 1} = 1$
Given number: 2

38. Equation: $\frac{2(1 - x)}{x^2 - 3} = 0$
Given number: 1

For Exercises 39–46, write each sentence as an equation using x as a variable. Do not attempt to find solutions.

39. The sum of a number and ten is 15.

40. Eight plus twice a number is five.

41. The product of six and a number is 11.

42. The product of a number and seven is 20.

43. Twelve divided by twice a number is 34.

44. The difference of the square of a number and two is nine.

45. The product of ten and the square of a number is five.

46. The product of the square of a number and nine is 23.

SAY IT IN WORDS

47. In your own words, explain the difference between a numerical expression and an algebraic expression.

48. What is a variable?

ENRICHMENT EXERCISES

1. Write the following sentence as an equation using n to represent the smallest of four consecutive positive even integers.

Twice the sum of the two smallest integers is six more than the sum of the two largest integers.

2. What is the solution to the equation in Exercise 1?

3. What are four consecutive even integers that satisfy the sentence in Exercise 1?

Answers to Enrichment Exercises begin on page A.1.

1.7 Properties of Real Numbers

OBJECTIVES

- *To identify the commutative, associative, and distributive properties of real numbers*
- *To use these properties to simplify numerical expressions*

LEARNING ADVANTAGE

As you progress through the course, it is important to continually review previous material. New concepts are based on mathematics studied earlier, and, therefore, a sound foundation is very important. To help you review the concepts from previous sections, review problems occur at the end of each exercise set starting with Chapter 2. There are also review exercise sets at the end of each chapter.

Some basic properties of the real number system will be stated and illustrated in this section. These properties will be used throughout the book.

From your past experience with numbers, you know that the order makes no difference when adding two numbers. The sum $2 + 3$ is the same as $3 + 2$. Furthermore, order is not important when multiplying two numbers. The product $4(7)$ is the same as $7(4)$. We say that both addition and multiplication are commutative operations.

PROPERTY

The Commutative Property of Addition

If a and b are numbers, then

$$a + b = b + a$$

PROPERTY

The Commutative Property of Multiplication

If a and b are numbers, then

$$a \cdot b = b \cdot a$$

Example 1 **(a)** $3 + x = x + 3$ is an example of the commutative property of addition.
(b) $z \cdot 10 = 10z$ is an example of the commutative property of multiplication.

The next two properties deal with grouping or associating numbers. We find the sum of several numbers by adding two numbers at a time. For example, in the sum

$$46 + 25 + 75$$

we usually do the addition from left to right. Symbolically,

$$(46 + 25) + 75 = 71 + 75 = 146$$

However, for these three numbers, it would have been easier to add 25 and 75 first.

$$46 + (25 + 75) = 46 + 100 = 146$$

Notice that the sum is 146 no matter how we group, or *associate,* the numbers. With this example we illustrated the associative property of addition.

PROPERTY

The Associative Property of Addition

For any three numbers a, b, and c,

$$(a + b) + c = a + (b + c)$$

Similarly, the product of three or more numbers remains unchanged no matter how we group or associate the numbers.

PROPERTY

The Associative Property of Multiplication

For any three numbers a, b, and c,

$$(a \cdot b) \cdot c = a \cdot (b \cdot c)$$

For example, $(2 \cdot 3) \cdot 4 = 6 \cdot 4 = 24$, also $2 \cdot (3 \cdot 4) = 2 \cdot 12 = 24$, and, therefore,

$$(2 \cdot 3) \cdot 4 = 2 \cdot (3 \cdot 4)$$

We show in the next example how use of the associative property allows us to simplify algebraic expressions.

Example 2 Simplify the following expressions.

(a) $3 + (8 + x) = (3 + 8) + x$ * Associative property of addition.

$= 11 + x$

(b) $4(3y) = (4 \cdot 3)y$ Associative property of multiplication.

$= 12y$

(c) $2\left(\frac{3}{2}t\right) = \left(2 \cdot \frac{3}{2}\right)t$ Associative property of multiplication.

$= 3t$

We illustrate the next property with an example:

Four friends plan to attend Dornway Amusement Park for the day. The cost per person is a \$5 admission fee plus \$20 for an all-day pass. We may determine the total amount paid by the four people using two methods.

Method 1. Each person pays $5 + 20 = \$25$. Since there are four people, the total amount is

$$4 \cdot (5 + 20) = 4 \cdot 25 = \$100$$

Method 2. Four admission fees are paid plus four all-day passes. Therefore, the total amount is given by

$$4 \cdot 5 + 4 \cdot 20 = 20 + 80 = \$100$$

Use of this example illustrated the distributive property. Namely,

$$4 \cdot (5 + 20) = 4 \cdot 5 + 4 \cdot 20$$

PROPERTY

The Distributive Property

Let a, b, and c be numbers, then

$$a \cdot (b + c) = a \cdot b + a \cdot c \quad \text{and} \quad (b + c) \cdot a = b \cdot a + c \cdot a$$

We say that multiplication *distributes* over addition.

*A dashed box indicates a step that you should eventually do mentally.

Example 3 Simplify using the distributive property.

(a) $3(x + 22)$ **(b)** $6(a - 7)$

(c) $-(5 + z)$ **(d)** $(3 + x)(8)$

Solution **(a)** $3(x + 22) = 3 \cdot x + 3 \cdot 22$ The distributive property.

$= 3x + 66$

(b) $6(a - 7) = 6a + 6(-7)$

$= 6a - 42$

We see in this problem that multiplication also distributes over subtraction.

(c) $-(5 + z) = (-1)(5 + z)$ $-a = (-1)a$.

$= (-1)5 + (-1)z$ Distributive property.

$= -5 - z$

(d) $(3 + x)(8) = 3 \cdot 8 + x \cdot 8$

$= 24 + 8x$ $x \cdot 8 = 8 \cdot x$. The commutative property of multiplication.

In Example 3(c), we saw that $-(5 + z) = -5 - z$. In general, we have the following rule.

RULE

For any two numbers a and b,

$$-(a + b) = -a - b$$

Here are some more examples of simplifying algebraic expressions using properties of real numbers. In actual practice, we would not justify every step, but we realize that the underlying properties allow us to quickly and efficiently simplify expressions.

Example 4 Simplify each expression.

(a) $12 + 2(4x - 3)$ **(b)** $4 - (z - 1)$ **(c)** $2(1 + 3x + y)$

Solution **(a)** $12 + 2(4x - 3) = 12 + 2(4x) - 2(3)$ Distributive property.

$= 12 + 8x - 6$

$= 6 + 8x$

(b) $4 - (z - 1) = 4 - z + 1$ $-(a + b) = -a - b$.

$= 5 - z$

(c) The distributive property still holds when there is a sum of more than two terms. Therefore, we distribute the number two over the three terms.

$$2(1 + 3x + y) = 2 \cdot 1 + 2(3x) + 2 \cdot y$$
$$= 2 + 6x + 2y$$

Example 5 Erica Delaney had \$387 in her checking account. During the following week, she wrote checks for \$27, \$432, and \$212 and made deposits of \$402 and \$106. How much is in her checking account at the end of this week?

Solution We want to find the quantity

$$387 - 27 - 432 - 212 + 402 + 106$$

The associative and commutative properties allow us to group the positive numbers together and the negative numbers together. We then add each group.

$$387 + 402 + 106 = 895$$
$$-(27 + 432 + 212) = -671$$

Subtracting 671 from 895, we obtain a balance of \$224.

The two numbers zero and one have unique properties that we regularly use in algebra. We have already used these properties without giving them official names.

PROPERTY

Additive Identity Property

There exists a unique number 0 with the property that for any number a,

$$a + 0 = a \quad \text{and} \quad 0 + a = a$$

That is, the number zero preserves the identity of any number under the operation of addition.

PROPERTY

Multiplicative Identity Property

There exists a unique number 1 such that for any number a,

$$a \cdot 1 = a \quad \text{and} \quad 1 \cdot a = a$$

That is, the number one preserves the identity of any number under the operation of multiplication.

PROPERTY

Additive Inverse Property

For each number a, there exists a unique number $-a$ such that

$$a + (-a) = 0 \quad \text{and} \quad (-a) + a = 0$$

That is, the sum of a number and its opposite is 0.

PROPERTY

Multiplicative Inverse Property

For every number a, except 0, there exists a unique number $\frac{1}{a}$ such that

$$a\left(\frac{1}{a}\right) = 1 \quad \text{and} \quad \left(\frac{1}{a}\right)a = 1$$

That is, the product of a nonzero number and its reciprocal is one.

We show in the next example how the various properties of real numbers can be used to change algebraic expressions.

Example 6 State the property that is illustrated in each equation.

Equation	*Property*
(a) $3 + 5 = 5 + 3$	Commutative property of addition.
(b) $\left(-\frac{1}{2}\right)(-2) = 1$	The multiplicative inverse.
(c) $7(4 - 9) = 7 \cdot 4 - 7 \cdot 9$	The distributive property.
(d) $(x + 3)[a + (-a)] = (x + 3)(0)$	The additive inverse.
(e) $1 \cdot (x + y) = x + y$	The multiplicative identity.

The properties of real numbers discussed in this section will be used throughout the book to simplify algebraic expressions. Most of the time, however, we will not state the name of the property being used.

WARMING UP

Answer true or false.

1. $2 + 3 = 3 + 2$
2. $2 \cdot 4 = 4 \cdot 2$
3. $4 \div 1 = 1 \div 4$
4. Addition is a commutative operation.
5. $3 + (6 + x) = 9 + x$
6. $-2(4y) = -8y$
7. $3(2x + 4) = 6x + 4$
8. $(y - 1)2 = 2y - 2$
9. $-(x + 4) = -x + 4$
10. $2 + 4(1 - 2x) = 6 - 8x$

EXERCISE SET 1.7

For Exercises 1–30, simplify using either the associative property of addition or multiplication.

1. $3 + (2 + x)$
2. $4 + (5 + 2x)$
3. $(x + 1) + 7$
4. $(3x + 6) + 2$
5. $-2 + (1 + 3x)$
6. $-5 + (3 + y)$
7. $1 + (-4 + a)$
8. $10 + (-8 + 2y)$
9. $3(2x)$
10. $4(5x)$
11. $(-2)(6y)$
12. $(-4)(2a)$
13. $3(-7c)$
14. $2(-6z)$
15. $(-5)(-3x)$
16. $(-8)(-2b)$
17. $(3z - 1) - 2$
18. $(6x - 5) - 1$
19. $2\left(\frac{1}{2}x\right)$
20. $5\left(\frac{1}{5}y\right)$
21. $3\left(\frac{2}{3}a\right)$
22. $4\left(\frac{3}{4}u\right)$
23. $\frac{2}{3}\left(\frac{3}{2}x\right)$
24. $\frac{3}{5}\left(\frac{5}{3}y\right)$
25. $-3\left(-\frac{1}{3}a\right)$
26. $-7\left(-\frac{1}{7}v\right)$
27. $-2\left(-\frac{3}{2}z\right)$
28. $-4\left(-\frac{3}{4}c\right)$
29. $-\frac{3}{2}\left(-\frac{2}{3}t\right)$
30. $-\frac{4}{5}\left(-\frac{5}{4}x\right)$

For Exercises 31–50, simplify using the distributive property.

31. $2(3 + x)$
32. $3(1 + 3x)$
33. $(2 + a)(4)$
34. $(5 + x)(2)$
35. $6(1 - 2y)$
36. $7(2 - 3c)$
37. $2(-1 + y)$
38. $3(-4 + 3z)$
39. $-4(-y + 1)$
40. $-5(-2c + 4)$
41. $-(2 + x)$
42. $-(3 + 2a)$
43. $-(1 - y)$
44. $-(2 - u)$
45. $\frac{1}{2}(2x + 4)$
46. $\frac{1}{3}(3y + 12)$
47. $\frac{1}{4}(4 - 8b)$
48. $\frac{1}{6}(6t - 18)$

49. $\frac{2}{3}\left(6 + \frac{3}{2}s\right)$

50. $\frac{5}{3}\left(\frac{3}{5}x - 3\right)$

For Exercises 51–60, simplify.

51. $2(3x - 1) + 4$

52. $4(2y + 2) - 6$

53. $-12 + 3(1 - x)$

54. $8 + 2(-3 - y)$

55. $-(3 + x) + (-2)^2$

56. $(-4)^2 + 2(4z - 8)$

57. $\frac{2}{3}(9 - 6x) - (-2)^3$

58. $(-5)^2 + \frac{2}{5}(-30 + 5a)$

59. $0.34(1.2 - 4.7x)$

60. $5.7(9.2y - 8.1)$

For Exercises 61–74, state the property of real numbers that the given equation illustrates.

61. $(3 + 2) + 5 = 3 + (2 + 5)$

62. $1 \cdot 1 = 1$

63. $2 \cdot \left(\frac{1}{2}\right) = 1$

64. $4 \cdot \left(\frac{1}{4}\right) = \left(\frac{1}{4}\right) \cdot 4$

65. $1.7 + (-1.7) = 0$

66. $-12 + 8 = 8 + (-12)$

67. $1 \cdot x = x$

68. $5\left(\frac{1}{5}\right) = 1$

69. $19 + (-19) = (-19) + 19$

70. $-34 + 34 = 0$

71. $(0 + 2z) + 1 = 2z + 1$

72. $1 \cdot (4 + 3a) - 2 = (4 + 3a) - 2$

73. $-3(5 - a) = (-3)5 + (-3)(-a)$

74. $\left(\frac{1}{3} + 1\right)(-2) = \left(\frac{1}{3}\right)(-2) + 1(-2)$

For Exercises 75–80, use the stated property to complete the equality.

75. $2x + 4 =$ ________ The commutative property of addition.

76. $a \cdot 3 =$ ________ The commutative property of multiplication.

77. $1 \cdot (x + y) =$ ________ The multiplicative identity property.

78. $(5z - a) + 0 =$ ________ The additive identity property.

79. $(a - b) - (a - b) =$ ________ The additive inverse property.

80. $\left(\frac{1}{2} \cdot 2\right)a =$ ________ The multiplicative inverse property.

For Exercises 81–84, state which instructions are commutative.

81. Instruction 1: Start the car.
Instruction 2: Get into the car.

82. Instruction 1: Dry your hair.
Instruction 2: Wash your hair.

83. Instruction 1: Put on your right glove.
Instruction 2: Put on your left glove.

84. Instruction 1: Watch TV.
Instruction 2: Do your homework.

SAY IT IN WORDS

85. Explain why variables, and not just numbers, are used in algebra.

86. Explain why $2(x + 3) \neq 2x + 3$.

ENRICHMENT EXERCISES

1. A set S is **closed** with respect to an operation such as addition, if for any two members a and b of S, then $a + b$ is a member of S. For example, the set of real numbers is closed with respect to addition and multiplication. Is the set

$$\{\ldots, -3, -2, -1, 0, 1, 2, 3, \ldots\}$$

closed with respect to addition? multiplication?

2. Is subtraction a commutative operation? Explain.

Answers to Enrichment Exercises begin on page A.1.

CHAPTER 1 Summary and review

Fractions (1.1)

To **write a fraction in lowest terms,** use the property

$$\frac{a \cdot c}{b \cdot c} = \frac{a}{b}, \qquad c \neq 0$$

Example: $\frac{4}{6} = \frac{2 \cdot 2}{3 \cdot 2} = \frac{2}{3}$

To **multiply two fractions,** multiply the numerators and then multiply the denominators, then express the result in lowest terms. In symbols,

$$\frac{a}{b} \cdot \frac{c}{d} = \frac{a \cdot c}{b \cdot d}$$

Example: $\frac{5}{2} \cdot \frac{3}{4} = \frac{5 \cdot 3}{2 \cdot 4} = \frac{15}{8}$

To **divide two fractions,** multiply the first by the reciprocal of the second. In symbols,

$$\frac{a}{b} \div \frac{c}{d} = \frac{a}{b} \cdot \frac{d}{c}$$

Example: $\frac{7}{15} \div \frac{4}{3} = \frac{7}{15} \cdot \frac{3}{4} = \frac{7}{20}$

To **add or subtract two fractions with the same denominator,** add or subtract numerators keeping the same denominator. In symbols,

$$\frac{a}{b} + \frac{c}{b} = \frac{a + c}{b}$$

$$\frac{a}{b} - \frac{c}{b} = \frac{a - c}{b}$$

Examples:

$\frac{4}{3} + \frac{1}{3} = \frac{4 + 1}{3} = \frac{5}{3}$

$\frac{4}{3} - \frac{2}{3} = \frac{4 - 2}{3} = \frac{2}{3}$

Examples

To **add or subtract fractions with unlike denominators,** first rewrite each fraction so that they have a common denominator by using the following property:

$\frac{1}{2} + \frac{3}{4} = \frac{2}{4} + \frac{3}{4} = \frac{5}{4}$

$$\frac{a}{b} = \frac{a \cdot c}{b \cdot c}, \qquad c \neq 0$$

Basic symbols (1.2)

Example	Symbol	Meaning
$\frac{2}{8} = \frac{1}{4}$	$a = b$,	a is equal to b
$2 + 2 \neq 5$	$a \neq b$,	a is not equal to b
$6 < 10$	$a < b$,	a is less than b
$3 \leq 3$	$a \leq b$,	a is less than or equal to b
$9 > 5$	$a > b$,	a is greater than b
$3 \geq 3$	$a \geq b$,	a is greater than or equal to b

Exponents (1.2)

Exponents are used to indicate repeated multiplication. For example, $4 \cdot 4 \cdot 4$ is written as 4^3. The number 3 tells us that 4 is used as a factor 3 times. The number 3 is called the **exponent** and 4 is called the **base.**

Order of operations (1.2)

When simplifying an expression that has no grouping symbols, follow the order of operations convention.

Order of Operations Convention

$$3^2 + 4 \cdot 2 = 9 + 8$$
$$= 17$$

1. Simplify all powers.
2. Perform all multiplications and divisions from left to right.
3. Perform all additions and subtractions from left to right.

If an expression contains grouping symbols, use the order of operations convention within each grouping, starting with the innermost, and work outward until you reach a final answer.

The real numbers (1.3)

The set of real numbers can be classified according to the following subsets.

Counting numbers: $\{1, 2, 3, \ldots\}$
Whole numbers: $\{0, 1, 2, 3, \ldots\}$
Integers: $\{\ldots, -3, -2, -1, 0, 1, 2, 3, \ldots\}$

Examples of rational numbers:

$\frac{2}{3}$, 4, and 0.95

Rational numbers: the set of all real numbers that can be written as a ratio of integers.

Examples

Examples of irrational numbers:

$\sqrt{2}$, $-\sqrt{3}$, and π

Irrational numbers: the set of all real numbers that cannot be written as a ratio of integers.

$|7| = 7, \quad |-2| = 2$

The **absolute value** of a number x, denoted by $|x|$, is the distance that it is located from the origin.

Addition and subtraction (1.4)

The number -6 is the opposite of 6 and $6 + (-6) = 0$.

If a is a number, $-a$ is called the **opposite** or **additive inverse** of a. The sum of a number and its opposite is zero.

$$a + (-a) = 0$$

Rules for Adding Two Numbers

1. If the numbers are both positive or both negative, add their absolute values. The sum has the same sign as the two numbers.
2. If one number is positive and the other is negative, subtract the smaller absolute value from the larger absolute value. The result has the sign of the number with the larger absolute value.

Rule for Subtraction

For any two numbers a and b, the **difference** of a and b is the same as the sum of a and the additive inverse of b. That is,

$6 - 8 = 6 + (-8) = -2$

$$a - b = a + (-b)$$

Multiplication and division (1.5)

Rules for Multiplying Signed Numbers

$2 \cdot 3 = 6$

$(-2)(-3) = 6$

$2 \cdot (-3) = -6$

$(-2) \cdot 3 = -6$

$$(\text{positive})(\text{positive}) = \text{positive}$$
$$(\text{negative})(\text{negative}) = \text{positive}$$
$$(\text{positive})(\text{negative}) = \text{negative}$$
$$(\text{negative})(\text{positive}) = \text{negative}$$

Two multiplication properties are the following:

If a is a number, then

$(-4) \cdot 0 = 0$

$$a \cdot 0 = 0$$

$(-1)(5) = -5$

and

$$(-1) \cdot a = -a$$

The fractions $\frac{3}{4}$ and $\frac{4}{3}$ are reciprocals, since

$\left(\frac{3}{4}\right)\left(\frac{4}{3}\right) = 1$

If a is a nonzero number, then a and $\frac{1}{a}$ are called **reciprocals** of each other, since their product is one:

$$a \cdot \frac{1}{a} = 1$$

Rules for Division

For all real numbers a and b, where b is not zero,

$$a \div b = a \cdot \frac{1}{b}$$

The rules for division of signed numbers correspond to the rules for multiplication of signed numbers. The quotient of two positive or two negative numbers is positive. If one number is positive and the other is negative, then the quotient is negative.

Remember that division *into* zero is zero, but division *by* zero is undefined.

Examples

$$\frac{24}{6} = 4$$

$$\frac{-24}{6} = -4$$

$$\frac{24}{-6} = -4$$

$$\frac{-24}{-6} = 4$$

$$\frac{3}{4} \div 6 = \frac{3}{4} \cdot \frac{1}{6} = \frac{1}{8}$$

Variables (1.6)

A **variable** is a symbol that stands for a number or numbers. A **constant** is a symbol whose value is fixed.

An **algebraic expression** contains constants and/or variables that are combined using the four operations of addition, subtraction, multiplication, and division.

Algebraic expressions:

$3x - 1$, $\frac{5ab - c}{a}$

When a specific value of the variable is selected, we can **evaluate** an algebraic expression to obtain a number.

Evaluating an expression: When $x = -4$,

$$3x + 2 = 3(-4) + 2 = -10$$

An **equation** is formed by placing an equal sign between two numerical or variable expressions, called **sides** of the equation.

An equation:

$3x - 1 = 2$

Properties of real numbers (1.7)

	Addition	***Multiplication***
Commutative	$a + b = b + a$	$a \cdot b = b \cdot a$
Associative	$a + (b + c) = (a + b) + c$	$a \cdot (b \cdot c) = (a \cdot b) \cdot c$
Identity	$a + 0 = a$	$a \cdot 1 = a$
	$0 + a = a$	$1 \cdot a = a$
Inverse	$a + (-a) = 0$	$a\left(\frac{1}{a}\right) = 1$
	$(-a) + a = 0$	$\left(\frac{1}{a}\right)a = 1$
Distributive	$a \cdot (b + c) = a \cdot b + a \cdot c$	
	$(b + c) \cdot a = b \cdot a + c \cdot a$	

CHAPTER 1 REVIEW EXERCISE SET

Section 1.1

For Exercises 1–4, write each fraction in lowest terms.

1. $\frac{7}{14}$ **2.** $\frac{5}{25}$ **3.** $\frac{18}{15}$ **4.** $\frac{21}{49}$

For Exercises 5–8, find the product or quotient and write the answer in lowest terms.

5. $\frac{2}{3} \cdot \frac{1}{2}$ **6.** $\frac{4}{5} \cdot \frac{10}{3}$ **7.** $\frac{5}{9} \div \frac{2}{3}$ **8.** $\frac{11}{2} \div \frac{22}{3}$

For Exercises 9–14, add or subtract as indicated. Write the answer in lowest terms.

9. $\frac{3}{4} + \frac{3}{4}$ **10.** $\frac{1}{3} + \frac{2}{3}$ **11.** $\frac{6}{7} - \frac{1}{7}$

12. $\frac{3}{4} - \frac{1}{4}$ **13.** $\frac{2}{3} - \frac{1}{6} + \frac{1}{2}$ **14.** $\frac{7}{16} + \frac{1}{4} - \frac{3}{8}$

Section 1.2

For Exercises 15–17, symbolize each statement.

15. x is greater than 2. **16.** z is less than or equal to 9. **17.** a is not equal to 5.

For Exercises 18–22, simplify each numerical expression.

18. $(7 - 5)(2 - 1)$ **19.** $(4 + 3 - 1)^2 + 15$ **20.** $(-3)^2 - (4 - 10)$

21. $\frac{30 + 2 \cdot 10}{3 \cdot 2}$ **22.** $\frac{2 \cdot (4 + 5) + 3^2}{(1 + 2)^2}$

Section 1.3

23. Draw a number line, then graph the following numbers.

(a) -3 **(b)** $-2\frac{1}{2}$ **(c)** 0.25 **(d)** 1.5 **(e)** 2 **(f)** π

For Exercises 24–28, place either $<$ or $>$ between the two numbers to make the statement true.

24. -3 $<$ 1 **25.** $\frac{4}{3}$ $<$ $\frac{3}{2}$ **26.** 4.2 $<$ 4.5

27. $-\frac{1}{4}$ $>$ $-\frac{1}{2}$ **28.** 0.9065 $>$ 0.9056

For Exercises 29–33, find the absolute value of each number.

29. 7 **30.** 5.65 **31.** -8 **32.** $-\frac{1}{4}$ **33.** $-\pi$

Section 1.4

For Exercises 34–36, simplify each expression.

34. $-4 + (-2) - (-9)$

35. $|-4| - 4$

36. $|2 - 6| - |4 - 10|$

37. Subtract -3 from -2.

38. Find the difference of -5 and 3.

Section 1.5

For Exercises 39–42, find the product or quotient. If the division is not defined, write "not defined."

39. $(-1)(-5.2)$

40. $63 \div (-9)$

41. $0 \div 2$

42. $2 \div 0$

43. Simplify: $-2(-6) - |3 \cdot 4 - 18| \div 2$

Section 1.6

For Exercises 44–46, find the value of each algebraic expression when $x = 2$.

44. $3 + 5x$

45. $x^2 - 2x + 6$

46. $\frac{|3 - x|}{x^3}$

47. Evaluate $7x + 6y + 8$, when $x = -5$ and $y = 3$.

For Exercises 48–50, write each word phrase as an algebraic expression.

48. Twelve less than a number y.

49. The sum of a number z and nine.

50. Six times the difference of two and a number x.

For Exercises 51–53, write each sentence as an equation. Use x as the variable.

51. The difference of a number and six is ten.

52. The product of a number and four is 36.

53. Thirty divided by the square of a number is five.

54. Determine solutions of $x^2 + 10 = 3x$ from the set $\{1, 2, 5\}$.

For Exercises 55–57, write each sentence as an equation using x as a variable. Then find all solutions from the set $\{0, 1, 2, 3\}$.

55. The sum of twice a number and three is seven.

56. The difference of three and the absolute value of a number is one.

57. The product of four and a number is three more than the number.

Section 1.7

For Exercises 58–60, simplify using either the associative property of addition or the associative property of multiplication.

58. $3(6y)$

59. $-1 + (3 - 7b)$

60. $-\frac{3}{5}\left(-\frac{5}{3}a\right)$

61. Simplify: $10 - (3 + y) - 5(-2)$

For Exercises 62–66, state the property that each equation illustrates.

62. $(-2)(5) = (5)(-2)$

63. $1 \cdot (x + z) = x + z$

64. $2 + (a + 5) = 2 + (5 + a)$

65. $\left[\left(\frac{1}{2}\right)2\right]x = x$

66. $3(a - 4) = 3a - 12$

CHAPTER 1 TEST

1. Write the fraction $\frac{20}{25}$ in lowest terms.

For Problems 2 and 3, combine as indicated and write the answer in lowest terms.

2. $\frac{2}{15} \div \frac{7}{3}$

3. $\frac{2}{3} + \frac{1}{4}$

4. Write the statement using symbols: The sum of x and 3 is greater than or equal to 5.

5. Find the value of 2^3.

For Problems 6 and 7, simplify each numerical expression.

6. $4 + 2 \cdot 6$

7. $\frac{2(4 + 1) - 1}{3^2}$

8. Represent the quantity as either a positive or negative number: Frank withdrew $120 from his checking account.

For Problems 9 and 10, place either $<$ or $>$ between the two numbers to make the statement true.

9. $\frac{3}{5} \quad \frac{4}{7}$

10. $4.12 \quad 4.2$

11. Find the absolute value of -4.1.

For Problems 12–14, simplify each expression.

12. $-7 + (-3) + 5$

13. $\frac{|-5 + 2| + 3}{|-2 + (-8)|}$

14. $-6 - [13 - (2 - 4)]$

15. Subtract -2 from 10.

For Problems 16–19, find the product or quotient if the division is not defined, write "not defined."

16. $(-3)(7)$

17. $\left(\frac{2}{5}\right)\left(\frac{15}{16}\right)$

18. $\frac{2}{7} \div 2$

19. $\frac{3}{4} \div 0$

20. Simplify: $\left[\frac{4}{7}(-14) + \frac{1}{3} \cdot 2\right] \div \frac{11}{12}$

For Problems 21 and 22. Find the value of each algebraic expression when $x = 3$.

21. $6 - 2x$

22. $x^2 + 3x - 5$

23. Evaluate $x^2 - (4 + y)^2 + 3x$ for $x = -2$ and $y = 1$.

24. Write the sentence as an equation using x as the variable: The difference of twice a number and three is 11.

25. State the property of real numbers that the equation illustrates:
$x + (5 + y) = x + (y + 5)$

26. Simplify using the distributive property: $4(x - 5)$

For Problems 27 and 28, simplify.

27. $-(3 - x) + 7(-1)$

28. $4\left(u - \frac{5}{4}\right) + 3$

Linear Equations and Inequalities

CHAPTER 2

CONNECTIONS

If peanuts cost one dollar per pound and cashews cost five dollars per pound, how much of each should be used for a mixture of peanuts and cashews costing four dollars per pound? See Exercise 44 in Section 2.5.

Overview

In this chapter, we will see how algebraic expressions are used to construct linear equations and inequalities. In Sections 2.2 and 2.3, we will investigate properties of equality that will be used to solve linear equations. Next, we develop the skill of translating word problems to mathematics. In this chapter, to solve a word problem means solving either a linear equation or a linear inequality. In the final section, we look at formulas and their applications.

2.1 Algebraic Expressions

OBJECTIVES

- *To identify like terms in an algebraic expression*
- *To simplify an expression by combining like terms*

LEARNING ADVANTAGE

Read slowly the section or sections in the book that are assigned, not once but twice—more often if you have trouble understanding it. Read it first in class if there is time available; read it again later in the day. Each time you read it, you will learn more or find questions to ask.

In order to solve an equation, it is frequently necessary to simplify it first. In this section, we will develop techniques for simplifying an equation.

First, we start with the concept of *term.*

DEFINITION

A **term** is an algebraic expression that is the product and/or quotient of numbers and variables.

For example, the expression $3x^2y$ is a term, since it is the product $3 \cdot x \cdot x \cdot y$ of a constant and variables. Other examples of terms are

$$21, \quad -4x, \quad ab, \quad \frac{3}{z^4}, \quad \text{and} \quad \frac{x+y}{2}$$

In this chapter, we will study only terms that are connected by multiplication. In Chapter 6, we will work with terms involving division.

In the term $6xy^2$, 6 is called the *numerical coefficient.* The **numerical coefficient,** or simply **coefficient,** of a term is the number by which the variable or variables are multiplied. Examples of terms, variables, and coefficients are given in the following table.

Term	Variable Part	Coefficient
$3y$	y	3
x	x	1
$-0.4str$	str	-0.4
$-a^5b^3c$	a^5b^3c	-1

The expression $4x - 3y$ can be considered as $4x + (-3y)$ and consists of two terms. The first term, $4x$, has a coefficient of 4 and the second term, $-3y$, has a coefficient of -3.

Example 1 State the variable part and the coefficient of the terms.

(a) x **(b)** $-x$ **(c)** $\frac{az^2}{2}$

Solution **(a)** Since $x = 1 \cdot x$, the variable part is x and the coefficient is 1.

(b) Since $-x = (-1)x$, the variable part is x and the coefficient is -1.

(c) Since $\frac{az^2}{2} = \frac{1}{2}az^2$, the variable part is az^2 and the coefficient is $\frac{1}{2}$.

In the expression $3x + 7x$, there are two terms, $3x$ and $7x$. Both terms have the same variable part and differ only in their coefficients. They are examples of *like terms.*

DEFINITION

Like (or **similar**) **terms** are terms whose variable parts are the same.

Examples of pairs of like terms are the following:

$$3x^2 \text{ and } -x^2, \quad 1.4st \text{ and } 5st, \quad q^3p \text{ and } q^3p$$

Examples of pairs of **unlike terms** are the following:

$$2x \text{ and } 2x^2, \quad \frac{1}{2}a \text{ and } ac, \quad qp^3 \text{ and } q^3p$$

NOTE *The two terms, $2x$ and $2x^2$, are* unlike terms *since the exponents are not the same.*

Like terms in an algebraic expression can be **combined** by using addition and subtraction to obtain a single term. As an example, consider the expression $3x + 7x$. Using the distributive property, we have

$$3x + 7x = (3 + 7)x \quad \text{or} \quad ba + ca = (b + c)a$$
$$= 10x$$

To combine like terms, we need not use the distributive property each time. Simply *combine the coefficients* of the terms using addition or subtraction as indicated.

Example 2 Simplify by combining like terms.

(a) $16y + 9y - 4y$ **(b)** $x^2 - 3x^2 + 5 - 14 + 8x^2$

(c) $7p^2qr - pq^2r + pqr^2$

Solution **(a)** This expression has three like terms.

$$16y + 9y - 4y = (16 + 9 - 4)y$$ Distributive property.
$$= 21y$$

(b) By using the commutative and associative properties of addition, we can reorder the terms so that like terms are together.

$$x^2 - 3x^2 + 5 - 14 + 8x^2 = x^2 - 3x^2 + 8x^2 + 5 - 14$$
$$= (1 - 3 + 8)x^2 + (5 - 14)$$
$$= 6x^2 - 9$$ Combine like terms.

(c) This expression contains no like terms and therefore cannot be simplified.

If an expression contains parentheses, we first use the distributive property to remove them before combining like terms.

Example 3 Remove the parentheses by using the distributive property, then combine like terms.

(a) $7(x - 2) - 2x + 14$ **(b)** $15x^2 - a - 3(x^2 - 5a)$

(c) $(3)(s + 2t) - (s + t)$

Solution **(a)** $7(x - 2) - 2x + 14 = 7x - 14 - 2x + 14$ Distributive property.
$= 5x$ Combine like terms.

(b) $15x^2 - a - 3(x^2 - 5a) = 15x^2 - a - 3x^2 + 15a$ Distributive property.
$= 12x^2 + 14a$ Combine like terms.

(c) We first use the distributive property on $(3)(s + 2t)$. Next, the expression $-(s + t)$ is the same as $(-1)(s + t)$, and, therefore, we distribute -1 through the s and t.

$$(3)(s + 2t) - (s + t) = 3s + 6t - s - t \quad \text{Distributive property.}$$
$$= 2s + 5t \quad \text{Combine like terms.}$$

In the last chapter we converted word phrases to algebraic expressions. The next step is to simplify the resulting expression.

Example 4 Write each word phrase as an algebraic expression. Then, simplify the expression.

(a) Three times the sum of n and five.

(b) Four minus twice the sum of c and seven.

Solution **(a)** The word phrase is written as $3(n + 5)$, which simplifies to $3n + 15$ by the distributive property.

(b) The word phrase is written as $4 - 2(c + 7)$. Next, we simplify.

$$4 - 2(c + 7) = 4 - 2c - 14 \quad \text{Distributive property.}$$
$$= -10 - 2c \quad \text{Combine like terms.}$$

Example 5 Write each word phrase as an algebraic expression choosing a letter for the variable. Then, simplify the expression.

(a) A number subtracted from seven times the sum of the number and three.

(b) The difference of 14 and a number, the result added to three times the sum of four and the number.

Solution **(a)** Let n be the variable. Then, the word phrase can be written as $7(n + 3) - n$. Using the distributive property and then combining like terms,

$$7(n + 3) - n = 7n + 21 - n$$
$$= 6n + 21$$

(b) Let x be the variable. Then, the word phrase can be written as $3(4 + x) + (14 - x)$. We now simplify this expression.

$$3(4 + x) + (14 - x) = 12 + 3x + 14 - x \quad \text{Distributive property.}$$
$$= 2x + 26 \quad \text{Combine like terms.}$$

WARMING UP

Answer true or false.

1. The coefficient of $2x^3$ is 3.

2. The variable part of x^2y is x^2y.

3. The coefficient of $-t$ is -1.

4. The coefficient of $\frac{uv^2}{5}$ is $\frac{1}{5}$.

5. $2x + 5x = 7x$

6. $x^2 + 2x^2 = 3x^2$

7. $3y^2 + 2y^2 = 5y^4$

8. $\frac{v}{2} + \frac{v}{2} = v$

9. $3(x - 2) + x - 1 = 4x - 7$

10. $5a^2b + 2ab^2 = 7a^3b^3$

EXERCISE SET 2.1

For Exercises 1–12, state the variable part and the coefficient of the term.

1. $2x$
2. $3y$
3. $-3x^2$
4. $-6y^3$
5. y
6. s^2t^2
7. $-x^2y$
8. $-u^2v^3$
9. $\frac{a}{3}$
10. $\frac{z}{4}$
11. $\frac{3x}{5}$
12. $-\frac{2y}{3}$

For Exercises 13–38, simplify by combining like terms.

13. $5x + 3x$
14. $4y + 5y$
15. $4t - 2t$
16. $9a - 3a$
17. $7c - 6c$
18. $5x - 4x$
19. $x + 7x$
20. $3y + y$
21. $z - 9z$
22. $-6y + y$
23. $-b - b$
24. $-3u + u$
25. $3x - 5 + 2x + 1$
26. $4a + 3 - 5a + 2$
27. $8 - 3x^2 + 2 - 5x^2$
28. $-4 + 7a^3 + 5 - 6a^3$
29. $xy + a - 2xy + 2a$
30. $3u^2v - 2t + 4t - 4u^2v$
31. $3x - 3 + \frac{2x}{2} - 1$
32. $4 + \frac{3a}{3} - 5 + 2a$
33. $\frac{z}{2} + 1 + \frac{z}{2} - 7$
34. $3 - \frac{a}{5} - 4 - \frac{4a}{5}$
35. $\frac{y^2}{2} - x - \frac{y^2}{2} + \frac{x}{2}$
36. $\frac{m^3}{4} - \frac{n^2}{2} + \frac{m^3}{3} + n^2$
37. $2.32x - 7.91x - 5.71 + 8.64$
38. $7.29y^2 + 9.07 - 5.87y^2 + 9.92$

For Exercises 39–54, remove parentheses using the distributive property, then simplify.

39. $6x - 2(5x - 3)$
40. $3(1 - 2a) + 7$
41. $7 - 3(2 + 4y^2) - y^2$
42. $5ab + 4(1 - ab) + 4$
43. $2 - (x^2 - 3) + 2x^2 - 3$
44. $4(x^2y^3 - 4) - x^2y^3 + 1$

45. $2(x - 2y) - (x + 3y)$

46. $3(a - b) - (b - a)$

47. $-3(4y - 3x^2) + 4(y - x^2)$

48. $2(3x - y) + 5(x + 2y)$

49. $3(x + 2y - 1) - 2 + 2(y + 1)$

50. $-6 + 4(2a - b + 2) - (2a + 3b)$

51. $2[u + (-4)] + (-3u)$

52. $4t - [3t + (-5)] - 2$

53. $6.62(3.5a - 7.1)$

54. $58(32 - 64z^2) - 61(24 + 30z^2)$

For Exercises 55–58, write each word phrase as an algebraic expression. Then, simplify the expression.

55. Three times the sum of x and four.

56. Ten times the sum of y and seven.

57. Three times the sum of a and 12, the result subtracted from six.

58. Four times the sum of b and 15, the result subtracted from five.

For Exercises 59–62, write each word phrase as an algebraic expression choosing a letter for the variable. Then simplify the expression.

59. A number is subtracted from six times the sum of two and the number.

60. Negative two times the difference of one and twice a number.

61. Two times the difference of a number and five, the result subtracted from three times the number.

62. Three times the difference of four and twice a number, the result subtracted from seven.

SAY IT IN WORDS

63. Explain like terms and give examples of two like terms and two terms that are unlike.

64. Make up a word phrase that can be written as an algebraic expression.

REVIEW EXERCISES

The following exercises review parts of Section 1.6.

65. Evaluate $4x - 2y$, when $x = 1$ and $y = 7$.

66. Evaluate $|2a + b| - ab$, when $a = -1$ and $b = 1$.

For Exercises 67–69, determine any solutions of each equation from the set $\{-1, 1, 2\}$.

67. $2x - 4 = 0$

68. $|3 - 4x| = 1$

69. $2x^2 - 5x + 3 = 0$

ENRICHMENT EXERCISES

Simplify the following expressions.

1. $1 - \frac{2}{3}[6 - 3(1 - x)]$

2. $2 - \{3 - [x - (4y - 2x) + 1] - 1 + 4(1 - 8y)\}$

3. $x[5 - 6(4 - 3x)]$

4. Pick any number and follow these steps in order:
(a) Multiply the number by 5.
(b) Subtract 2.
(c) Multiply by 3.
(d) Add 6.
(e) Divide by 15.
Is your final number the same as the one you picked? Show that this will always happen.

Answers to Enrichment Exercises begin on page A.1.

2.2 Equations and Properties of Equality

OBJECTIVES

▶ *To solve equations using the addition-subtraction property of equality*

▶ *To solve equations using the multiplication-division property of equality*

LEARNING ADVANTAGE

When reading the text, write down questions that need to be answered. You can get more help from your instructor or friends if you ask specific questions. Read examples of a new procedure and try to do them without looking at the book.

In this section, we will introduce properties of equality that will be used to solve an equation. To **solve an equation,** such as $5x + 7 = -3$, means to find all numbers that make the equation a true statement when used in place of x. The collection of these numbers is called the **solution set** of the equation.

For example, the equation $5x + 7 = -3$ has $\{-2\}$ as its solution set, since the equation becomes the true statement $5(-2) + 7 = -3$, or $-3 = -3$, when x is replaced by -2. We say that -2 *satisfies* the equation.

The goal in solving an equation is to convert the equation to a simpler equation that has the same solution set. We give equations that have the same solution set a special name.

DEFINITION

Two or more equations are **equivalent** if they have the same solution set.

Example 1

(a) $5x + 7 = -3$ and $x = -2$ are equivalent equations, since they both have $\{-2\}$ as the common solution set.

(b) $3y - 2 = 7$, $3y = 9$, and $y = 3$ are equivalent, since they each have the solution set $\{3\}$. ◀

In order to solve an equation, we will produce equivalent equations by using properties of equality until we obtain an *obvious equation* of the form

"x = a constant" or "a constant = x." For example, as shown before, the equation $5x + 7 = -3$ is equivalent to the equation $x = -2$, from which it is obvious that the solution set is $\{-2\}$. Therefore, our goal is to *isolate the variable* to one side of the equation.

STRATEGY

Solving an Equation

Use properties of equality to isolate the variable to one side of the equation.

To solve an equation, we have two properties of equality. The first one allows us to add (or subtract) the same number to both sides of an equation without changing the solution set. The second allows us to multiply (or divide) both sides of an equation by the same nonzero number without changing the solution set.

PROPERTIES

Addition-Subtraction Property of Equality

Add (or subtract) the same quantity on *both* sides of the equation.

Multiplication-Division Property of Equality

Multiply (or divide) *both* sides of an equation by the same *nonzero* quantity.

Example 2 Solve the equation and check your answer.

$$x - 4 = -13$$

Solution Our goal is to isolate x by itself on the left side of this equation. If we add 4 to the left side, then $-4 + 4$ has a sum of zero. Therefore, we add 4 to *both* sides of the equation.

$$x - 4 = -13$$

$$x - 4 + 4 = -13 + 4 \quad \text{Add 4 to both sides.}$$

$$x + 0 = -9 \quad \text{Simplify.}$$

$$x = -9$$

To check our answer, we replace x by -9 in the original equation.

$$x - 4 = -13$$

$$-9 - 4 \stackrel{?}{=} -13$$

$$-13 = -13$$

The answer checks since the statement is true if x has the value of -9. The solution set is $\{-9\}$.

Example 3 Solve the equation $y + 5 = -9$ and check your answer.

Solution We use the addition-subtraction property of equality by subtracting 5 from both sides.

$$y + 5 = -9$$
$$y + 5 - 5 = -9 - 5$$
$$y + 0 = -14$$
$$y = -14$$

To check our answer, replace y by -14 in the original equation.

$$y + 5 = -9$$
$$-14 + 5 \stackrel{?}{=} -9$$
$$-9 = -9$$

The answer checks, and the solution set is $\{-14\}$.

Example 4 Solve $3x + 8 = 2x + 10$.

Solution Our plan is to isolate x to the left side. We start by subtracting 8 from both sides

$$3x + 8 = 2x + 10$$
$$3x + 8 - 8 = 2x + 10 - 8 \quad \text{Subtract 8 from both sides.}$$
$$3x = 2x + 2 \quad \text{Simplify both sides.}$$
$$3x - 2x = 2x + 2 - 2x \quad \text{Subtract } 2x \text{ from both sides.}$$
$$x = 2 \quad \text{Simplify both sides.}$$

Check As a check, replace x by 2 in the original equation.

$$3x + 8 = 2x + 10$$
$$3 \cdot 2 + 8 \stackrel{?}{=} 2 \cdot 2 + 10 \quad \text{Replace } x \text{ by 2.}$$
$$6 + 8 \stackrel{?}{=} 4 + 10$$
$$14 = 14.$$

The solution to the equation is 2.

Example 5 Solve $\frac{x}{3} = -2$ and check your answer.

Solution Since the left side is $\frac{x}{3}$, we multiply by 3 to get $3 \cdot \frac{x}{3} = x$. Therefore, multiply both sides of the equation by 3.

$$\frac{x}{3} = -2$$
$$3 \cdot \frac{x}{3} = 3(-2) \quad \text{Multiply each side by 3.}$$
$$x = -6 \quad \text{Simplify.}$$

Check To check, replace x by -6.

$$\frac{x}{3} = -2$$

$$\frac{-6}{3} \stackrel{?}{=} -2$$

$$-2 = -2$$

The solution of the equation is -6.

Example 6 Solve $-5x = 15$.

Solution If we divide $-5x$ by -5, we obtain x: $\frac{-5x}{-5} = x$. Therefore, we divide each side of the equation by -5.

$$-5x = 15$$

$$\frac{-5x}{-5} = \frac{15}{-5} \quad \text{Divide both sides by } -5.$$

$$x = -3 \quad \text{Simplify.}$$

Check

$$-5x = 15$$

$$-5(-3) \stackrel{?}{=} 15$$

$$15 = 15$$

The solution set is $\{-3\}$.

The next example shows how both properties of equality are sometimes used to solve an equation.

Example 7 Solve $4t - 3 = 7t - 1$.

Solution We start by adding 3 to each side

$$4t - 3 = 7t - 1$$

$$4t - 3 + 3 = 7t - 1 + 3 \quad \text{Add 3 to both sides.}$$

$$4t = 7t + 2 \quad \text{Simplify.}$$

$$4t - 7t = 7t + 2 - 7t \quad \text{Subtract } 7t \text{ from each side.}$$

$$-3t = 2 \quad \text{Simplify.}$$

$$\frac{-3t}{-3} = \frac{2}{-3} \quad \text{Divide each side by } -3.$$

$$t = -\frac{2}{3} \quad \text{Simplify.}$$

Check that $\left\{-\frac{2}{3}\right\}$ is the solution set.

For our final example, we verbally describe an unknown number that can be found using the techniques of this section.

Example 8 Write an equation for the given sentence, then solve the equation.

The sum of four times a number and five is three times the number.

Solution Letting x be the variable, the sentence can be written as

$$4x + 5 = 3x$$

Using the properties of equality, we solve this equation.

$$4x + 5 - 5 - 3x = 3x - 5 - 3x$$
$$x = -5$$

Check Does -5 satisfy the conditions of the problem?

Four times -5 plus 5 is $-20 + 5$ or -15.

Three times -5 is $3(-5)$ or -15.

Therefore, our answer does satisfy the stated conditions.

WARMING UP

Answer true or false.

1. $x + 2 = 5$ and $x = 7$ are equivalent equations.
2. $y - 3 = -1$ and $y = 2$ are equivalent equations.
3. $2t = 4$ and $t = 2$ are equivalent equations.
4. $3z = -2$ and $z = -6$ are equivalent equations.
5. The solution set of $x - 7 = -4$ is $\{3\}$.
6. The solution set of $k + 4 = -1$ is $\{-4\}$.
7. The solution set of $3u = -3$ is $\{-1\}$.
8. The solution set of $\frac{x}{3} = -1$ is $\{-3\}$.
9. $\left\{-\frac{1}{2}\right\}$ is the solution set of $-\frac{x}{2} = 1$.
10. Two equations are equivalent if they have the same solution set.

EXERCISE SET 2.2

For Exercises 1–48, solve the equation and check your answer.

1. $x + 3 = 2$
2. $x + 5 = 3$
3. $y + 8 = 5$

4. $z + 10 = 7$

5. $x - 2 = 6$

6. $x - 7 = 4$

7. $z - \frac{3}{4} = -\frac{5}{2}$

8. $s + \frac{2}{5} = -\frac{7}{10}$

9. $5 = -12 + d$

10. $-7 = 6 + r$

11. $4x + 2 = 3x + 1$

12. $5y - 3 = 4y + 2$

13. $2m - 7 = m - 5$

14. $7t + 6 = 6t + 9$

15. $\frac{x}{3} = 1$

16. $\frac{x}{2} = 4$

17. $\frac{x}{4} = -2$

18. $\frac{y}{5} = -1$

19. $-\frac{s}{6} = 1$

20. $-\frac{v}{2} = -5$

21. $2x = -14$

22. $3z = -9$

23. $5t = 2$

24. $4u = 3$

25. $-3m = 12$

26. $-6c = 18$

27. $4x = 6$

28. $8x = -10$

29. $\frac{1}{3}x = \frac{2}{9}$

30. $\frac{1}{2}x = \frac{1}{4}$

31. $\frac{1}{4}x = -\frac{3}{8}$

32. $\frac{1}{5}y = -\frac{7}{10}$

33. $\frac{2}{3}s = \frac{4}{9}$

34. $\frac{3}{4}c = \frac{9}{16}$

35. $\frac{4}{3} = \frac{m}{12}$

36. $-\frac{9}{14} = \frac{3}{7}x$

37. $-x = 7$

38. $-a = -12$

39. $3x - 5 = 15 + 5x$

40. $2y - 3 = 5y + 3$

41. $4z + 7 = 8z - 1$

42. $8 - 2v = 3v - 7$

43. $6 - 3x = 5x + 2$

44. $4y - 1 = -3 - 2y$

45. $0.32x = 0.128$

46. $84y = 1{,}848$

47. $1.9x + 5.7 = 7.79$

48. $5.31a + 38.7 = -25.02$

For Exercises 49–54, write an equation for the given sentence, then solve the equation.

49. The sum of five times a number and seven is ten less than four times the number.

50. The sum of twice a number and nine is one less than the number.

51. Twelve minus six times a number is the difference of 11 and five times the number.

52. Twenty plus five times a number is the sum of 15 and six times the number.

53. A number divided by three is five.

54. Twice a number is negative eight.

SAY IT IN WORDS

55. Explain the addition-subtraction property of equality and how it is used.

56. Explain the multiplication-division property of equality and how it is used.

REVIEW EXERCISES

The following exercises review parts of Section 2.1. Doing these problems will help prepare you for the next section.

For Exercises 57–62, simplify each expression.

57. $2x - 4x + 1 - 7$

58. $5 - 12 + 3y - 10y + 4$

59. $3(1 - 2a + 3) + 4a$

60. $5z - 2(3z - 8 + z) + 1$

61. $5(1 - 3t) - (4 - 3t)$

62. $\frac{1}{2}[4 - 6(1 - z)]$

ENRICHMENT EXERCISES

1. Write an equation for the given sentence, then solve the equation.

Five times the sum of four and twice a number is three times the difference of three times the number and two.

2. Find a, if $x = -2$ is equivalent to $ax + 2 = 4$.

3. Find a, if $x = -1$ is equivalent to $ax - 3 = 5$.

4. Solve $3x + 5 = 14$ by two methods:

Method 1 First use the addition-subtraction property, then the multiplication-division property.

Method 2 First use the multiplication-division property, then the addition-subtraction property.

Which method did you prefer? Why?

Answers to Enrichment Exercises begin on page A.1.

2.3 Solving Linear Equations

OBJECTIVE

▶ *To apply techniques for solving linear equations*

LEARNING ADVANTAGE

Don't be afraid to talk to your instructor. If you are having a problem, make an appointment to see him or her. Instructors are busy throughout the week, but if contacted early enough, a meeting can be arranged. It doesn't take long to get most mathematics questions answered.

The equations we solved in the last section are examples of *linear equations.*

DEFINITION

A **linear equation** is an equation that can be put into the standard form

$$ax + b = 0$$

where a and b are constants with $a \neq 0$.

Most linear equations that we will encounter will *not* be in this standard form. For example,

$$2x + 1 = 4x \qquad \text{and} \qquad 2(1 - 3x) = 7x + 5$$

are linear equations. Each one, however, *could* be put into the standard form.

The following steps are guidelines to be used for solving linear equations.

STRATEGY

Solving a Linear Equation

Step 1 **Clear the equation of any fractions or decimals,** if desired, by multiplying by the LCD of the fractions appearing or, in the case of decimals, by an appropriate power of ten.

Step 2 **Combine like terms on each side,** if necessary, to simplify. You may have to use the distributive property first to separate terms.

Step 3 **Use the addition property of equality,** if necessary, to move all terms containing a variable to one side and all constant terms to the other side.

Step 4 **Use the multiplication property of equality,** if necessary, to make the coefficient of the variable term equal to one. The equation is now in the form

$$x = \text{the solution} \qquad \text{or} \qquad \text{the solution} = x$$

Step 5 **(Optional) Check your answer:** Substitute it into the original equation.

Step 6 **Write the solution set.**

Example 1 Solve the equation

$$3x - 5 = 11$$

Then, check your answer.

Solution Steps 1 and 2 These do not apply.

Step 3 Add 5 to both sides and simplify.

$$3x - 5 + 5 = 11 + 5$$
$$3x = 16$$

Step 4 Divide both sides by 3.

$$\frac{3x}{3} = \frac{16}{3}$$
$$x = \frac{16}{3}$$

Step 5 We check our answer by replacing x by $\frac{16}{3}$ in the original equation to determine if the resulting statement is true.

Check $\frac{16}{3}$:

$$3x - 5 = 11$$
$$3\left(\frac{16}{3}\right) - 5 \stackrel{?}{=} 11$$
$$16 - 5 \stackrel{?}{=} 11$$
$$11 = 11$$

Step 6 The solution set is $\left\{\frac{16}{3}\right\}$.

Example 2 Solve the equation.

$$3 - 2x = 4(1 + x) - 13$$

Solution We use the strategy for solving a linear equation.

Step 1 This does not apply.

Step 2 Use the distributive property, then combine like terms on both sides.

$$3 - 2x = 4(1 + x) - 13$$
$$3 - 2x = 4 + 4x - 13 \qquad \text{Distributive property.}$$
$$3 - 2x = -9 + 4x$$

Step 3 Subtract 3 and $4x$ from both sides; then simplify.

$$3 - 2x - 3 - 4x = -9 + 4x - 3 - 4x$$
$$-6x = -12$$

Step 4 Divide both sides by -6 and simplify.

$$\frac{-6x}{-6} = \frac{-12}{-6}$$
$$x = 2$$

Step 5 On your own, check the answer 2.

Step 6 The solution set is $\{2\}$.

Example 3 Solve the equation

$$0.3y + 0.7 = 0.2(y + 3)$$

Solution Steps 1 and 2 Clear the equations of decimals by multiplying both sides by 10; then, simplify.

$$0.3y + 0.7 = 0.2(y + 3)$$
$$10(0.3y + 0.7) = 10(0.2)(y + 3) \quad \text{Multiply each side by 10.}$$
$$10(0.3y) + 10(0.7) = 10(0.2)(y + 3) \quad \text{Distribute the 10.}$$
$$3y + 7 = 2(y + 3) \quad \text{Simplify.}$$
$$3y + 7 = 2y + 6$$

Step 3 Subtract 7 and $2y$ from both sides.

$$3y + 7 - 7 - 2y = 2y + 6 - 7 - 2y$$
$$y = -1 \quad \text{Simplify.}$$

Step 4 Does not apply.

Step 5

Check -1:

$$0.3y + 0.7 = 0.2(y + 3)$$
$$0.3(-1) + 0.7 \stackrel{?}{=} 0.2(-1 + 3)$$
$$-0.3 + 0.7 \stackrel{?}{=} 0.2(2)$$
$$0.4 = 0.4$$

Step 6 The solution set is $\{-1\}$.

The next example shows that the solution of a linear equation need not be a simple number.

Example 4 Solve the equation:

(a) $3(1 + 2x) = 2(3x + 1)$

(b) $6 - 4x = 2(-2x + 3)$

Solution **(a)**

$$3(1 + 2x) = 2(3x + 1)$$
$$3 + 6x = 6x + 2$$
$$3 + 6x - 6x = 6x + 2 - 6x$$
$$3 = 2$$

What happened? Since the work is correct, the given equation has the same solution set as the equation $3 = 2$. But the equation $3 = 2$ is satisfied by no real number. The solution set is the set with no elements or the empty set $\varnothing$.

(b)

$$6 - 4x = 2(-2x + 3)$$
$$6 - 4x = -4x + 6$$
$$6 - 4x + 4x = -4x + 6 + 4x$$
$$6 = 6$$

The variables again vanish from the equation. In this case, however, the resulting equation, $6 = 6$, is true for all real numbers. The solution set is the set of all real numbers.

Example 5 Write an equation for the following sentence. Then solve the equation.

The sum of three times a number and five is negative eight.

Solution Let n be the unknown number. Then the sentence can be written as

$$3n + 5 = -8$$

We now solve this equation.

$$3n + 5 = -8$$
$$3n + 5 - 5 = -8 - 5 \qquad \text{Subtract 5 from each side.}$$
$$3n = -13$$
$$n = \frac{-13}{3} \qquad \text{Divide each side by 3.}$$

Therefore, the number is $-\frac{13}{3}$.

WARMING UP

Answer true or false.

1. The solution set of $2(x + 1) = 6$ is $\{2\}$.
2. The solution set of $3(1 - x) = 0$ is $\{1\}$.
3. $4(1 + x) = 1$ and $4 + x = 1$ are equivalent equations.
4. $1 - 2(3 - x) = 0$ and $-5 + 2x = 0$ are equivalent equations.
5. $0.23 + 0.1x = 0.59$ and $23 + x = 59$ are equivalent equations.

6. $\frac{1}{2} + \frac{x}{4} = -2$ and $2 + x = -8$ are equivalent equations.

7. The solution set of $2(1 + x) = 2x + 5$ is $\varnothing$.

8. The solution set of $6x - 4 = 2(3x - 2)$ is the set of real numbers.

EXERCISE SET 2.3

For Exercises 1–44, solve the equation using the strategy on page 90.

1. $3x + 7 = -2$

2. $2y - 1 = 3$

3. $4z - 3 = -1$

4. $5t + 6 = 2$

5. $6 - 3x = 4 - x$

6. $3 - 4a = 9 - 2a$

7. $-5 + 6u = 2u - 11$

8. $1 + 7m = m - 1$

9. $5 + 4z - 7z = 6 - 4$

10. $12 - 3x + 14 = 3x - 5x$

11. $4x - 11 + 3x = 2x + 9$

12. $c - 5c + 6 - 8 = 3c - 1 - 15$

13. $-4t + 9t - 7 = 3 - t + 6$

14. $12s + s - 19 + 5 = 2s - s$

15. $\frac{1}{3}a - \frac{5}{6}a = \frac{1}{6} - \frac{5}{3}$

16. $\frac{3}{4} - \frac{1}{8}w = \frac{7}{8} + \frac{1}{4}w$

17. $\frac{2}{5}y - \frac{1}{10} = 1 - \frac{7}{10}y$

18. $\frac{3}{7} - \frac{5}{14}n = \frac{1}{2} - \frac{1}{7}n$

19. $3.2m - 5.9 + 4.9m = -1.5 + 3.7$

20. $-0.6z + 0.5z + 1.9 = 2.8$

21. $3(5 + 2x) = 31 - 10$

22. $2(3z - 8) = 15 - 1$

23. $-2 + 3(4 - x) = 2x$

24. $7 - 3b = 5(b - 5)$

25. $3 - (2 - y) = -5y$

26. $7 - (2 - v) = -17$

27. $a + 4 = -3(7 - a) - 7a$

28. $23 + 3y = -2y + 3(2y + 4)$

29. $-(2x - 5) + 13 = 2(x + 2)$

30. $10 - (a - 3) = 3(5 - 3a)$

31. $1 - \frac{1}{3}(3 - t) = \frac{1}{3}(2t + 5)$

32. $\frac{1}{4}(s - 9) = \frac{1}{4}(3s - 1) - 3$

33. $m - \frac{3(1 + 2m)}{2} = \frac{m - 3}{4}$

34. $\frac{2(x + 1)}{3} = 1 - \frac{x - 3}{6}$

35. $0.7x + 0.8 = 0.6x + 0.9$

36. $0.2y - 0.1 = 0.6 + 0.3y$

37. $2x - 1 = 2(x + 1)$

38. $3 - 6y = 2(1 - 3y)$

39. $4(6 + 3z) = 3(4z + 8)$

40. $15w - 6 = 3(-2 + 5w)$

41. $0.525w + 0.4w - 0.045 = 1.25$

42. $4.08z - 9.903 = 5(6.4187 - 1.03z)$

43. $3(8790 + 875a) = 71(36a - 209)$

44. $25(-386.6 + 5y) = 4(-3940.5 + 14.5y)$

For Exercises 45–50, write an equation for each sentence. Then, solve the equation.

45. The sum of four times a number and five is three.

46. The sum of five times a number and seven is negative five.

47. The sum of twice a number and four thirds is six.

48. Ten times a number plus 25 is -80.

49. The sum of two and three-fourths times a number is negative one-fourth.

50. The sum of one and two-ninths times a number is negative one-third.

SAY IT IN WORDS

51. Explain your strategy for solving a linear equation.

REVIEW EXERCISES

The following exercises review parts of Section 1.6. Doing these exercises will help prepare you for the next section.

For Exercises 52–57, write each word phrase as an algebraic expression. Choose a letter for the variable.

52. The sum of a number and 25.

53. The difference of a number and 14.

54. Three times the sum of four and a number.

55. The quotient of a number and ten.

56. The difference of one and a number, the result divided by two.

57. The product of six and a number.

ENRICHMENT EXERCISES

1. Solve the following equation:

$$2[3 - 2(3x - 1)] = 3(2 - 3x)$$

For Exercises 2–5, solve for x.

2. $ax + b = 0$

3. $ax + b = cx + d,\ a \neq c$

4. $x + 3d = 4x$

5. $\dfrac{x - 3}{2} = ax$

Answers to Enrichment Exercises begin on page A.1.

2.4 Solving Word Problems

OBJECTIVES

- *To convert word problems into mathematical problems*
- *To solve word problems*

LEARNING ADVANTAGE

When solving a word problem, the information must be read carefully and slowly. Make sure you understand each word. Do not attempt to digest the information too quickly; you may miss an important point.

Throughout the previous sections, we have touched upon translating word phrases and sentences into algebraic expressions and equations. In this section, we will convert a greater variety of sentences into equations. Then, we will learn how to solve word problems.

Before we solve word problems, we first look at some ways that word phrases are converted into algebraic expressions.

Word Phrase	***Algebraic Expression***
Addition	
The sum of a number and five	$x + 5$
Eight more than a number	$x + 8$
A number increased by 2.3	$x + 2.3$
Subtraction	
A number minus three	$x - 3$
Five less than a number	$x - 5$
A number decreased by four	$x - 4$
A number subtracted from three	$3 - x$
Ten fewer than a number	$x - 10$
The difference of a number and nine	$x - 9$
The difference of nine and a number	$9 - x$

Word Phrase	Algebraic Expression
Multiplication	
Ten times a number	$10x$
Twice a number	$2x$
A number multiplied by six	$6x$
Three fourths of a number	$\frac{3}{4}x$
Twenty-eight percent of a number	$0.28x$
Division	
A number divided by 8	$\frac{x}{8}$
The quotient of 12 and a number	$\frac{12}{x}$
The quotient of a number and 12	$\frac{x}{12}$
The reciprocal of a number	$\frac{1}{x}$

Example 1 Convert the following word phrases into algebraic expressions. Use x for the variable.

(a) A number subtracted from two *means* $2 - x$.

(b) The quotient of twice a number and five *means* $\frac{2x}{5}$.

(c) Three-eighths of the sum of a number and seven *means* $\frac{3}{8}(x + 7)$.

(d) Eighteen less than 0.57 times a number *means* $0.57x - 18$.

(e) The quotient of a number increased by two and four times the number *means* $\frac{x + 2}{4x}$.

(f) Thirty-five percent of the sum of a number and 170 *means* $0.35(x + 170)$.

The rest of the chapter is devoted to solving word problems. You may be tempted to solve some of them by "trial and error." However, it is important to develop a strategy for solving word problems, since not all of them will have simple solutions.

The following is a plan or strategy for solving word problems.

STRATEGY

Solving Word Problems

1. **Read** the problem carefully. Take note of what is being asked and what information is given.
2. **Plan** a course of action. Represent the unknown number by a letter. If there are two or more unknowns, represent one of them by a letter and express the others in terms of the letter.
3. **Create** an equation from the given information.
4. **Solve** this equation.
5. **Check** your solution against the words of the original problem.
6. **Answer** the original question. Read again the problem to make sure you have answered the question.

We conclude this section with two examples of number problems. As you read through these examples, keep in mind the strategy for solving word problems. In the next section, we will solve more complex word problems using the same strategy.

Example 2 The sum of three times a number and ten is negative two times the number. Find the number.

Solution From reading the problem, we want to find a number that satisfies the given information. Let

$$x = \text{the unknown number}$$

Now, we read the problem again with the idea of converting the information into an equation. From the given information,

The sum of 3 times the number and 10	is	−2 times the number
$3x + 10$	$=$	$-2x$

We now solve this equation.

$$3x + 10 = -2x$$

$$3x + 10 - 10 + 2x = -2x - 10 + 2x \quad \text{Add } -10 + 2x \text{ to both sides.}$$

$$5x = -10 \quad \text{Simplify each side.}$$

$$x = -2$$

To check our solution, we determine if -2 satisfies the original problem. The sum of three times -2 and 10 is $3(-2) + 10$ or 4, and -2 times -2 is 4. Our answer checks, and, therefore, the number we wanted to find is -2.

NOTE *The information in Example 2 is the sentence, "The sum of three times a number and ten is negative two." To convert the sentence to an equation, the word "is" becomes the equal sign. Any word or group of words that means* the same as *triggers the equal sign. For example, phrases like "is equal to" and "the result is (equal to)" translate as an equal sign (=).*

Example 3 Twenty percent of the sum of a number and 12 is six. Find the number.

Solution Let

$$x = \text{the unknown number}$$

20% of the sum of x and 12 is 6 can be written as

$$0.20(x + 12) = 6$$

We can solve this equation by first multiplying both sides by 10 to clear the equation of decimals.

$$0.20(x + 12) = 6$$
$$(10)(0.20)(x + 12) = (10)(6)$$
$$2(x + 12) = 60$$
$$\frac{2(x + 12)}{2} = \frac{60}{2} \quad \text{Divide both sides by 2.}$$
$$x + 12 = 30$$
$$x + 12 - 12 = 30 - 12 \quad \text{Subtract 12 from both sides.}$$
$$x = 18$$

The number is 18.

Example 4 The $85 selling price of a cassette player is $20 more than twice the cost. Find the cost.

Solution Since we are to find the unknown cost, let

$$x = \text{the cost}$$

We read the problem again to convert the information into an equation.

the selling price	is	20 more than twice the cost
85	=	$2x + 20$

We now solve this equation.

$$85 - 20 = 2x + 20 - 20$$
$$65 = 2x$$
$$\frac{65}{2} = \frac{2x}{2}$$
$$32.50 = x$$

The cost is $32.50.

WARMING UP

Answer true or false.

1. The sum of a number x and 12 can be written as $x + 12$.

2. Seven divided by a number x is $\frac{x}{7}$.

3. The product of -5 and a number x is $-5x$.

4. The difference of a number x and 1 is $1 - x$.

5. Forty-five percent of a number x is $45x$.

6. The sum of twice a number x and 3 is $2x + 3$.

7. The quotient of 3 times a number x and 4 is $\frac{4x}{3}$.

8. Ten times the difference of a number x and 2 is $10(x - 2)$.

EXERCISE SET 2.4

For Exercises 1–14, convert the word phrase into a mathematical expression using x as the variable.

1. The sum of a number and 6.

2. The sum of 2 and a number.

3. A number decreased by 7.

4. The difference of 4 and a number.

5. The product of 4 and a number.

6. A number divided by negative four.

7. Five times the sum of a number and 1.

8. Three-tenths of the sum of 4 and a number.

9. Sixty-five percent of a number.

10. Thirty-two percent of a number.

11. The reciprocal of the sum of a number and 25.

12. Nine more than twice a number.

13. Four times the difference of a number and 7.

14. The difference of twice a number and 3, multiplied by 2.

For Exercises 15–32, find the number described.

15. The sum of five times a number and seven is negative three.

16. One-half increased by three-fourths times a number is two.

17. The difference of 4 and a number is 6.

18. A number subtracted from 1.6 is -3.9.

19. Sixteen fewer than a number is twice the number.

20. Negative two times the difference of a number and 5 is 8.

21. Three increased by 4 times a number is 27.

22. Five more than twice a number is the same as the product of 12 and the number.

23. Twenty more than four times a number is six times the number.

24. Three times the sum of twice a number and one is the number reduced by 12.

25. Four times the sum of a number and 0.2 is 0.8.

26. The sum of twice a number and 2.1 is -4.3.

27. The sum of a number and 14 is divided by four. The result is the difference of twice the number and seven.

28. The sum of 4 times a number and 3 is divided by 2. The result is the difference of the number and 1.

29. The sum of 40% of a number and 20 is 40.

30. The sum of 60% of a number and 2 is 20.

31. The quotient of three times a number and five is 75 more than 45% of the number.

32. The difference of twice a number and 36 is 140% of the number.

For Exercises 33–38, solve by using the strategy for solving word problems.

33. Three times Fernando's age decreased by 12 is 18. How old is he?

34. The difference of 4 times Karen's age and 32 is 48. Find her age.

35. Four times the number of coins in a parking meter, reduced by 12, is 52. Find the number of coins.

36. Six pounds less than twice Fred's weight is 234 pounds. What is Fred's weight?

37. The $300 selling price of a car CD system is $27 more than 3 times the cost. Find the cost.

38. Three less than 4 times the temperature is $-23°$. Find the temperature.

SAY IT IN WORDS

39. Explain your strategy for solving words problems.

$20 + 4x = 6x$

REVIEW EXERCISES

The following exercises review parts of Section 1.6. Doing these exercises will help prepare you for the next section.

For Exercises 40–43, write an algebraic expression from the given information.

40. The cost of x pounds of grapes at $2.40 per pound.

41. The number of feet in y yards.

42. The sales tax on a purchase that cost z dollars, if the sales tax is 6%.

43. The amount of money received for working t hours at $14 per hour.

ENRICHMENT EXERCISES

1. Find two consecutive even integers whose sum is 94.
2. Find two consecutive odd integers whose sum is -76.
3. Find three consecutive odd integers whose sum is 57.

Answers to Enrichment Exercises begin on page A.1.

2.5 More Word Problems

OBJECTIVES

- *To solve more word problems*
- *To solve coin problems*
- *To solve mixture problems*

NOTE *When solving a word problem, sometimes the information is not given in an organized form. If this is the case, organize the data by constructing a table. The exercise set has problems designed to help you make a table.*

In this section, we solve more word problems. The six-step strategy for solving word problems will still be used to help you.

Example 1 Ann is planning to buy a VCR which costs \$650. This is \$72 more than twice the amount she saved last month. How much did she save last month?

Solution Let

$$x = \text{the amount of money that Ann saved last month}$$

From the information, we have

$$650 = 2x + 72$$

We solve this equation.

$$650 - 72 = 2x + 72 - 72$$
$$578 = 2x$$
$$\frac{1}{2} \cdot 578 = \frac{1}{2} \cdot 2x$$
$$289 = x$$

We check our solution by determining if \$72 more than twice \$289 is equal to \$650.

$$72 + 2(289) = 72 + 578 = 650$$

Our answer checks, so Ann saved \$289 last month.

Example 2 A pair of shoes is on sale at 25% off the original price. If the sale price is \$45, what was the original price?

Solution Let

$$x = \text{the original price of the shoes}$$

Since the pair of shoes is on sale at 25% off the original price,

$$\text{sale price} = \text{original price} - 25\% \text{ of original price}$$

Therefore,

$$45 = x - 0.25x$$

We solve this equation.

$$45 = 0.75x$$

$$\frac{45}{0.75} = \frac{0.75}{0.75}x$$

$$60 = x$$

The original price is \$60. We now check our answer. Does \$60 minus 25% of \$60 equal \$45?

$$60 - 0.25(60) = 60 - 15 = 45 \qquad \text{yes}$$

Alternate Method. If the price of the shoes is marked down at 25% off the original price x, then the sale price is 75% of the original price. Therefore,

$$0.75x = 45$$

$$x = \frac{45}{0.75}$$

$$x = 60$$

◀

In some word problems, there are two unknowns that are related. For example, suppose that the sum of two numbers is 20. If we denote one of the numbers by x, then the other can be expressed as $20 - x$, since $x + (20 - x) = 20$.

Example 3 Represent one number by a variable. Then express the other number in terms of the variable.

(a) The sum of two numbers is 5.

(b) Two consecutive integers.

(c) Two consecutive even integers.

(d) Stephanie's age now and her age three years from now.

Solution

(a) If we denote one of the numbers by x, then the other number is $5 - x$.

(b) If the first integer is n, then the next consecutive integer is $n + 1$.

(c) If the first integer is n, then the next consecutive even integer is $n + 2$.

(d) If Stephanie's age now is x, then her age three years from now is $x + 3$.

◀

The next few examples involve word problems that contain two unknowns.

Example 4 The sum of two consecutive odd integers is 80. Find the two integers.

Solution Let

$$n = \text{the first odd integer}$$

then, since the two *odd* integers are consecutive,

$$n + 2 = \text{the second odd integer}$$

Since their sum is 80,

$$n + (n + 2) = 80$$

We solve this equation.

$$2n + 2 = 80 \quad \text{Simplify the left side.}$$

$$2n = 78 \quad \text{Subtract 2 from both sides.}$$

$$n = 39 \quad \text{Divide both sides by 2.}$$

Therefore, the two consecutive odd numbers are 39 and 41. To check, the sum, $39 + 41$, is equal to 80.

Example 5 Jill must cut a 5-foot board into two pieces, so that the larger piece has a length that is three-halves the length of the smaller piece. Find the lengths of the two pieces.

Solution Let

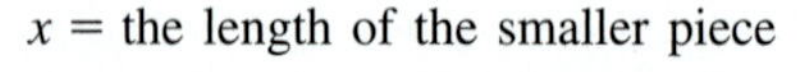

$$x = \text{the length of the smaller piece}$$

Since the total length of the board is 5 feet,

$$5 - x = \text{the length of the larger piece}$$

The length of the larger piece is three-halves the length of the smaller piece. Therefore,

$$5 - x = \frac{3}{2}x$$

Solving this equation,

$$5 - x + x = \frac{3}{2}x + x \quad \text{Addition property of equality.}$$

$$5 = \frac{5}{2}x \quad \text{Simplify each side.}$$

$$\frac{2}{5}(5) = \frac{2}{5}\left(\frac{5}{2}x\right) \quad \text{Multiplication property of equality.}$$

$$2 = x \quad \text{Simplify.}$$

Therefore, the smaller piece is 2 feet and the longer piece is $5 - 2$ or 3 feet.

As a check, is the length of the larger piece equal to three-halves the length of the smaller piece? The answer is yes, since $\frac{3}{2} \cdot 2 = 3$, the length of the larger piece. ◀

We finish the section with a coin problem and a mixture problem.

Example 6 (Coin problem) Reggie has \$2.35 in nickels and dimes in his pocket. If he has 7 more dimes than nickels, how many of each coin does he have in his pocket?

Solution Let

$$x = \text{the number of nickels}$$

Since he has 7 more dimes than nickels,

$$x + 7 = \text{the number of dimes}$$

Since each nickel is worth 5 cents, $5x$ is the value of the x nickels.
Since each dime is worth 10 cents, $10(x + 7)$ is the value of the $x + 7$ dimes.

We summarize this information in the table.

	Nickels	*Dimes*
Number	x	$x + 7$
Value	$5x$ cents	$10(x + 7)$ cents

Since the total value of the coins is \$2.35, we obtain the equation

$$\begin{array}{ccccc} \text{The value of the } x \text{ nickels} & + & \text{The value of the } x + 7 \text{ dimes} & = & \text{total value} \\ 5x & + & 10(x + 7) & = & 235 \end{array}$$

Solving this equation,

$$5x + 10x + 70 = 235 \qquad \text{Distributive property.}$$

$$15x + 70 - 70 = 235 - 70$$

$$15x = 165$$

$$x = \frac{165}{15}$$

$$x = 11$$

Therefore, Reggie has 11 nickels and 11 + 7 or 18 dimes. To check our answers, we determine if the value of 11 nickels and 18 dimes is \$2.35.

$$5(11) + 10(18) = 55 + 180 = 235 \qquad \text{or} \qquad \$2.35$$

Therefore, 11 nickels and 18 dimes are the correct numbers. ◀

Example 7 (Mixture problem) A solution containing 8% salt is to be mixed with a solution containing 16% salt to obtain a 50-liter mixture with 10% salt. How many liters of each solution should be used?

Solution Let

$$x = \text{the number of liters of the first (8\%) solution}$$

Since the total mixture is 50 liters, the amount of the second (16%) solution is given by $50 - x$. (See figure below).

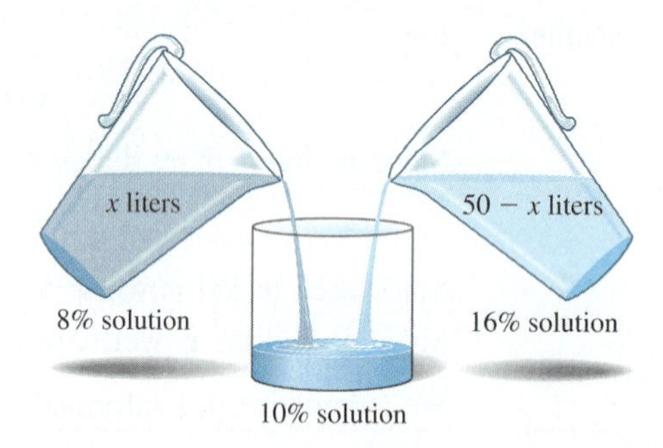

Now, the amount of salt in the x liters of the 8% solution is $0.08x$, and the amount of salt in the $50 - x$ liters of the 16% solution is $0.16(50 - x)$. We summarize this information in the table:

	8% Solution	16% Solution
Amount of solution	x	$50 - x$
Amount of salt	$(0.08)x$	$(0.16)(50 - x)$

Since the amount of salt in the final mixture is to be 10%,

$$\begin{array}{ccccc} \text{Amount of salt in the 8\% solution} & + & \text{Amount of salt in the 16\% solution} & = & \text{Total amount of salt} \\ 0.08x & + & 0.16(50 - x) & = & 0.10(50) \end{array}$$

We solve this equation.

$$0.08x + 0.16(50 - x) = 0.10(50)$$

$$0.08x + 8 - 0.16x = 5 \qquad \text{Simplify each side.}$$

$$8 - 0.08x = 5$$

$$8 - 0.08x - 8 = 5 - 8 \qquad \text{Subtract 8 from both sides.}$$

$$-0.08x = -3$$

$$8x = 300 \qquad \text{Multiply both sides by } -100.$$

$$x = \frac{300}{8} \quad \text{or} \quad 37.5 \text{ liters}$$

Therefore, we should use 37.5 liters of the 8% solution and 50 − 37.5 or 12.5 liters of the 16% solution. As a check,

$$\begin{array}{ccccc} 8\% \text{ of } 37.5 \text{ liters} & + & 16\% \text{ of } 12.5 \text{ liters} & = & 10\% \text{ of } 50 \text{ liters} \\ (0.08)(37.5) & + & (0.16)(12.5) & = & (0.10)(50) \\ 3 & + & 2 & = & 5 \end{array}$$

TEAM PROJECT

(3 or 4 Students)

MAKING UP PROBLEMS FROM NEWSPAPER ADVERTISEMENTS

Materials Needed: Newspaper advertisements.

Course of Action: As a team, find a newspaper ad of a sale on clothes. Make up a word problem similar to Example 2 of this section. Write up the solution of this problem. Trade problems with another team. Solve their problem. Then correct each other's problem solutions.

As a team, find a newspaper ad for a sale on a car. Make up a word problem. Write up the solution of this problem. Trade problems with another team. Solve their problem. Then correct each other's problem solutions.

As a team, make a chart of all the pennies, nickels, and dimes that each team member has today. Make up a word problem similar to Example 6 of this section. Write up the solution of this problem. Trade problems with another team. Solve their problem. Then correct each other's problem solutions.

Team Report: How many problems did your team solve? What difficulties, if any, did the team encounter?

WARMING UP

Answer true or false.

1. The number of dollars in x 20 dollar bills is $20x$.

2. The number of yards in x feet is $3x$.

3. The cost of y paperback books at \$4.95 each is $\dfrac{y}{4.95}$.

4. If the sum of two numbers is 8 and x is one of them, then $8 - x$ is the other one.

5. If a 12-foot board is cut into two pieces and the length of one piece is x feet, then the length of the other piece is $12 - x$ feet.

6. Thirty-two percent of n dollars is $3.2n$ dollars.

7. Fifteen percent of x quarts of a liquid is $0.15x$ quarts.

8. If Meredith is n years old today, then she will be $n + 5$ years old in 5 years.

EXERCISE SET 2.5

For Exercises 1–14, represent one number by a variable. Then, express the other number in terms of the variable. Be sure to clearly state what the variable represents.

1. The sum of two numbers is 6.

2. The sum of two numbers is 12.

3. The sum of two numbers is −7.

4. The sum of two numbers is −10.

5. The selling price of a chair and the cost, where the cost is 55% of the selling price.

6. The selling price of a lamp and the cost, where the cost is 30% of the selling price.

7. The regular price and the sale price of a video camera, if the sale price is 20% off the regular price.

8. The regular price and the sale price of a dress, if the sale price is 30% off the regular price.

9. A wire that is 14 inches long is cut into two pieces.

10. A 9-foot board is cut into two parts.

11. The number of apples and oranges in a produce display totals 50.

12. Twenty flashlights are separated into two groups, defective and nondefective.

13. The combined income of Mr. and Mrs. Abbot is $110,000 a year.

14. A parking meter contains 32 coins consisting of dimes and quarters.

For Exercises 15–18, solve the word problem. See Example 1.

15. Marcy is planning to buy a racing bike that costs $598. This is $50 more than four times the amount she saved last month. How much did she save last month?

16. Juan is planning to buy a keyboard that costs $300. This is $60 more than three times the amount he saved last month. How much did he save last month?

17. The selling price of a sofa is $500. This is $80 more than twice the cost. What is the cost of the sofa?

18. The selling price of a video game machine is $120. This is $15 less than three times the cost. What is the cost of the machine?

For Exercises 19–22, solve the word problem. See Example 2.

19. Carpets Galore is selling rugs at 30% off. If the sale price is $140, what was the original price?

20. A ceiling fan is on sale at 20% off. If the sale price is \$56, what was the original price?

21. A picnic table is on sale at 25% off. If the sale price is \$60, what was the original price?

22. Warren's Jewelers has watches on sale at 40% off. If the sale price is \$210, what was the original price?

For Exercises 23–26, solve the word problem. See Example 4.

23. Find two consecutive even integers whose sum is 78.

24. Find two consecutive odd integers whose sum is 168.

25. Find three consecutive odd integers whose sum is -39.

26. Find three consecutive even integers whose sum is -66.

For Exercises 27–30, solve the word problem. See Example 5.

27. A 15-foot board is to be cut into two so that one part is twice the length of the other part. Find the lengths of the two pieces.

28. A 24-inch glass rod is to be cut into two parts so that the length of one piece is three times the length of the other piece. Find the lengths of the two parts.

29. Luigi the plumber must cut a 16-foot-long pipe into two parts. If one piece is to be 3/5 the length of the other piece, find the lengths of the two parts.

30. A rubber tube that is 30 centimeters long must be cut into two pieces so that one piece is 2/3 the length of the other piece. Find the length of the two parts.

For Exercises 31–34, solve the word problem. See Example 6.

31. Faith has five dollars in dimes and quarters. If she has six more quarters than dimes, how many of each coin does she have?

	Dimes	Quarters
Number	x	
Value		

32. A store has \$320 in ten- and twenty-dollar bills in its cash register. If there are four more twenty-dollar bills than ten-dollar bills, how many bills of each kind does the store have?

	\$10 Bills	\$20 Bills
Number	x	
Value		

33. Deirdre has \$125 in five- and ten-dollar bills. If she has a total of 16 bills, how many of each kind does she have?

34. Victor has one dollar in nickels and dimes. If he has a total of 14 coins, how many coins of each type does he have?

For Exercises 35–40, solve the word problem. See Example 7.

35. A 20% alcohol solution is mixed with a 30% alcohol solution to obtain 25 gallons of a solution with 26% alcohol. How many gallons of each solution should be used?

	20% Solution	30% Solution
Amount of solution	x	
Amount of alcohol		

36. A 40% salt solution is mixed with a 25% salt solution to obtain 24 liters of a 30% salt solution. How many liters of each solution should be used?

	40% Solution	25% Solution
Amount of solution	x	
Amount of salt		

37. Milk that is 1% butterfat is mixed with milk that is 4% butterfat to make 12 gallons of milk that is 2% butterfat. How many gallons of each type is needed?

38. A solution containing 24% iodine is mixed with a solution containing 40% iodine to make 32 liters of a solution with 35% iodine. How many liters of each are needed?

39. How many liters of a 2% zinc solution must be added to 20 liters of a 6% zinc solution to make a solution containing 3% zinc?

40. How many gallons of a 5% quinine solution must be added to 10 gallons of a 12% quinine solution to make a solution containing 7% quinine?

For Exercises 41–44, solve the word problem.

41. Eddy makes \$250 per week plus 20% on sales. How much must he sell in a week to make a total of \$3,000?

42. Thirty-six students are separated into two groups. The number in the first group is three less than twice the number in the second group. How many students are in each group?

43. Sherri has 14 fish in her aquarium this year. This is two less than twice the number of fish she had last year. How many fish did she have last year?

44. A store is making a mixture of peanuts and cashews. If three pounds of peanuts that cost one dollar per pound are used, how many pounds of cashews costing five dollars per pound should be used if the total mixture is to cost four dollars per pound?

REVIEW EXERCISES

The following exercises review parts of Section 1.3. Doing these exercises will help prepare you for the next section.

For Exercises 45–50, place either < or > between the two numbers to make a true statement.

45. $3 \quad 1$

46. $-3 \quad -1$

47. $\frac{3}{4} \quad \frac{1}{2}$

48. $\frac{1}{2} \quad \frac{1}{3}$

49. $-\frac{2}{5} \quad -\frac{1}{4}$

50. $-0.1 \quad -0.01$

ENRICHMENT EXERCISES

1. Find the number of nickels, dimes, and quarters, if the value of x nickels, $4x$ dimes, and $2x - 1$ quarters is \$3.55.

2. The sum of five consecutive odd integers is -5. Find the integers.

Answers to Enrichment Exercises begin on page A.1.

2.6 Solving Inequalities

OBJECTIVES

- ▶ *To solve linear inequalities using properties of inequalities*
- ▶ *To solve linear inequalities using the multiplication-division property of inequality*
- ▶ *To solve general linear inequalities*

LEARNING ADVANTAGE

The instructor may give you some time to work on one or more problems in class. Don't waste this time by allowing your mind to wander. Whether you believe the problems are too simple and not worth doing or too difficult that you have no chance of succeeding, it is important to make your best effort during class time.

The inequality symbols introduced in Section 1.2 to compare numbers can also be used to compare algebraic expressions. In this section we will solve **linear inequalities** such as

$$3x - 1 \geq 5 \qquad \text{and} \qquad 6 - \frac{5x - 2}{4} < 7x + 12$$

When solving a *linear equation,* we obtained a *single* number that made the equation a true statement. However, a *linear inequality,* in general, will have a solution set containing *many* numbers. For example, consider the linear inequality

$$2x + 4 \geq 8$$

If we replace x by 2,

$$2(2) + 4 \overset{?}{\geq} 8$$
$$4 + 4 \overset{?}{\geq} 8$$
$$8 \geq 8$$

Therefore 2 is a solution to the given inequality.

If we replace x by 2.5,

$$2(2.5) + 4 \overset{?}{\geq} 8$$
$$5 + 4 \overset{?}{\geq} 8$$
$$9 \geq 8$$

Therefore, 2.5 is another solution to the given inequality. In fact, the solution set of

$$2x + 4 \geq 8$$

is the set of all numbers greater than or equal to 2. We can write these numbers as

$$x \geq 2$$

The inequality $x \geq 2$ is an example of a "basic" linear equality. The following is a list of basic linear inequalities and their meanings.

Basic Linear Inequalities

Inequality	*Meaning*
$x < 3$	All numbers less than 3.
$x \leq -1$	All numbers less than or equal to -1.
$x > 4$	All numbers greater than 4.
$x \geq 0$	All numbers greater than or equal to 0.
$-5 < x < 7$	All numbers greater than -5 and less than 7.
$3 < x \leq 6$	All numbers greater than 3 and less than or equal to 6.
$4 \leq x < 5$	All numbers greater than or equal to 4 and less than 5.
$-9 \leq x \leq -5$	All numbers greater than or equal to -9 and less than or equal to -5.

NOTE *The inequality $7 < x$ means all numbers greater than 7. It is frequently written as $x > 7$.*

> **COMMON ERROR**
>
> Symbolism like $3 < x < 2$ is never used. This would mean all numbers greater than 3 *and* less than 2 *at the same time*. No such numbers exist.

Example 1 Let x be the number of people waiting in line at a theater. Write an inequality for each statement.

(a) There are more than 15 people waiting in line.
(b) There are no fewer than 10 people waiting in line.
(c) There are no more than 12 people waiting in line.

Solution

(a) If there are more than 15 people waiting in line, then the number x is greater than 15; that is, $x > 15$.
(b) No fewer than 10 people means 10 or more people are waiting in line. Therefore, the correct inequality is $x \geq 10$.
(c) No more than 12 people waiting in line means that 12 or less people are waiting in line. Therefore, the correct inequality is $x \leq 12$.

A linear inequality such as $x < 2$ will have infinitely many solutions. On a number line, all numbers to the left of 2 are solutions. We can draw the **graph** of the solution set as shown.

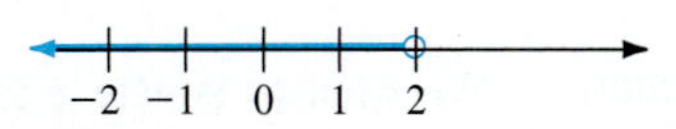

We use an *open circle* when the inequality is either $<$ or $>$ and a *solid circle* when the inequality is either $\leq$ or $\geq$. Another standard notation for graphing inequalities is to use *parentheses* for either $<$ or $>$ and *brackets* for either $\leq$ or $\geq$. The two notations are shown in the next example.

Example 2 Graphing inequalities:

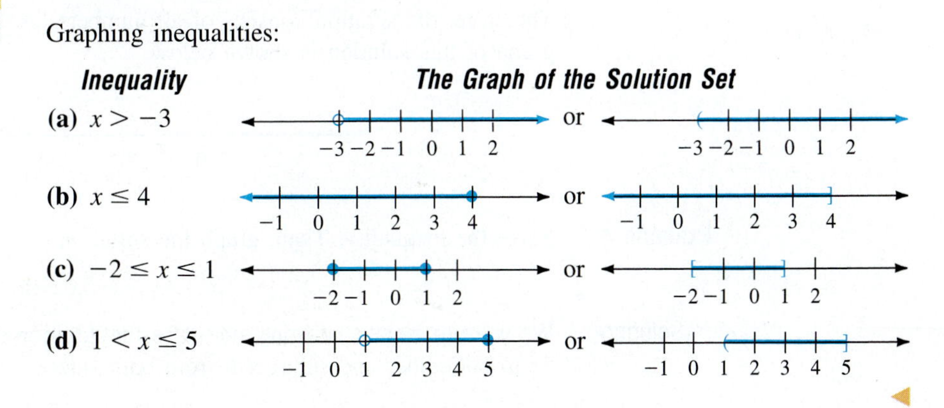

Solving a linear inequality requires properties that are similar to the properties for solving linear equations.

PROPERTY

The solution set of an inequality does not change, if we apply any of the following properties:

Addition-Subtraction Property of Inequality

Adding or subtracting the same quantity on both sides of the inequality.

Multiplication-Division Property of Inequality

Multiplying or dividing both sides by the same *positive* number.
Multiplying or dividing both sides by the same *negative* number and changing the direction of the inequality.

These properties are true for any of the four inequalities: $<$, $>$, $\leq$, $\geq$.

Notice that the multiplication-division property for a negative number requires that the direction of the inequality be changed. For example, if we multiply both sides of $-2 < 3$ by -4, then the rule states that

$$(-4)(-2) > (-4)(3) \qquad \text{or} \qquad 8 > -12,$$

which *is* a true statement.

Example 3 Solve the inequality. Then, graph the solution.

$$x - 12 \leq -9$$

Solution We want to isolate x to one side of the inequality. To achieve this, use the addition-subtraction property of inequality and add 12 to both sides.

$$x - 12 \leq -9$$

$$x - 12 + 12 \leq -9 + 12 \qquad \text{Add 12 to both sides.}$$

$$x \leq 3 \qquad \text{Simplify.}$$

Therefore, the solution consists of all numbers less than or equal to 3. The graph of this solution is shown below.

Example 4 Solve the inequality. Then, graph the solution.

$$6 - 4x > -5x - 1$$

Solution We want to isolate x to one side. Use the addition-subtraction property to add $5x$ to both sides and subtract 6 from both sides.

$$6 - 4x > -5x - 1$$

$$6 - 4x + 5x - 6 > -5x - 1 + 5x - 6 \quad \text{Add } 5x \text{ and subtract 6.}$$

$$x > -7 \quad \text{Simplify.}$$

Therefore, the solution consists of all numbers greater than -7. The graph of this solution is shown below.

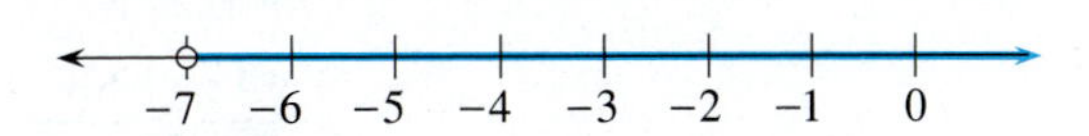

Example 5 Solve the inequality. Then, graph the solution set.

$$-1 \leq x + 6 < 11$$

Solution There are two inequalities in this problem.

$$-1 \leq x + 6 \qquad \text{and} \qquad x + 6 < 11$$

However, we can solve both of them at the same time. We subtract 6 from each part.

$$-1 \leq x + 6 < 11$$

$$-1 - 6 \leq x + 6 - 6 < 11 - 6 \quad \text{Subtract 6.}$$

$$-7 \leq x < 5 \quad \text{Simplify each part.}$$

Therefore, the solution consists of all numbers greater than or equal to -7 and less than 5. The graph of the solution set is shown below.

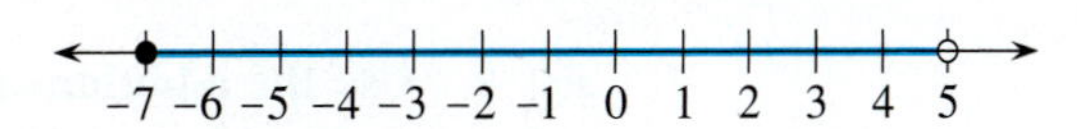

Example 6 Solve the inequality and graph the solution set.

$$2x - 1 > -7$$

Solution

$$2x - 1 > -7$$

$$2x - 1 + 1 > -7 + 1 \quad \text{Add 1 to both sides.}$$

$$2x > -6 \quad \text{Simplify.}$$

$$\frac{2x}{2} > \frac{-6}{2} \quad \text{Divide both sides by (positive) 2.}$$

$$x > -3$$

Therefore, the solution set consists of all numbers greater than -3 and is graphed below.

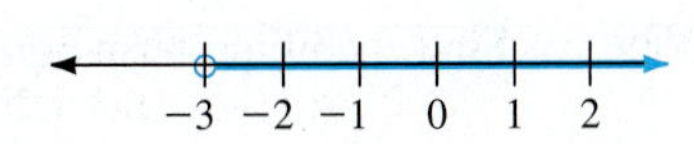

Example 7 Solve the inequality and graph the solution set.

$$3 - 4x \leq 11$$

Solution

$$3 - 4x \leq 11$$

$$3 - 4x - 3 \leq 11 - 3 \qquad \text{Subtract 3 from both sides.}$$

$$-4x \leq 8 \qquad \text{Simplify.}$$

$$\frac{-4x}{-4} \geq \frac{8}{-4} \qquad \text{Divide by } -4 \text{ and change the direction of the inequality.}$$

$$x \geq -2$$

The graph of the solution set is shown below.

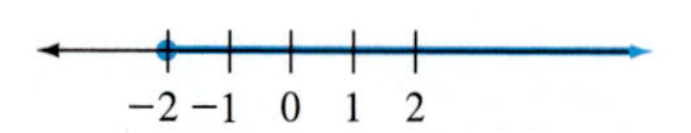

The following steps are guidelines for solving linear inequalities.

STRATEGY

To Solve a Linear Inequality

Step 1 **Combine like terms on each side.** Use the distributive property first, if necessary, to separate terms.

Step 2 **Use the addition-subtraction property of inequality,** if necessary, to move all terms containing the variable to one side and all constant terms to the other side.

Step 3 **Use the multiplication-division property of inequality,** if necessary, to make the coefficient on the variable term equal to one. The inequality now has one of the basic forms:

$$x < b \qquad x \leq b$$

$$x > a \qquad x \geq a$$

or a combination of these such as $a < x \leq b$.

Example 8 Solve the inequality and graph the solution.

$$3x - 5 + 2x - 7 \leq 8x + x - 12 + 2$$

Solution First, combine terms on both sides.

$$3x - 5 + 2x - 7 \leq 8x + x - 12 + 2$$

$$5x - 12 \leq 9x - 10 \qquad \text{Combine terms.}$$

$$5x - 12 + 12 - 9x \leq 9x - 10 + 12 - 9x \qquad \text{Add 12 and subtract } 9x.$$

$$-4x \leq 2 \qquad \text{Combine terms on each side.}$$

$$\frac{-4x}{-4} \geq \frac{2}{-4} \qquad \text{Divide by } -4 \text{ and change the direction of the inequality.}$$

$$x \geq -\frac{1}{2} \qquad \text{Simplify.}$$

Therefore, the solution set consists of all numbers greater than or equal to $-\dfrac{1}{2}$.

The graph is shown below.

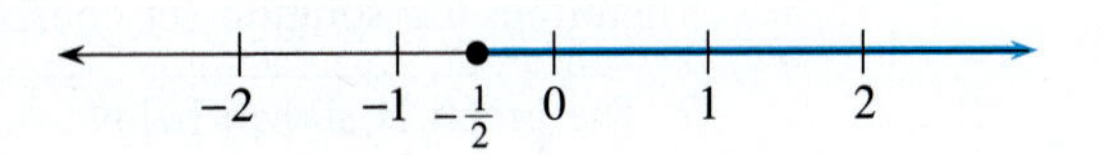

Example 9 Solve the inequality and graph the solution.

$$-5 < 7 - 3t < 10$$

Solution To isolate t in the middle, first subtract 7 from all parts.

$$-5 < 7 - 3t < 10$$

$$-5 - 7 < 7 - 3t - 7 < 10 - 7 \qquad \text{Subtract 7.}$$

$$-12 < -3t < 3 \qquad \text{Simplify.}$$

$$\frac{-12}{-3} > \frac{-3t}{-3} > \frac{3}{-3} \qquad \text{Divide by } -3. \text{ Do not forget to change the direction of each inequality.}$$

$$4 > t > -1 \qquad \text{Simplify.}$$

This answer can be written as $-1 < t < 4$. This second form is easier to visualize, although both are correct. Therefore, the solution set consists of all numbers greater than -1 and less than 4. The graph is shown below.

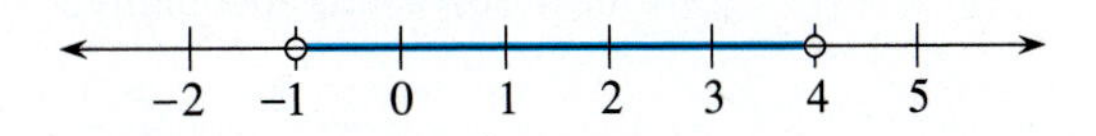

Example 10 Solve the inequality and graph the solution.

$$3 < \frac{3y + 4}{2} < 5$$

Solution We start by multiplying by 2 to clear the fraction.

$$3 < \frac{3y + 4}{2} < 5$$

$$2 \cdot 3 < (2)\frac{3y + 4}{2} < 2 \cdot 5 \qquad \text{Multiply each part by 2.}$$

$$6 < 3y + 4 < 10 \qquad \text{Simplify.}$$

$$6 - 4 < 3y + 4 - 4 < 10 - 4 \qquad \text{Subtract 4 from each part.}$$

$$2 < 3y < 6 \qquad \text{Simplify.}$$

$$\frac{2}{3} < \frac{3y}{3} < \frac{6}{3} \qquad \text{Divide each part by 3.}$$

$$\frac{2}{3} < y < 2 \qquad \text{Simplify.}$$

Therefore, the solution set consists of all numbers greater than $\frac{2}{3}$ and less than 2. The graph is shown below.

We finish this section with word problems involving inequalities.

Example 11 Find all numbers satisfying the given information.

A number subtracted from two is greater than the sum of three and twice the number.

Solution Let x be this unknown number. Reading the information again, we obtain the following inequality.

a number subtracted from 2	is greater than	the sum of 3 and twice the number
$2 - x$	$>$	$3 + 2x$

We now solve this inequality.

$$2 - x > 3 + 2x$$

$$2 - x + x - 3 > 3 + 2x + x - 3 \qquad \text{Add } x \text{ and subtract 3.}$$

$$-1 > 3x \qquad \text{Simplify.}$$

$$-\frac{1}{3} > x \qquad \text{Divide by 3.}$$

Therefore, the solution consists of all numbers less than $-\frac{1}{3}$.

Example 12 Mike received grades of 82%, 91%, and 84% on his first three algebra tests. What possible scores on the fourth test will give him a final average between 80% and 89%, inclusive?

Solution Let x represent Mike's grade on the fourth test. The average of the four tests is given by

$$\frac{82 + 91 + 84 + x}{4}$$

This average must fit between 80 and 89, *inclusive.* Therefore,

$$80 \leq \frac{82 + 91 + 84 + x}{4} \leq 89$$

Solving for x,

$$80 \leq \frac{82 + 91 + 84 + x}{4} \leq 89$$

$$4 \cdot 80 \leq 4\,\frac{82 + 91 + 84 + x}{4} \leq 4 \quad 89 \qquad \text{Multiply by 4.}$$

$$320 \leq 257 + x \leq 356 \qquad \text{Simplify.}$$

$$320 - 257 \leq 257 + x - 257 \leq 356 - 257 \qquad \text{Subtract 257.}$$

$$63 \leq x \leq 99 \qquad \text{Simplify.}$$

Therefore, he must score between 63% and 99%, inclusive, on the fourth test to achieve a final average that falls in the given range.

WARMING UP

Answer true or false.

1. If $x + 1 < 3$, then $x < 2$.
2. $x - 3 \geq 4$ and $x \geq 1$ have the same solution set.
3. If $2x > x$, then $x > 1$.
4. $x + 2 < -1$ and $x < -3$ have the same solution set.
5. We can subtract the same number from both sides of an inequality and not change the solution set.
6. If $-3 < x - 4 \leq 5$, then $-7 < x \leq 1$.
7. If $-1 \leq x + 1 < 0$, then $-2 < x \leq -1$.
8. If $0 \leq x + 3 \leq 3$, then $-3 \leq x \leq 0$.

EXERCISE SET 2.6

For Exercises 1–6, let x be the number of people waiting in line at a theater. Write an inequality for each statement.

1. There are more than ten people waiting in line.

2. There are fewer than six people waiting in line.

3. There are at least nine people waiting in line.

4. There are no more than 12 people waiting in line.

5. There are seven or fewer people waiting in line.

6. There are 14 or more people waiting in line.

For Exercises 7–16, graph the given inequality.

7. $x < 7$

8. $x > -2$

9. $x \leq 3$

10. $x > 4$

11. $x \geq -2$

12. $x \leq 0$

13. $1 < x$

14. $-1 \geq x$

15. $-1 < x < 1$

16. $-2 < x < 3$

For Exercises 17–52, solve the inequality. Then, graph the solution set.

17. $x - 3 \leq 7$

18. $x - 5 < 5$

19. $x + 2 > 2$

20. $x + 3 \geq 2$

21. $x + 30 < 50$

22. $x - 10 \geq -20$

23. $2 + 3x < 2x + 1$

24. $-5 + 6x \geq 5x + 2$

25. $-5 < x - 3 < 0$

26. $-1 < x - 2 \leq 0$

27. $7 \leq x + 7 < 10$

28. $6 < x + 11 \leq 12$

29. $-3 < x - 1 < -1$

30. $-1 \leq x + 2 \leq 4$

31. $4x \geq -8$

32. $3x < 12$

33. $2x < 12$

34. $5x \geq -15$

35. $-3x \geq 9$

36. $-2x \leq 8$

37. $-x < -1$

38. $-4x > -4$

39. $\frac{x}{2} < -1$

40. $\frac{y}{3} > -2$

41. $-\frac{t}{3} \geq -1$

42. $-\frac{u}{2} < 1$

43. $2x - 5 + 3x > -5x + 15$

44. $16x - 21x + 7 \leq 3x + 14 + 2x + 8$

45. $y + 3(2y + 1) \leq 12 + y$

46. $4 + 3(x - 3) \geq x - 6$

47. $4 < x + 7 < 5$

48. $-1 < y - 3 < 2$

49. $3 < 1 - 4x < 5$

50. $-3 < 2 - 5x < -1$

51. $-1 \leq \frac{2x - 1}{3} < 1$

52. $0 \leq \frac{2x - 4}{5} < 2$

For Exercises 53–64, solve the inequality.

53. $-4x \geq 15$

54. $-2z < 9.6$

55. $-9 \leq -3w < 5$

56. $22 < -4v < 28$

57. $1 < 5 - 2x < 4$

58. $4 < 3 - 2t < 6$

59. $-3 < 1 - 4y \leq -1$

60. $-6 \leq -1 - 3x < 2$

61. $0.0912 - 0.88x < -0.4x + 0.1392$

62. $0.057 - 0.76z > 0.19(z - 0.3)$

63. $27(323 - 17n) \leq 8262$

64. $0.875 \leq \frac{-0.0805 + 0.35t}{0.43} < 1.085$

For Exercises 65–68, write an inequality from the given information. Then, solve it and graph the solution set.

65. The sum of a number and 2 is less than 1.

66. The sum of a number and -3 is not less than -2.

67. The difference of a number and 3 is not more than 1.

68. The difference of a number and 3 is not less than -5.

69. Sharon's goal is to have a bowling average of at least 180 for the tournament. Her scores so far are 166 and 189. What must she bowl the last game to achieve this goal?

70. Roberto received grades of 66%, 72%, and 48% on his first three history tests. What possible scores on the fourth test will give a final average of at least 70%?

71. A company that makes frozen chicken nuggets makes a profit of $0.45 per box sold. How many boxes must be sold to have a profit of at least $90,000?

72. A company that produces conveyer belts makes a profit of $250 per belt. How many belts must be sold to have a profit of at least $100,000?

For Exercises 73–76, does the given inequality statement make sense? Why?

73. $-4 > x > -1$

74. $5 < x < 3$

75. $2 > x > 5$

76. $-3 < x < -7$

SAY IT IN WORDS

77. Explain the addition-subtraction property of inequality.

78. Explain the multiplication-division property of inequality.

REVIEW EXERCISES

The following exercises review parts of Section 1.6. Doing these problems will help prepare you for the next section.

79. Evaluate $2x + 2y$ for $x = 3$ and $y = 4$.

80. Evaluate $(s + t)(s + 2t)$ for $s = -1$ and $t = 2$.

81. Evaluate $\frac{a + b}{2}$ for $a = 5$ and $b = 3$.

82. Evaluate $\frac{1}{2}bh$ for $b = 2.4$ and $h = 10$.

83. Evaluate πr^2 for $r = 2$. Use 3.14 to approximate π.

84. Evaluate $\frac{4}{3}\pi r^3$ for $r = 3$. Use 3.14 to approximate π.

ENRICHMENT EXERCISES

1. Solve $|x| < 2$. (*Hint:* Consider the two cases, where Case 1 is $x \geq 0$ and Case 2 is $x < 0$.)

Solve for x.

2. $|x| < 3$

3. $|x - 2| < 1$

4. $|x + 4| \leq 5$

5. What is the solution set of $|x| < -1$?

6. What is the solution set of $|x| > 0$?

Answers to Enrichment Exercises begin on page A.1.

2.7 Formulas and Applications

OBJECTIVES

- *To use formulas in solving word problems*
- *To solve a formula for one variable in terms of the other variables*

LEARNING ADVANTAGE *To solve a word problem, avoid the temptation to start with the first fact stated. Many times the key to the solution may appear in the last sentence or in a question asked at the end after all the information has been given. The appropriate place to begin may not be apparent at first.*

Some problems use a formula in their solution. A **formula** is an equation that contains two or more variables.

For example, when a car travels at a constant speed, or rate, for a given time, then the distance traveled is the product of the rate and the time. This is described by the formula

$$d = rt$$

where d is the distance traveled, r is the (constant) rate, and t is the time. For example, if a car is being driven at a rate of 55 miles per hour (mph) for two hours, then the distance traveled is given by $d = 55 \cdot 2$ or 110 miles.

Example 1 Tom and Joe leave school at the same time and bicycle in opposite directions. If Tom rides his bicycle at 4 mph and Joe rides his bicycle at 5 mph, how long will it take for them to be 3 miles apart?

Solution When solving a motion problem, we first draw a diagram as shown below.

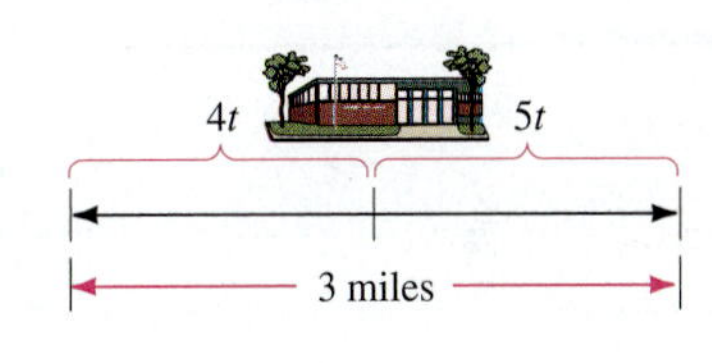

Let t be the time, in hours, it takes for Tom and Joe to be 3 miles apart. We can organize the information in a table.

	Rate	*Time*	*Distance*
Tom	4	t	$4t$
Joe	5	t	$5t$

Now, from the diagram,

$$\begin{array}{ccccc} \text{Tom's distance} & + & \text{Joe's distance} & = & \text{3 miles} \\ 4t & + & 5t & = & 3 \end{array}$$

Next, we solve this equation for t:

$$4t + 5t = 3$$
$$9t = 3$$
$$t = \frac{3}{9} \text{ or } \frac{1}{3} \text{ hour}$$

Therefore, in $\frac{1}{3}$ hour, or 20 minutes, Tom and Joe will be three miles apart. ◀

Example 2 One morning, Marla leaves for work driving 40 mph. Fifteen minutes later, her husband discovers that she forgot some important papers. He drives 50 mph to catch up to her. When does he catch up to her?

Solution Let

t = the time he travels to catch up to Marla

Since Marla started for work 15 minutes $\left(\frac{1}{4} \text{ hour}\right)$ earlier, she has been traveling $t + \frac{1}{4}$ hours. We now fill in the table:

	Rate	*Time*	*Distance*
Marla	40	$t + \frac{1}{4}$	$40\left(t + \frac{1}{4}\right)$
Her husband	50	t	$50t$

When Marla's husband catches up to her, they both will have traveled the same distance. Therefore,

$$\text{Marla's distance} = \text{Her husband's distance}$$

$$40\left(t + \frac{1}{4}\right) = 50t$$

We now solve this equation.

$$40t + 40\left(\frac{1}{4}\right) = 50t \qquad \text{Distribute the 40.}$$

$$40t + 10 = 50t$$

$$40t + 10 - 10 - 50t = 50t - 10 - 50t \qquad \text{Subtract 10 and } 50t \text{ from both sides.}$$

$$-10t = -10$$

$$t = 1 \qquad \text{Divide both sides by } -10.$$

Therefore, it takes one hour for Marla's husband to catch up to her.

We next consider an application in business. If you deposit an amount P of money, called the **principal,** in a savings account, the bank will pay you **interest** on the principal. The interest I is determined by the **interest rate** r, expressed as a decimal, and the time t, in years, that the money remains on deposit. The **(simple) interest** is given by the formula

$$I = Prt$$

Example 3 How much interest is made on \$100 deposited in a savings account for six months that pays 7% interest?

Solution To compute the interest at the end of six months, we change 7% to the decimal 0.07 and convert six months to $\frac{6}{12}$ or $\frac{1}{2}$ year. Next, we evaluate the formula $I = Prt$,

$$\begin{aligned} I &= Prt \\ &= (100)(0.07)\left(\frac{1}{2}\right) \end{aligned}$$

$$= 7\left(\frac{1}{2}\right)$$

$$= \$3.50$$

Therefore, the \$100 earned \$3.50 in interest during the six months.

The next example shows how the interest formula can be used to solve an investment problem.

Example 4 Judy plans to invest \$10,000, part at 12% and the other part at 8%. She wants the total interest for one year to be \$950. How much should she invest at each interest rate?

Solution Let x be the amount she invests at 12%. Since the total amount for investment is \$10,000, then $10{,}000 - x$ is the amount that she invests at 8%.

We can now represent the interest earned by each part using the formula $I = Prt$. For the investment of x dollars at 12% for one year, $I = x(0.12)(1) = 0.12x$. For the investment of $10{,}000 - x$ dollars at 8% for one year, $I = (10{,}000 - x)(0.08)(1) = (0.08)(10{,}000 - x)$. Now, the sum of these two interests must be \$950,

$$\text{interest at 12\%} + \text{interest at 8\%} = \text{total interest}$$
$$0.12x + (0.08)(10{,}000 - x) = 950$$

Next, we solve this equation.

$$\begin{aligned} 0.12x + 800 - 0.08x &= 950 \quad \text{Distributive property.} \\ 0.04x &= 150 \\ x &= 3750 \end{aligned}$$

Therefore, Judy should invest $x = \$3750$ at 12% and $10{,}000 - x = 10{,}000 - 3750 = \$6250$ at 8%.

There are many formulas associated with geometric figures. The inside front cover lists geometric figures and formulas. We may divide figures into two types: *two dimensional* and *three dimensional.*

Two dimensional figures, such as a rectangle, have length and width. The figures enclose an *area,* which is measured in square units. Examples of two dimensional figures are the following:

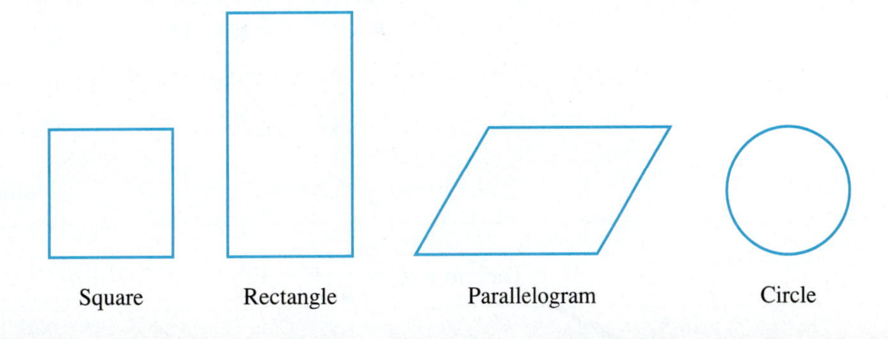

The **perimeter** is the distance around the figure. Common units to measure distance are inches, feet, centimeters, and meters. The **area** of a figure is the enclosed region measured in units such as square inches, square feet, square centimeters, and square meters.

Example 5 A rectangular flower garden is being planned to have a length that is 3 feet more than twice the width. If the perimeter is to be 42 feet, find the length and width of the garden.

Solution We let x be the width of the garden, then $2x + 3$ is the length (see the figure). Now, the perimeter is 42 feet and the formula for perimeter is

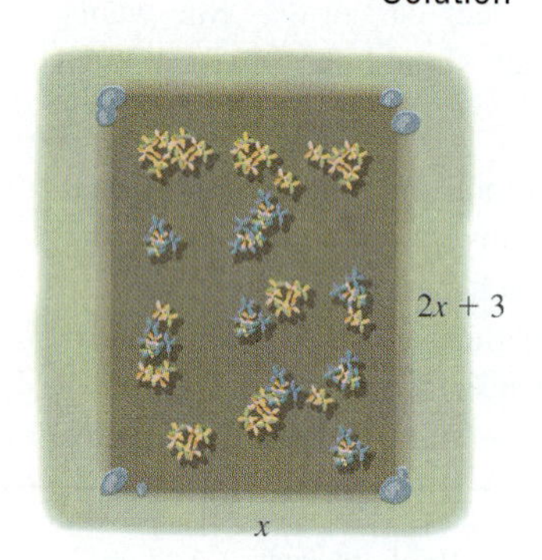

$$2(\text{width}) + 2(\text{length}) = \text{perimeter}.$$

Therefore,

$$2(\text{width}) + 2(\text{length}) = 42$$

Since width $= x$ and length $= 2x + 3$, we have

$$2x + 2(2x + 3) = 42$$

Next, we solve this equation.

$$2x + 4x + 6 = 42 \quad \text{Distribute the 2.}$$
$$6x + 6 = 42 \quad \text{Simplify.}$$
$$6x = 36 \quad \text{Subtract 6 from both sides.}$$
$$x = 6 \quad \text{Divide both sides by 6.}$$

Therefore, the width is 6 feet and the length is $2(6) + 3$, or 15 feet.

In the next example, we solve for one of the variables in a formula.

Example 6 The formula to convert temperatures from Celsius (C) to Fahrenheit (F) is given by $F = \frac{9}{5}C + 32$. Solve for C.

Solution To solve for C, we must isolate it on one side of the equation. We start by subtracting 32 from both sides, and then multiply by $\frac{5}{9}$.

$$F = \frac{9}{5}C + 32$$

$$F - 32 = \frac{9}{5}C + 32 - 32 \quad \text{Subtract 32 from both sides.}$$

$$F - 32 = \frac{9}{5}C$$

$$\frac{5}{9}(F - 32) = \frac{5}{9}\left(\frac{9}{5}C\right) \quad \text{Multiply both sides by } \frac{5}{9}.$$

Therefore, $C = \frac{5}{9}(F - 32)$.

NOTE *The two formulas $F = \frac{9}{5}C + 32$ and $C = \frac{5}{9}(F - 32)$ both determine the same relationship between the variables C and F. The first one, however, is used to convert from Celsius to Fahrenheit and the second one is used to convert from Fahrenheit to Celsius.*

Example 7 Solve $4y - 3x + 2 = 0$ for y.

Solution To solve for y, first isolate the term containing y on the left side by using the addition property of equality.

$$4y - 3x + 2 = 0$$

$$4y - 3x + 2 + 3x - 2 = 0 + 3x - 2 \qquad \text{Add } 3x - 2 \text{ to both sides.}$$

$$4y = 3x - 2 \qquad \text{Simplify.}$$

$$\frac{1}{4}(4y) = \frac{1}{4}(3x - 2) \qquad \text{Multiply both sides by } \frac{1}{4}.$$

$$y = \frac{3}{4}x - \frac{1}{2} \qquad \text{Simplify.}$$

TEAM PROJECT

(3 or 4 Students)

TO DETERMINE THE PERCENTAGE OF FAT IN A FOOD PRODUCT

The percentage of fat in a product at a grocery store can be found by using the information on the package or can. Here is the formula:

$$\text{percentage of fat} = \frac{(920)(\text{grams of fat})}{\text{total calories}}$$

Course of Action: Select five products, such as yogurt, cottage cheese, and bacon, that have both regular and low-fat versions. For each product, make the following table:

PRODUCT ______________________________

	Grams of Fat	*Total Calories*	*Percentage of Fat (from the formula)*
Regular			
Low-fat			

For each product, take the difference of the percentage in fat of the regular and the low-fat versions. Compare your results with another team. Write a report of your team's results. How do you feel about the significance of regular versus low-fat products? Make any recommendations concerning the label of ''low-fat'' on a product.

WARMING UP

Answer true or false.

1. If a train travels for three hours at 50 mph, it has gone 150 miles.
2. A car traveling at 20 mph for 30 minutes covers 600 miles.
3. If $100 is invested at 6% for two years, then the (simple) interest is $12.
4. The perimeter of a rectangle 2 feet in width and 4 feet in length is 8 feet.
5. The area of a rectangle that is 12 inches by 2 inches is 24 square inches.

EXERCISE SET 2.7

For Exercises 1–6, solve the distance-rate-time problem. See Example 1.

1. Bert and Bill leave their home at the same time, traveling in opposite directions. Bert drives at 35 mph and Bill at 45 mph. How long will it take for them to be 80 miles apart?

	Rate	Time	Distance
Bert			
Bill			

2. Two trains leave the Reading Railroad Terminal at the same time, one traveling west and the other traveling east. The first train travels at 50 mph and the second train travels at 60 mph. In how many hours will the trains be 220 miles apart?

	Rate	Time	Distance
First train			
Second train			

3. Two planes leave an airport at the same time, one flying north and the other flying south. If the first plane flies at 80 mph and the second plane flies at 60 mph, how long will it take them to be 210 miles apart?

	Rate	Time	Distance
First plane			
Second plane			

4. Two bicyclists, Pat and Ephraim, pass each other on a bike path going in opposite directions. If Pat is pedaling at 8 mph and Ephraim is pedaling at 12 mph, how long will it take them to be 10 miles apart?

	Rate	Time	Distance
Pat			
Ephraim			

5. Two all-terrain vehicles are 285 miles apart in a desert and traveling toward each other. The first vehicle is traveling at 45 mph and the second one is traveling at 50 mph. How much time passes before they meet?

6. Two boats, 45 nautical miles apart, travel towards each other. If the first boat is traveling at 8 knots (nautical miles per hour) and the second boat is traveling at 10 knots, how long will it take them to meet?

	Rate	Time	Distance
First vehicle			
Second vehicle			

	Rate	Time	Distance
First boat			
Second boat			

For Exercises 7–10, solve the distance-rate-time problem. See Example 2.

7. A freight train left the depot traveling at 40 mph. One-half hour later, a passenger train left the depot in the same direction but on a parallel track traveling 60 mph. In how many hours after the passenger train starts will the two trains meet?

	Rate	Time	Distance
Freight train			
Passenger train			

8. A car left Portland traveling 45 mph. A second car left from the same place one hour later on the same road traveling 55 mph. How long does it take the second car to catch up to the first car?

	Rate	Time	Distance
First car			
Second car			

9. A motorcyclist left Cedar Rapids traveling at 50 mph. One and one-half hours later, a car left Cedar Rapids using the same highway and traveling at 65 mph. How long will it take the car to catch up to the motorcyclist?

	Rate	Time	Distance
Motorcyclist			
Car			

10. Brad left the house traveling 25 mph. One-fifth hour later, his sister discovers that he had forgotten his medicine. She leaves the house traveling 55 mph to catch up to him. How long does it take?

	Rate	Time	Distance
Brad			
His sister			

For Exercises 11–14, solve the simple interest problem. See Example 3.

11. How much interest is made on $2,000 deposited for two years in a savings account that pays 3% interest?

12. Find the interest made on $800 deposited in a savings account for three years that pays 2% interest.

13. Find the interest made on a $4,000 investment that pays 3.5% interest for five years.

14. How much interest is made on $10,000 deposited in an account for six months that pays 4%?

For Exercises 15–18, solve the investment problem. See Example 4.

15. Roman plans to invest his lottery winnings of $8,000, part at 4% and the other part at 5%. He wants the total interest for one year to be $350. How much should he invest at each interest rate?

16. Sally wants to invest $10,000, part at 5% and part at 6%. She wants the total interest for one year to be $560. How much should she invest at each interest rate?

17. Ridge has $20,000 to invest in two securities, one paying 5% and the other paying 8%. How much should he invest at each interest rate to realize a yearly interest payment of $1,240?

18. Perez is planning to invest $12,000 for one year, part at 8% and part at 6%. If he wants to obtain $860 in combined interest, how much should he invest at each interest rate?

For Exercises 19–24, solve the geometric problem. See Example 5.

19. A rectangular garden is to have a length that is 1 foot less than twice the width. If the perimeter is designed to be 28 feet, find the width and length.

20. A rectangular sign is to be made so that the length is 4 inches less than three times the width. If the perimeter must be 56 inches, find the width and the length.

21. For display purposes, the front of a rectangular cereal box is to have a length that is 3 inches more than the width. If the perimeter is designed to be 42 inches, find the width and the length.

22. A rectangular poster has a width that is 1 foot less than one-half its length. If the perimeter is 25 feet, find the width and the length.

23. The length of a rectangular sign is 12 feet. If the area is 72 square feet, find the width of the sign.

24. The width of a rectangular sheet of metal is 8 inches. If the area is 112 square inches, find the length.

For Exercises 25–30, solve the formula for the indicated variable.

25. $C = 2\pi r$ for r. Circumference of a circle.

26. $I = Prt$ for t. Simple interest.

27. $E = \dfrac{I}{d^2}$ for I. Illuminance of a light source.

28. $V = \pi r^2 h$ for h. Volume of a right circular cylinder.

29. $P = 2l + 2w$ for w. Perimeter of a rectangle.

30. $h = vt - 16t^2$ for v. Height of a projectile.

For Exercises 31–38, solve the equation for x.

31. $y = 3x + 1$

32. $y = -5x + 3$

33. $y = \dfrac{2x - 1}{3}$

34. $y = \dfrac{3 - x}{2}$

35. $4x - 2y - 1 = 0$

36. $-6x + 3y + 2 = 0$

37. $y = mx + b$

38. $y = mx - 4$

For Exercises 39–44, solve the equation for y.

39. $2x - 3y + 1 = 0$

40. $x + 5y + 3 = 0$

41. $y - 1 = 3(x - 2)$

42. $y + 3 = -2(x - 5)$

43. $\dfrac{x}{3} - \dfrac{y}{4} = 1$

44. $\dfrac{x}{2} - \dfrac{y}{2} = 1$

45. If $2x - 3y = 6$, find y if $x = 0$.

46. If $-4x + 3y = 12$, find x if $y = 0$.

47. If $\dfrac{x}{5} - \dfrac{y}{3} = 1$, find x if $y = 6$.

48. If $\dfrac{y}{2} + \dfrac{x}{4} = -2$, find y if $x = -4$.

49. Find the area of a triangle if $b = 3.51$ feet and $h = 12.80$ feet.

50. A rectangle has a length that is the sum of 4.23 times the width and 5.545. If the perimeter is 18.412 inches, find the length and width.

51. Fran plans to invest $23,600, part at 11% and part at 9.5% for one year. If she wants to obtain a total interest of $2410, how much should she invest at each interest rate?

SAY IT IN WORDS

52. Explain how you solve a word problem that uses a formula such as $d = r \cdot t$.

ENRICHMENT EXERCISES

For Exercises 1 and 2, solve for y.

1. $Ax + By + C = 0,\ B \neq 0$

2. $\dfrac{x}{a} + \dfrac{y}{b} = 1$

3. Solve $\dfrac{V_2}{T_2} = \dfrac{V_3}{T_3}$ for T_3.

Answers to Enrichment Exercises begin on page A.1.

CHAPTER 2 Summary and review

Algebraic expressions (2.1)

A **term** is an algebraic expression that is comprised of the product and/or quotient of numbers and variables.

In the term $3x^3y$, the number 3 is called the **coefficient** of the term.

Like terms are terms whose variable parts are the same.

Examples — Like terms:

xy, $-2xy$, and $5xy$

Sums and differences of like terms can be combined into a single term.

$$2x - 5x = (2 - 5)x = -3x$$

Equations (2.2)

To **solve an equation** means to find all numbers that make the equation a true statement when used in place of the variable. The collection of these numbers is called the **solution set** of the equation.

Two equations are **equivalent** if they have the same solution set.

Equivalent equations:

$3x = 6$ and $x = 2$

There are two properties of equalities that are used to solve equations.

The Addition-Subtraction Property of Equality

Add (or subtract) the same quantity on both sides of the equation.

The Multiplication-Division Property of Equality

Multiply (or divide) both sides of an equation by the same nonzero quantity.

We use these properties to solve linear equations.

Solving linear equations (2.3)

A **linear equation** is an equation that can be put into the standard form

$$ax + b = 0$$

where a and b are constants with $a \neq 0$.

Linear equations:

$2x + 3 = 0$, $x = -4$, and $3(x - 1) = x$

These steps can be used as a guideline for solving linear equations.

Strategy for Solving a Linear Equation

Step 1 **Clear the equation of any fractions or decimals,** if desired, by multiplying by the LCD of the fractions appearing or, in the case of decimals, by an appropriate power of ten.

Step 2 **Combine like terms on each side,** if necessary, to simplify. You may have to use the distributive property first to separate terms.

Step 3 **Use the addition property of equality,** if necessary, to move all terms containing a variable to one side and all constant terms to the other side.

Solve:

$$2(1 - 3x) = 1$$
$$2 - 6x = 1$$
$$-2 + 2 - 6x = -2 + 1$$
$$-6x = -1$$
$$\frac{-6x}{-6} = \frac{-1}{-6}$$
$$x = \frac{1}{6}$$

Examples

Step 4 **Use the multiplication property of equality,** if necessary, to make the coefficient of the variable term equal to one. The equation is now in the form

$$x = \text{the solution} \qquad \text{or} \qquad \text{the solution} = x$$

Step 5 **Check your answer** by substituting it into the original equation.

Solving word problems (2.4) and (2.5)

Here is a strategy for solving word problems.

1. **Read** the problem carefully. Take note of what is being asked and what information is given.
2. **Plan** a course of action. Represent the unknown number by a letter. If there is more than one unknown, represent one of them by a letter and express the others in terms of the letter.
3. **Create** an equation from the given information.
4. **Solve** this equation.
5. **Check** your solution in the original problem.
6. **Answer** the original question. Read again the problem to make sure you have answered the question.

Review the word problem examples in Sections 2.4 and 2.5.

The addition-subtraction property of inequality (2.6)

Adding or subtracting the same quantity does not change the solution set.

The multiplication-division property of inequality (2.6)

The solution set does not change, if we

(a) Multiply or divide both sides by the same positive number.
(b) Multiply or divide both sides by the same negative number and change the direction of the inequality.

Solving a linear inequality (2.6)

Solve:

$$3x - 1 > 5x$$
$$3x - 1 + 1 - 5x > 5x + 1 - 5x$$
$$-2x > 1$$
$$\left(-\frac{1}{2}\right)(-2x) < \left(-\frac{1}{2}\right)1$$
$$x < -\frac{1}{2}$$

1. Combine terms on each side. This may entail using the distributive property first to separate terms.
2. Use the addition property of inequality, if necessary, to move all terms containing the variable to one side and all constant terms to the other side.
3. Use the multiplication property of inequality, if necessary, to make the coefficient of the variable term equal to one. The inequality now has one of the basic forms $x < b$, $x \leq b$, $x > a$, $x \geq a$, or a combination of these such as $a \leq x < b$.

Formulas (2.7)

To solve $V = \pi r^2 h$ for h, divide both sides by πr^2.

$$\frac{V}{\pi r^2} = h$$

A **formula** is an equation that contains two or more variables. We may solve a formula for a specified variable by using the properties of equality.

CHAPTER 2 REVIEW EXERCISE SET

Section 2.1

For Exercises 1–3, simplify each algebraic expression by combining like terms.

1. $20y - 15y$

2. $st + 4st - 3st$

3. $a^2 - 2 + \frac{1}{2} - a^2$

For Exercises 4–6, remove the parentheses by using the distributive property, then combine like terms.

4. $6(5z + 6) - 15$

5. $3x^2yz - 2(5 - 9x^2yz)$

6. $ab^2 - 2(a^2b + 5ab^2) + a^2b$

Section 2.2

For Exercises 7–15, solve each equation.

7. $x + 6 = 7$

8. $-4y + 11 = 17 - 3y$

9. $13 + 3n - 7n + 1 = -14n + 9n$

10. $2(3z - 1) = 3(z + 2) - 5$

11. $\frac{1}{4}y = -3$

12. $6x = -84$

13. $\frac{n}{5} = -6$

14. $-z = 7$

15. $\frac{9}{7}x = 1$

16. Write an equation for the given sentence, then solve it.

The sum of six times a number and three is five times the number.

Section 2.3

For Exercises 17–20, solve each equation.

17. $5y + 7 = -3$

18. $7t + t - 5 = 2t + 3t - 20$

19. $2(1 - 3x) + 3(4x - 1) = 5$

20. $5 - 2b = 3(b + 1)$

Section 2.4

For Exercises 21–27, convert each word phrase into an algebraic expression using x as the variable.

21. The difference of a number and 17.

22. Five times a number, the result subtracted from 5.

23. The product of 3 and a number, increased by nine.

24. The quotient of three times a number and 5.

25. The reciprocal of the sum of a number and ten.

26. The sum of the reciprocal of a number and four.

27. Twelve percent of a number, the result subtracted from the number.

28. A number is reduced by five. If the result is multiplied by four, it is equal to five subtracted from three times the number. Find the number.

Section 2.5

For Exercises 29–31, represent one of the numbers by a variable. Then, express the other number in terms of the variable.

29. The sum of two numbers is 17.

30. Two consecutive odd integers.

31. One number is 50 more than another number.

For Exercises 32–36, solve the problem.

32. A dress is on sale at 35% off the original price. If the sale price is \$162.50, what was the original price?

33. Eric sold a vase for \$72. This is \$15 less than three times the amount he paid for it. How much did Eric pay for the vase?

34. The sum of two consecutive even integers is -54. Find the two integers.

35. Mario must cut an 18-foot-long pipe into two parts. If one part must be twice the length of the other part, find the lengths of the two pieces.

36. A 14% alcohol solution is mixed with a 22% alcohol solution to obtain 20 gallons of a solution with 16% alcohol. How many gallons of each solution should be used?

Section 2.6

For Exercises 37–40, let x be the number of pennies in a jar. Symbolize each statement with an inequality.

37. There are fewer than 250 pennies in the jar.

38. There are at least 300 pennies in the jar.

39. There are no more than 200 pennies in the jar.

40. There are not less than 100 pennies in the jar.

For Exercises 41–43, graph each inequality.

41. $x \leq 4$

42. $x > -\frac{3}{4}$

43. $-1 < x < 3$

For Exercises 44–50, solve the inequality. Then graph the solution set.

44. $x + 4 < 6$

45. $-1 < x - 3 \leq 2$

46. $x - 2 \geq -3$

47. $3x + 2 > -4$

48. $-3 \le 4x - 3 < 1$

49. $4(1 - 2x) + 1 < -3$

50. $-2 \le \dfrac{2t - 6}{5} \le 0$

51. Write an inequality from the given information. Then, solve and graph the solution set.

The sum of twice a number and six is not less than the sum of the number and three.

52. Find all numbers satisfying the given information. Two times the sum of four and a number is subtracted from seven. The result is less than one fifth times the sum of twice the number and one.

53. For each of the first three weeks in July, Ted earned \$20, \$25, and \$36 mowing lawns. If he wants to average at least \$30 a week for the entire month, how much must he earn during the last week in July?

Section 2.7

54. Two boats, 5 miles apart, are traveling toward each other. If one boat travels at a rate of 7 mph and the other travels at a rate of 13 mph, how long will it take the two boats to meet?

55. How much interest is made on \$800 invested at a rate of 16% for nine months?

56. Ellen is planning to invest \$20,000, part at 10% and the other part at 6%, for one year. If she wants the total interest for the year to be \$1600, how much should she invest at each interest rate?

57. A car left Kansas City traveling 45 mph. A second car left from the same place one hour later on the same road traveling 60 mph. How long does it take the second car to catch up to the first car?

58. A rectangular piece of wood has a length that is 1 foot longer than twice the width. If the perimeter is 20 feet, find the width and length.

59. Solve the formula $A = \dfrac{1}{2}bh$ for b.

60. Solve $y = -5x + 20$ for x.

CHAPTER 2 TEST

1. Simplify by combining like terms: $2x^2 - 8 + 10x - 3x + 4x^2 + x$

2. Remove parentheses using the distributive property, then simplify: $3(2x^5 + 1) - 4(x^5 - 3)$

3. Write the given word phrase as an algebraic expression, choosing a letter for the variable. Then, simplify the expression. The difference of six and twice a number, the result is multiplied by -2.

For Problems 4 and 5, solve each equation.

4. $x - 3 = -9$

5. $5x - 1 = 6(x + 3)$

6. Write an equation for the given sentence, then solve it. The sum of twice a number and five is two less than the number.

For Problems 7–10, solve each equation.

7. $4n = -12$

8. $\frac{x}{10} = 3$

9. $2 - 9z = 11$

10. $4(2 - a) = 12 - 6a$

For Problems 11 and 12, convert each word phrase into an algebraic expression using x as a variable.

11. Five less than a number.

12. Fifteen percent of a number.

13. The sum of two numbers is 18. One number is three less than six times the other number. Find the numbers.

14. Graph the inequality: $-5 < x \leq 3$.

15. Solve the given inequality. Then, graph the solution.
$8x - 4 \geq 6x$

For Problems 16 and 17, solve the inequality.

16. $-5 \leq 3x + 10 < 22$

17. $-\frac{2}{3}x > -12$

18. How many liters of a 5% zinc solution must be added to 4 liters of an 8% zinc solution to make a solution containing 6% zinc?

19. The length of a rectangular sign is 2 feet less than three times the width. If the perimeter is 20 feet, find the width and the length.

20. Solve $-8x + 4y = -2$ for y.

Linear Equations and Their Graphs

CHAPTER 3

CONNECTIONS

The limited edition of Stephen King's *The Stand* sold for $200 in 1990. By 1994, copies of this book were selling for $350. Assuming that the value of the book increases linearly with time, what is the predicted value of the book for 1996? See Exercise 45 in Section 3.2.

Overview

Analytic geometry has its foundation based on associating points on a line with the set of real numbers. We continue this idea by associating points in a plane with pairs of real numbers. This in turn enables us to describe paths in a plane with equations. There are many applications that involve describing a path or curve. A communications satellite, for example, travels a path around the earth. Once we have the connection between paths and mathematics, we can apply the powerful tool of mathematical analysis to help describe and understand the physical world.

3.1 Linear Equations and Their Graphs

OBJECTIVES

- *To determine the coordinates of a point in a plane*
- *To plot a point, given its coordinates*
- *To draw the graph of a linear equation*
- *To find the x- and y-intercepts of a linear equation*

LEARNING ADVANTAGE

Why study mathematics? *Mathematics is not merely learning formulas, but looking for results and determining a repeated sequence of events that can be described by a mathematical formula. Too frequently, a person mistakes arithmetic for mathematics. Arithmetic and algebra are only the start toward using mathematics to solve problems.*

In Chapter 2, we solved equations in one variable. For example, the equation $2x - 5 = 3$ has $x = 4$ as its solution, and we can graph this solution on a number line:

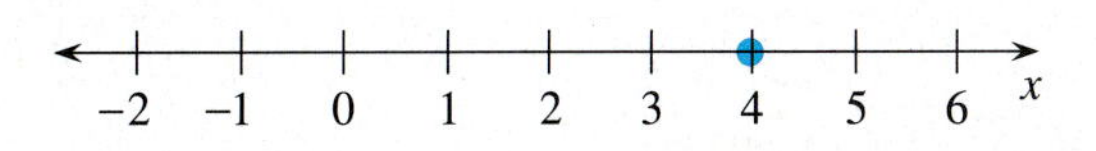

In this chapter, we will study equations in *two* variables such as $2x + y = 8$. By a **solution,** we mean *two* numbers—a value for x and a value for y—that make the equation true. For example, $x = 1$ and $y = 6$ is a solution, since $2(1) + 6 = 8$. Also, $x = 2$ and $y = 4$ is a solution, since $2(2) + 4 = 8$. In fact, there are infinitely many solutions to the equation. We will develop these concepts in more detail in the next section. Notice that for the equation in one variable, $2x - 5 = 3$, we graphed the solution set using a number line. For an equation in two variables, however, a number line no longer works for graphing the solution set since a single solution consists of two numbers. In this section, we develop the *rectangular coordinate system* that will be used for graphing the solution set of these new equations. We start our discussion with the following situation.

Suppose you are attending a rock concert in an auditorium. Your ticket indicates that you are to sit in the seventh seat of the tenth row. Notice that two numbers are needed to find your place. The **ordered pair** of numbers to locate your place can be written as (7, 10).

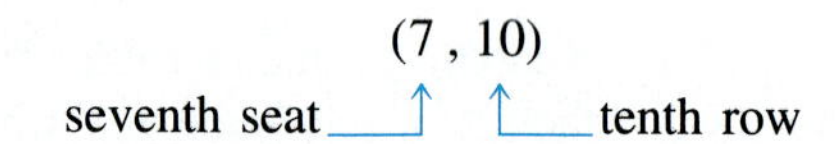

Notice that the *order* in which the two numbers are written is important. The ordered pair (7, 10) means the seventh seat in the tenth row, whereas (10, 7) means the tenth seat in the seventh row.

We will use ordered pairs of numbers to locate points in a plane. Recall, we have used numbers to locate points on a number line. For example, the point P on the number line below has the number 2.5 as its coordinate.

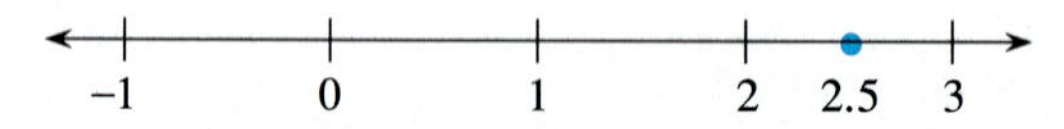

To locate a point in a plane, we need an ordered pair of numbers. To do this, we first draw a **rectangular coordinate system.**

STRATEGY

To Construct a Rectangular Coordinate System

Step 1 Draw a *horizontal* number line.

Step 2 At the point with coordinate 0, draw a *vertical* number line. Each line has the same zero point, called the **origin.**

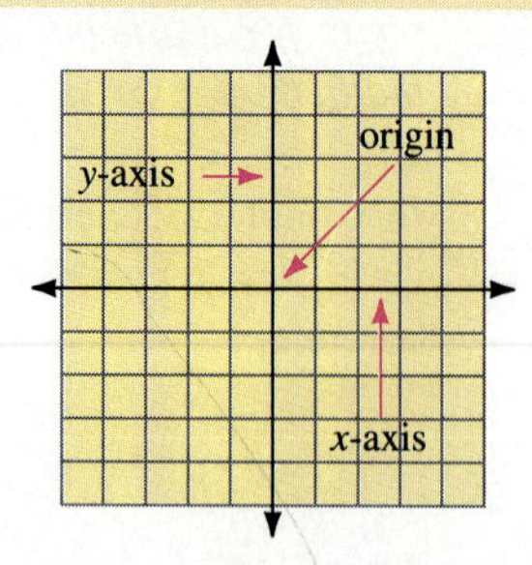

The horizontal number line is called the ***x*-axis.** The *positive* direction is to the *right* on the horizontal axis. The vertical number line is called the ***y*-axis.** The *positive* direction is *upward* on the vertical axis.

We use the coordinate system to locate points that lie in a plane. For example, the origin is given by the ordered pair (0, 0). The point P in the *coordinate plane* as shown in the figure below is labeled with the ordered pair (3, 4). Three is called the ***x*-coordinate** and 4 is called the ***y*-coordinate** of P. The coordinate plane gives us a way to label a geometric point P with an ordered pair of numbers (x, y). René Descartes, (1596–1650), a French mathematician, is given credit for this idea.

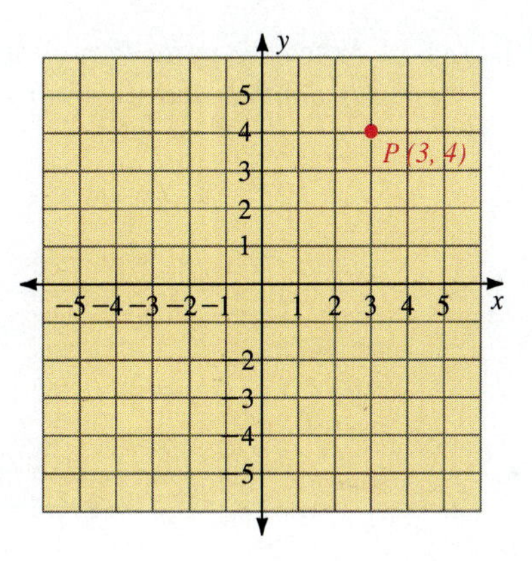

Example 1 Find the ordered pair for each point shown in the coordinate plane.

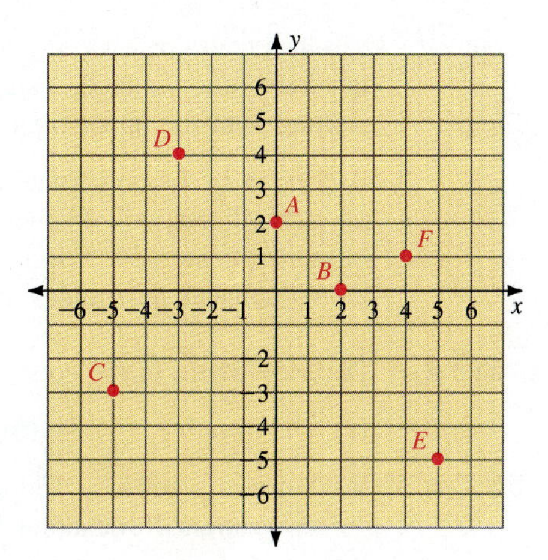

Solution Be sure to always write the x-coordinate as the first entry of each ordered pair (x, y).

Point	Ordered Pair
A	$(0, 2)$
B	$(2, 0)$
C	$(-5, -3)$
D	$(-3, 4)$
E	$(5, -5)$
F	$(4, 1)$

NOTE *When we write $P(x, y)$, we mean the point P with coordinates (x, y). Usually P is omitted and we simply write (x, y).*

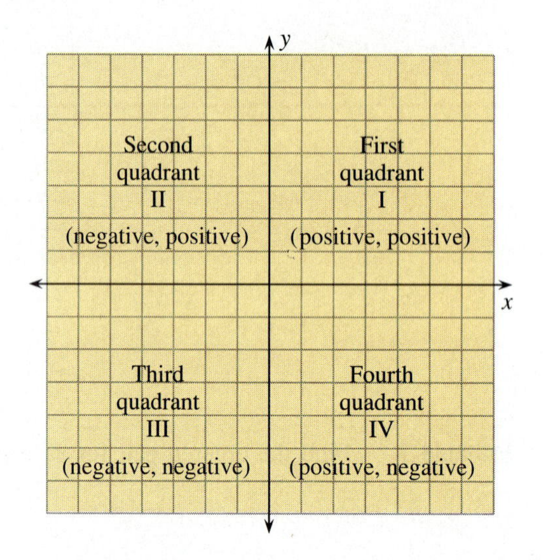

The coordinate axes divide the plane into four parts or **quadrants.** These quadrants are numbered using Roman numerals in a counterclockwise direction starting with the upper right one.

Any point in the coordinate plane either lies in one of the four quadrants or lies on a coordinate axis. If one of the coordinates of a point is zero, the point lies on a coordinate axis. For a point not on a coordinate axis, the sign (+ or −) of the coordinates determines in which quadrant the point lies.

Example 2 Describe where in the coordinate plane the point (x, y) would be located.

(a) x is negative. **(b)** $x = 0$.

Solution **(a)** Since x is negative, the point (x, y) must lie to the left of the y-axis, either in quadrant II, quadrant III, or on the (negative) x-axis.

(b) $x = 0$ means that the point (x, y) is on the y-axis.

Given a point such as $A(3, 5)$, we can **plot** or **graph** it by locating A in the coordinate plane.

Example 3 Plot the points $(-2, 4)$, $(0, -4)$, $(-4, 0)$, and $(5, -3)$.

Solution Since the x-coordinate of $(-2, 4)$ is -2, start at the origin and move two units to the left. The y-coordinate is 4, so we move upward four units. Since the x-coordinate of $(0, -4)$ is 0, stay at the origin. The y-coordinate is -4 so we move down four units. The four points are graphed in the figure that follows.

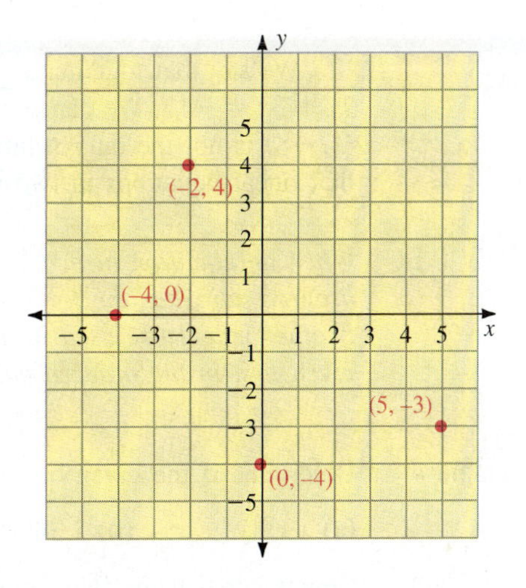

In the previous chapter, we studied equations such as

$$3x + 1 = 4 \qquad \text{and} \qquad 4t - 5 = 2$$

These equations have exactly one variable. We will now study equations having two variables. Examples of *linear* equations in *two* variables are

$$3y - 2x = 6, \qquad y = -\frac{1}{2}x + 1, \qquad \text{and} \qquad y - 2 = 3(x - 4)$$

DEFINITION

A **linear equation in two variables** is an equation that can be put in the standard form

$$Ax + By = C$$

where A, B, and C are numbers, A and B are not both zero.

Examples of linear equations written in standard form are

$$2x + 3y = 5 \qquad (A = 2, B = 3, C = 5)$$
$$x - y = -2 \qquad (A = 1, B = -1, C = -2)$$
$$x = 4 \qquad (A = 1, B = 0, C = 4)$$

A linear equation in one variable, such as $3x + 1 = 2$, has a solution that is one number. However, two numbers are required for a solution to a linear equation in two variables. For example, the equation $7x + 3y = -17$ becomes a true statement when x is replaced by 1 and y is replaced by -8, since

$$\begin{aligned} 7x + 3y &= -17 \\ 7(1) + 3(-8) &\stackrel{?}{=} -17 \\ 7 - 24 &\stackrel{?}{=} -17 \\ -17 &= -17 \end{aligned}$$

We say that together $x = 1$ and $y = -8$ is a **solution** of the equation $7x + 3y = -17$. We can write this solution as the ordered pair $(1, -8)$. In fact, $(1, -8)$ is not the only solution of the equation. We will see in the examples that linear equations in two variables will have many solutions.

NOTE *When an equation in two variables involves x and y, a solution, written as an ordered pair, has the form (x, y). That is, the x value of a solution is given first. If other letters are used in an equation, it will be made clear which variable is given first in the ordered pair form of a solution.*

Example 4 Determine if the given ordered pair is a solution of the equation $3x + 2y = 12$.

(a) $(2, 3)$ **(b)** $(-4, 0)$

Solution Remember that the first number in an ordered pair is the x value and the second number is the y value.

(a) Replace x by 2 and y by 3 in the equation.

$$\begin{aligned} 3x + 2y &= 12 \\ 3(2) + 2(3) &\stackrel{?}{=} 12 \\ 6 + 6 &= 12 \end{aligned}$$

Since replacing x by 2 and y by 3 resulted in the true statement, $12 = 12$, the ordered pair $(2, 3)$ is a solution of the equation.

(b) Replace x by -4 and y by 0.

$$\begin{aligned} 3x + 2y &= 12 \\ 3(-4) + 2(0) &\stackrel{?}{=} 12 \\ -12 + 0 &\neq 12 \end{aligned}$$

Since replacing x by -4 and y by 0 resulted in the false statement, $-12 = 12$, the ordered pair $(-4, 0)$ is not a solution of the equation. ◀

Example 5 Find the ordered pairs that are solutions of $-7x + 3y = 14$ when

(a) $x = -1$ **(b)** $y = 0$

Solution **(a)** Replace x by -1 in the equation and solve for y.

$$-7x + 3y = 14$$
$$-7(-1) + 3y = 14$$
$$7 + 3y = 14$$
$$3y = 7 \quad \text{Subtract 7 from both sides.}$$
$$y = \frac{7}{3} \quad \text{Divide by 3.}$$

Therefore, the ordered pair $\left(-1, \frac{7}{3}\right)$ is a solution.

(b) Replace y by 0 in the equation and solve for x.

$$-7x + 3y = 14$$
$$-7x + 3(0) = 14$$
$$-7x = 14$$
$$x = -2 \quad \text{Divide by } -7.$$

Therefore, $(-2, 0)$ is a solution.

The solution set of a linear equation is a collection of ordered pairs (x, y) that make the equation true. This solution set can be visualized by graphing these ordered pairs in a coordinate plane.

In the next example, we generate and then graph several solutions of a linear equation.

Example 6 Complete the table below for the equation $y = 2x - 3$. Then, graph the resulting ordered pairs in a coordinate plane.

x	y
-1	
0	
1	
2	
3	

Solution Replacing x by -1, we have

$$y = 2(-1) - 3$$
$$= -2 - 3$$
$$= -5$$

Replacing x by 0, 1, 2, and 3, respectively, in the equation, we complete the table.

x	y
-1	-5
0	-3
1	-1
2	1
3	3

From this table we get five solutions to the equation; namely, $(-1, -5)$, $(0, -3)$, $(1, -1)$, $(2, 1)$, and $(3, 3)$. Next, we draw a coordinate plane and graph these ordered pairs as shown in the figure below.

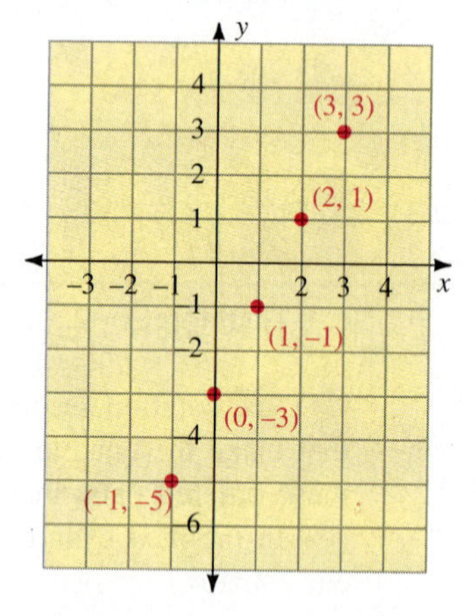

Notice the pattern in the figure above. The points all lie on a straight line. In fact, all solutions of the equation in Example 6 do lie on this same line.

DEFINITION

The **graph** of the linear equation $Ax + By = C$ is the set of points in the coordinate plane with coordinates (x, y) that are solutions of the equation.

As indicated by the graph in the figure on the previous page, we have the following statement.

The graph of the linear equation $Ax + By = C$ is a straight line.

A straight line is easy to draw because of the following statement from geometry.

Two points determine a straight line in the plane.

Therefore, to graph a linear equation we need to find only two ordered pairs that are solutions. However, to serve as a check, we will also find a third pair.

Example 7 Draw the graph of $4x - 3y = 12$.

Solution Since it is of the form $Ax + By = C$, the equation is linear. We know that the graph of any linear equation is a straight line.

1. We may choose any convenient points that are solutions of the equation. Setting $x = 0$,

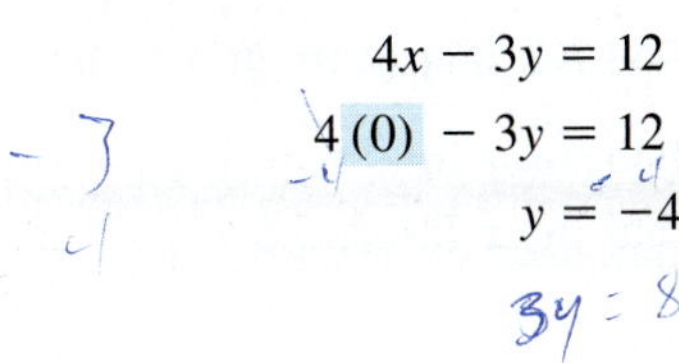

$$4x - 3y = 12$$
$$4(0) - 3y = 12$$
$$y = -4$$

x	y
0	

Therefore, $(0, -4)$ is a solution.

2. Next, setting $y = 0$,

$$4x - 3(0) = 12$$
$$4x = 12$$
$$x = 3$$

x	y
0	−4
	0

Therefore, $(3, 0)$ is another solution.

3. As a check point, set $x = 6$. Then

$$4x - 3y = 12$$
$$4(6) - 3y = 12$$
$$24 - 3y = 12$$
$$-3y = -12$$
$$y = 4$$

x	y
0	−4
−3	0
6	

Therefore, our check point is $(6, 4)$. We now plot these three points in a coordinate plane and draw the line through these points. See the figure below.

x	y
0	−4
3	0
6	4

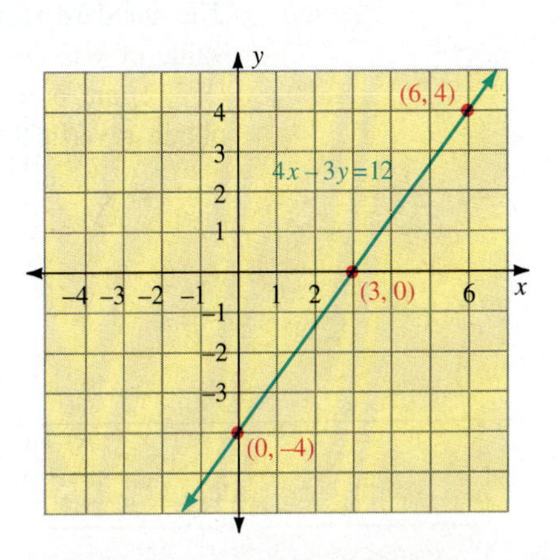

Notice that the graph in the figure above crosses the y-axis at $y = -4$. The number -4 is called the **y-intercept** of the line. Notice also that the graph crosses the x-axis at 3. The number 3 is called the **x-intercept** of the line. It is often convenient to graph linear equations by finding the intercepts.

STRATEGY

To Find the Intercepts of a Line

To find the **y-intercept,** set $x = 0$ in the equation and solve for y.

To find the **x-intercept,** set $y = 0$ in the equation and solve for x.

Example 8 Draw the graph of $y = -2x$.

Solution Setting $x = 0$, $y = -2(0) = 0$ and so $(0, 0)$ lies on the line. Notice, if we set $y = 0$, then $0 = -2x$, giving $x = 0$ and yielding the same ordered pair $(0, 0)$. To find other points on the line, set $x = 1$, giving $y = -2$ and so the point $(1, -2)$ lies on the line. As a check point, we choose $x = -1$, giving $y = 2$ and so $(-1, 2)$ also lies on the line. We use these three ordered pairs to graph the line as shown in the figure below.

x	y
0	0
1	−2
−1	2

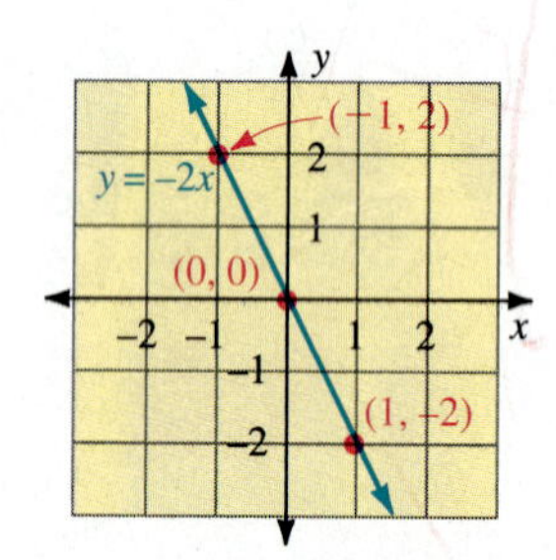

In the next two examples, we draw the graphs of two special types of linear equations.

Example 9 Draw the graph of the equation $x = -3$.

Solution The standard form for this equation is $x + 0y = -3$. Therefore, no matter what value of y is chosen, the x-coordinate is always -3. For example, $(-3, 0)$ and $(-3, 2)$ are both solutions. Plotting these two points in the figure below, we obtain a vertical line.

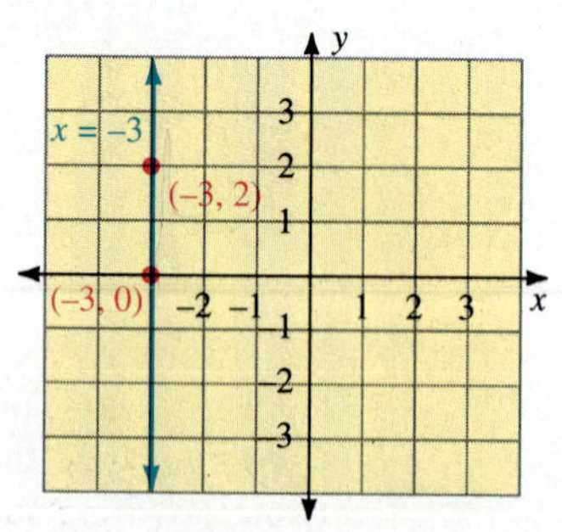

Example 10 Draw the graph of $y = 10$.

Solution This equation in standard form is $0x + y = 10$. Therefore, no matter what value of x is chosen, $y = 10$. For example, (0, 10) and (5, 10) are both points on the graph. Plotting these two points, we obtain a horizontal line as shown in the figure below.

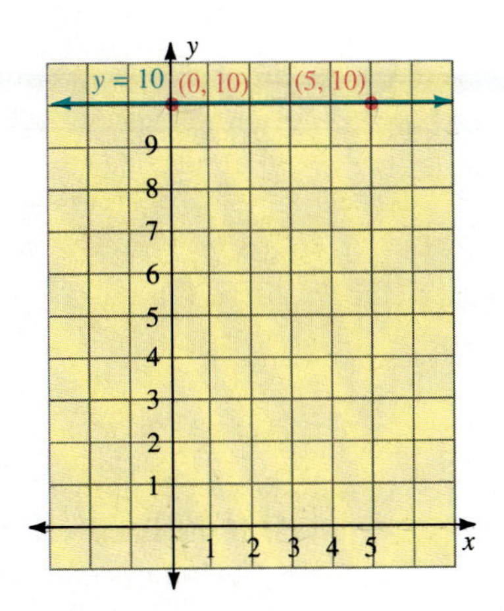

The special linear equations $x = a$ and $y = b$ are illustrated in Examples 9 and 10.

The graph of the equation $x = a$ is a vertical line passing through the x-axis at a.

The graph of the equation $y = b$ is a horizontal line passing through the y-axis at b.

These two special graphs, (a) and (b), are shown in the figure below.

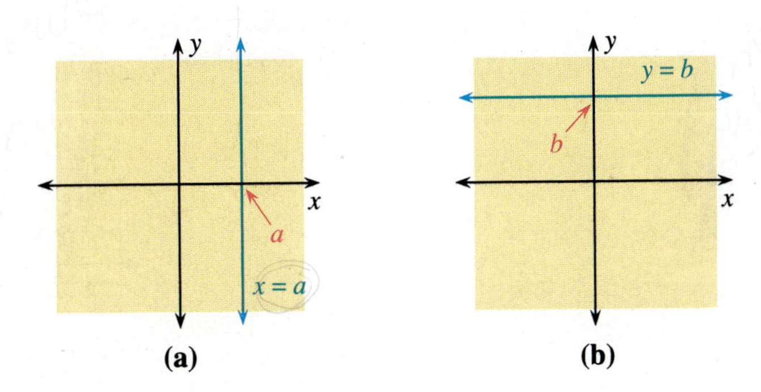

We summarize the results obtained in the examples in this section with the following graphing strategy.

STRATEGY

Graphing Linear Equations

1. If the linear equation is of the form $x = a$, then plot the point $(a, 0)$ and draw a vertical line through that point.

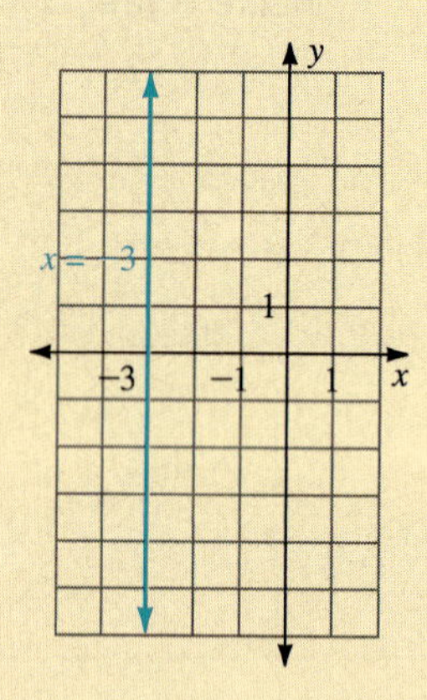

2. If the linear equation is of the form $y = b$, then plot the point $(0, b)$ and draw a horizontal line through that point.

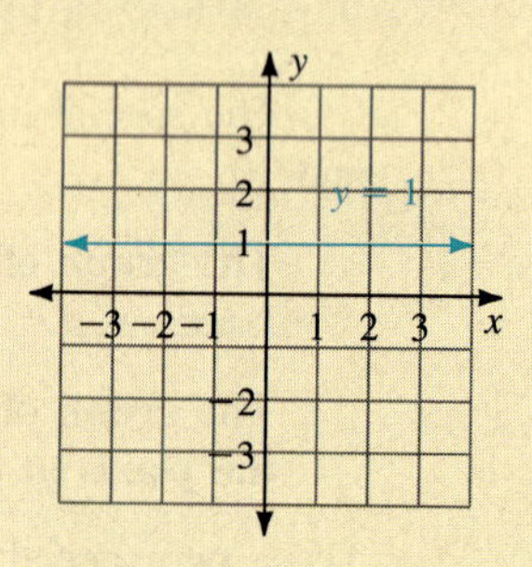

3. If the linear equation is of the form $y = mx$, the line passes through $(0, 0)$. Obtain other points that lie on the line by choosing values for x other than zero.

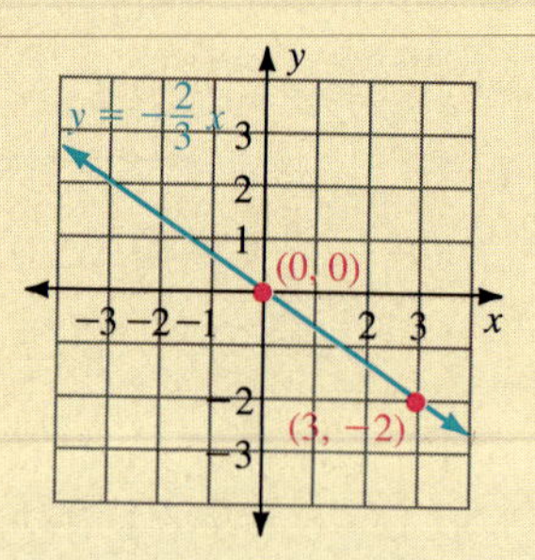

4. If the linear equation is of the form $Ax + By = C$ that is not covered in the first three cases, find any two points that lie on the line. In particular, you may choose to use the intercepts: set $y = 0$ and solve for x, then set $x = 0$ and solve for y.

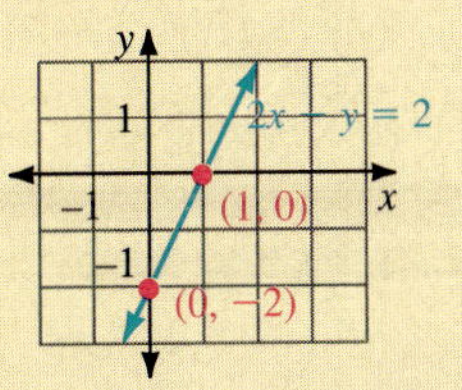

GRAPHING CALCULATOR CROSSOVER

A graphing calculator can be used to graph linear equations. We must first construct a rectangular coordinate system, a portion of which will then be displayed on the rectangular screen when graphing a linear function. The coordinate system is defined by numbers called "range values." The term RANGE is commonly used to refer to these values. The range screen will look like the figure below:

Range	
Xmin: −10	The smallest value on the x-axis
Xmax: 10	The largest value on the x-axis
Xscl: 1	The value of each tic mark on the x-axis
Ymin: −10	The smallest value on the y-axis
Ymax: 10	The largest value on the y-axis
Yscl: 1	The value of each tic mark on the y-axis

The coordinate system with these range values is shown in the following figure.

The RANGE values can be changed if desired. Check your manual for details.

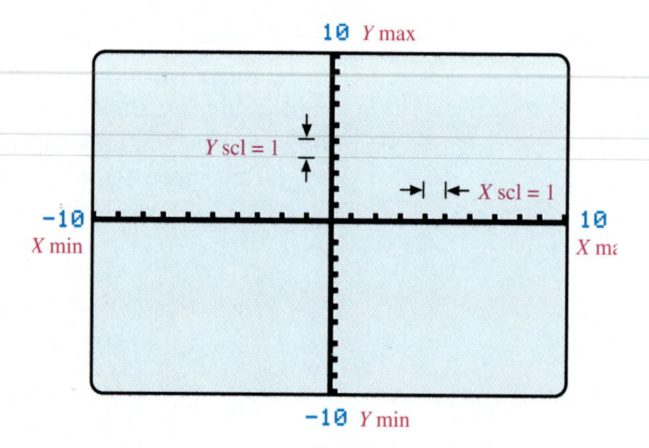

Example 11 Use a graphing calculator or computer to graph the linear equation

$$y = x + 1.$$

Solution We set the RANGE values as follows: Xmin = −10, Xmax = 10, Xscl = 1, Ymin = −10, Ymax = 10, Yxcl = 1. Next, use GRAPH to enter the equation $y = x + 1$. The graph appearing on your screen should look like this:

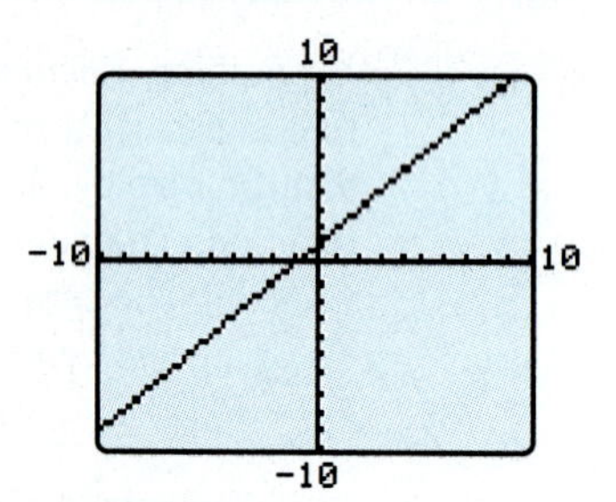

Example 12 Graph $x + y = 1$ using a graphing calculator or computer.

Solution For graphing purposes, we solve the equation for y: $y = 1 - x$. Next, using the GRAPH procedure, we graph the equation as shown in the figure.

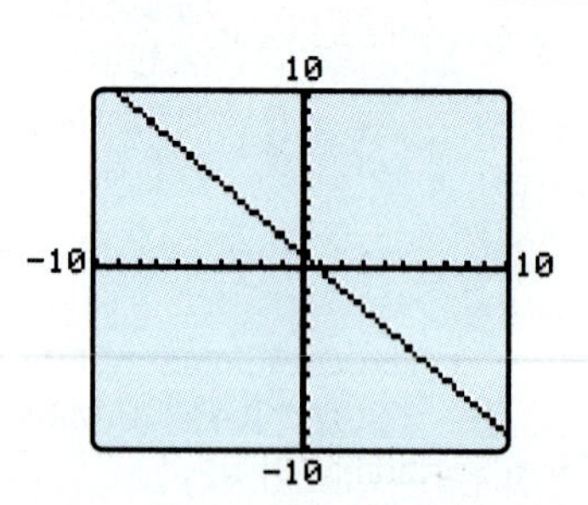

TEAM PROJECT

(3 or 4 Students)

FINDING THE COORDINATES OF A CITY

Materials Needed: Transparencies, markers, rulers, and state maps.

Course of Action: Alternate jobs so everyone participates.

Job 1: Using a marker and a ruler, draw a set of axes on the transparency. Mark the horizontal axis with an x and the vertical axis with a y. Mark the origin as the point $(0, 0)$.

Job 2: Using a marker and a ruler, mark a scale on each of the axes.

Job 3: From the state map, select a city. Place the map under the transparency so that your city is at the origin marked on the transparency.

Job 4: Select a city in the first quadrant.

Job 5: What are the coordinates of that city?

Job 6: Select a city in the second quadrant.

Job 7: What are the coordinates of that city?

Job 8: Select a city in the third quadrant and one in the fourth quadrant.

Job 9: What are the coordinates of these two cities?

As a team, take turns selecting six cities and finding the coordinates of each. Next, have each person write down a pair of coordinates and swap with another person. Find the city associated with those coordinates. Continue swapping coordinates until time is up.

WARMING UP

Answer true or false.

1. The point $(3, 2)$ lies in the first quadrant.

2. The point $(3, 0)$ lies in the y-axis.

3. The point (x, y) lies in the x-axis if $y = 0$.

4. The point $(-12, -2)$ is the quadrant IV.

5. The coordinates of $(-2, 3)$ satisfy $4x + 6y = 0$.

6. The origin lies in all four quadrants.

7. The point $(5, -4)$ lies in quadrant IV.

8. The coordinates of $(5, 7)$ satisfy $y = 2x - 3$.

9. The graph of $x = 3$ is a horizontal line.

10. The y-intercept of $2x + 3y = 6$ is 2.

EXERCISE SET 3.1

1. Find the ordered pair for each point shown in the following coordinate plane.

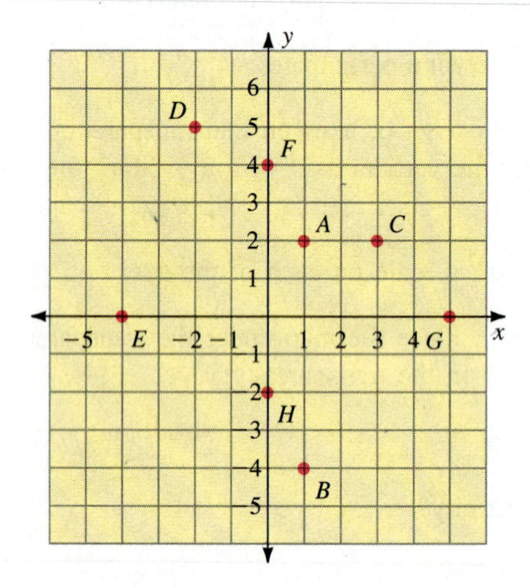

2. Find the ordered pair for each point shown in the following coordinate plane.

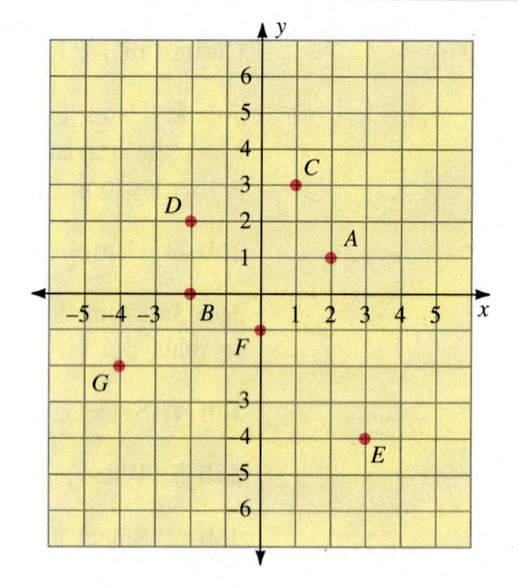

For Exercises 3–6, describe where in the coordinate plane the point (x, y) would be located.

3. x is positive and y is positive.

4. $y = 0$.

5. $x = 0$ and $y = 0$.

6. x is positive and y is negative.

7. Plot the points $A(-3, 2)$, $B(1, 0)$, $C(3, 5)$, and $D(-4, -1)$.

8. Plot the points $A(3, 3)$, $B(-2, 0)$, $C(-1, -5)$, and $D(0, -1)$.

9. Plot the points $A(50, 100)$, $B(50, -100)$, $C(-50, 100)$, and $D(-50, -100)$.

10. Plot the points $A(10, 10)$, $B(-10, 10)$, $C(10, -10)$, and $D(-10, -10)$.

For Exercises 11–14, determine if the given ordered pair is a solution to the equation $4x - 3y = -1$.

11. $\left(-\frac{1}{4}, 0\right)$ **12.** $(-1, 0)$ **13.** $(-1, -2)$ **14.** $(2, 3)$

For Exercises 15 and 16, find the ordered pair (x, y) that is a solution to $8x - 3y = 12$ when

15. $x = 0$

16. $y = 0$

17. Complete the following table of x and y values for the equation $3x + y = -5$.

x	y
0	
	0
−2	
	4

18. Complete the following table of x and y values for the equation $x - 4y = 20$.

x	y
0	
	0
4	
	−8

For Exercises 19–40, use the graphing strategy for linear equations to draw the graph of the linear equation.

19. $3x - y = 9$

20. $2x + 3y = 12$

21. $y = -4x + 5$

22. $y = 2x - 6$

23. $x + y = 50$

24. $x - y = 100$

25. $x = 2y + 10$

26. $x = -5y + 20$

27. $2x + 7y = 14$

28. $5x + 2y = 10$

29. $x - 2y = 1$

30. $-2x + y = 4$

31. $y = 2x$

32. $y = 3x$

33. $y = -3x$

34. $y = -\frac{1}{2}x$

35. $x = 4$

36. $y = 2$

37. $x = 0$

38. $y = 0$

39. $x = 40$

40. $y = -25$

For Exercises 41–44, write an equation from the given information. Then, draw the graph of the equation.

41. The sum of the y-coordinate and twice the x-coordinate is negative two.

42. The difference of the x-coordinate and the y-coordinate is 12.

43. The y-coordinate is twice the difference of the x-coordinate and five.

44. The sum of the x-coordinate and y-coordinate, reduced by one, is zero.

For Exercises 45–48, use a graphing calculator or computer to draw the graph.

45. $y = x - 1$

46. $y = 2 - x$

47. $2x + y = 0$

48. $-3x + y = -2$

SAY IT IN WORDS

49. What is the difference between (x, y) and $\{x, y\}$?

50. Is (x, y) ever the same as (y, x)? Explain.

REVIEW EXERCISES

The following exercises review parts of Section 1.1. Doing these exercises will help prepare you for the next section.

For Exercises 51–54, simplify.

51. $\dfrac{7 - 6}{5 - 3}$

52. $\dfrac{3 - 2}{4 - 1}$

53. $\dfrac{5 - 3}{4 - 2}$

54. $\dfrac{7 - 3}{2 - 4}$

ENRICHMENT EXERCISES

1. Three vertices of a square are $A(0, 0)$, $B(0, 3)$, and $C(3, 0)$. What are the coordinates of the fourth vertex?

2. Three vertices of a rectangle are $A(2, 1)$, $B(6, 1)$, and $C(6, 7)$. What are the coordinates of the fourth vertex?

3. Three vertices of a rectangle are $A(-3, 2)$, $B(5, -4)$, and $C(-3, -4)$. What are the coordinates of the fourth vertex?

4. Find the value of m, if the graph of $y = mx + 5$ passes through $(3, 10)$.

5. Determine c so that the point $(-2, c)$ lies on the graph of $2x + 3y = c$.

Answers to Enrichment Exercises begin on page A.1.

3.2 The Slope of a Line

OBJECTIVES

- ▶ *To find the slope of a line from the formula*
- ▶ *To find the slope of a line from the graph*
- ▶ *To understand the difference between zero slope and undefined slope*
- ▶ *To understand the geometric meaning of positive and negative slope*

LEARNING ADVANTAGE

Why study mathematics? *To attain mathematical maturity means to be able to use mathematics as a language. Mathematical language is used to determine the truth of a statement as well as being able to predict the outcome of events. The language of mathematics is acquired through courses in elementary and intermediate algebra.*

The slope of a line is suggested by the slope of a roof of a house.

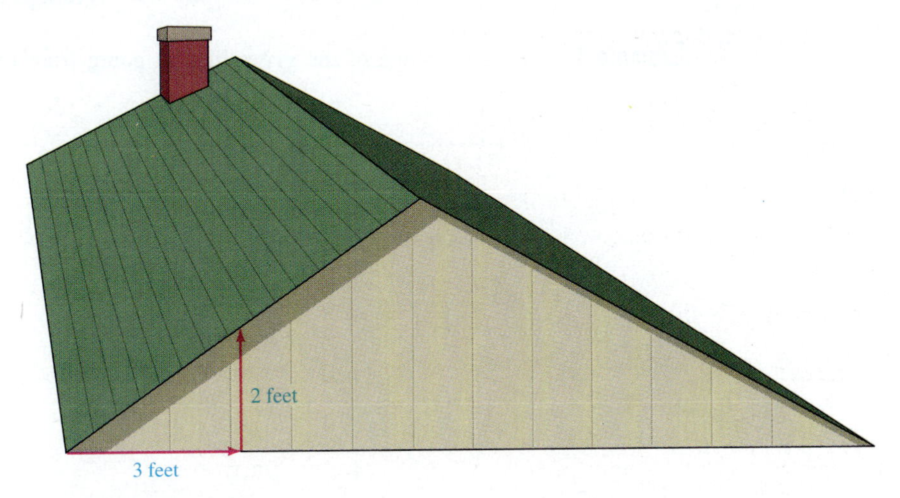

The roof in the figure rises 2 feet for every 3 feet in the level direction. A homebuilder considers the slope of the roof to be $\frac{2}{3}$.

In a similar way, each nonvertical line has associated with it a number called the *slope* of the line. The slope is defined in such a way to give us two important properties of the line. One is the steepness of the line and the other is whether the line rises or falls.

Consider the line shown in the figure below.

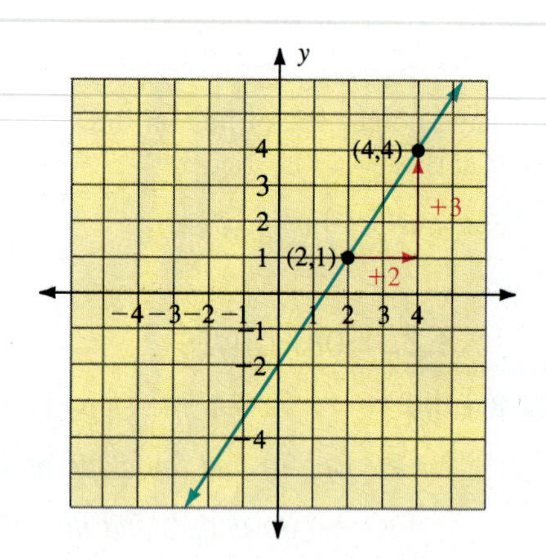

To move from (2, 1) to (4, 4), we go 2 units to the right then 3 units upward. That is, to move from (2, 1) to (4, 4), there is a +3 change in the y-coordinate and a +2 change in the x-coordinate. The slope of the line is defined as $\frac{3}{2}$. In general,

The slope of a line is

$$\textbf{slope} = \frac{\textbf{change in } y}{\textbf{change in } x}$$

Example 1 Find the slope of the given line by going from point A to point B.

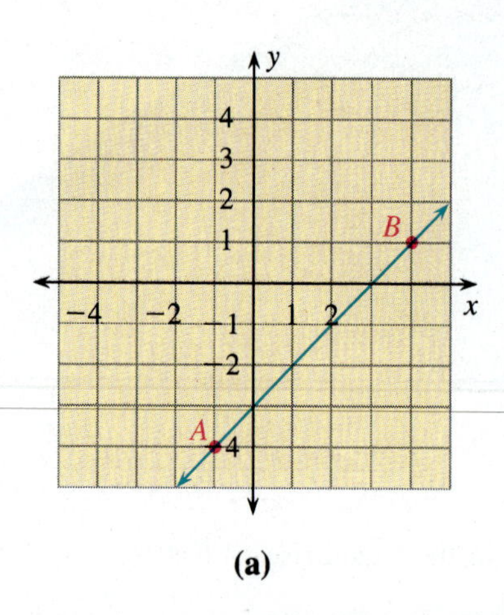

(a)

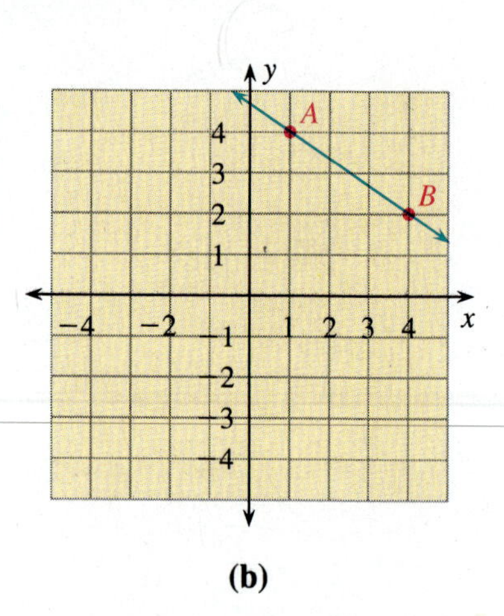

(b)

Solution (a) Starting at A, move to the right 5 units, then up 5 units, we arrive at point B. The slope of the line is therefore, $\frac{5}{5}$ or 1.

(b) If we start at point A, move 3 units to the right, then *down* 2 units, we arrive at point B. The slope of the line is therefore, $\frac{-2}{3}$ or $-\frac{2}{3}$.

As shown in Example 1, the change in x as well as the change in y can be either positive or negative. When going from A to B,

The change in x is $\begin{cases}\textbf{positive, if you move to the right.}\\ \textbf{negative, if you move to the left.}\end{cases}$

and

The change in y is $\begin{cases}\textbf{positive, if you move up.}\\ \textbf{negative, if you move down.}\end{cases}$

The slope of a line can be found if we know the coordinates of two points on the line. The (formal) definition of slope gives us the formula to use.

DEFINITION

Let (x_1, y_1) and (x_2, y_2) be any two points on a nonvertical line. The **slope** of the line, denoted by m, is

$$m = \frac{\text{change in } y}{\text{change in } x} = \frac{y_2 - y_1}{x_2 - x_1}$$

That is, the slope is found by taking the difference of the y-coordinates and dividing by the difference (in the same order) of the x-coordinates. See the figure below.

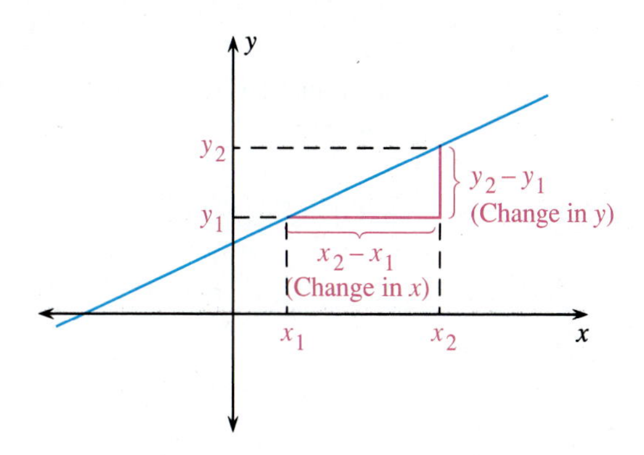

NOTE *The subscript notation such as x_1 is read "x sub-one." The variables x_1 and x_2 simply mean two choices for the x-coordinate.*

Example 2 Find the slope of the line passing through $(-2, 1)$ and $(3, 4)$.

Solution Let $(-2, 1)$ be (x_1, y_1) and $(3, 4)$ be (x_2, y_2) in the slope formula. Then, the slope m is given by

$$\begin{aligned} m &= \frac{y_2 - y_1}{x_2 - x_1} \\ &= \frac{4 - 1}{3 - (-2)} \\ &= \frac{3}{5} \end{aligned}$$

The order in which the two points are used in the slope formula is unimportant. If, we had interchanged the role of the two points by taking $(3, 4)$ as (x_1, y_1) and $(-2, 1)$ as (x_2, y_2), the slope formula would still yield the same number, $\frac{3}{5}$. The line together with the two points is shown in the figure below.

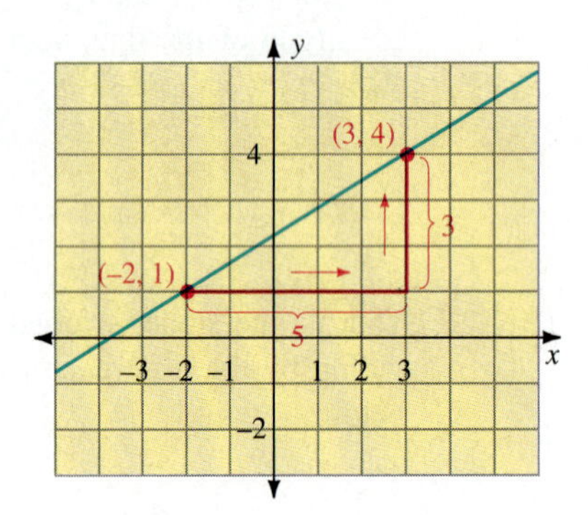

Notice that the line in the figure above has a *positive* slope and slants *up* to the right.

Example 3 Find the slope of the line passing through the points $(-3, 1)$ and $(2, -1)$.

Solution Letting $(x_1, y_1) = (-3, 1)$ and $(x_2, y_2) = (2, -1)$ in slope formula,

$$\begin{aligned} m &= \frac{y_2 - y_1}{x_2 - x_1} \\ &= \frac{-1 - 1}{2 - (-3)} \\ &= -\frac{2}{5} \end{aligned}$$

The line is shown in the figure below. Notice that the line has a *negative* slope and slants *down* to the right.

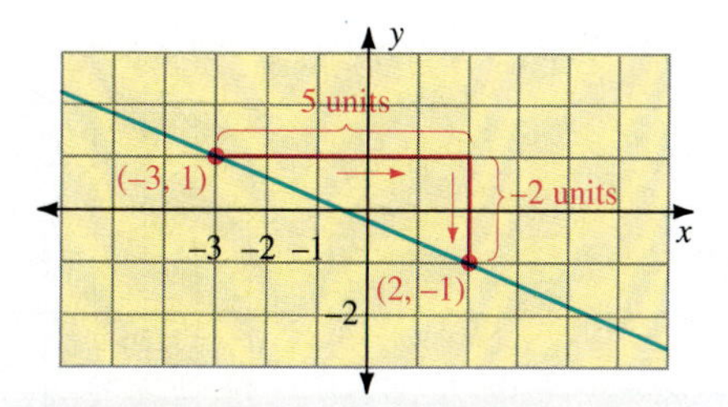

It is illustrated in Examples 2 and 3 that **positive slope** means an *upward slant* and **negative slope** means a *downward slant* of the line as the graph is scanned from left to right.

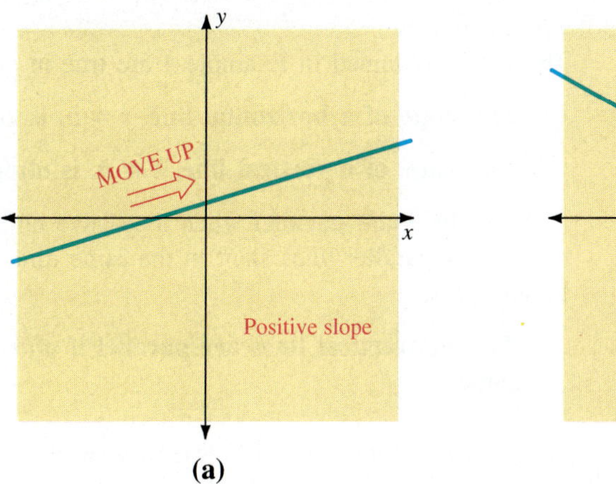

(a)

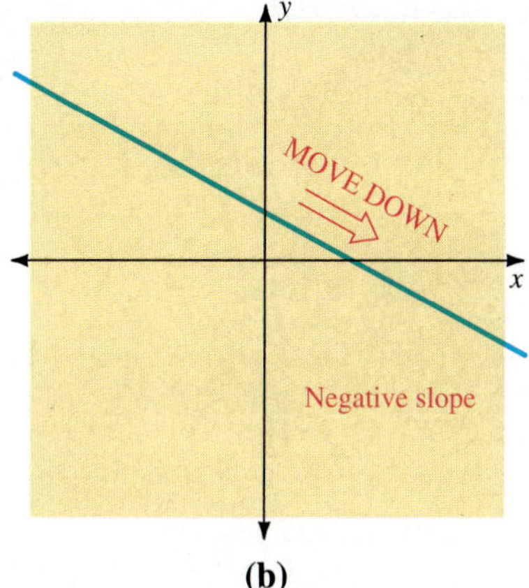

(b)

In the next two examples, we look at slope as it relates to horizontal and vertical lines.

Example 4 Find the slope of the line passing through

(a) $(-3, -2)$ and $(2, -2)$ **(b)** $(4, 1)$ and $(4, 3)$.

Solution **(a)** The graph of this line is shown in figure (a) on the following page. Note that the line is horizontal; that is, it is parallel to the x-axis. Using the slope formula,

$$m = \frac{-2 - (-2)}{2 - (-3)} = \frac{0}{5} = 0$$

(b) The graph of this line is shown in figure (b) on the following page. Note that this line is vertical; that is, it is perpendicular to the x-axis. Using the slope formula,

$$m = \frac{3 - 1}{4 - 4} = \frac{2}{0}$$

Recall that division by zero is undefined, so the slope of a vertical line is undefined.

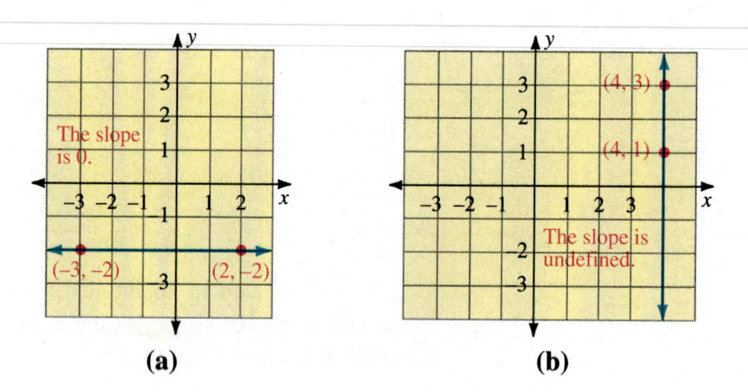

(a) (b)

The results obtained in Example 4 are true in general.

The slope of a horizontal line, $y = b$, is zero.

The slope of a vertical line, $x = a$, is undefined.

Two lines are **parallel** when they have no points in common. Since two nonvertical parallel lines slant in the same direction, they have the *same* slope. In summary,

Two nonvertical lines are parallel if and only if they have the same slope.

NOTE *The phrase "if and only if" is a short way to say two different things:*

(1) If the two lines are parallel, then they have the same slope.

(2) If two lines have the same slope, then they are parallel.

Example 5 By checking their slopes, determine if the line through the first pair of points is parallel to the line through the second pair of points.

(a) First pair of points: $(0, 5)$ and $(2, 4)$
Second pair of points: $(-2, 2)$ and $(2, 0)$

(b) First pair of points: $(0, 1)$ and $(-1, 3)$
Second pair of points: $(3, 2)$ and $(0, 4)$

Solution **(a)** We use the slope formula to compute the slope of the two lines:

The slope of the line through $(0, 5)$ and $(2, 4)$ is

$$\frac{4-5}{2-0} = -\frac{1}{2}$$

The slope of the line through $(-2, 2)$ and $(2, 0)$ is

$$\frac{0-2}{2-(-2)} = \frac{-2}{4}$$
$$= -\frac{1}{2}$$

Since the two slopes are the same, the two lines are parallel. (See Part (a) of the figure below.)

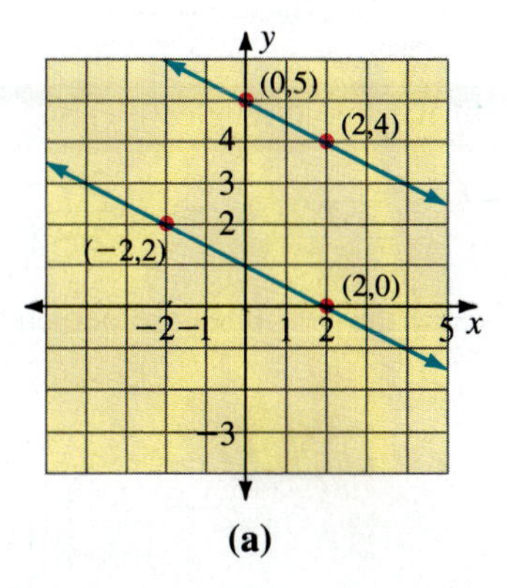

(a)

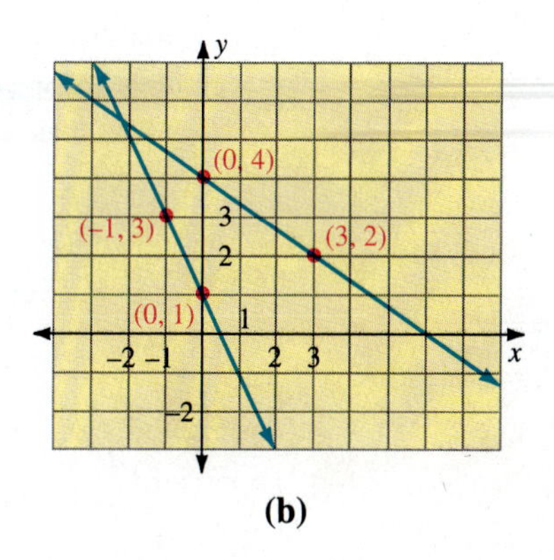

(b)

(b) The slope of the line through (0, 1) and (−1, 3) is

$$\frac{3-1}{-1-0} = \frac{2}{-1}$$
$$= -2$$

The slope of the line through (3, 2) and (0, 4) is

$$\frac{4-2}{0-3} = \frac{2}{-3}$$
$$= -\frac{2}{3}$$

Since the slopes are not the same, $-2 \neq -\frac{2}{3}$, the two lines are not parallel. (See Part (b) of the figure above.)

Two lines are **perpendicular** if they intersect at a right (90°) angle.

Two nonvertical lines with slopes m_1 and m_2 are perpendicular if and only if

$$m_2 = -\frac{1}{m_1}$$

We say that m_1 and m_2 are *negative reciprocals* of each other.

Example 6 By checking their slopes, determine if the line through the first pair of points is perpendicular to the line through the second pair of points.

First pair of points: $(-1, 4)$ and $(1, 0)$
Second pair of points: $(0, 1)$ and $(2, 2)$

Solution The slope m_1 of the line through $(-1, 4)$ and $(1, 0)$ is

$$m_1 = \frac{0 - 4}{1 - (-1)} = \frac{-4}{2} = -2$$

The slope m_2 of the line through $(0, 1)$ and $(2, 2)$ is

$$m_2 = \frac{2 - 1}{2 - 0} = \frac{1}{2}$$

Since $m_2 = -\dfrac{1}{m_1}$, the two lines are perpendicular. (See the figure below.)

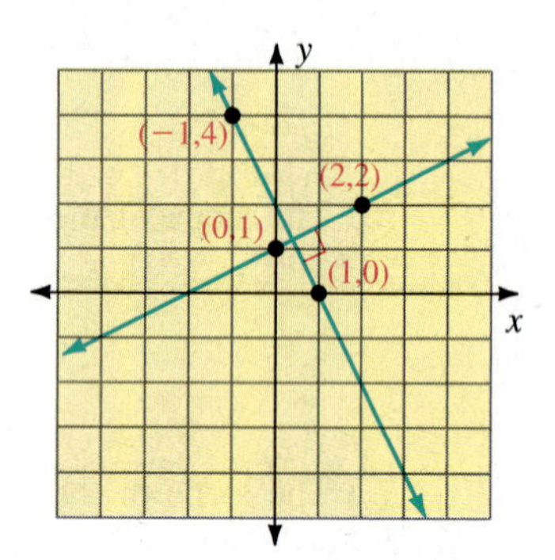

WARMING UP

Answer true or false.

1. The slope of a line is always positive.
2. The slope of the line passing through $(1, 0)$ and $(2, 3)$ is 3.
3. The slope of a vertical line is 0.
4. The slope of the line passing through $(-1, 3)$ and $(-2, -4)$ is $\frac{1}{7}$.
5. The slope of a line parallel to a line with slope $\frac{1}{2}$ is $\frac{1}{2}$.
6. Two lines with slopes 3 and $-\frac{1}{3}$, respectively, are perpendicular.
7. Slopes of parallel lines are the negative reciprocals of each other.
8. If the slope of a line is undefined, then the line is vertical.

EXERCISE SET 3.2

For Exercises 1–8, find the slope, if it is defined, of each line.

1.

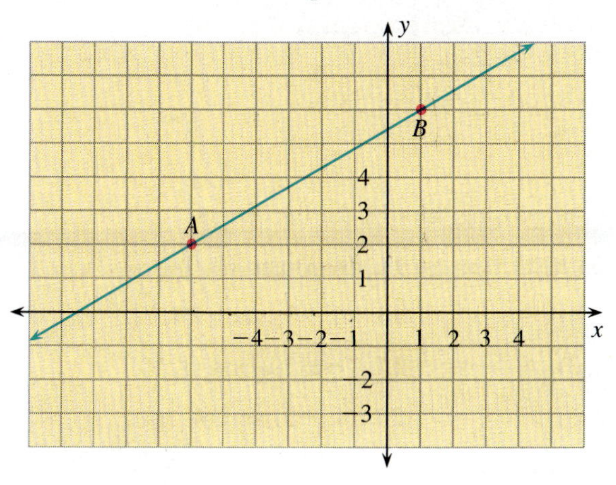

2.

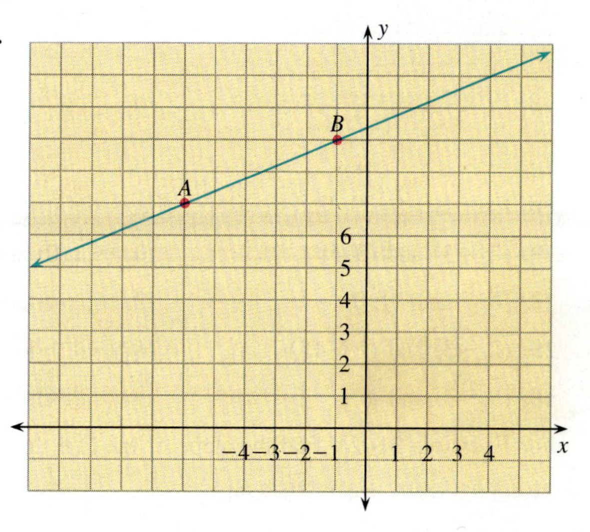

3.

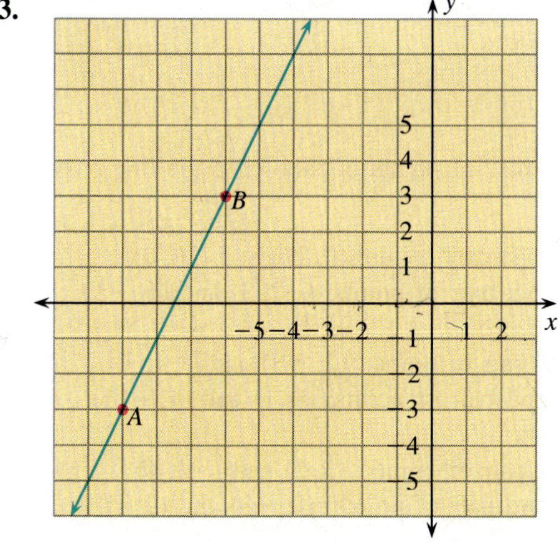

4.

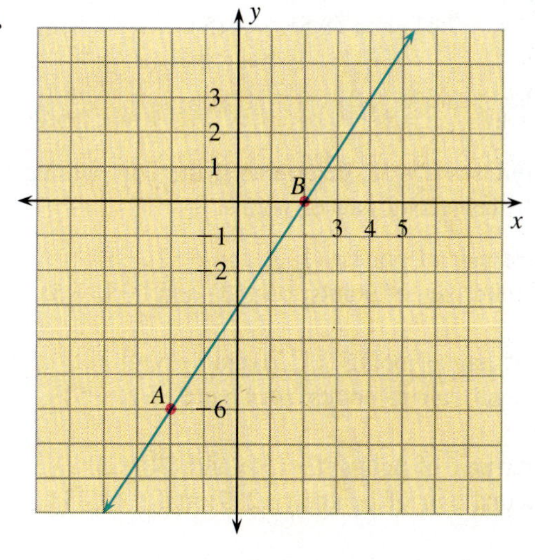

5.

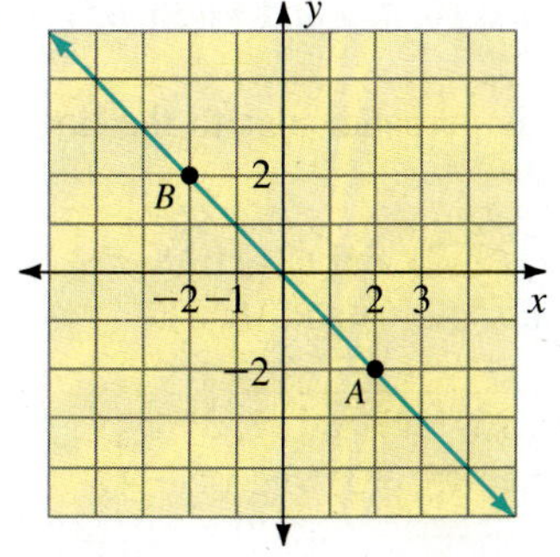

6.

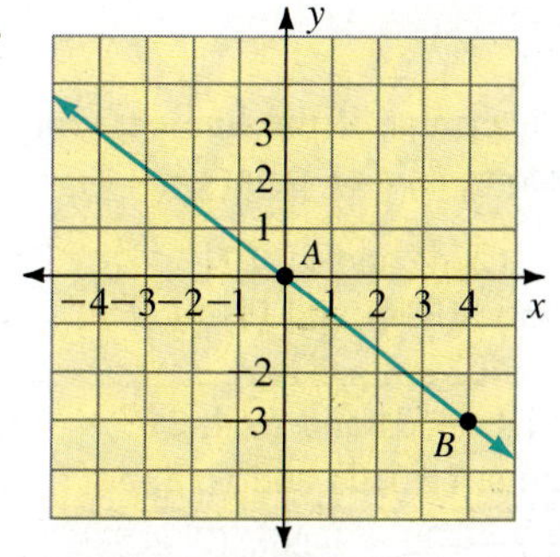

7.

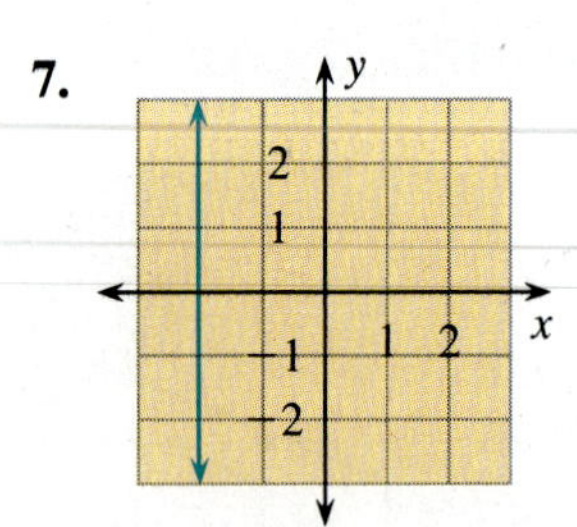

8.

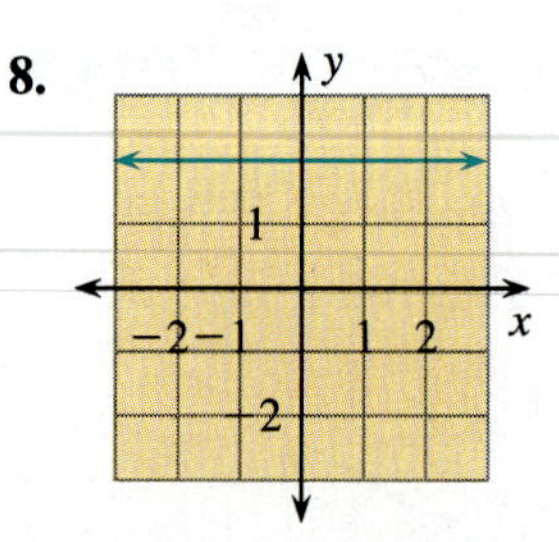

For Exercises 9–20, find the slope, if it is defined, of the line passing through the two points.

9. $(2, -5)$ and $(3, 4)$

10. $(-1, 0)$ and $(3, 5)$

11. $(0, 3)$ and $(2, 0)$

12. $(4, 1)$ and $(1, 0)$

13. $(6, -1)$ and $(-2, 4)$

14. $(6, -3)$ and $(-4, 7)$

15. $(2, -3)$ and $(-6, 15)$

16. $(-4, -3)$ and $(2, 3)$

17. $(-2, 10)$ and $(1, 10)$

18. $(5, -3)$ and $(-4, -3)$

19. $(-2, 4)$ and $(-2, 6)$

20. $(4, -3)$ and $(4, -4)$

For Exercises 21–24, find the slope of the line through the two points.

21. $(5.72, 6.69)$ and $(3.10, 2.94)$

22. $(-1.06, 3.98)$ and $(4.95, -10.21)$

23. $(4.14, -7.31)$ and $(8.55, -9.15)$

24. $\left(0, \frac{1}{2}\right)$ and $\left(\frac{1}{2}, 0\right)$

For Exercises 25–32, determine if the line passing through the first pair of points is parallel to the line passing through the second pair of points.

25. First pair of points: $(1, 3)$ and $(-2, -3)$
Second pair of points: $(0, -3)$ and $(-1, -5)$

26. First pair of points: $(0, 10)$ and $(10, 0)$
Second pair of points: $(-2, 3)$ and $(6, -5)$

27. First pair of points: $(1, 3)$ and $(-5, 9)$
Second pair of points: $(4, 1)$ and $(-2, -5)$

28. First pair of points: $(2, -3)$ and $(-1, 3)$
Second pair of points: $(0, 1)$ and $(4, -11)$

29. First pair of points: $(-1, 4)$ and $(3, -2)$
Second pair of points: $(-2, 5)$ and $(-4, 11)$

30. First pair of points: $(1, 7)$ and $(-1, 4)$
Second pair of points: $(0, -3)$ and $(2, 1)$

31. First pair of points: $(-1, 5)$ and $(-1, -8)$
Second pair of points: $(2, -5)$ and $(2, 7)$

32. First pair of points: $(-20, -3)$ and $(10, -3)$
Second pair of points: $(0, 45)$ and $(-33, 45)$

For Exercises 33–38, determine if the line passing through the first pair of points is perpendicular to the line passing through the second pair of points.

33. First pair of points: $(2, 4)$ and $(1, 2)$
Second pair of points: $(3, 4)$ and $(1, 5)$

34. First pair of points: $(3, 7)$ and $(2, 3)$
Second pair of points: $(2, 6)$ and $(6, 5)$

35. First pair of points: $(-3, 3)$ and $(-4, 2)$
Second pair of points: $(-5, 2)$ and $(-8, 5)$

36. First pair of points: $(6, -2)$ and $(1, 3)$
Second pair of points: $(5, -3)$ and $(4, -4)$

37. First pair of points: $(-2, -8)$ and $(-1, -5)$
Second pair of points: $(-3, 4)$ and $(-6, 3)$

38. First pair of points: $(-9, -3)$ and $(-1, 5)$
Second pair of points: $(-3, 2)$ and $(1, 6)$

39. What is the slope of a line that is perpendicular to a line with slope $\frac{2}{3}$?

40. What is the slope of a line that is perpendicular to a line with slope $-\frac{5}{4}$?

41. In a coordinate plane, plot the point $(3, 4)$. Then plot two other points so that all three points lie on a line with a slope of $-\frac{1}{2}$.

42. In a coordinate plane, plot the point $(5, 2)$. Then plot two other points so that all three points lie on a line with a slope of $\frac{3}{2}$.

43. In a coordinate plane, plot the point $(-1, 3)$. Then plot two other points so that all three points lie on a line with a slope of -2.

44. In a coordinate plane, plot the point $(3, -1)$. Then plot two other points so that all three points lie on a line with a slope of -1.

45. The limited edition of Stephen King's *The Stand* sold for \$200 in 1990. By 1994, copies of this book were selling for \$350. Assuming that the value of the book increases linearly with time, what is the predicted value of the book for 1996?

46. The National Transportation Safety Board reported 550 near collisions by airplanes in this country in 1982 and steadily increased to 790 in 1990. Assuming linearity, how many near collisions are predicted for 1995?

SAY IT IN WORDS

47. Explain in a short paragraph the concept of slope. Include the ideas of positive, negative, zero, and undefined slope.

REVIEW EXERCISES

The following exercises review parts of Section 2.9. Doing these exercises will help prepare you for the next section.

For Exercises 48–51, solve the equation for y.

48. $3x - 2y = 6$

49. $y - 3 = 2(x + 1)$

50. $\dfrac{y}{10} + \dfrac{x}{5} = 1$

51. $5x - 3y = \dfrac{3}{4}$

ENRICHMENT EXERCISES

For Exercises 1 and 2, find the slope of the line passing through the two points.

1. $\left(\frac{1}{3}, -1\right)$ and $\left(2, \frac{2}{3}\right)$

2. $\left(-2, \frac{1}{4}\right)$ and $\left(-\frac{1}{2}, \frac{1}{2}\right)$

3. Find the value of k so that the line through $(3, k)$ and $(2, -k)$ has a slope of 4.

4. The shallow end of a swimming pool is 3 feet deep and the other end is 12 feet deep as shown in the figure. If the pool is 81 feet long, what is the slope of the bottom of the pool?

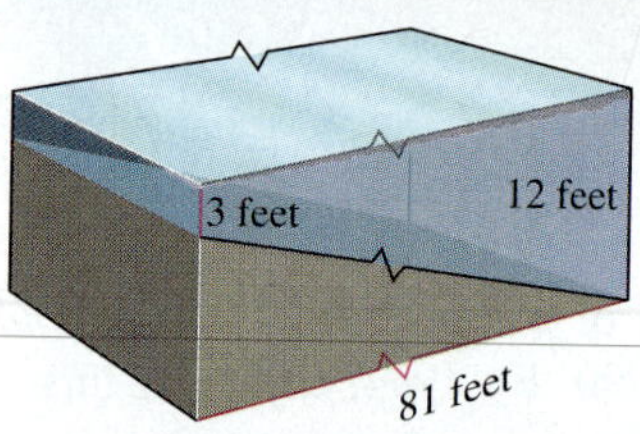

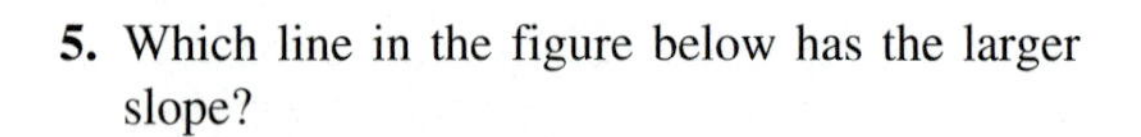

5. Which line in the figure below has the larger slope?

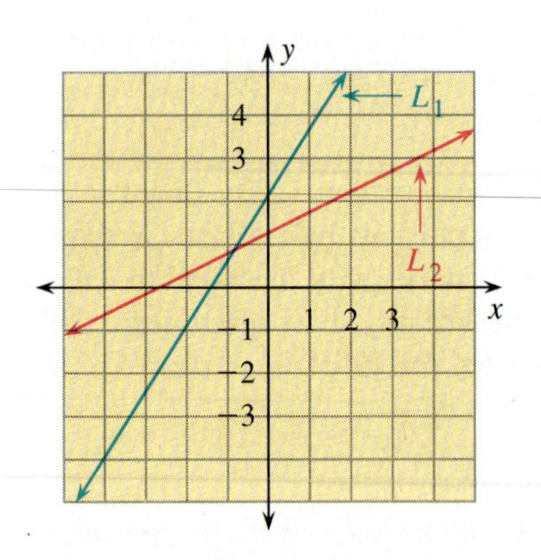

Answers to Enrichment Exercises begin on page A.1.

3.3 The Equation of a Line

OBJECTIVE

▶ *To find the equation of a line given*
(a) the slope and y-intercept
(b) the slope and a point on the line
(c) two points on the line

LEARNING ADVANTAGE

Why study mathematics? *Mathematics is used in areas such as pollution control. Mathematics was used to discover the relationship between the quantity of pollution that can be dissipated by an average sized tree. This in turn can be used to estimate the number of trees needed to counterbalance the exhaust from commuter cars during the rush hour in cities such as Seattle, Washington.*

Many parts of environmental science utilize mathematics. The problems of pollution and toxic waste, for example, cannot be quantified and studied without a knowledge of mathematics.

There is a relationship between the slope of a line and the equation of the line. For example, consider the line given by the equation $y = 3x + 2$. Two points on this line can be found by setting x equal to 0 and 4. When $x = 0$, $y = 3(0) + 2 = 2$. When $x = 4$, $y = 3(4) + 2 = 14$. Therefore, $(0, 2)$ and $(4, 14)$ are two points satisfying the equation $y = 3x + 2$. We now find the slope of the line using the slope formula with these two points.

$$\begin{aligned} m &= \frac{y_2 - y_1}{x_2 - x_1} \\ &= \frac{14 - 2}{4 - 0} \\ &= \frac{12}{4} \\ &= 3 \end{aligned}$$

Notice that the slope of the line, 3, is the same number as the coefficient of x in the equation $y = 3x + 2$.

The y-intercept is 2, since when $x = 0$, $y = 2$. Notice that 2 is also the constant term in the equation $y = 3x + 2$. The line is shown in the figure below.

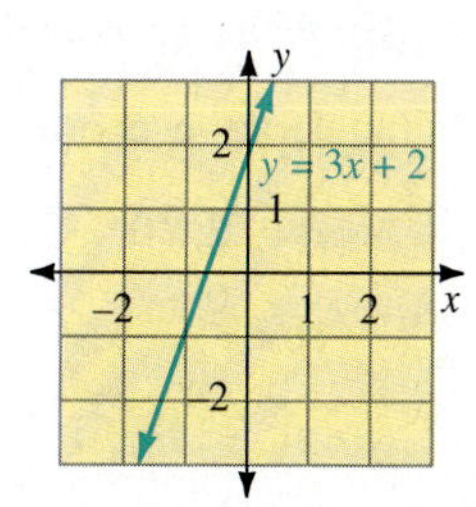

In summary, given the line with equation $y = 3x + 2$, the coefficient of x, 3, is the slope, and the constant term, 2, is the y-intercept. We can generalize this result.

RULE

When the equation of a line is written as

$$y = mx + b$$

then the slope of the line is m and the y-intercept is b.

The equation $y = mx + b$ is called the **slope-intercept form** for the equation of a line.

To determine the slope and y-intercept from the equation of a line, the equation must first be solved for y to obtain the slope-intercept form $y = mx + b$. We illustrate this technique in Example 1.

Example 1 Find the slope and y-intercept of the line given by $5x + 3y = 6$.

Solution We solve the equation for y:

$$5x + 3y = 6$$

$$3y = -5x + 6 \quad \text{Subtract } 5x \text{ from both sides.}$$

$$y = -\frac{5}{3}x + 2 \quad \text{Divide by 3.}$$

The equation is now in the slope-intercept form $y = mx + b$. Therefore, the slope is $-\frac{5}{3}$ and the y-intercept is 2.

As shown in Example 1, given the equation of a line, we can find the slope and y-intercept. Now, suppose we are given the slope and y-intercept of a line. From this information, we can draw the line and also write an equation for this line.

Example 2 A line has a slope of $\frac{3}{4}$ and a y-intercept of -2.

(a) Draw the line.

(b) Write an equation for the line.

Solution **(a)** Draw a coordinate plane as shown in the figure below. Since the y-intercept is -2, $(0, -2)$ is a point on the line. Plot this point. Next, the slope is $\frac{3}{4}$, so,

$$\frac{\text{change in } y}{\text{change in } x} = \frac{3}{4}$$

This means that for each (horizontal) change in x to the right of four units, there is a corresponding (vertical) change in y upward of three units. Starting at the point $(0, -2)$, move to the right four units, then upward three units. The movement gives the point (4, 1), which therefore lies on the line. We draw the line determined by (0, −2) and (4, 1) as shown in the figure below.

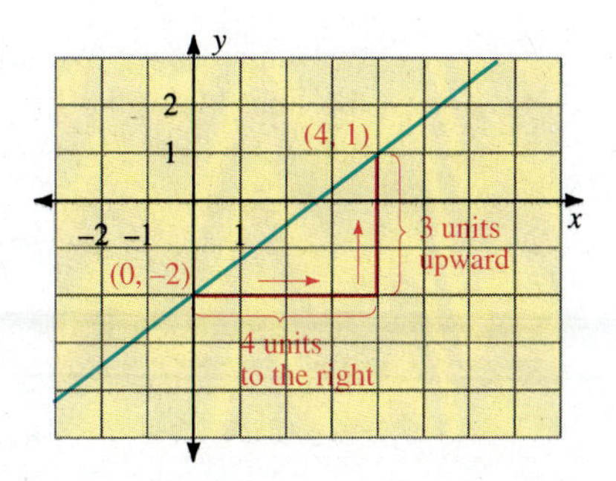

(b) We use the slope-intercept form, $y = mx + b$, with $m = \frac{3}{4}$ and $b = -2$.

Therefore, the equation of the line is $y = \frac{3}{4}x + (-2)$ or simply

$$y = \frac{3}{4}x - 2$$

NOTE *In Example 2, there is another way to use the slope $\frac{3}{4}$ to find another point on the line. Starting at (0, −2), first move upward three units, then to the right four units. Notice that we still arrive at the point (4, 1) as shown in the figure below.*

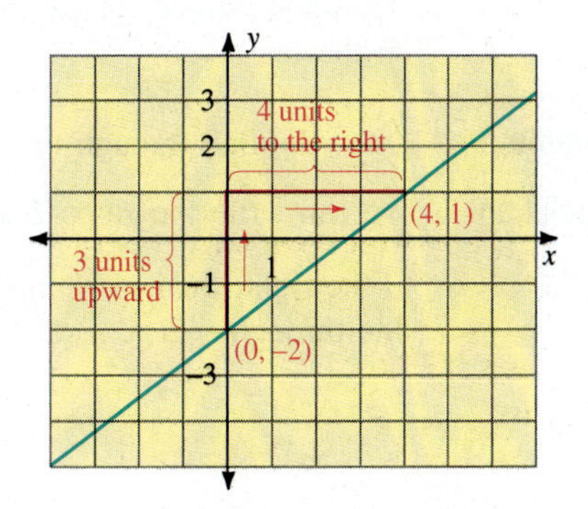

Example 3 A line has a slope of $-\frac{2}{3}$ and a y-intercept of 1.

(a) Draw the line.

(b) Write an equation for the line.

Solution **(a)** Since the y-intercept is 1, (0, 1) lies on the line. The slope $-\frac{2}{3}$ can be written as $\frac{-2}{3}$. Therefore, starting at (0, 1), move three units to the right, then two units *down.* We arrive at the point $(3, -1)$, which is another point on the line. Draw the line between (0, 1) and $(3, -1)$ as shown in the figure below.

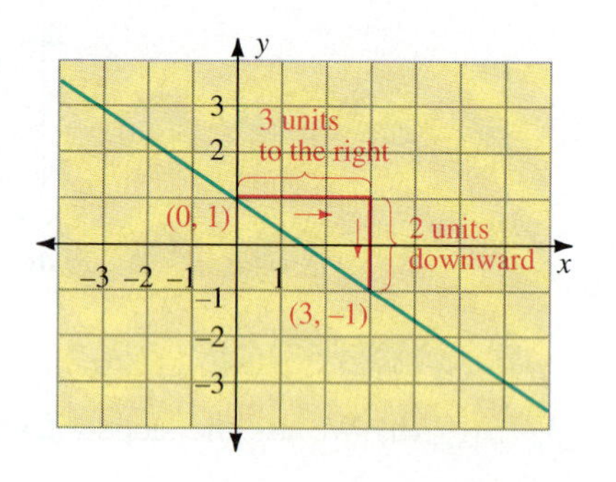

(b) Using the slope-intercept form, $y = mx + b$, where $m = -\frac{2}{3}$ and $b = 1$, we have

$$y = -\frac{2}{3}x + 1$$

NOTE *In Example 3, the slope* $-\frac{2}{3}$ *could be written as* $\frac{2}{-3}$. *In this form, move three units to the left, then upward two units, arriving at a third point on the line.*

The method that was used in Examples 2 and 3 to draw a line given its slope and y-intercept can be used to draw a line given its slope and *any* point on the line.

Example 4 Draw the line through $(-1, -3)$ with slope 2.

Solution We can write the slope 2 as $\frac{2}{1}$. Starting at the point $(-1, -3)$, move one unit to the right, then two units upward. We arrive at the point $(0, -1)$. Draw the line through $(-1, -3)$ and $(0, -1)$ as shown in the figure below.

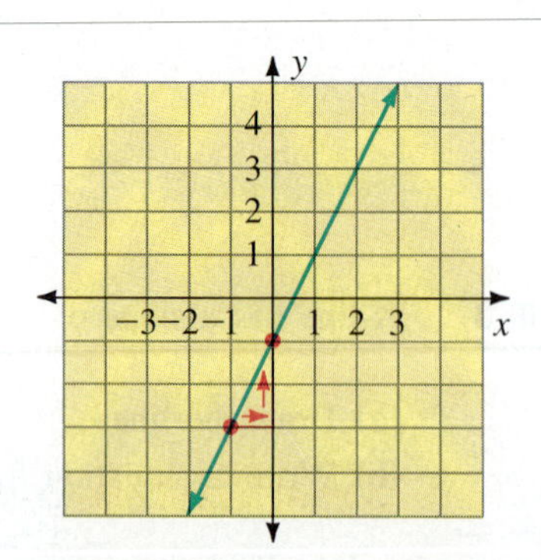

If we are given a point (x_1, y_1) and the slope m, we may find the equation of the line. Let (x, y) be any other point on the line. Using the slope formula with (x, y) replacing (x_2, y_2)

$$m = \frac{y - y_1}{x - x_1}$$

Multiplying both sides by the quantity $x - x_1$, we obtain

$$m(x - x_1) = y - y_1$$

that is,

$$y - y_1 = m(x - x_1)$$

This last equation is called the **point-slope form** of the equation of a line.

Example 5 Use the point-slope form to find the equation of the line through the point $(-1, 2)$ with slope -3. Write the answer in the form $Ax + By = C$.

Solution Letting $(-1, 2)$ be (x_1, y_1) and -3 be m in the point-slope form, we have

$$y - y_1 = m(x - x_1)$$
$$y - 2 = -3[x - (-1)]$$

Next, simplify the equation.

$$y - 2 = -3(x + 1)$$
$$y - 2 = -3x - 3 \quad \text{Distributive property.}$$
$$3x + y = -1 \quad \text{Add } 3x + 2 \text{ to both sides.}$$

Recall that two points determine a line. In the next example we show how to find the equation of this line.

Example 6 Use the point-slope form to find the equation of the line through $(-2, 3)$ and $(-5, 4)$. Write the answer in the form $Ax + By = C$, where A, B, and C are integers.

Solution We must find the slope of the line in order to use the point-slope form. Since we are given two points, we can find the slope using the slope formula.

$$m = \frac{y_2 - y_1}{x_2 - x_1}$$
$$= \frac{4 - 3}{-5 - (-2)}$$
$$= \frac{1}{-3} \quad \text{or} \quad -\frac{1}{3}$$

We now use the point-slope form using either given point and $-\frac{1}{3}$ as the slope.

We arbitrarily choose $(-2, 3)$.

$$y - 3 = -\frac{1}{3}[x - (-2)]$$

$$y - 3 = -\frac{1}{3}(x + 2) \qquad -(-2) = 2.$$

$$3(y - 3) = 3\left(-\frac{1}{3}\right)(x + 2) \qquad \text{Multiply by 3 to clear the fraction.}$$

$$3y - 9 = -x - 2 \qquad \text{Distributive property.}$$

$$x + 3y = 7 \qquad \text{Add } 9 + x \text{ to both sides and simplify.}$$

Therefore, the line through the two given points has $x + 3y = 7$ as its equation.

In the next example we show how the slope-intercept form may also be used to find the equation of a line when two points are given.

Now we show a summary, together with examples, of the types of linear equations we encountered in this section.

Example	***Linear Equations: Ax + By = C***
1. $2x - 3y = 12$ slope $= \frac{2}{3}$ y-intercept $= -4$ x-intercept $= 6$	**1.** If neither A nor B is zero, then the slope is $-\frac{A}{B}$, the y-intercept is $\frac{C}{B}$, and the x-intercept is $\frac{C}{A}$.
2. $y = 3$ slope $= 0$ y-intercept $= 3$	**2.** If $A = 0$, the linear equation has the form $y = b$. The slope is 0 and the y-intercept is b. The graph is a horizontal line.
3. $x = -2$ slope is undefined x-intercept $= -2$	**3.** If $B = 0$, the linear equation has the form $x = a$. The slope is undefined, the x-intercept is a, and the graph is a vertical line.

STRATEGY

To Write the Equation of a Line

1. The **slope-intercept form:** $y = mx + b$, where m is the slope and b is the y-intercept.
2. The **point-slope form:** $y - y_1 = m(x - x_1)$, where m is the slope and (x_1, y_1) is a point on the line.

WARMING UP

Answer true or false.

1. The slope of $y = 2x + 3$ is 3.
2. The y-intercept of $y = -3x + 1$ is 1.
3. The slope of $y = -x + 2$ is -1.
4. The y-intercept of $y = x$ is 0.
5. The equation of the line with slope 1 and y-intercept -4 is $y = x - 4$.
6. For the line $y = 4$, its slope is 4.
7. The y-intercept of $2x - 4y = 8$ is -2.
8. The line given by $x + 7 = 0$ is vertical.

EXERCISE SET 3.3

For Exercises 1–18, find the slope and y-intercept of the line.

1. $y = 3x + 2$
2. $y = 2x + 5$
3. $y = -4x - 1$
4. $y = -3x + 3$
5. $y = x + 6$
6. $y = x - 7$
7. $y = -x - 4$
8. $y = -x + 10$
9. $y = x$
10. $y = -x + 1$
11. $y - x = 0$
12. $y + x = 2$
13. $y + 3x = -6$
14. $y - 2x = 0$
15. $y + 5x - 1 = 0$
16. $y + 3x + 7 = 0$
17. $2y = x - 8$
18. $3y = -2x + 9$
19. Does the graph of $x = 3$ have a slope? A y-intercept? Why?
20. Does the graph of $y = -1$ have a slope? A y-intercept? Why?

For Exercises 21–32, the slope and y-intercept of a line are given.

(a) Draw the line. **(b)** Write an equation for the line.

21. The slope is $\frac{1}{2}$ and the y-intercept is -1.
22. The slope is $\frac{1}{3}$ and the y-intercept is 1.

23. The slope is $\frac{3}{4}$ and the y-intercept is 1.

24. The slope is $\frac{2}{3}$ and the y-intercept is -2.

25. The slope is $-\frac{4}{3}$ and the y-intercept is 4.

26. The slope is $-\frac{3}{2}$ and the y-intercept is 3.

27. The slope is 2 and the y-intercept is 0.

28. The slope is 3 and the y-intercept is -2.

29. The slope is $-\frac{2}{3}$ and the y-intercept is 0.

30. The slope is -3 and the y-intercept is 1.

31. The slope is 0 and the y-intercept is 3.

32. The slope is 0 and the y-intercept is -2.

For Exercises 33–40, draw the line through the given point P and having slope m.

33. $P(0, 0)$; $m = \frac{3}{2}$

34. $P(1, 0)$; $m = \frac{1}{2}$

35. $P(-2, 3)$; $m = -\frac{3}{5}$

36. $P(-1, 4)$; $m = -\frac{3}{4}$

37. $P(-1, -2)$; $m = 2$

38. $P(-3, -3)$; $m = 3$

39. $P(-2, 4)$; $m = -4$

40. $P(0, 3)$; $m = -2$

For Exercises 41–46, find the equation of a line through the point P and having slope m. Write your answer in the form $Ax + By = C$, where A, B, and C are integers.

41. $P(1, 5)$; $m = 3$

42. $P(2, -1)$; $m = 2$

43. $P(-2, -5)$; $m = \frac{1}{2}$

44. $P(-7, -1)$; $m = \frac{1}{3}$

45. $P(2, 3)$; $m = -4$

46. $P(-1, 1)$; $m = -3$

For Exercises 47–52, find the equation of a line through the two points P and Q. Write your answer in the form $Ax + By = C$, where A, B, and C are integers.

47. $P(1, 5)$ and $Q(-1, 6)$

48. $P(-3, 4)$ and $Q(2, 1)$

49. $P(5, 7)$ and $Q(5, -2)$

50. $P(-3, -1)$ and $Q(4, -1)$

51. $P(-4, -3)$ and $Q(0, -1)$

52. $P(3, -6)$ and $Q(-7, -10)$

For Exercises 53–58, find the equation of a line through the two points P and Q. Write your answer in the form of $y = mx + b$.

53. $P(0, 0)$ and $Q(2, 4)$

54. $P(-1, -3)$ and $Q(0, 0)$

55. $P(3, 2)$ and $Q(5, 1)$

56. $P(1, -2)$ and $Q(3, -1)$

57. $P(7, -1)$ and $Q(-4, -1)$

58. $P(3, 6)$ and $Q(-2, 6)$

For Exercises 59–62, find the equation of the line from the given information. Write your answer in the form $y = mx + b$.

59. The slope is -7.069 and the line passes through $(-12.1, 10.3)$.

60. The slope is 10.92 and the line passes through $(-7.95, -3.82)$.

61. The line passes through $(1.35, -2.47)$ and $(3.90, 1.06)$.

62. The line passes through $(-4.56, -3.19)$ and $(2.07, -4.48)$.

For Exercises 63–66, determine whether or not the two lines of the given equations are parallel.

63. $y = -\frac{1}{3}x + 2$ and $x + 3y = 1$

64. $-2x + y = -1$ and $-x + 2y = 2$

65. $x + 3y = 3$ and $x - 3y = -6$

66. $y = \frac{1}{2}x - 7$ and $3x - 6y = 5$

SAY IT IN WORDS

67. Explain how to find the equation (in standard form) of a line if you are given two points on the line.

68. If you are given the equation of a line, $Ax + By = C$, where $B \neq 0$, explain how you would obtain the slope-intercept form.

REVIEW EXERCISES

The following exercises review parts of Section 2.5. Doing these exercises will help prepare you for the next section.

For Exercises 69–72, solve the inequality.

69. $4x < 20$

70. $-3y \geq 12$

71. $3 \leq 2 + 3x < 14$

72. $4 < 1 + 2t \leq 9$

ENRICHMENT EXERCISES

1. Find an equation of the vertical line through the origin.

2. Find an equation of the horizontal line through the origin.

3. Find an equation of the line through $(2, 3)$ and parallel to the line $2x + y = 1$.

4. Find an equation of the line through $(-1, 2)$ and perpendicular to the line $-3x + y = 2$.

5. Find an equation of the line parallel to the line $y = 3$ and passing through $(3, -2)$.

6. Find an equation of the line parallel to the line $x = 1$ and passing through $(-5, 3)$.

7. Find an equation of the line perpendicular to the line $x = -4$ and passing through $(5, -1)$.

8. Find an equation of the line perpendicular to the line $y = -5$ and passing through $(3, 0)$.

Answers to Enrichment Exercises begin on page A.1.

3.4 Applications of Linear Equations

OBJECTIVE

▶ *To use linear equations in applications*

LEARNING ADVANTAGE

Why study mathematics? *Mathematics is an increasingly important tool for jobs of the future. Using mathematics in problem solving will be necessary for the typical college graduate. No longer can we say that the knowledge of math-*

ematics is nice, but not necessary. To be a success in the world of tomorrow requires being able to use the powerful tool of mathematics.

A linear equation in the slope-intercept form, $y = mx + b$, is a formula that expresses y in terms of x. We say that y is *linearly dependent upon x.* By choosing a value for x, we can use the equation to find the corresponding value for y. We say that x is the **independent variable** and y is the **dependent variable.**

Example 1 Suppose that y is linearly dependent upon x and is given by the equation

$$y = 2x + 3, \text{ where } -3 \leq x \leq 2$$

Graph this equation.

Solution The graph of the linear equation $y = 2x + 3$ is a straight line. However, since x is restricted to be between -3 and 2, inclusive, we make the following table by choosing x to be -3 and 2 and then finding the corresponding values of y from the equation.

x	y
-3	-3
2	7

Next, plot the two points $(-3, -3)$ and $(2, 7)$ and draw the line segment between these two points.

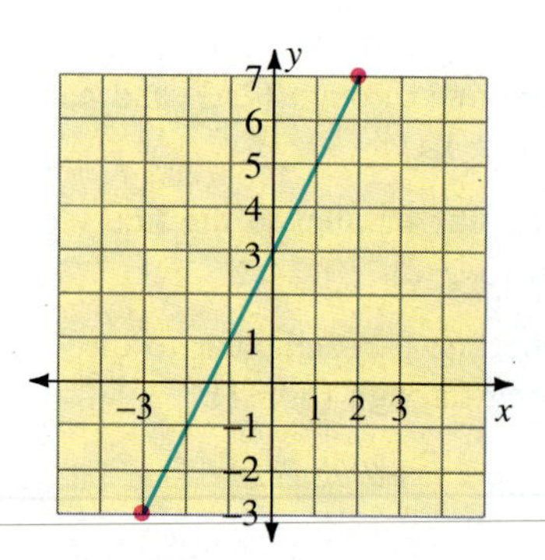

There are many relationships in the real world where one variable is linearly dependent upon another variable.

Example 2 Let C be the cost of installing x feet of fencing. The cost is \$40 plus \$1 per foot.

(a) Express C in terms of x.

(b) Draw the graph of the equation found in Part (a).

Solution **(a)** The cost is the sum of two things, the variable cost and the fixed cost.

$$\text{cost} = \text{variable cost} + \text{fixed cost}$$

The variable cost is the product of the cost per foot and the number of feet. The fixed cost is \$40. Therefore,

$$\text{cost} = (\text{cost per foot})(\text{number of feet}) + \text{fixed cost}$$
$$C = x + 40$$

(b) Since x represents the number of feet of fencing, $x \geq 0$. Choosing $x = 0$ and $x = 10$, we find the corresponding values for C. The graph is shown below.

x	C
0	40
10	50

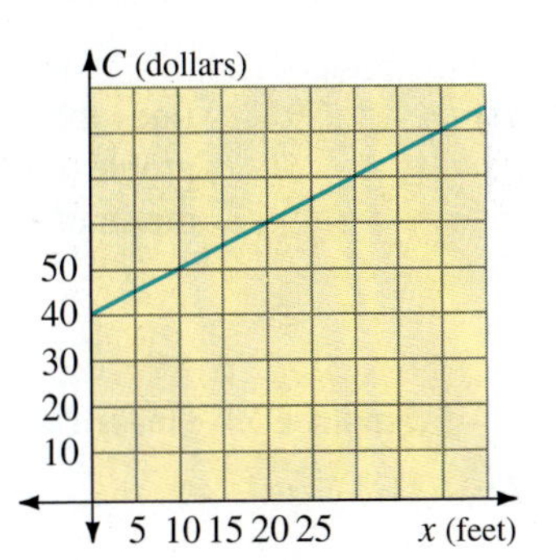

Notice that in Example 2, we used the letter C instead of y to represent the dependent variable, since C reminds us that the dependent variable means the cost.

Example 3 Jeremy drives his Porsche at 60 mph for t hours, where $0 \leq t \leq 5$.

(a) Express the distance d he has traveled in terms of time t.

(b) Draw the graph of this equation.

Solution **(a)** We use the formula for distance-rate-time: $d = rt$ and replace r by 60. The desired equation is

$$d = 60t,\ 0 \leq t \leq 5$$

(b) Since the independent variable is t and the dependent variable is d, we draw a coordinate system with the horizontal axis labeled t and the vertical axis labeled d. Two points are needed to draw the graph, we choose $t = 0$ and $t = 5$:

t	d
0	0
5	300

We plot the two points (0, 0) and (5, 300) and draw the line segment between them.

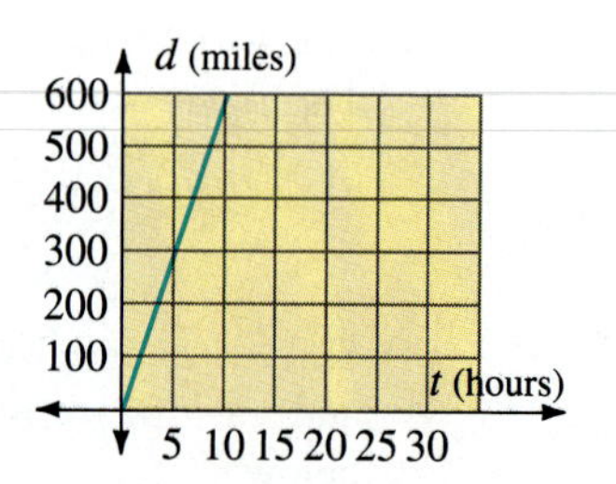

When two variables are related linearly, we can find the equation that defines the relationship, if we are given two points. This is the same kind of problem as finding the slope-intercept equation of a line given two points on the line.

Example 4 Linear Depreciation Green Acres farm buys a corn husker for 60 thousand dollars. The value of the machine decreases with time. At the end of 10 years, the machine has a salvage value of 20 thousand dollars. For tax purposes, assume that the value V of the machine is linearly dependent upon the number of years t after purchase.

(a) Write the linear equation that expresses V in terms of t.

(b) Draw the graph of this linear equation.

Solution **(a)** Since we assume that V is linearly dependent upon t, there are two numbers m and b so that

$$V = mt + b$$

From the information given, we can find two points (t, V). The machine was purchased for 60 thousand dollars, so then $t = 0$, $V = 60$. We write this information as (0, 60). Ten years later, the machine is worth 20 thousand dollars, which gives the point (10, 20).

Now, the problem is to find m and b so that the linear equation $V = mt + b$ contains the two points (0, 60) and (10, 20). Using the slope formula,

$$m = \frac{20 - 60}{10 - 0} = \frac{-40}{10} = -4$$

The number b is the V-intercept, and when $t = 0$, $V = 60$. Therefore, b is 60, and the equation is

$$V = -4t + 60$$

(b) Notice that the interval of values for the independent variable t is $0 \leq t \leq 10$. Therefore, we plot the two points (0, 60) and (10, 20) and draw the line segment between them as shown in the figure.

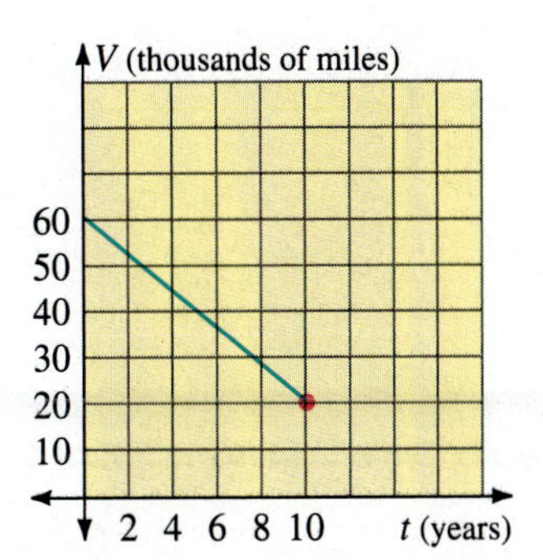

WARMING UP

Answer true or false.

1. $C = 2x + 7$ is a linear equation.

2. $d = 4r^2$ is a linear equation.

3. If $V = 3x + 5$, when $x = 4$, then $V = 17$.

4. If $C = 6x + 40$ gives the cost C of producing x items, then C is linearly dependent upon x.

EXERCISE SET 3.4

For Exercises 1–14, graph the equation.

1. $y = 2x - 1, \ -2 \leq x \leq 3$

2. $y = 3x - 4, \ -1 \leq x \leq 3$

3. $y = 2x, \ 0 \leq x \leq 3$

4. $y = -2x + 5, \ -2 \leq x \leq 2$

5. $C = -2x + 10, 0 \leq x \leq 5$

6. $C = -x + 8, 0 \leq x \leq 6$

7. $V = x + 4, 0 \leq x \leq 4$

8. $V = 2x, 0 \leq x \leq 3$

9. $d = 30t, 0 \leq t \leq 3$

10. $d = 20t, 0 \leq t \leq 4$

11. $C = x + 1, x \geq 0$

12. $C = 2x + 2, x \geq 0$

13. $N = s - 1,\ s \geq 1$

14. $p = 2r - 4,\ r \geq 2$

15. Let C be the cost of installing x feet of fencing. The cost is \$20 plus \$10 per foot.

(a) Express C in terms of x.

(b) Draw the graph of the equation found in Part (a).

16. Elaine wants x yards of wooden fencing installed on her property. The cost is \$25 plus \$5 per yard.

(a) Express C in terms of x.

(b) Draw the graph of the equation found in Part (a).

17. Taro drives a car at 20 mph for t hours, where $0 \leq t \leq 3$.

(a) Express the distance traveled d in terms of time t.

(b) Draw the graph of this equation.

18. A car is driven at 30 mph for t hours, where $0 \leq t \leq 2$.

(a) Express the distance d that the car travels in terms of time t.

(b) Draw the graph of this equation.

19. A laundry firm buys a new press for 50 thousand dollars. This press has a useful life of 20 years with a salvage value of 10 thousand dollars. Assume that the value V of the press is linearly dependent upon time t.

(a) Write the linear equation that expresses V in terms of t.

(b) Draw the graph of this linear equation.

20. The Clear Cola Company buys a new capping machine for 30 thousand dollars. The machine has a useful life of 20 years, at which time it has a scrap value of 10 thousand dollars. Assume that the value V of the machine is linearly dependent upon time t.

(a) Find the linear equation of V in terms of t.

(b) Draw the graph of this linear equation.

SAY IT IN WORDS

21. When graphing an equation, explain how you decide on the scale for the horizontal axis and for the vertical axis.

22. In your own words, explain dependent and independent variables.

REVIEW EXERCISES

The following exercises review parts of Section 3.3. Doing these exercises will help prepare you for the next section.

For Exercises 23–28, draw the graph of the linear equation.

23. $-2x + 3y = 6$

24. $x + 3y = 9$

25. $y = -2x + 3$

26. $x = 4$

27. $y = -2$

28. $x + y = 8$

ENRICHMENT EXERCISES

For Exercises 1–4, draw the graph.

1. $y = \begin{cases} 2x - 1, & \text{if } -2 \leq x \leq 0 \\ -x - 1, & \text{if } 0 < x \leq 2 \end{cases}$

2. $y = \begin{cases} 2, & \text{if } -3 \leq x < 0 \\ 1, & \text{if } x = 0 \\ -2, & \text{if } 0 < x \leq 3 \end{cases}$

3. $y = |x|, \ -3 \leq x \leq 3$

4. $y = |x - 1|, \ 0 \leq x \leq 2$

Answers to Enrichment Exercises begin on page A.1.

3.5 Functions

OBJECTIVES

- *To understand the definitions of relation and function*
- *To determine when a relation is also a function*
- *To determine the domain and the range of a function*
- *To use the $f(x)$ notation*

LEARNING ADVANTAGE

Why study mathematics? *Mathematics enables us to quantify real life problems. It allows us to investigate relationships that occur and attempt to find patterns that logically describe them. The power of mathematics is translated in terms of time saved as well as savings in money and manpower used to solve a problem.*

There are many examples of dependence of one quantity upon another. The price of soybeans depends upon the supply. The level of blood sugar after a meal depends upon the amount of insulin present. The height of a rocket depends upon the time elapsed since ignition. In mathematics, such relationships are called functions.

Before we study functions, we start with the definition of relation.

DEFINITION

A **relation** is any set of ordered pairs of numbers. The **domain** of the relation is the set of all first coordinates of the ordered pairs, and the **range** is the set of all second coordinates of the ordered pairs.

Example 1 Consider the relation $\{(0, 1), (0, 2), (1, 5), (-1, 0)\}$. The domain is $\{-1, 0, 1\}$ and the range is $\{0, 1, 2, 5\}$.

In addition to listing the elements in a relation, another way to specify a relation is to give a rule, or equation, for obtaining the set of ordered pairs.

Example 2 Consider the relation $\{(x, y) | x + y = 1\}$. It is the set of all ordered pairs whose sum of the coordinates equals one. Some members of this relation are $(1, 0)$, $\left(\frac{1}{2}, \frac{1}{2}\right)$, and $(3, -2)$. We cannot list all of the members of the relation, since there are infinitely many solutions to $x + y = 1$. The domain is the set of all real numbers and the range is also the set of all real numbers.

Since a relation is a set of ordered pairs, these ordered pairs can be graphed in a coordinate plane.

DEFINITION

The **graph** of a relation is the set of points $P(x, y)$ in the coordinate plane whose coordinates belong to the relation.

Example 3 The graph of $\{(0, 1), (0, 2), (1, 5), (-1, 0)\}$ is shown in the figure below. We simply plot the four ordered pairs of the relation.

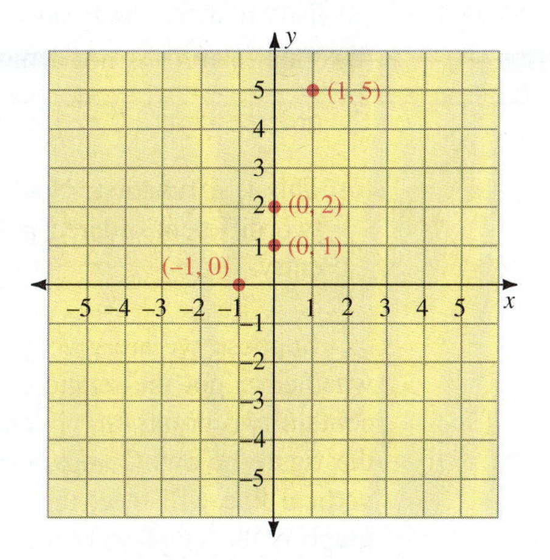

Example 4 The graph of the relation $\{(x, y) \mid x + y = 1\}$ is the straight line $x + y = 1$ as shown in the figure below.

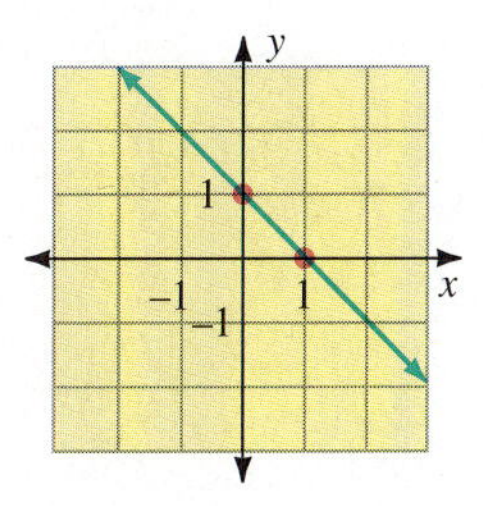

A very important concept in mathematics is that of a *function.* A function is a special kind of relation.

DEFINITION

A **function** is a relation in which no two different ordered pairs have the same first entry.

Example 5 Determine which of the following relations are functions.

(a) $\left\{(-3, 0), (-2, 1), \left(1, \frac{1}{2}\right), \left(2, \frac{1}{4}\right)\right\}$

(b) $\left\{\left(0, \frac{4}{3}\right), (2, 3), (-1, 1), (0, 1)\right\}$

(c) $\{(2, 1), (3, 1), (4, -1)\}$

Solution **(a)** This is a function, since no first entry is repeated.

(b) This relation is not a function, since 0 is the first entry in the two different ordered pairs $\left(0, \frac{4}{3}\right)$ and $(0, 1)$.

(c) This is a function. Note that a function may have the same *second* entry in two different ordered pairs. In this example, 1 is used twice as the second entry.

Suppose we are given the graph of some relation and we want to determine whether or not the relation is also a function. The graph of a function cannot contain two points (a, y_1) and (a, y_2), where $y_1 \neq y_2$, as shown in Part (a) of the figure below. Compare this to the graph in Part (b) of the figure. Here any vertical line will meet the graph in no more than one point. Therefore, this graph is the graph of a function.

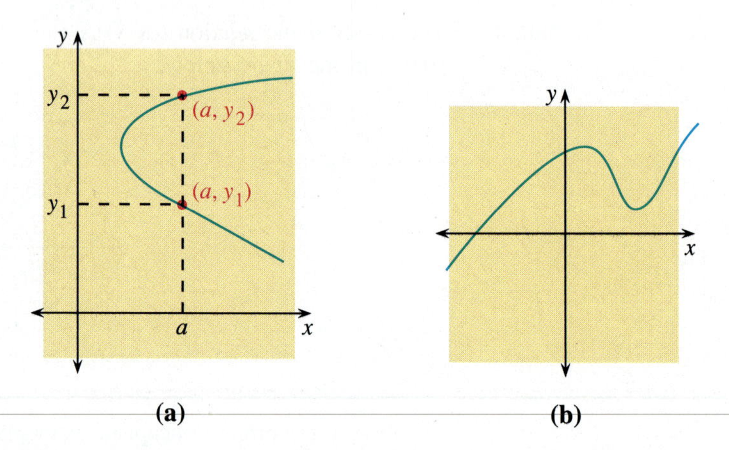

(a) (b)

We state these results as the vertical line test.

Vertical Line Test

1. If any vertical line intersects the graph of a relation in more than one point, the relation is not a function.
2. If any vertical line intersects the graph of a relation in at most one point, the relation is a function.

Example 6 Use the vertical line test to check if each relation is a function.

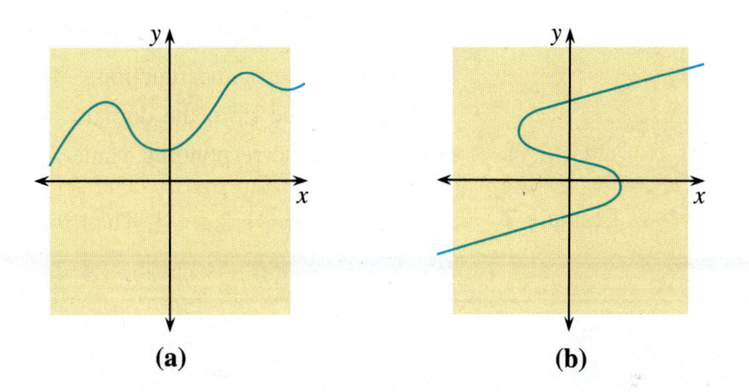

(a) **(b)**

Solution **(a)** Using the vertical line test, any vertical line will cross the graph in precisely one point and therefore this is the graph of a function.

(b) The vertical line test for this graph indicates that the graph is not the graph of a function, since a vertical line could cross the graph in more than one point.

As we have seen so far, many relations and functions are given in equation form. It is easier to just write the equation rather than use the set-builder notation. For example, the function $\{(x, y) \mid y = x + 1\}$ can be stated as the function defined by $y = x + 1$ and we say that *y is a function of x*. We call x the **independent variable** and y the **dependent variable,** and say that y depends upon x. Recall in the last section that we talked of linear dependence. To say that y is linearly dependent upon x means that y is a function of x and this function is given by a linear equation.

Suppose that a function is given by an equation. If the domain is not specified, it is assumed to be the largest set of replacements for x so that the equation is defined as a real number. We illustrate this concept in the next example.

So far, we have worked with functions without giving them names. Frequently, we use a letter like f or g to denote a function. For example, the function $y = 3x - 2$ is written as

$$f(x) = 3x - 2$$

where $f(x)$ is read "f of x." The symbolism $f(x)$ is an alternate way to write y in the equation defining the function. It is a convenient way to find the value of y given a value for x. For example, when $x = 4$, we can find the corresponding value for y using the equation $f(x) = 3x - 2$:

$$f(4) = 3(4) - 2 \quad \text{Replace } x \text{ by 4.}$$
$$= 12 - 2$$
$$= 10$$

Therefore, $f(4) = 10$, which is read "f of 4 is equal to 10." This means that $y = 10$ when $x = 4$; that is, $f(4)$ is the member of the range of the function f that corresponds to the domain value of 4. In summary, here are three important observations about the function notation.

1. f is the name of the function.
2. x is the value from the domain of f.
3. $f(x)$ is the corresponding value from the range of f.

Example 7 Let $f(x) = -3x^2 + 2x - 4$. Then,

(a) $f(0) = -3(0)^2 + 2(0) - 4$ Replace x by 0.
$= -4$

(b) $f(1) = -3(1)^2 + 2(1) - 4$ Replace x by 1.
$= -5$

(c) $f(-2) = -3(-2)^2 + 2(-2) - 4$ Replace x by -2.
$= -20$

(d) $f(a) = -3a^2 + 2a - 4$ Replace x by a.

WARMING UP

Answer true or false.

1. $\{(1, 2), (1, 3)\}$ is a function
2. $\{(0, 0), (1, 1), (2, 4)\}$ is a function
3. If $f(x) = 2x - 5$, then $f(0) = -3$.
4. If $g(x) = 6 - 3x$, then $g(0) = 6$.
5. If $f(x) = x^2 - x + 1$, then $f(1) = 1$.
6. If $g(x) = 2x^2 + x - 5$, then $g(-1) = -8$.

EXERCISE SET 3.5

For Exercises 1–8, state the domain and range of the relation and determine if it is also a function.

1. $\{(2, 3), (-3, 2), (-1, -2), (1, -1)\}$
2. $\{(6, -2), (2, -6), (0, 7), (1, 5)\}$
3. $\{(2, -2), (3, -5), (3, -6), (7, -1)\}$
4. $\{(3, -4), (-4, 6), (3, -5), (0, 0)\}$
5. $\{(-5, 1), (4, -1), (-3, 0), (6, -3)\}$
6. $\{(9, 8), (0, 1), (-4, 3), (-4, 0)\}$
7. $\{(-1, -3), (-3, 2), (-2, -2), (-5, 10)\}$
8. $\{(10, 9), (-9, 10), (-10, 9), (9, -10)\}$

For Exercises 9–12, write the relation S whose graph is given. Determine the domain and range of S and whether S is also a function.

9.

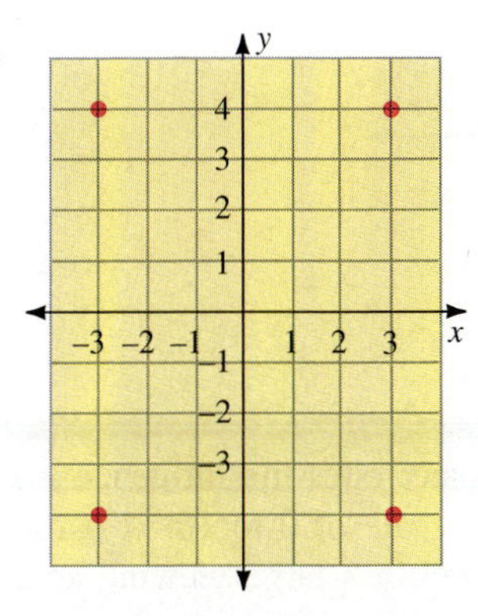

10.

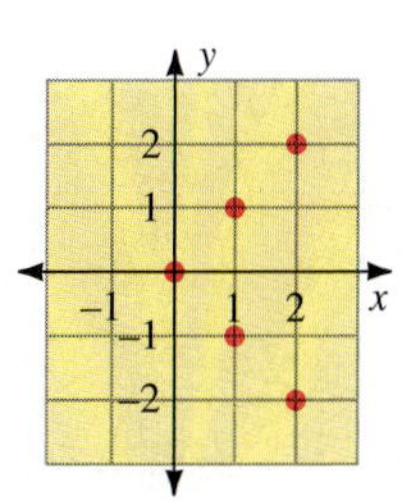

11.

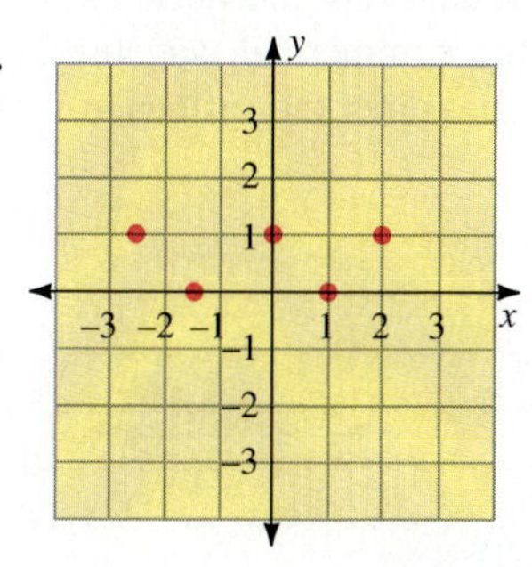

12.

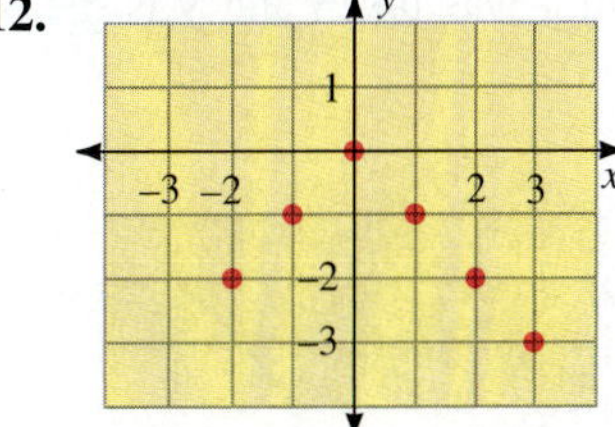

For Exercises 13–18, use the vertical line test to determine if the relation is also a function.

13.

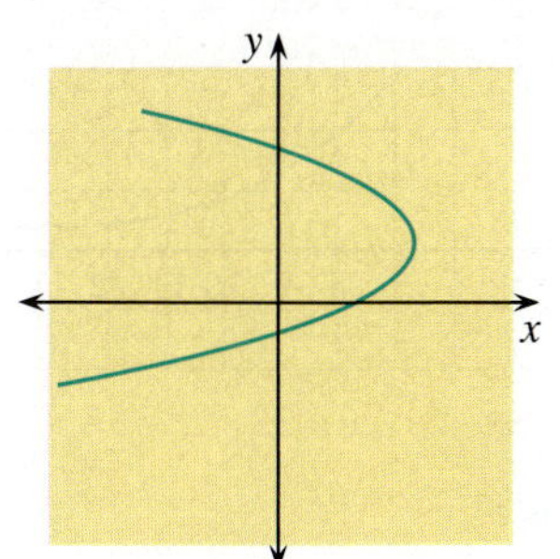

14.

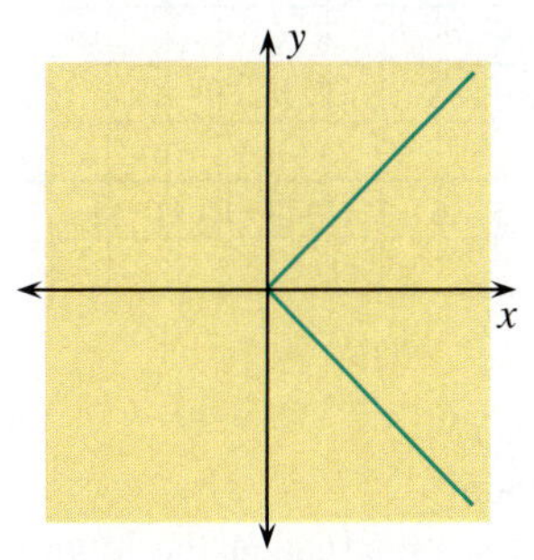

15.

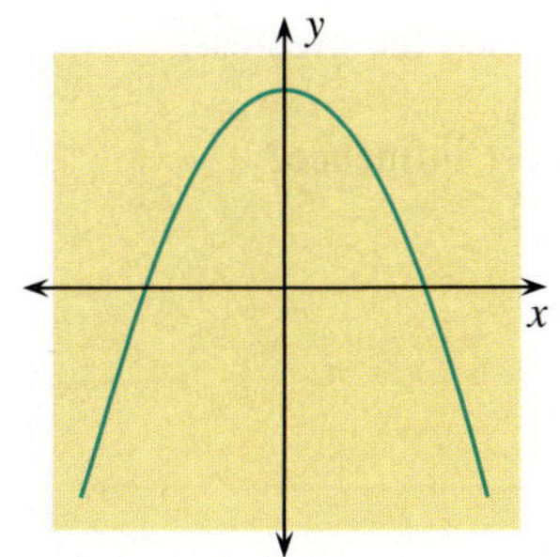

16.

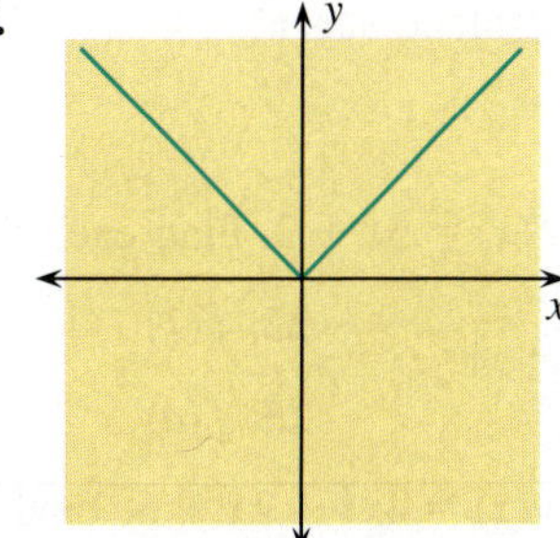

17.

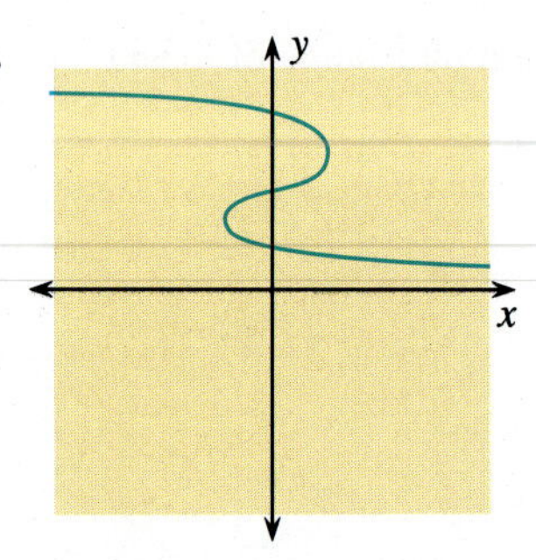

18.

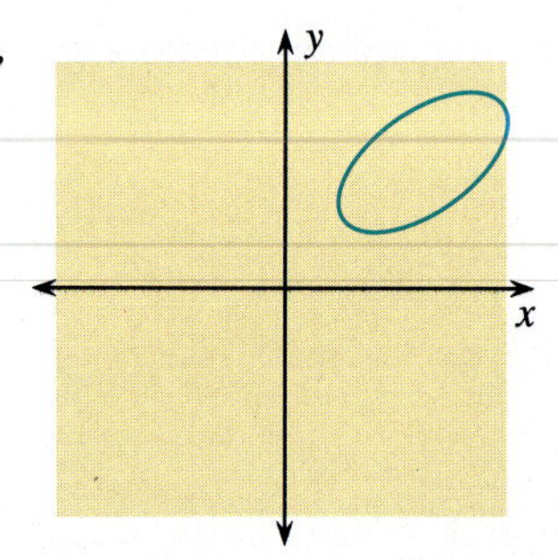

19. The blood sugar of a diabetic is regulated by a mixture of time-released insulin. The following graph shows the relationship between the amount of blood sugar y at time t, where t is the number of hours after the insulin was taken and y is measured in mg/dl.

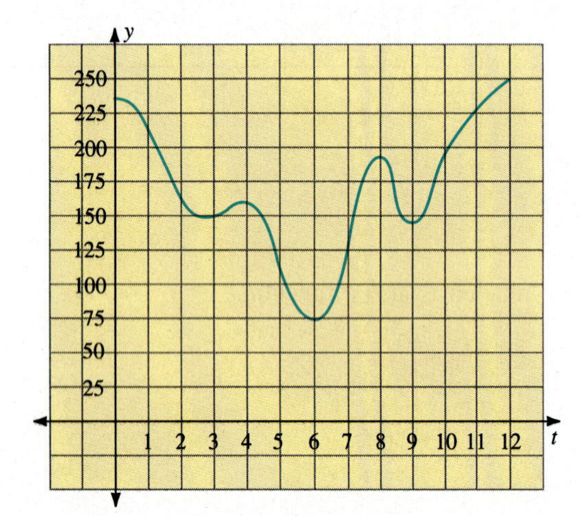

(a) Is y a function of t?
(b) What is the domain?
(c) At what time is the blood sugar the lowest?
(d) What is the lowest blood sugar?

20. An advertising company rates the influence a TV commercial has on a scale of 0 to 10. A particular commercial is aired in a large viewing area and a graph constructed so that the horizontal axis is time t days since the commercial was aired, and the vertical axis measures the influence y of the commercial.

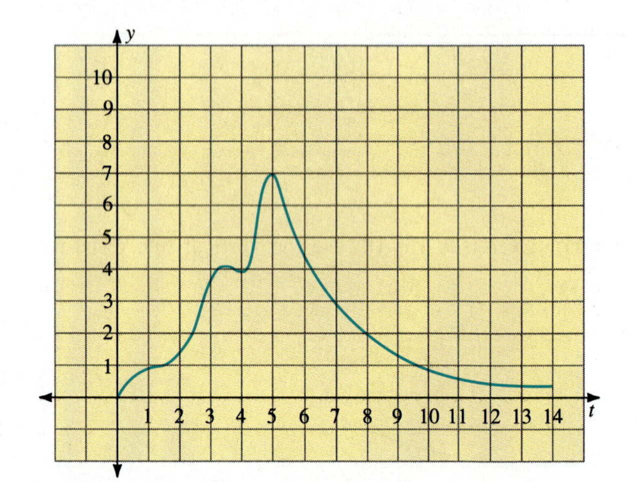

(a) Is y a function of t?
(b) What is the domain?
(c) When is the influence the greatest?
(d) What is the greatest influence?

For Exercises 21–26, let $f(x) = 3x + 2$. Find each of the following.

21. $f(0)$ **22.** $f(1)$ **23.** $f(3)$

24. $f(-2)$ **25.** $f(a)$ **26.** $f(t)$

For Exercises 27–32, let $g(x) = 4x^2 - 2x + 5$. Find each of the following.

27. $g(0)$ **28.** $g(2)$ **29.** $g(-1)$

30. $g\left(-\frac{1}{2}\right)$ **31.** $g(s)$ **32.** $g(a)$

SAY IT IN WORDS

33. Explain the difference between relation and function.

34. Explain the "f of x" notation.

REVIEW EXERCISES

The following exercises review parts of Section 1.2. Simplify.

35. 3^2 **36.** 2^3 **37.** $\frac{1}{2^2}$

38. 2^{-2} **39.** $2^3 \cdot 2^2$ **40.** 2^{3+2}

41. $(2 + 3)^2$ **42.** $2^2 + 3^2$

ENRICHMENT EXERCISES

If $f(x) = \frac{x}{x+1}$, find

1. $f(0)$ **2.** $f(1)$ **3.** $f(-2)$

4. $f(a)$ **5.** $f\left(\frac{1}{2}\right)$ **6.** $f\left(\frac{1}{3}\right)$

Answers to Enrichment Exercises begin on page A.1.

CHAPTER 3 Summary and review

Examples

The points (0, 3), (3, 0), (2, 4), (2, −4), (−2, 4), and (−2, −4) are graphed in the following coordinate plane.

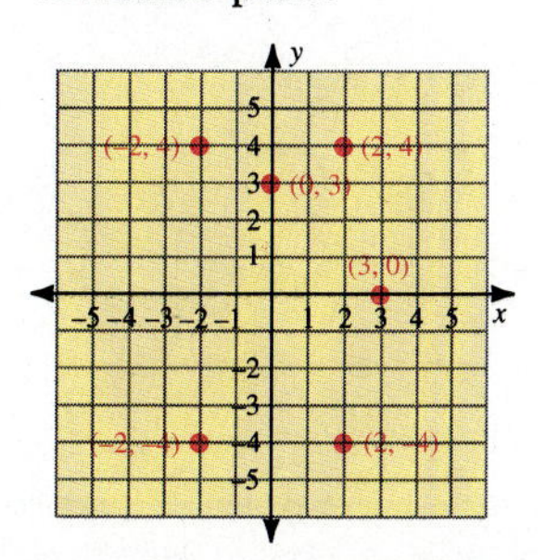

Plotting points in the plane (3.1)

To plot (or graph) a point (x, y) in a coordinate plane, start at the origin and consider the x-coordinate.

(a) If $x > 0$, move right x units.
(b) If $x < 0$, move left $|x|$ units.
(c) If $x = 0$, stay at the origin.

Next, consider the y-coordinate.

(a) If $y > 0$, move up y units.
(b) If $y < 0$, move down $|y|$ units.
(c) If $y = 0$, do not move.

The slope of a line (3.2)

The slope m of a line containing the two points (x_1, y_1) and (x_2, y_2) is given by

$$m = \frac{\text{change in } y}{\text{change in } x} = \frac{y_2 - y_1}{x_2 - x_1}, \quad x_1 \neq x_2$$

Examples

The slope of the line containing the points (2, 8) and (5, −4) is

$$m = \frac{-4 - 8}{5 - 2} = \frac{-12}{3} = -4$$

Linear equations (3.1 and 3.3)

1. $Ax + By = C$ is the general form. If neither A nor B is zero,

   slope is $-\dfrac{A}{B}$

   y-intercept is $\dfrac{C}{B}$

   x-intercept is $\dfrac{C}{A}$

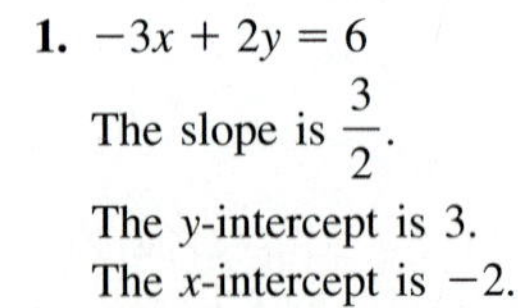

1. $-3x + 2y = 6$

   The slope is $\dfrac{3}{2}$.

   The y-intercept is 3.
   The x-intercept is −2.

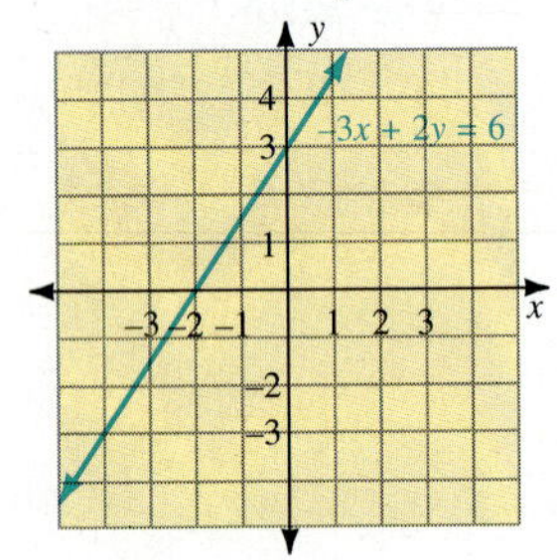

2. If $A = 0$, the linear equation has the form $y = b$. The slope is 0 and the y-intercept is b. The graph is a horizontal line.

2. $y = -2$
   The slope is 0.
   The y-intercept is −2.

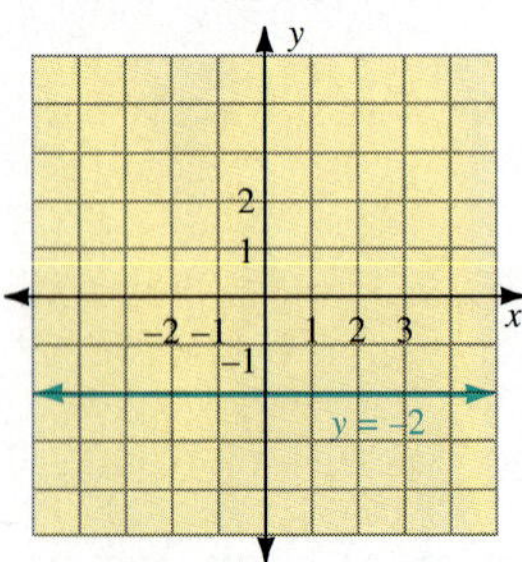

3. If $B = 0$, the linear equation has the form $x = a$. The slope is undefined, the x-intercept is a, and the graph is a vertical line.

3. $x = 2$
   The slope is undefined.
   The x-intercept is 2.

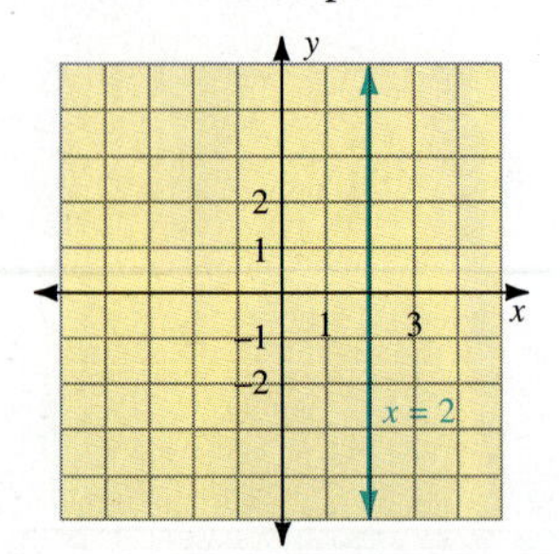

The slope-intercept form of the equation of a line (3.3)

A line that has slope m and y-intercept b is given by

$$y = mx + b$$

Examples

The equation of a line with slope 5 and y-intercept -1 is

$$y = 5x - 1.$$

The point-slope form of the equation of a line (3.3)

A line with slope m passing through the point (x_1, y_1) is given by

$$y - y_1 = m(x - x_1)$$

The equation of a line with slope 3 and passing through $(-1, 2)$ is given by $y - 2 = 3(x + 1)$, which simplifies to

$$y = 3x + 5.$$

Applications of linear equations (3.4)

In Section 3.4, we saw applications involving linear depreciation, linear cost equations, linear demand equations, and linear supply equations. Check the section for details.

The Viking Company makes x computer hard disks each day with a cost per disk of \$800 and a fixed cost of \$350 per day. If C is the daily total cost, then

$$C = 800x + 350.$$

Functions (3.5)

A **relation** is any set of ordered pairs of numbers. The **domain** is the set of all first coordinates and the **range** is the set of all second coordinates.

Relation:

$S = \{(-1, -1), (-1, 2), (1, 0), (0, 1)\}$

Domain:

$$\{-1, 0, 1\}$$

Range:

$$\{-1, 0, 1, 2\}$$

This relation is not a function.

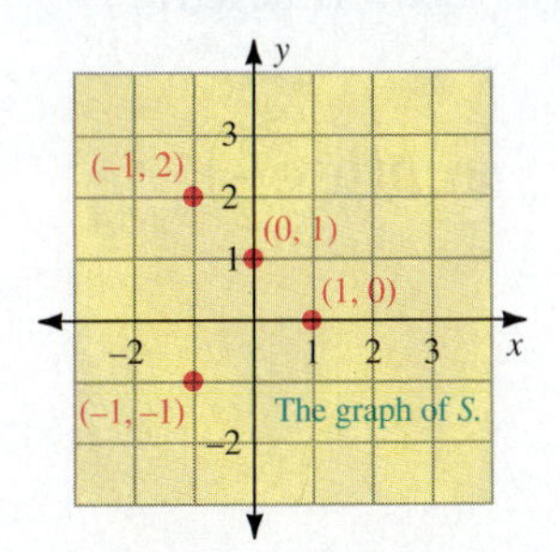

A **function** is a relation in which no two different ordered pairs have the same first entry.

The **graph** of a relation is the set of points $P(x, y)$ in the coordinate plane whose coordinates belong to the relation.

A function f:

$$f(x) = -2x + 3$$

For example,

$$f(0) = -2(0) + 3 = 3$$

and

$$f(1) = -2(1) + 3 = 1$$

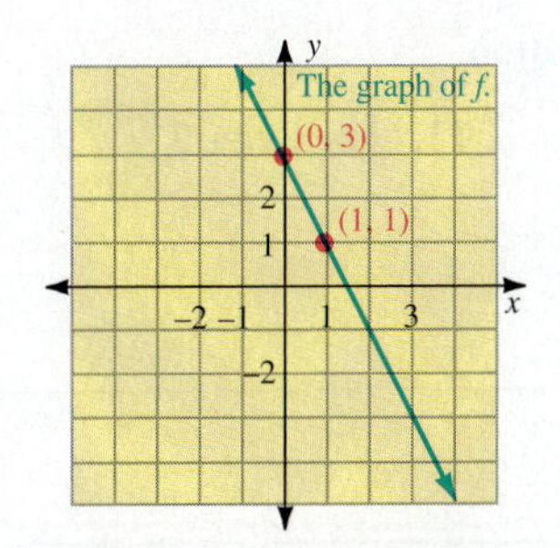

CHAPTER 3 REVIEW EXERCISE SET

Section 3.1

1. Find the ordered pair for each point shown in the following coordinate plane.

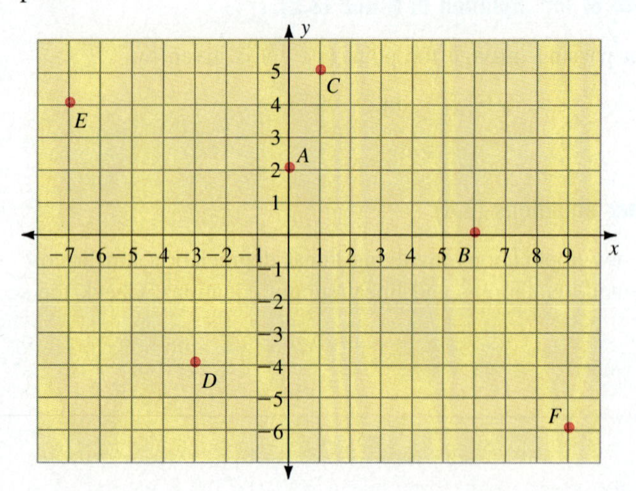

For Exercises 2 and 3, describe where in the coordinate plane the point (x, y) would be located.

2. y is positive

3. x is negative

4. Plot each point in the same coordinate plane.

(a) $A(6, 4)$ **(b)** $B(-2, 0)$ **(c)** $C(-4, -1)$ **(d)** $D\left(0, -\frac{5}{2}\right)$

For Exercises 5 and 6, determine if the given ordered pair is a solution of the equation $4x - 3y = 18$.

5. $(0, -6)$

6. $(-2, -4)$

For Exercises 7–9, find the ordered pair that is a solution of $-5x + 2y = 10$ when

7. $x = 0$

8. $y = 0$

9. $x = 2$

For Exercises 10–13, draw the graph of the given equation.

10. $2x - 5y = 10$

11. $3x + y = 9$

12. $y = -6$

13. $x = 4$

Section 3.2

For Exercises 14–17, find the slope, if it exists, for each line through the two points.

14. $(0, -1)$ and $(6, 4)$

15. $(-2, 6)$ and $(-2, 3)$

16. $(-5, 10)$ and $(-1, 10)$

17. $(-3, 2)$ and $(-4, 4)$

For Exercises 18 and 19, determine if the line through the first pair of points is parallel to the line through the second pair of points.

18. First pair of points: $(0, 1)$ and $(-3, -3)$
Second pair of points: $(6, 7)$ and $(3, 0)$

19. First pair of points: $(-2, 2)$ and $(-2, 7)$
Second pair of points: $(-3, 4)$ and $(-3, 0)$

Section 3.3

For Exercises 20 and 21, find the slope and the y-intercept of the line from the given equation.

20. $y = \frac{7}{2}x - 12$

21. $-5x + 3y = -18$

22. A line has a slope of $\frac{2}{3}$ and y-intercept of -3.

(a) Draw the line.

(b) Write an equation for the line.

23. A line has a slope of -1 and y-intercept of 1.

(a) Draw the line.

(b) Write an equation for the line.

24. Use the point-slope form to find the equation of the line through $(2, 7)$ with slope 1. Write the answer in the form $Ax + By = C$.

25. Use the point-slope form to find the equation of the line through $(-3, 5)$ and $(1, -4)$. Write the answer in the form $Ax + By = C$.

26. Use the slope-intercept form to find the equation of the line through $(1, 2)$ and $(-4, -3)$.

Section 3.4

27. Draw the graph of $y = -3x + 12$, $0 \leq x \leq 3$

28. Draw the graph of $V = t + 3$, $t \geq 0$

29. The sanitation department of Whitehall Township buys a water-carrying street sweeper for $28,000. The useful life of the sweeper is 10 years at which time it has a salvage or trade-in value of $8000. Assume that the value V of the machine is linearly dependent upon the number of years t after purchase.

(a) Write a linear equation that expresses V in terms of t.

(b) Draw the graph of this linear equation. (*Note:* Since the useful life is 10 years, draw the line only for $0 \leq t \leq 10$.)

30. Custom Design Corporation makes x hundred radiator enclosures each day with a cost per one hundred enclosures of $25 and a fixed cost of $275 per day.

(a) Express the daily total cost C in terms of x.

(b) Draw the graph of the cost equation.

Section 3.5

31. Let S be the relation $\{(-2, -1), (-1, 0), (0, 2), (0, 3)\}$

(a) Find the domain of S.

(b) Find the range of S.

(c) Is S a function?

(d) Draw the graph of S.

32. Consider the function f defined by $f(x) = -4x^2 + 1$. Find

(a) $f(0)$

(b) $f(-1)$

(c) $f(a)$

CHAPTER 3 TEST

1. Plot each point in the same coordinate plane.

(a) $A(-2, 1)$ **(b)** $B(2, -3)$ **(c)** $C(-2, -3)$ **(d)** $D(0, 2)$

2. Find the ordered pair for each point in the coordinate plane below.

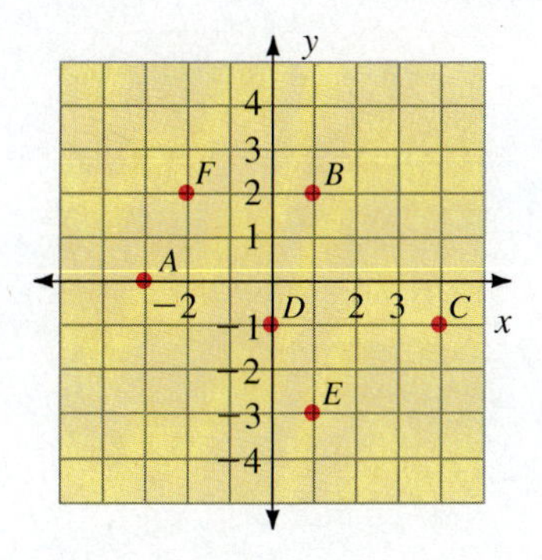

For Problems 3–6, draw the graph of the given equation.

3. $3x + 4y = 12$

4. $y = 3$

5. $2x - y = 4$

6. $y = -2x$

For Problems 7–9, find the slope, if it exists, of the line passing through the two points.

7. $(3, -1)$ and $(-2, -4)$

8. $(2, 6)$ and $(2, 4)$

9. $(-1, 3)$ and $(-2, 6)$

10. Is the line passing through $(2, -3)$ and $(4, 5)$ parallel to the line passing through $(3, -2)$ and $(4, 2)$?

For Problems 11 and 12, find the slope and y-intercept of the line.

11. $y = -x + 5$

12. $2x - 3y = 6$

13. Find the equation of a line through $(3, -1)$ with slope -2. Write the answer in the form $Ax + By = C$.

14. Find the equation of a line through $(5, 4)$ and $(10, 5)$. Write the answer in the form $y = mx + b$.

15. Find the equation of the line with x-intercept -2 and y-intercept 3. Write the answer in the form $y = mx + b$.

16. Draw the graph of $y = -2x + 8$, $0 \leq x \leq 4$.

17. Draw the graph of $V = 5n + 15$, $n \geq 0$.

18. The Sport Supply Company makes ice skates and has a fixed cost of \$110 with a variable cost of \$50 per ice skate. If the company makes x ice skates each day, express the daily total cost C in terms of x.

19. Let S be the relation $\{(-4, -3), (-2, 1), (0, 3), (1, 3), (2, -1)\}$.
(a) Find the domain of S.
(b) Find the range of S.
(c) Is S a function?
(d) Draw the graph of S.

20. Let $f(x) = x^3 - 3x^2 + 4$. Find each of the following:
(a) $f(0)$
(b) $f(-2)$
(c) $f(t)$

CHAPTER 4 Exponents and Polynomials

CONNECTIONS

The city of Allentown is planning to expand the area of a rectangular rose garden by increasing the width by x feet and the length by $2x$ feet. If the original garden measures 20 feet by 40 feet, express the additional area of the expanded garden in terms of x. See Exercise 59 in Section 4.5.

Overview

We begin this chapter by looking at properties of exponents, which leads to the algebra of polynomials. By developing our skills at working with these algebraic expressions, we will be able to solve certain quadratic equations in the next chapter.

4.1 Properties of Exponents

OBJECTIVES

- *To simplify expressions using $a^m a^n = a^{m+n}$*
- *To simplify expressions using $(a^m)^n = a^{m \cdot n}$*
- *To simplify expressions using division properties with exponents*

LEARNING ADVANTAGE

The main purposes of assignments are (1) to help reinforce what you learned in class, (2) to help identify those parts of the lesson that you do not understand, and (3) to increase the number of correctly solved problems to which you can refer when reviewing for a test.

Exponents in algebra are used in the same way as exponents in arithmetic. For example, $3^2 = 3 \cdot 3 = 9$ and $2^3 = 2 \cdot 2 \cdot 2 = 8$.

DEFINITION

If a is any real number and n is a positive integer,

$$a^n = \underbrace{a \cdot a \cdot a \cdots a}_{n \text{ factors of } a}$$

a^n is called the ***n*th power of *a*.**

a is called the **base.**

n is the **exponent.**

Example 1

(a) $2^5 = 2 \cdot 2 \cdot 2 \cdot 2 \cdot 2 = 32$ The base is 2 and the exponent is 5.

(b) $(-1)^2 = (-1)(-1) = 1$ The base is -1 and the exponent is 2.

(c) $(-4)^3 = (-4)(-4)(-4) = -64$ The base is -4 and the exponent is 3.

(d) $-5^2 = -5 \cdot 5 = -25$ The base is 5 and the exponent is 2. ◀

COMMON ERROR

Be sure not to mistake -5^2 for $(-5)^2$. The expression -5^2 means $-(5^2)$ or -25, whereas $(-5)^2$ means $(-5)(-5)$ or 25.

An expression such as $a^3 \cdot a^2$ can be simplified. Since a^3 means $a \cdot a \cdot a$ and a^2 means $a \cdot a$, then

$$a^3 \cdot a^2 = \underbrace{\overbrace{(a \cdot a \cdot a)}^{\text{3 factors}}\overbrace{(a \cdot a)}^{\text{2 factors}}}_{\text{5 factors}} = a^5$$

Therefore,

$$a^3 \cdot a^2 = a^{3+2} = a^5$$

That is, to multiply powers with the same base, use the common base and add the exponents.

RULE

Exponent Rule 1

$$a^m \cdot a^n = a^{m+n}$$

To find the product of two powers, each with the same base, use the common base raised to the sum of the two exponents.

Example 2 Simplify each of the following:

(a) $x^4 \cdot x^5$

(b) $b^6b^9b^2$

(c) $x^2 \cdot x^{16} \cdot y^7 \cdot y^3$

(d) ab^8a^7b

Solution (a) We use Exponent Rule 1 with x as the base a, $m = 4$, and $n = 5$.

$$x^4 \cdot x^5 = x^{4+5}$$
$$= x^9$$

(b) Exponent Rule 1 can be extended to simplify the product of three or more powers with the same base: Use the common base with the new exponent being the sum of the original exponents.

$$b^6b^9b^2 = b^{6+9+2}$$
$$= b^{17}$$

(c) Here we have two different bases. By the associative property, group the factors with a common base together, then apply Rule 1 to each grouping.

$$\begin{aligned} x^2 \cdot x^{16} \cdot y^7 \cdot y^3 &= (x^2 \cdot x^{16})(y^7 \cdot y^3) \\ &= x^{2+16} \cdot y^{7+3} \quad \text{Rule 1 used twice.} \\ &= x^{18} \cdot y^{10} \quad \text{or} \quad x^{18}y^{10} \end{aligned}$$

(d) First, using the commutative and associative properties of multiplication, group together powers with the same base, then apply Rule 1.

$$\begin{aligned} ab^8a^7b &= (aa^7)(b^8b) \quad \text{Group according to common base.} \\ &= a^{1+7}b^{8+1} \quad \text{Rule 1 used twice.} \\ &= a^8b^9 \end{aligned}$$

COMMON ERROR

Be careful not to mistake $a^6 \cdot a$ as a^6. Recall that a means a^1. Therefore,

$$\begin{aligned} a^6 \cdot a &= a^6 \cdot a^1 \\ &= a^{6+1} \\ &= a^7 \end{aligned}$$

If the rule $a^m a^n = a^{m+n}$ is to hold for $n = 0$, then

$$\begin{aligned} a^m a^0 &= a^{m+0} \\ &= a^m \end{aligned}$$

If $a \neq 0$, we may divide both sides by a^m.

$$\frac{a^m a^0}{a^m} = \frac{a^m}{a^m}$$

$$\begin{aligned} a^0 &= \frac{a^m}{a^m} \\ &= 1 \end{aligned}$$

Therefore, the following definition makes sense.

DEFINITION

If a is any nonzero real number,

$$a^0 = 1$$

NOTE *Any* nonzero *quantity raised to the zero power is one. For example,*

$$(4)^0 = 1, \qquad (3a)^0 = 1, \qquad (xy^3)^0 = 1$$

We are assuming that a, x, and y are each nonzero.

The next rule is used to simplify the nth power of a number that is raised to a power. For example, the expression $(5^2)^3$ means $5^2 \cdot 5^2 \cdot 5^2$. But,

$$5^2 \cdot 5^2 \cdot 5^2 = 5^{2+2+2}$$
$$= 5^6$$

Therefore,

$$(5^2)^3 = 5^{2 \cdot 3}$$

RULE

Exponent Rule 2

$$(a^m)^n = a^{m \cdot n}$$

The expression a^m raised to the power of n is the same as the base a raised to the product mn of the two exponents.

Example 3 Simplify each of the following:

(a) $(x^5)^7$ **(b)** $(x^3)^5(y^7)^2$

Solution

(a) $(x^5)^7 = x^{5 \cdot 7}$ Exponent Rule 2.
$= x^{35}$

(b) $(x^3)^5(y^7)^2 = x^{3 \cdot 5}y^{7 \cdot 2}$ Apply Exponent Rule 2 twice.
$= x^{15}y^{14}$

Consider the expression $(ab)^3$. This may be written as $(ab)(ab)(ab)$. But, by regrouping the terms,

$$(ab)(ab)(ab) = aaabbb$$
$$= a^3b^3$$

In general, we have Exponent Rule 3.

RULE

Exponent Rule 3

$$(ab)^n = a^nb^n$$

The product ab raised to the power of n is the same as the product of a^n and b^n.

Example 4 Simplify each of the following:

(a) $(xy)^5$ **(b)** $(2t)^3$

(c) $(xyz)^8$ **(d)** $(-a^6)^3$

Solution

(a) $(xy)^5 = x^5y^5$ Exponent Rule 3.

(b) $(2t)^3 = 2^3t^3$ Exponent Rule 3.

$= 8t^3$

(c) Exponent Rule 3 works for a product of three numbers: $(abc)^n = a^nb^nc^n$. Therefore,

$$(xyz)^8 = x^8y^8z^8$$

(d) $(-a^6)^3 = [(-1)a^6]^3$ Recall $-a = (-1)a$.

$= (-1)^3(a^6)^3$ $(ab)^n = a^nb^n$.

$= (-1)^3a^{6\cdot 3}$ $(a^m)^n = a^{mn}$.

$= -a^{18}$ $(-1)^3 = -1$.

We want the rule $a^ma^n = a^{m+n}$ to hold for *negative* integers as well as positive integers. If we set $m = -n$, then

$$a^{-n}a^n = a^{-n+n}$$
$$= a^0$$
$$= 1$$

Dividing both sides by a^n,

$$\frac{a^{-n}a^n}{a^n} = \frac{1}{a^n}$$
$$a^{-n} = \frac{1}{a^n}$$

This leads us to the next definition.

DEFINITION

$a^{-n} = \dfrac{1}{a^n}$, where a is any nonzero number

Example 5 Write each expression without negative exponents and simplify.

(a) 3^{-2} **(b)** 8^{-1}

(c) $(2x)^{-3}$ **(d)** $(u^2v)^{-4}$

Solution

(a) $3^{-2} = \dfrac{1}{3^2}$ Definition.

$= \dfrac{1}{3 \cdot 3}$

$= \dfrac{1}{9}$

(b) $8^{-1} = \dfrac{1}{8^1}$ Definition.

$= \dfrac{1}{8}$

(c) $(2x)^{-3} = \dfrac{1}{(2x)^3}$ Definition.

$= \dfrac{1}{2^3x^3}$ $(ab)^n = a^nb^n$.

$= \dfrac{1}{8x^3}$

(d) $(u^2v)^{-4} = \dfrac{1}{(u^2v)^4}$ Definition.

$= \dfrac{1}{(u^2)^4v^4}$ $(ab)^n = a^nb^n$.

$= \dfrac{1}{u^{2\cdot 4}v^4}$ $(a^m)^n = a^{m\cdot n}$.

$= \dfrac{1}{u^8v^4}$

We next develop a rule for the division of a^m by a^n. For example, consider the quotient $\dfrac{a^5}{a^2}$. This quotient can be written as a product.

$$\frac{a^5}{a^2} = a^5\left(\frac{1}{a^2}\right)$$

$$= a^5a^{-2} \qquad \frac{1}{a^n} = a^{-n}.$$

$$= a^{5-2} \qquad a^ma^n = a^{m+n}.$$

So,

$$\frac{a^5}{a^2} = a^{5-2} \quad \text{or} \quad a^3$$

This answer is correct, since

$$\frac{a^5}{a^2} = \frac{a \cdot a \cdot a \cdot \not{a} \cdot \not{a}}{\not{a} \cdot \not{a}}$$

$$= a \cdot a \cdot a$$

$$= a^3$$

Consider the quotient $\dfrac{a^3}{a^5}$. Notice that the exponent in the denominator, 5, is larger than the exponent in the numerator, 3. Nonetheless, we can write this quotient as a product.

$$\frac{a^3}{a^5} = a^3\left(\frac{1}{a^5}\right)$$

$$= a^3a^{-5} \qquad \frac{1}{a^n} = a^{-n}.$$

$$= a^{3-5} \qquad a^m a^n = a^{m+n}.$$

So, $\frac{a^3}{a^5} = a^{3-5}$ or a^{-2}, which can be written as $\frac{1}{a^2}$. This answer is correct, since

$$\frac{a^3}{a^5} = \frac{a \cdot a \cdot a}{a \cdot a \cdot a \cdot a \cdot a}$$

$$= \frac{1}{a \cdot a}$$

$$= \frac{1}{a^2}$$

This leads us to the next rule.

RULE

Exponent Rule 4

$$\frac{a^m}{a^n} = a^{m-n}$$

where $a \neq 0$, m and n are integers.

The expression a^m divided by a^n can be simplified by replacing the quotient by the common base a raised to the $m - n$ power.

Example 6 Simplify and write your answer without negative exponents.

(a) $\frac{5^6}{5^4}$ **(b)** $\frac{x^3}{x^5}$ **(c)** $\frac{2b^3}{b^{-2}}$

Solution **(a)** $\frac{5^6}{5^4} = 5^{6-4}$ Exponent Rule 4.

$= 5^2$

$= 25$

(b) $\frac{x^3}{x^5} = x^{3-5}$ Exponent Rule 4.

$= x^{-2}$

$= \frac{1}{x^2}$ $\quad a^{-n} = \frac{1}{a^n}$.

(c) $\dfrac{2b^3}{b^{-2}} = 2\dfrac{b^3}{b^{-2}}$

$= 2b^{3-(-2)}$ Exponent Rule 4.

$= 2b^{3+2}$

$= 2b^5$

COMMON ERROR

The quotient $\dfrac{a}{a^7}$ is *not* $\dfrac{1}{a^7}$. If we realize that $a = a^1$, then

$$\frac{a}{a^7} = \frac{a^1}{a^7} = a^{1-7} = a^{-6} = \frac{1}{a^6}$$

Consider the expression $\left(\dfrac{a}{b}\right)^3$. It means $\left(\dfrac{a}{b}\right)\left(\dfrac{a}{b}\right)\left(\dfrac{a}{b}\right)$. Therefore,

$$\left(\frac{a}{b}\right)^3 = \left(\frac{a}{b}\right)\left(\frac{a}{b}\right)\left(\frac{a}{b}\right)$$

$$= \frac{a \cdot a \cdot a}{b \cdot b \cdot b}$$

$$= \frac{a^3}{b^3}$$

In general, we have Exponent Rule 5.

RULE

Exponent Rule 5

If n is an integer,

$$\left(\frac{a}{b}\right)^n = \frac{a^n}{b^n}, \qquad b \neq 0$$

The quotient $\dfrac{a}{b}$ raised to an integer n is the same as the quotient of a^n and b^n.

Next, consider a negative exponent like $n = -2$.

$$\left(\frac{a}{b}\right)^{-2} = \frac{1}{\left(\dfrac{a}{b}\right)^2}$$

$$= \frac{1}{\frac{a^2}{b^2}}$$

$$= \frac{b^2}{a^2}$$

$$= \left(\frac{b}{a}\right)^2$$

This is an example of Exponent Rule 6.

RULE

Exponent Rule 6

If n is an integer,

$$\left(\frac{a}{b}\right)^{-n} = \left(\frac{b}{a}\right)^n, \qquad a \neq 0, b \neq 0$$

To simplify the expression $\left(\frac{a}{b}\right)^{-n}$, form the reciprocal of $\frac{a}{b}$, which is $\frac{b}{a}$, and raise it to the nth power.

Example 7 Simplify and write the answer without negative exponents.

(a) $\left(\frac{1}{3}\right)^{-3}$ **(b)** $\left(\frac{8}{9}\right)^2$

(c) $\left(\frac{t}{s}\right)^{-6}$ **(d)** $\left(\frac{7}{z}\right)^{-2}$

Solution **(a)** $\left(\frac{1}{3}\right)^{-3} = \left(\frac{3}{1}\right)^3$ Exponent Rule 6.

$= (3)^3$ $\frac{3}{1}$ is 3.

$= 27$

(b) $\left(\frac{8}{9}\right)^2 = \frac{8^2}{9^2}$ Exponent Rule 5.

$= \frac{64}{81}$

(c) $\left(\frac{t}{s}\right)^{-6} = \left(\frac{s}{t}\right)^6$ Exponent Rule 6.

$= \frac{s^6}{t^6}$ Exponent Rule 5.

(d) $\left(\frac{7}{z}\right)^{-2} = \left(\frac{z}{7}\right)^{2}$ Exponent Rule 6.

$= \frac{z^2}{7^2}$ Exponent Rule 5.

$= \frac{z^2}{49}$

For convenience, we list here all the definitions and rules of exponents. Assume that m and n are integers, and a and b are any real numbers.

Definitions and Rules of Exponents	
Definition	$a^0 = 1$, provided $a \neq 0$
Definition	$a^{-n} = \frac{1}{a^n}$, provided $a \neq 0$
Exponent Rule 1	$a^m a^n = a^{m+n}$
Exponent Rule 2	$(a^m)^n = a^{m \cdot n}$
Exponent Rule 3	$(ab)^n = a^n b^n$
Exponent Rule 4	$\frac{a^m}{a^n} = a^{m-n}$, provided $a \neq 0$
Exponent Rule 5	$\left(\frac{a}{b}\right)^n = \frac{a^n}{b^n}$, provided $b \neq 0$
Exponent Rule 6	$\left(\frac{a}{b}\right)^{-n} = \left(\frac{b}{a}\right)^n$, provided $a \neq 0$ and $b \neq 0$.

We show in the next example how two or more properties of exponents are sometimes needed to simplify an expression.

Example 8 Simplify and write the answer without negative exponents.

(a) $(x^2)^3(y^{-4}z)^5$ **(b)** $\left(\frac{s^2}{t^3}\right)^{-6}$ **(c)** $\frac{(a^{-5})^{-2}}{a^7}$

Solution **(a)** $(x^2)^3(y^{-4}z)^5 = x^{2 \cdot 3}y^{(-4) \cdot 5}z^{1 \cdot 5}$ Exponent Rules 2 and 3.

$= x^6y^{-20}z^5$

$= x^6\left(\frac{1}{y^{20}}\right)z^5$ Definition of negative exponent.

$= \frac{x^6z^5}{y^{20}}$

(b) $\left(\frac{s^2}{t^3}\right)^{-6} = \left(\frac{t^3}{s^2}\right)^6$ Exponent Rule 6.

$= \frac{(t^3)^6}{(s^2)^6}$ Exponent Rule 5.

$$= \frac{t^{3\cdot 6}}{s^{2\cdot 6}}$$ Exponent Rule 2 used twice.

$$= \frac{t^{18}}{s^{12}}$$

(c) $$\frac{(a^{-5})^{-2}}{a^7} = \frac{a^{(-5)(-2)}}{a^7}$$ Exponent Rule 2.

$$= \frac{a^{10}}{a^7}$$

$$= a^{10-7}$$ Exponent Rule 4.

$$= a^3$$

WARMING UP

Answer true or false.

1. $9^1 = 9$ **2.** $2^4 = 8$ **3.** $(-2)^2 = 4$ **4.** $x^5 \cdot x^5 = x^{10}$

5. $(k^2)^3 = k^5$ **6.** $(2x)^{-3} = \frac{8}{x^3}$ **7.** $\left(\frac{1}{3}\right)^{-2} = 9$ **8.** $(xy^2)^{-2} = \frac{1}{x^2y^4}$

EXERCISE SET 4.1

For Exercises 1–14, find the value of each numerical expression.

1. 2^1 **2.** 3^3 **3.** 6^2 **4.** 7^3

5. $(-3)^2$ **6.** $(-4)^3$ **7.** $(-1)^7$ **8.** $(-1)^8$

9. $\left(\frac{2}{3}\right)^3$ **10.** $\left(\frac{3}{5}\right)^2$ **11.** $(-6)^2$ **12.** $(-7)^2$

13. -6^2 **14.** -7^2

For Exercises 15–38, simplify. Assume that the variables are nonzero.

15. $x^2 \cdot x^7$ **16.** $b^3 \cdot b^4$ **17.** $y^{12}y^4zz^{25}$ **18.** $a^{20}a^3b^{10}b$

19. $(ax)^3(ax)^4$ **20.** $(bz)^2(bz)^6$ **21.** 2^0 **22.** $(st)^0$

23. $(x^2)^3$ **24.** $(a^5)^{10}$ **25.** $(b^4)^4$ **26.** $(z^3)^5$

27. $(5a)^2$ **28.** $(2b)^4$ **29.** $(x^2)^4x^3$ **30.** $a^7(a^3)^6$

31. $(-x)^5$ **32.** $(-t)^4$ **33.** $-(-a)^4$ **34.** $-(-r)^5$

35. $(3w^2)^3$ **36.** $(2v^3)^4$ **37.** $(2mn)^2$ **38.** $(4x^2y)^3$

For Exercises 39–48, write the expression without negative exponents and simplify.

39. 4^{-2} **40.** 3^{-3} **41.** 10^{-1} **42.** 9^{-1}

43. a^{-2} **44.** t^{-3} **45.** $(uv)^{-8}$ **46.** $(ax)^{-9}$

47. $(st^2)^{-3}$ **48.** $(4c)^{-2}$

For Exercises 49–80, simplify and write your answer without negative exponents.

49. $\frac{3^{10}}{3^8}$ **50.** $\frac{7^7}{7^6}$ **51.** $\frac{a^5}{2a}$

52. $\frac{4y^3}{y^2}$ **53.** $\frac{12^2}{12^3}$ **54.** $\frac{26^3}{26^4}$

55. $\frac{s}{s^2}$ **56.** $\frac{3p}{p^3}$ **57.** $\frac{18n^5}{6n^7}$

58. $\frac{21z^6}{3z^{10}}$ **59.** $\frac{m^3}{m^{-3}}$ **60.** $\frac{x^5}{x^{-6}}$

61. $\frac{x^3}{3^{-2}}$ **62.** $\frac{pq^2}{12^{-1}}$ **63.** $\frac{(az)^4}{(az)^5}$

64. $\frac{(bc^2)^{10}}{(bc^2)^{13}}$ **65.** $\left(\frac{2}{7}\right)^{-2}$ **66.** $\left(\frac{3}{2}\right)^{-3}$

67. $\left(\frac{1}{4}\right)^{-1}$ **68.** $\left(\frac{1}{9}\right)^{-2}$ **69.** $\left(\frac{a}{b}\right)^{-5}$

70. $\left(\frac{c}{n}\right)^{-7}$ **71.** $\left(\frac{2x}{z}\right)^{-3}$ **72.** $\left(\frac{3y}{u}\right)^{-2}$

73. $(x^{-2})^{-3}$ **74.** $(a^{-3})^{-4}$ **75.** $(5^{-3})^{-1}$

76. $(4^{-1})^{-2}$ **77.** $(y^3)^2(z^{-2}x)^4$ **78.** $(rv^3)^5(r^2v^{-1})^2$

79. $(t^{-2}w)^{-3}(tw^4)^{-1}$ **80.** $(m^2n^{-2})^{-3}mn^2$

For Exercises 81–86, give a reason why the statement is incorrect.

81. $2^5 \cdot 2^3 = 4^8$ **82.** $\frac{3^8}{3^2} = 3^4$ **83.** $(x^3)^4 = x^7$

84. $x^2 \cdot x^3 = x^6$ **85.** $\frac{x^6}{x^3} = x^2$ **86.** $x^3 + x^5 = x^8$

For Exercises 87–90, evaluate.

87. $[8(2.1)]^3$ **88.** $[-3.12(-5)^2]^2$

89. $-[(-1.6)(-27)2]^2$ **90.** $\frac{(3.2)^2}{4(-4)^2}$

SAY IT IN WORDS

91. State the six rules of exponents.

92. Explain why it makes sense to define $a^0 = 1$, where $a \neq 0$.

REVIEW EXERCISES

The following exercises review parts of Section 1.4. Doing these exercises will help prepare you for the next section.

For Exercises 93–96, multiply or divide.

93. 10^2 **94.** 10^3 **95.** $\frac{1}{10}$ **96.** $\frac{1}{100}$

ENRICHMENT EXERCISES

Simplify, where n is a positive integer.

1. $x^{2n}x^{6n}$

2. $(x^3)^n(x^4)^n$

3. $(ab^2)^n(a^2b)^n$

4. $\left(\frac{u^3}{v^2}\right)^{-n}$

Answers to Enrichment Exercises begin on page A.1.

4.2 Scientific Notation

OBJECTIVES

- *To write numbers in scientific notation*
- *To simplify numerical expressions using scientific notation*

LEARNING ADVANTAGE *When doing your assignment, write it all down; it is sometimes easier to do a mathematics problem in your head and then jot down a few of the steps and the answer. But the more you have on paper, the easier it is for another person to figure out what it is you are doing wrong and what it is you are doing right. So if others will be helping you later, they will need to be able to read your assignment.*

Many disciplines deal with numbers that are either very large or very small. For example, the Earth is approximately 93,000,000 miles from the sun. The probability of winning the Pennsylvania lottery is 0.00000009. It is inconvenient to use numbers containing large numbers of zeros in calculations. We can avoid this inconvenience by writing these types of numbers in *scientific notation.*

DEFINITION

A number is in **scientific notation** when it is of the form

$$a \times 10^n, \quad \text{where } 1 \leq a < 10 \text{ and } n \text{ is an integer}$$

We use the multiplication sign (×) to symbolize multiplication, since the raised dot (·) could be mistaken for a decimal point.

Example 1 Write the following in standard notation.

(a) $2.9 \times 10^3 = 2.9 \times 1{,}000 = 2{,}900$

(b) $7.6 \times 10^{-2} = 7.6 \times \frac{1}{10^2} = 7.6 \times \frac{1}{100}$

$$= \frac{7.6}{100} = 0.076$$

Example 2 Write 53,710 in scientific notation.

Solution First, locate the decimal point. It is understood to be to the right of the 0. To be in scientific notation, the decimal point must be to the right of the first nonzero digit, which for this number is between the 5 and the 3. When we multiply 5.371 by 10^4, we obtain the original number 53,710. The details look like this:

Place the decimal point here.

$$53{,}710 = 5.371 \times 10^4$$

Move 4 places to the *left*.

The exponent indicates the number of places to the *left* that we moved the decimal point.

Example 3 Write 0.00083 in scientific notation.

Solution We want the decimal point to be placed to the right of the first nonzero digit; that is, between the 8 and the 3. This means moving the decimal point four places to the right. If we multiply 8.3 by 10^{-4}, we revert back to the original form, since

$$8.3 \times 10^{-4} = 8.3 \times \frac{1}{10^4} = 8.3 \times \frac{1}{10{,}000}$$

$$= \frac{8.3}{10{,}000} = 0.00083$$

Here is the pattern.

Place the decimal point here.

$$0.00083 = 8.3 \times 10^{-4}$$

Move 4 places to the *right*.

The exponent gives the number of places to the *right* that we moved the decimal point.

Additional examples of numbers written in standard form and again in scientific notation are given in the following table. Verify that the two numbers on each line are equal.

The Number in Standard Notation	The Same Number in Scientific Notation
913,000	9.13×10^5
80,510	8.051×10^4
1,700	1.7×10^3
400	4×10^2
16	1.6×10^1 or 1.6×10
5.7	5.7×10^0 or 5.7
0.3002	3.002×10^{-1}
0.07	7×10^{-2}
0.0018	1.8×10^{-3}
0.0009	9×10^{-4}
0.0000162	1.62×10^{-5}

NOTE *Any number between 1 and 10, expressed as a decimal, is already in scientific notation. For example, the number 5.7 is the same in both columns of the preceding table.*

Writing numbers in scientific notation before evaluating a numerical expression makes the computation easier. This point is illustrated in the next example.

Example 4 Use scientific notation to find the value of

$$\frac{(0.004)(90{,}000)}{(0.0002)}$$

Solution We convert each of the three numbers into scientific notation, then use rules of exponents to simplify.

$$\begin{aligned}\frac{(0.004)(90{,}000)}{(0.0002)} &= \frac{(4 \times 10^{-3})(9 \times 10^4)}{2 \times 10^{-4}} \\ &= \frac{4 \cdot 9}{2} \times \frac{10^{-3}10^4}{10^{-4}} \\ &= 18 \times 10^{-3+4-(-4)} \qquad \text{Exponent rules.} \\ &= 18 \times 10^5 \\ &= 1{,}800{,}000\end{aligned}$$

TEAM PROJECT

(3 or 4 Students)

WRITING NUMBERS IN SCIENTIFIC NOTATION

Course of Action: Make a list of very large numbers. Each team member should contribute two numbers to this list. Take turns rewriting each number in scientific notation. Check each answer on a calculator if possible.

Now make a list of very small numbers. Each team member should contribute two numbers to this list. Take turns rewriting each number in scientific notation. Use a calculator to check each answer.

As a group, make a list of a mixture of very large and very small numbers. Each person should contribute two numbers. Rewrite each number in scientific notation.

Trade lists of numbers with another team. Solve the other team's problems. Correct each other's solutions using a calculator.

Group Report: How successful was the team in rewriting very large and very small numbers in scientific notation? What difficulties, if any, did the team encounter?

WARMING UP

Answer true or false.

1. $4 \times 10^3 = 4{,}000$

2. $3.2 \times 10^4 = 3{,}200$

3. $5.6 \times 10^{-1} = 0.56$

4. $7.92 \times 10^{-3} = 0.000792$

5. $10^6 \cdot 10^2 = 10^{12}$

6. $(2 \times 10^3) \times (6 \times 10^2) = 1.2 \times 10^6$

7. 42×10^{-3} is in scientific notation.

8. 9.7×10^6 is in scientific notation.

EXERCISE SET 4.2

For Exercises 1–14, write each number in standard notation.

1. 3×10^4

2. 7×10^3

3. 3×10^{-4}

4. 7×10^{-3}

5. 5.92×10^{-5}

6. 1.28×10^{-4}

7. 1.003×10^5

8. 6.027×10^4

9. 69.26×10^{-2}

10. 552.3×10^{-3}

11. 0.029×10^1

12. 9.001×10^1

13. 4.6×10^0

14. 92.8×10^0

For Exercises 15–30, write each number in scientific notation.

15. 3591

16. 8273

17. 23,981

18. 73,917

19. 7.4

20. 5.9

21. 0.0021

22. 0.00042

23. 0.00004

24. 0.003

25. 81.03

26. 9.002

27. 1,293,002

28. 394,001

29. 830.04

30. 90.3026

For Exercises 31–40, evaluate each term using scientific notation. Write your answer in standard notation.

31. $(0.0009)(20{,}000)$

32. $(800{,}000)(0.004)$

33. $\dfrac{9000}{0.0003}$

34. $\dfrac{0.0012}{60}$

35. $\dfrac{(0.0018)(1400)}{(60{,}000)(0.07)}$

36. $\dfrac{(800{,}000)(0.00003)}{(0.00004)(3000)}$

37. $\dfrac{(0.002)(11{,}000)}{0.022}$

38. $\dfrac{(40)(0.015)}{(0.03)(200)}$

39. $\dfrac{(50)(0.0009)}{(0.00005)(0.01)}$

40. $\dfrac{(0.00026)(40{,}000)}{(13{,}000)(0.0002)}$

For Exercises 41–44, write each number in scientific notation.

41. In four months, a pair of houseflies could produce about 190,000,000,000,000,000,000 descendants, if all of them lived.

42. Child care in the United States costs between \$50,000,000,000 and \$100,000,000,000 a year.

43. The human thyroid contains about 0.0002822 ounce of iodine.

44. The mass of an oxygen molecule is 0.0000000000000000000531 mg.

For Exercises 45–48, explain why the number is not in scientific notation, then write it in scientific notation.

45. 34×10^2

46. 83×10^3

47. 0.2×10^{-3}

48. 0.5×10^{-2}

When a number is too large or too small to be displayed on a calculator, scientific notation is used. For example, some calculators display numbers in scientific notation as 1.54 E 06 or 1.54 06. Each one means 1.54×10^6 or 1,540,000. Many calculators have a button labeled EE that is used to convert to scientific notation. Check your calculator's manual for details.

For Exercises 49–51, multiply and divide as indicated.

49. (93,845,000,000)(0.0000004723)

50. $\dfrac{0.0000008103}{0.000000058}$

51. $\dfrac{(0.000000482)(803{,}912{,}000{,}000)}{(0.000003371)(34{,}892{,}000)}$

SAY IT IN WORDS

52. Explain where scientific notation is used.

REVIEW EXERCISES

The following exercises review parts of Section 4.1. Doing these exercises will help prepare you for the next section.

For Exercises 53–62, simplify.

53. $x^2 \cdot x^6$

54. $c^3 \cdot c^2$

55. $r^4r^2t^5t$

56. $(s^2)^3$

57. $(4x)^2$

58. $(y^3)^3$

59. $(2v)^2$

60. $\left(\dfrac{3}{2}t^4\right)^2$

61. $(-z)^3$

62. $(-3h^2)^3$

ENRICHMENT EXERCISE

1. A light-year is the distance that light travels in one year, which is about 5.88×10^{12} miles. A parsec is equal to 3.26 light-years. Our galaxy has a diameter at its longest of 3×10^4 parsecs. What is the length, in miles, of this diameter?

Answers to Enrichment Exercises begin on page A.1.

4.3 Multiplying and Dividing Monomials

OBJECTIVES

- *To multiply monomials*
- *To divide monomials*

LEARNING ADVANTAGE

When doing your assignment, keep it neat and organized. You are the one who will have to use it for review at the end of the chapter. If you are careful and complete, you will not have any regrets at the last minute.

In this section, we study a special type of term called a *monomial.* Some examples of monomials follow:

5 is a constant.
y is a variable.
$3y^5$ is a product of a constant and a variable.
$-6xy^3z^2$ is a product of a constant and several variables raised to positive integer exponents.

DEFINITION

A **monomial** is a single term that is either a number or a number times one or more variables raised to positive integer exponents.

Additional examples of monomials are the following:

$$\frac{5}{4}, \quad 7x, \quad -2a^2b, \quad xy^3z, \quad \text{and} \quad -x$$

Examples of expressions that are *not* monomials are

$$\frac{a}{b}, \quad zx^{-2}, \quad \frac{3}{5x}, \quad \text{and} \quad a^2 - a$$

The numerical factor in a monomial is called the **numerical coefficient** or simply the **coefficient** of the monomial.

Example 1 (a) The coefficient of $2x$ is 2.

(b) The coefficient of $\frac{3x^2y}{5}$ is $\frac{3}{5}$, since $\frac{3x^2y}{5} = \frac{3}{5}x^2y$.

(c) The coefficient of $-a$ is -1, since $-a = (-1)a$.

Just as we use the operations of addition, subtraction, multiplication, and division on real numbers, we will perform these operations on monomials. First, we start with the multiplication of two monomials such as $7x$ and $2x^2y$. The product $(7x)(2x^2y)$ can be simplified using the associative and commutative properties that allow us to group together the coefficients as well as powers with the same base. Then, we multiply the coefficients and multiply the powers with the same base using the exponent rule $a^m a^n = a^{m+n}$.

$$\begin{aligned}(7x)(2x^2y) &= 7 \cdot 2(xx^2)y \\ &= 14x^{1+2}y \qquad xx^2 = x^1x^2 = x^{1+2}. \\ &= 14x^3y\end{aligned}$$

STRATEGY

Steps for Multiplying Two Monomials

Step 1 Multiply the coefficients.

Step 2 Multiply the variables, grouping together powers with the same base, then use $a^m a^n = a^{m+n}$ to simplify.

Example 2 Multiply $-2a^3b^6c^2$ by $5a^2b^4$.

Solution We first group the coefficients together, as well as the powers with the same base.

$$\begin{aligned}(-2a^3b^6c^2)(5a^2b^4) &= (-2)(5)(a^3a^2)(b^6b^4)c^2 \\ &= -10a^{3+2}b^{6+4}c^2 \qquad \text{Multiply coefficients and add exponents.} \\ &= -10a^5b^{10}c^2\end{aligned}$$

COMMON ERROR

When multiplying two monomials like $3a^2$ and $5a^7$, the answer is *not* $3 \cdot 5a^{2 \cdot 7}$. Make sure that you multiply the coefficients, but *add* the exponents. The correct form is

$$(3a^2)(5a^7) = 3 \cdot 5 \cdot a^{2+7} = 15a^9$$

In the next example, we show how the various rules of exponents can be used to simplify products of monomials.

Example 3 Simplify each expression.

(a) $(3x^2yz)\left(\frac{7}{12}xyz^2\right)(8x^4y^2z)$ **(b)** $(st^3)^2(5s^7t^4)$

Solution **(a)** For three monomials, we still multiply the coefficients and group the variables accordingly.

$$(3x^2yz)\left(\frac{7}{12}xyz^2\right)(8x^4y^2z) = 3 \cdot \frac{7}{12} \cdot 8(x^2xx^4)(yyy^2)(zz^2z)$$

$$= 14x^7y^4z^4 \quad \text{Multiply coefficients and add exponents.}$$

(b) First use the exponent rule $(ab)^n = a^nb^n$ with $n = 2$ to simplify $(st^3)^2$.

$$(st^3)^2(5s^7t^4) = s^2(t^3)^2(5s^7t^4)$$

$$= s^2t^{3\cdot 2}(5s^7t^4) \quad (a^m)^n = a^{m\cdot n}.$$

$$(= 1s^2t^6(5s^7t^4)) \quad \text{This step could be done mentally.}$$

$$= 1 \cdot 5(s^2s^7)(t^6t^4) \quad \text{Grouping.}$$

$$(= 5s^{2+7}t^{6+4}) \quad a^ma^n = a^{m+n}.$$

$$= 5s^9t^{10}$$

When dividing a monomial by a monomial, first divide the coefficients. Next, divide the variables using rules of exponents to simplify powers with common bases. For example, suppose we divide $6x^4y^{12}$ by $3x^3y^8$. Then,

$$\frac{6x^4y^{12}}{3x^3y^8} = \left(\frac{6}{3}\right)\left(\frac{x^4}{x^3}\right)\left(\frac{y^{12}}{y^8}\right)$$

$$= 2x^{4-3}y^{12-8} \quad \text{Divide coefficients and use the exponent rule } \frac{a^m}{a^n} = a^{m-n}.$$

$$= 2xy^4$$

STRATEGY

Steps for Dividing Two Monomials

Step 1 Divide the coefficients.

Step 2 Divide the variables, grouping together powers with the same base, then use rules of exponents to simplify.

Example 4 Divide $s^2t^5r^3$ by $-st^2r^6$. Write the answer without negative exponents.

Solution We first group the coefficients together as well as the powers with the same base.

$$\frac{s^2t^5r^3}{-st^2r^6} = -\left(\frac{s^2}{s}\right)\left(\frac{t^5}{t^2}\right)\left(\frac{r^3}{r^6}\right)$$

$$= -s^{2-1}t^{5-2}r^{3-6}$$ Divide the coefficients and subtract exponents.

$$= -st^3r^{-3}$$

$$= -\frac{st^3}{r^3}$$ Use $a^{-n} = \frac{1}{a^n}$.

NOTE *The quotient of two monomials need not be a monomial as shown in Example 4.*

We show in the next example how the various rules of exponents can be used to simplify products and quotients of monomials.

Example 5 Simplify $\frac{(2a)(3a^2b^3)}{6a^2b^5}$. Write the answer without negative exponents.

Solution

$$\frac{(2a)(3a^2b^3)}{6a^2b^5} = \left(\frac{2 \cdot 3}{6}\right)\left(\frac{aa^2}{a^2}\right)\left(\frac{b^3}{b^5}\right)$$ Group coefficients and powers with common bases.

$$= 1 \cdot ab^{3-5}$$ Simplify.

$$= ab^{-2}$$

$$= \frac{a}{b^2}$$ $b^{-2} = \frac{1}{b^2}$.

WARMING UP

Answer true or false.

1. The numerical coefficient of $4x$ is 4.
2. The numerical coefficient of $-x$ is -1.
3. For the monomial x, the numerical coefficient is 0.
4. $x + 1$ is a monomial.
5. $(2y)(3y^2) = 6y^3$
6. $\frac{x}{4}$ is a monomial.
7. $\frac{6x^3}{6x^2} = x$
8. The quotient of two monomials is always a monomial.

EXERCISE SET 4.3

For Exercises 1–10, name the numerical coefficient for each monomial.

1. $-3x$
2. $2y$

3. $\frac{a}{2}$

4. $\frac{7z}{3}$

5. $-\frac{2a^2}{5}$

6. $-\frac{3xy}{4}$

7. $-u$

8. $-ac$

9. st

10. v

11. Multiply $3x^2y^3$ by $2xy^2$.

12. Multiply $4ab^3$ by $3a^4b$.

13. Multiply $4st^2$ by $-2s^2t$.

14. Multiply $-5b^3c$ by $-bc^2$.

15. Divide $4s^3y^2$ by $2s^2y$.

16. Divide $6ab^3$ by ab.

17. Divide $8u^2v^3$ by $-4uv^2$.

18. Divide $-9a^3t^2$ by $3at$.

For Exercises 19–24, simplify each expression.

19. $(2xy^2)(-9x^2y)$

20. $(-3ab^4)(5a^2b^3)$

21. $(4v^2w)(-2vw^2)^2$

22. $(-m^2n)^3(3mn^3)$

23. $\left(\frac{1}{2}ab^2c\right)(2a^2bc^3)$

24. $(6x^2y^3z)\left(\frac{1}{3}xy\right)$

For Exercises 25–32, simplify each expression. Write your answer without negative exponents.

25. $\frac{3x^3y^2}{4x^6y}$

26. $\frac{-2ab^3}{6a^2b}$

27. $\frac{8x^2yz^4}{2xyz}$

28. $\frac{12s^3t^2u^2}{6s^2t^2u}$

29. $\frac{(uv^2)^3w^4}{(u^3w)^2v}$

30. $\frac{(a^2c^3)^2}{(b^2ac)^2a}$

31. $\frac{10(s^3t^2)^3(st^2)}{12(s^2t^3)(st^3)^2}$

32. $\frac{14(xy^3)^2(x^2y)}{8(x^2y)^3(xy^2)}$

SAY IT IN WORDS

33. What is a monomial?

34. What is your method for multiplying two monomials?

REVIEW EXERCISES

The following exercises review parts of Section 2.1. Doing these exercises will help prepare you for the next section.

For Exercises 35–40, simplify.

35. $4x - 2x + x$

36. $3y - 9y + 4y$

37. $4az - 3(2az - 1) + 3$

38. $5x^2 - 3x + 2 + (3x^2 + 4x - 6)$

39. $9s - 3r - 2 - 4s + r - 2s$

40. $7y^2 - 4y + 8 - (3y^2 + 4y - 1)$

ENRICHMENT EXERCISES

Simplify.

1. $a^n a^{2n}$

2. $(y^n)^3(y^3)^n$

3. $(x^2y)^n(xy^2)^n$

4. $\dfrac{(s^2t)^n}{s^n t^{2n}}$

Answers to Enrichment Exercises begin on page A.1.

4.4 Addition and Subtraction of Polynomials

OBJECTIVES

- ▶ *To recognize types of polynomials*
- ▶ *To find the degree of a polynomial*
- ▶ *To simplify polynomials*
- ▶ *To add polynomials*
- ▶ *To subtract polynomials*

LEARNING ADVANTAGE

When doing your assignments, be sure to correct them. This is the place to make mistakes. But do not stop there—diagnose your difficulties, get help as soon as possible, and do the problems over again until you get them right.

In the last section we talked about monomials. Recall that a monomial is either a number, a variable, or a product of numbers and variables raised to positive integral powers. In this section, we study *polynomials.*

A **polynomial** is the sum of monomials. For example,

$$3x^6 + 2x^3 + x + 5$$

is a polynomial, since it is the sum of four monomials. These monomials are usually called **terms.** The last term, 5, is called the **constant term.** The **degree of a term,** ax^n, is the exponent n. For example, the degree of $2x^3$ is 3 and the degree of 5 is 0, since $5 = 5 \cdot 1 = 5x^0$.

The **degree of a polynomial** is the highest degree of its terms. For example, the degree of $3x^6 + 2x^3 + x + 5$ is 6, since $3x^6$ is the term of highest degree. Notice that this polynomial has the term of highest degree written first with the other terms written in decreasing order of degrees. We say that this polynomial is in **standard form.**

Another example of a polynomial is

$$x^4 + (-8x^3) + (-x^2) + x + (-6)$$

which is usually written as

$$x^4 - 8x^3 - x^2 + x - 6$$

This is a fourth-degree polynomial with five terms and is in standard form.

Example 1 Write the polynomial in standard form and find the degree.

(a) $x^2 + 2x^4 + 6$ **(b)** $2x + 3$ **(c)** $-2 + 4x^3 - x + 6x^5$

Solution **(a)** Standard form: $2x^4 + x^2 + 6$; the degree is 4.
(b) Standard form: $2x + 3$; the degree is 1.
(c) Standard form: $6x^5 + 4x^3 - x - 2$; the degree is 5.

As stated earlier, a polynomial with one term is called a monomial. If a polynomial has two terms, it is called a **binomial.** A polynomial with three terms is a **trinomial.**

Example 2 Identify as a monomial, binomial, or trinomial. Find the degree.

(a) $4x^3$ **(b)** $8x - 2$ **(c)** $x^2 - 3x + 9$

Solution **(a)** Monomial; the degree is 3.
(b) Binomial; the degree is 1.
(c) Trinomial; the degree is 2.

To add two polynomials, use the commutative and associative properties to group the like terms together, then simplify.

Example 3 Add: $(x^3 + 4x^2 + 5) + (-7x^2 + 3x + 9)$.

Solution We group together like terms, then simplify.

$$\begin{aligned}(x^3 + 4x^2 + 5) + (-7x^2 + 3x + 9) &= x^3 + (4x^2 - 7x^2) + 3x + (5 + 9)\\ &= x^3 - 3x^2 + 3x + 14\end{aligned}$$

Recall that $-a = -1 \cdot a$; that is, the opposite of a is -1 times a. We can apply this property when a is a polynomial.

Example 4 Simplify $-(2x^2 - 5x + 3)$.

Solution We write $-(2x^2 - 5x + 3)$ as $(-1)(2x^2 - 5x + 3)$ and then distribute the -1 through the polynomial.

$$\begin{aligned}-(2x^2 - 5x + 3) &= (-1)(2x^2 - 5x + 3)\\ &= (-1)2x^2 + (-1)(-5x) + (-1)3\\ &= -2x^2 + 5x - 3\end{aligned}$$

Therefore, the opposite of $2x^2 - 5x + 3$ is $-2x^2 + 5x - 3$.

NOTE *From Example 4, we see that to find the* opposite *of a polynomial, change the sign of* every *term in the polynomial.*

Recall from Chapter 1 that subtraction $a - b$ is defined as $a + (-b)$. We use this same idea for subtraction of polynomials.

Example 5 Subtract: $(7x^3 - x^2 + 2x) - (3x^3 - 4x^2 + 3x)$.

Solution We rewrite the subtraction problem as an addition problem.

$$(7x^3 - x^2 + 2x) - (3x^3 - 4x^2 + 3x) = (7x^3 - x^2 + 2x) + (-3x^3 + 4x^2 - 3x)$$

Change the sign of each term of the second polynomial.

$$= (7x^3 - 3x^3) + (-x^2 + 4x^2) + (2x - 3x)$$

Group together like terms.

$$= 4x^3 + 3x^2 - x$$

Combine like terms.

Example 6 Simplify and write the answer in standard form.

$$(3x^2 - x + 5) - (4x^2 - 3x + 7) + (x^3 - x^2)$$

Solution

$$(3x^2 - x + 5) - (4x^2 - 3x + 7) + (x^3 - x^2)$$

$$= (3x^2 - x + 5) + (-4x^2 + 3x - 7) + (x^3 - x^2)$$

Take the opposite of each term of the middle polynomial.

$$= x^3 + (3x^2 - 4x^2 - x^2) + (-x + 3x) + (5 - 7)$$

Group together like terms.

$$= x^3 - 2x^2 + 2x - 2$$

Simplify.

In Chapter 1, we found the value of an algebraic expression when the variable was replaced by a number. We can *evaluate a polynomial* in the same way when a value of the variable is given.

Example 7 Evaluate $3a^2 + 2a - 5$ for $a = -1$.

Solution Replacing a by -1,

$$3a^2 + 2a - 5 = 3(-1)^2 + 2(-1) - 5$$
$$= 3 - 2 - 5$$
$$= -4$$

Example 8 A baseball is thrown vertically upward so that the distance s (in feet) from the ground after t seconds is

$$s = 5 + 56t - 16t^2$$

How far above the ground is the ball after two seconds?

Solution We replace t by 2 in the formula to find the value of s.

$$s = 5 + 56(2) - 16(2)^2$$
$$= 5 + 112 - 64$$
$$= 53$$

Therefore, the baseball will be 53 feet above the ground after two seconds.

WARMING UP

Answer true or false.

1. $x + 2x^2 - 4x^3$ is in standard form.
2. The degree of $x^3 - 2$ is 3.
3. $y^2 - 3y - 2$ is a trinomial.
4. $-(x^3 - 2x^2 - 3) = x^3 + 2x^2 + 3$
5. $(x + 2) + (-2x + 1) = -x + 3$
6. $(2x^2 - x + 3) - (x^2 - 1) = x^2 - x + 4$
7. The degree of $5x + 1$ is 0.
8. $x^2 + 3x + 7$ is a binomial.

EXERCISE SET 4.4

For Exercises 1–8, write the polynomial in standard form and find the degree.

1. $3x^4 + x^5 - 1$
2. $-t - t^2 + 12$
3. $x^2 + 5 - x^3 + 3x^4$
4. $7a^2 + a^3 - 1$
5. $-y^2 + 2 + y$
6. $n - n^2$
7. z
8. $-t$

For Exercises 9–16, identify as a monomial, binomial, or trinomial. Find the degree.

9. $2x^3 - x^2 + 4$
10. $y^4 - y^3 + 14$
11. $1 - t$
12. $3x^2 - 2$
13. $6x$
14. $-2x^3$
15. 4
16. -3

For Exercises 17–34, add the polynomials.

17. $(2x + 3) + (4x + 1)$
18. $(5x + 2) + (2x + 6)$
19. $(3x^2 - 4) + (2x^2 + 7)$
20. $(5y^2 - 7) + (-4y^2 + 1)$
21. $(4a^2 - 3a - 1) + (3a^2 + 4a + 5)$
22. $(6t^2 + 2t - 5) + (t^2 - 4t + 2)$
23. $(-z^3 + 2z^2 - 3z) + (4z^3 - 3z^2 + 9z)$
24. $(6s^4 - 3s^2 + s) + (2s^4 - s^2 + 4s)$
25. $(x^3 - 2x^2 + 1) + (4x^2 + 3)$
26. $(2y^3 + 3y + 5) + (-5y + 1)$
27. $(2 - 3x) + (5x - 1)$
28. $(y - 2y^2) + (y^2 + 2y)$
29. $(x^2 - 2x + 7) + (-6 + x - 2x^2)$
30. $(c^2 + 3c - 4) + (4c - 5 - 2c^2)$
31. $(x^3 + x^2 - 2) + (x^2 + 2x)$
32. $(z^3 - z + 1) + (z^3 + 4z^2)$
33. $(2.78x^2 - 4.92x + 7.26) + (3.09x^2 + 6.29x - 9.94)$
34. $(7.64y^2 + 2.91y - 5.02) + (-9.81y^2 + 4.34y + 6.91)$

For Exercises 35–38, simplify.

35. $-(2x^2 - 3x + 2)$
36. $-(3y^3 - 2y^2 + y)$

37. $-(-y + 2)$

38. $-(-2a^2 - a)$

For Exercises 39–52, perform the indicated subtraction.

39. $(3x^2 + 2x + 3) - (2x^2 + x + 1)$

40. $(4y^2 + 3y + 7) - (3y^2 + y + 4)$

41. $(6s^2 - 3s + 1) - (4s^2 + s + 3)$

42. $(5z^2 + 2z - 4) - (2z^2 + 3z + 5)$

43. $(8x^2 + 4x + 6) - (5x^2 - 2x + 1)$

44. $(7x^2 + 5x + 3) - (3x^2 - 4x + 5)$

45. $(2a^3 - 3a^2 + 1) - (-a^3 + 4a^2 - 2)$

46. $(5y^4 + 3y^2 - 7y) - (-y^4 + 4y^2 + y)$

47. $(3x + 1) - (2x^2 - x - 2)$

48. $(5y^2 - 2) - (3y^2 - 2y)$

49. $(-m^2 - 2m + 3) - (-4m - 5)$

50. $(-2x^3 + 5x) - (3x^2 + 7x)$

51. $(3.82x^2 - 4.96x - 3.08) - (2.96x^2 + 2.17x - 8.77)$

52. $(7.19y^2 + 3.42y - 6.52) - (-1.91y^2 - 5.47y + 2.93)$

For Exercises 53–58, simplify. Write your answer in standard form.

53. $(3a^2 - a + 5) - (6 - 2a + a^2) + (7a^2 - 4a + 12)$

54. $(u^5 - 7u^4 + u^2) - (u^3 - 2u^2) - (4u^5 - u^4)$

55. $(r^5 - r^3 + 7r^2) - (5 - 3r^2) - (r^4 + 2)$

56. $(-5x^4 + 7x^3) - (-3x^4 - 2x^2) - (x^4 + 9x^3 + x^2)$

57. $(s^3 - 7s^2 + 5) - (4s^3 - 8s) - (5s^3 - 7s^2 + 7)$

58. $(-v^4 - 4v^5) - (6 - 5v^5) - (14 + 8v^4 + v^2)$

59. The sides of a triangle are $3x^2 + x + 1$, $2x + 1$, and $5x$ inches. Express the perimeter as a polynomial in x.

60. The perimeter of a triangle is $6x^3 + 12x^2 + 10x - 2$ meters. If one side is $2x^3 + 7x^2$ meters and another is $4x^3 + 7x + 3$ meters, express the third side as a polynomial in x.

61. Evaluate $3x^3 - x^4 + 1$ for $x = 2$.

62. Evaluate $z - 10 + z^3$ for $z = 3$.

63. Evaluate $\frac{1}{2}a^2 + \frac{3}{4}a - 6$ for $a = 4$.

64. Evaluate $1.6b^2 - 290 + 15b$ for $b = 10$.

Many calculators have store STO and recall RCL keys. They are helpful when evaluating polynomials. See your calculator's manual for details.

For Exercises 65–68, evaluate.

65. $-8.096 + 2.1x^2$ for $x = 7.2$.

66. $45 - 2a^3$ for $a = 4$.

67. $7s^2 - 6s^4 + 12$ for $s = 3$.

68. $(8.3 - 5b^2)^2$ for $b = 2.4$.

69. The sum S of the first n positive integers is given by the formula $S = \frac{1}{2}n^2 + \frac{1}{2}n$. Find the sum of the first 60 positive integers.

70. The daily cost C (in dollars) of making x storage buildings is given by

$$C = 500x - 12x^2 + \frac{1}{3}x^3 + 600$$

Find the cost, if 12 storage buildings are made in a day.

REVIEW EXERCISES

The following exercises review parts of Section 4.3. Doing these exercises will help prepare you for the next section.

For Exercises 71–76, multiply.

71. $x^2 \cdot x^4$

72. $a^5 \cdot a^3$

73. $(2x^4)(3x)$

74. $(5y^3)(2y^3)$

75. $(-3x^2)(6x)$

76. $(-ab^2)(-a^2b)$

ENRICHMENT EXERCISES

Simplify.

1. $-[6 - (2 - x)]$

2. $5a - [2a - (a + 1)]$

3. $4z - [2 - (2z - 2)]$

4. $14 - [5c + 4 - (c - 3)]$

Answers to Enrichment Exercises begin on page A.1.

4.5 Multiplying Two Polynomials

OBJECTIVES

- *To multiply a monomial and a polynomial*
- *To multiply two polynomials*

LEARNING ADVANTAGE

Each day of a new chapter brings you one day closer to a chapter test. Be prepared for it by saving your assignments, correcting your mistakes, saving copies of quizzes, and asking questions in and out of class. From the textbook, class notes, and corrected assignment sheets, you should have an abundant supply of solved problems to help prepare you for the test.

To multiply a monomial and a polynomial, we use the distributive property, then simplify.

Example 1 Multiply:

(a) $4x^2(2x + 1)$ **(b)** $-2z(z^3 - 3z + 4)$

Solution **(a)** Apply the distributive property.

$$\begin{aligned} 4x^2(2x + 1) &= 4x^2(2x) + 4x^2(1) \\ &= 8x^3 + 4x^2 \end{aligned}$$

(b) Use the distributive property.

$$\begin{aligned} -2z(z^3 - 3z + 4) &= -2z(z^3) - 2z(-3z) - 2z(4) \\ &= -2z^4 + 6z^2 - 8z \end{aligned}$$

The distributive property can be used to multiply two binomials.

Example 2 Multiply: $(x + 2)(x + 3)$

Solution First distribute $x + 2$ through the two terms of the second binomial, then simplify. Here are the details:

$$\begin{aligned}(x + 2)(x + 3) &= (x + 2)x + (x + 2)3 && \text{Distribute } x + 2.\\ &= x^2 + 2x + 3x + 6 && \text{Use the distributive property again.}\\ &= x^2 + 5x + 6 && \text{Combine like terms.}\end{aligned}$$

We can use the method in Example 2 to multiply a binomial and a trinomial.

Example 3 Multiply:

$$(x - 2)(x^2 + 3x - 4).$$

Solution

$$\begin{aligned}&(x - 2)(x^2 + 3x - 4)\\ &= (x - 2)x^2 + (x - 2)3x + (x - 2)(-4) && \text{Distribute } x - 2.\\ &= x^3 - 2x^2 + 3x^2 - 6x - 4x + 8 && \text{Distribute again (three times).}\\ &= x^3 + x^2 - 10x + 8 && \text{Combine like terms.}\end{aligned}$$

We can also multiply two polynomials using a column form.

Example 4 Multiply:

$$(2x - 3)(3x^2 - x + 2)$$

Solution

$$\begin{array}{rrrrl} & 3x^2 & - \; x & + 2 & \\ & & 2x & - 3 & \\ \hline & -\; 9x^2 & + 3x & - 6 & \quad -3 \text{ times the top row.}\\ 6x^3 & -\; 2x^2 & + 4x & & \quad 2x \text{ times the top row.}\\ \hline 6x^3 & - 11x^2 & + 7x & - 6 & \quad \text{Add like terms that are lined up in columns.}\end{array}$$

These examples point to the rule.

RULE

Multiplying Polynomials

When multiplying two polynomials, multiply each term of the first polynomial by each term of the second polynomial.

Example 5 Multiply: $(x + 3)(x + 2)$.

Solution We multiply each term in the first binomial by each term in the second.

$$\begin{aligned}(x + 3)(x + 2) &= x \cdot x + x \cdot 2 + 3 \cdot x + 3 \cdot 2\\ &= x^2 + 2x + 3x + 6\\ &= x^2 + 5x + 6 && \text{Combine like terms.}\end{aligned}$$

We can obtain the same answer for the product $(x + 3)(x + 2)$ using a technique known as the FOIL method.

Multiply the First terms $(x + 3)(x + 2)$: x^2
Multiply the Outer terms $(x + 3)(x + 2)$: $2x$
Multiply the Inner terms $(x + 3)(x + 2)$: $3x$
Multiply the Last terms $(x + 3)(x + 2)$: 6

Adding the results, we obtain the answer.

$$\begin{aligned}(x + 3)(x + 2) &= x^2 + 2x + 3x + 6\\ &= x^2 + 5x + 6\end{aligned}$$

We summarize the FOIL method using the following diagram:

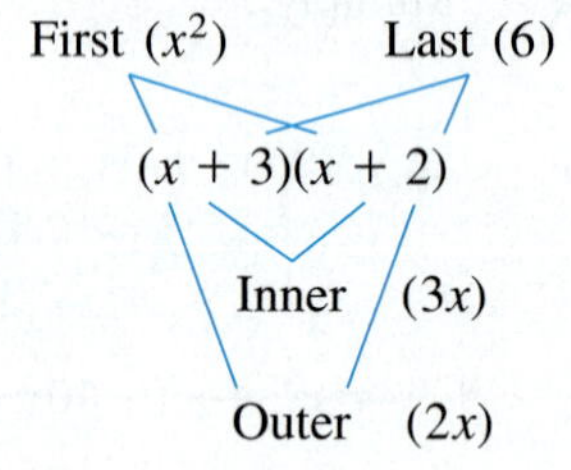

NOTE *The FOIL method can be used only when multiplying two* binomials.

The FOIL method gives us a system for obtaining all of the terms in a product of two binomials. Since multiplying binomials quickly will be important in factoring, we will do some of the multiplication mentally. We illustrate this technique in the next example.

Example 6 Multiply.

(a) $(x - 4)(x + 1)$ **(b)** $(3t - 2)(2t - 1)$
(c) $(5b + 3)(b - 2)$ **(d)** $(2x - 3y)(3x + 2y)$

Solution We multiply using the FOIL method.

(a) $(x - 4)(x + 1) = \overset{\text{F}}{x^2} + \overset{\text{O}}{x} - \overset{\text{I}}{4x} - \overset{\text{L}}{4}$
$= x^2 - 3x - 4$

(b) $(3t - 2)(2t - 1) = \overset{\text{F}}{6t^2} - \overset{\text{O}}{3t} - \overset{\text{I}}{4t} + \overset{\text{L}}{2}$
$= 6t^2 - 7t + 2$

(c) $(5b + 3)(b - 2) = \overset{\text{F}}{5b^2} - \overset{\text{O}}{10b} + \overset{\text{I}}{3b} - \overset{\text{L}}{6}$
$= 5b^2 - 7b - 6$

(d) $(2x - 3y)(3x + 2y) = \overset{\text{F}}{6x^2} + \overset{\text{O}}{4xy} - \overset{\text{I}}{9xy} - \overset{\text{L}}{6y^2}$
$= 6x^2 - 5xy - 6y^2$

With practice, you will be able to multiply two binomials by simply writing down the answer.

Example 7 Express the area of the shaded region in terms of x.

Solution The area A of the shaded region is the area of the larger rectangle minus the area of the smaller rectangle. The larger rectangle has a width of $8 + x$ inches and a length of $16 + x$ inches. The smaller rectangle is 8 inches by 16 inches. Therefore,

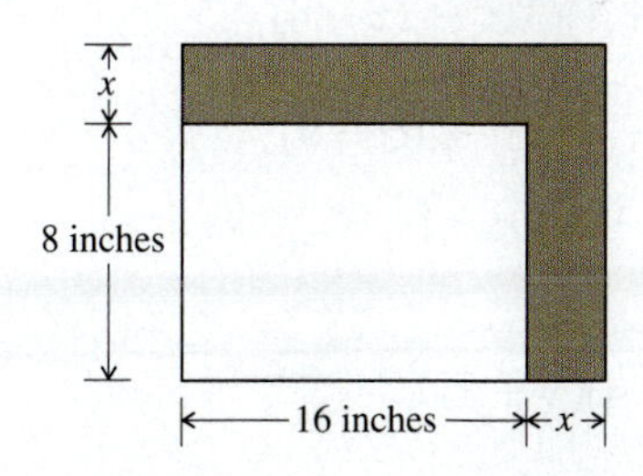

$$\begin{aligned} A &= \text{area of larger rectangle} - \text{area of smaller rectangle} \\ &= (8 + x)(16 + x) - 8 \cdot 16 \\ &= 8 \cdot 16 + 8x + 16x + x^2 - 8 \cdot 16 \\ &= 24x + x^2 \text{ square inches} \end{aligned}$$

WARMING UP

Answer true or false.

1. $x^2(1 + x)$ simplifies to $x^2 + x^3$.
2. The product of b and $b^2 + b - 2$ is $b^3 + b - 2$.
3. The answer to the multiplication problem $(x + 2)(x + 4)$ is $x^2 + 6x + 6$.
4. By the distributive property, $-2x^2(-x^2 + 3x - 2)$ is equal to $2x^4 - 6x^3 + 4x$.
5. The product of $y - 2$ and $y + 3$ is $y^2 + y - 6$.
6. The product of $2x + 1$ and $3x + 2$ is $6x^2 + 7x + 2$.

EXERCISE SET 4.5

For Exercises 1–22, use the distributive property to multiply.

1. $x^2(2x^3 - x + 2)$
2. $y^2(3y^2 - 2y + 1)$
3. $2a^3(a + 3)$
4. $4t(t^2 - 3t + 5)$
5. $-3r^2(2r^2 - 3r + 1)$
6. $-2z(-z^3 + 3z^2 + z)$
7. $-m(-m^4 + 2m^2 - 5)$
8. $-k^2(k^3 - 2k^2 + 9k)$
9. $(3x^2 - 4x + 2)(-4)$
10. $(3t^2 - 3t - 1)(-2)$
11. $(2x^2 - x + 9)(x)$
12. $(-3c^2 + c - 1)(2c)$
13. $(-2d^2 + 5d - 3)(-3d)$
14. $(3m^2 - 2m + 2)(-4m^2)$
15. $3t\left(\frac{1}{3}t^2 + 2\right)$
16. $4x^2\left(\frac{1}{4}x - 1\right)$
17. $10u(0.1u^2 - 2)$
18. $-10y(-0.2y + 1)$

19. $x(x^2 + y^2)$

20. $a(b - a)$

21. $3.2x^2(8.2x - 1.9)$

22. $2.9y(-3.4y^2 + 0.9y)$

For Exercises 23–28, multiply.

23. $(x + 4)(x^2 + 3x + 1)$

24. $(y + 2)(3y^2 + 3y + 4)$

25. $(2x + 1)(3x^2 - 2x - 4)$

26. $(3x + 4)(2x^2 + x - 5)$

27. $(3t - 4)(t^2 - 3t - 7)$

28. $(2z - 3)(2z^2 + 6z - 9)$

For Exercises 29–54, multiply.

29. $(x + 3)(x + 1)$

30. $(x + 5)(x + 2)$

31. $(y + 4)(y + 2)$

32. $(z + 3)(z + 6)$

33. $(a - 2)(a + 5)$

34. $(x + 4)(x - 1)$

35. $(t - 4)(t - 2)$

36. $(w - 7)(w - 3)$

37. $(x - 2)(x + 3)$

38. $(y - 5)(y - 4)$

39. $(a + 10)(a - 10)$

40. $(z - 6)(z + 6)$

41. $(t^2 - 7)(t^2 - 5)$

42. $(x^3 + 4)(x^3 + 2)$

43. $(2v + 3)(3v + 1)$

44. $(5y + 4)(6y + 2)$

45. $(3a + 4)(7a - 1)$

46. $(6n - 1)(6n + 2)$

47. $(3s - 1)(2s - 1)$

48. $(4r + 3)(r - 6)$

49. $(3x - 2)(3x + 2)$

50. $(2y + 3)(2y - 3)$

51. $(x + 3)(x + 3)$

52. $(y - 2)(y - 2)$

53. $(3z - 1)(3z - 1)$

54. $(2a + 5)(2a + 5)$

55. A rectangular rug has a width of $x + 2$ feet and a length of $2x + 1$ feet as shown in the figure. Express the area as a polynomial in x.

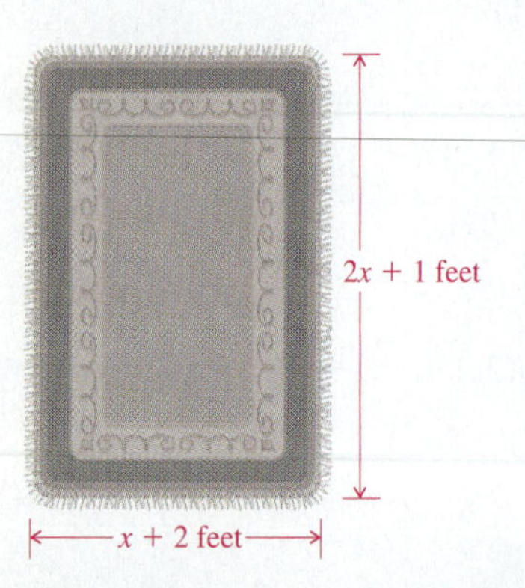

56. The triangular sail of a boat has a height of x feet and a base of $4x - 2$ feet as shown in the figure. Express the area as a polynomial in x.

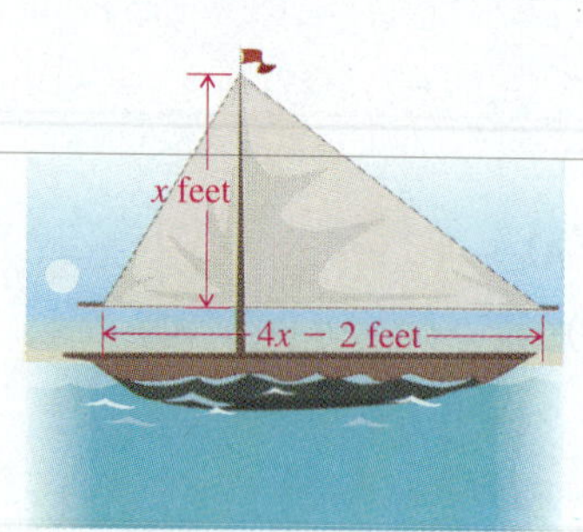

57. Find the area of the shaded region in the figure below.

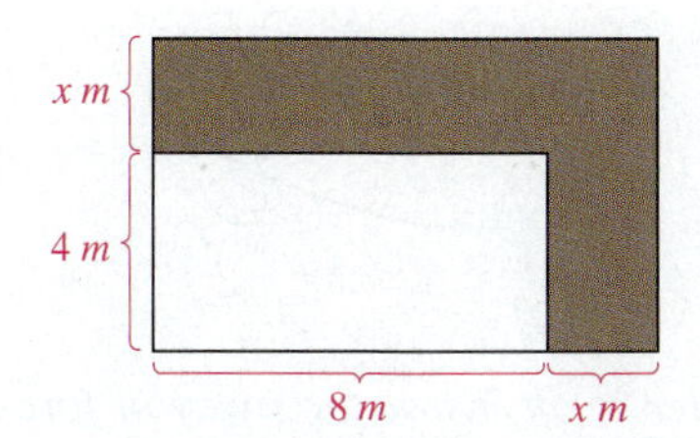

58. Find the area of the shaded region in the figure below.

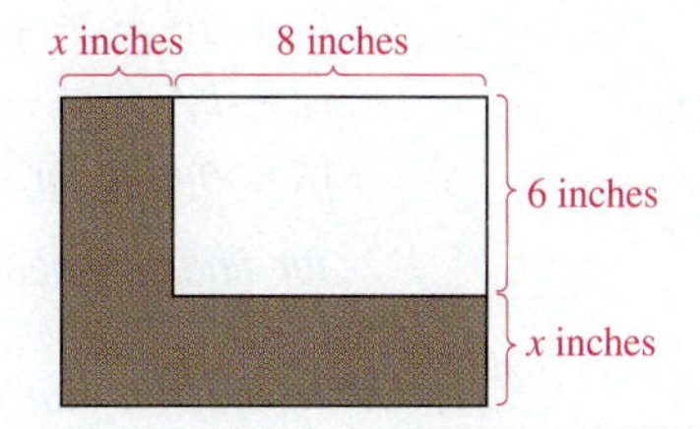

59. The city of Allentown is planning to expand the area of a rectangular rose garden by increasing the width by x feet and the length by $2x$ feet. If the original garden measures 20 feet by 40 feet, express the additional area of the expanded garden as a polynomial in x.

60. The Snyders have a rectangular back porch that measures 8 feet by 20 feet. They want to expand the porch by increasing the width by x feet and increasing the length also by x feet. Express the additional area of the expanded porch as a polynomial in x.

SAY IT IN WORDS

61. Explain your method for multiplying two binomials.

62. How do you multiply a binomial and a trinomial?

REVIEW EXERCISES

The following exercises review parts of Section 1.6. Doing these problems will help prepare you for the next section.

For Exercises 63–66, write each word phrase as an algebraic expression.

63. The sum of x and 2, the result squared.

64. Take the difference of a and b, then square the result.

65. The sum of the squares of x and y.

66. The square of the sum of x and y.

ENRICHMENT EXERCISES

For Exercises 1–5, multiply and simplify.

1. $(x + 1)^3$

2. $(x + y)^2$

3. $(x + y)(x - y) - (x - y)^2$

4. $\left(x + \frac{1}{2}\right)\left(x - \frac{1}{2}\right)$

5. $(x - y)(x^2 + xy + y^2)$

Answers to Enrichment Exercises begin on page A.1.

4.6 Special Products

OBJECTIVE

▶ *To memorize and use the rules*

$(a + b)^2 = a^2 + 2ab + b^2$
$(a - b)^2 = a^2 - 2ab + b^2$
$(a + b)(a - b) = a^2 - b^2$

for finding special products

LEARNING ADVANTAGE

Giving quizzes during the teaching of a chapter is an instructor's way of forcing you to organize everything you have studied so far. Both you and the instructor can measure how much of it you have learned. Although quizzes are not something to look forward to with pleasure, they generally have a long-term positive effect because they encourage you to review your notes and assignments each day.

In the last section, we learned how to multiply two polynomials. Certain special products are so common that methods for obtaining the answers have been developed. For example, the square of the sum of a and b, $(a + b)^2$, can be found using the FOIL method.

$$\begin{aligned} (a + b)^2 &= (a + b)(a + b) \\ &= a^2 + ab + ba + b^2 && \text{FOIL.} \\ &= a^2 + ab + ab + b^2 && ba = ab. \\ &= a^2 + 2ab + b^2 && \text{Combine like terms.} \end{aligned}$$

Thus, we can state the following rule.

RULE

The Square of a Sum

$$(a + b)^2 = a^2 + 2ab + b^2$$
$$(\text{first} + \text{last})^2 = (\text{first})^2 + 2(\text{first})(\text{last}) + (\text{last})^2$$

Example 1 Use the Square of a Sum Rule to find the following special products:

(a) $(a + 3)^2$ **(b)** $(2y + 5z)^2$

Solution **(a)** Use the Square of the Sum Rule with b replaced by 3.

	First term squared.		Twice first times last.		Last term squared.
$(a + 3)^2 =$	a^2	$+$	$2(a)(3)$	$+$	3^2

$$= a^2 + 6a + 9 \qquad \text{Simplify.}$$

(b) Replace a by $2y$ and b by $5z$.

$$\begin{array}{llll} & \text{First term squared.} & \text{Twice first times last.} & \text{Last term squared.} \\ (2y + 5z)^2 = & (2y)^2 & + \ 2(2y)(5z) & + \ (5z)^2 \\ & = 4y^2 + 20yz + 25z^2 & \text{Simplify.} & \end{array}$$

Using the FOIL method.

$$\begin{aligned} (a - b)^2 &= (a - b)(a - b) & \\ &= a^2 - ab - ab + b^2 & \text{FOIL.} \\ &= a^2 - 2ab + b^2 & \text{Combine like terms.} \end{aligned}$$

This gives us the next rule.

RULE

The Square of a Difference

$$(a - b)^2 = a^2 - 2ab + b^2$$

$$(\text{first} - \text{last})^2 = (\text{first})^2 - 2(\text{first})(\text{last}) + (\text{last})^2$$

Example 2 Use the Square of a Difference Rule to find the special products.

(a) $(x - 3)^2$ **(b)** $(3u - 2v)^2$

Solution **(a)** $(x - 3)^2 = x^2 - 2(x)(3) + 3^2$

$= x^2 - 6x + 9$

(b) $(3u - 2v)^2 = (3u)^2 - 2(3u)(2v) + (2v)^2$

$= 9u^2 - 12uv + 4v^2$

COMMON ERROR

It is incorrect to square $a + b$ as $a^2 + b^2$; that is,

$$(a + b)^2 \neq a^2 + b^2$$

To see why this is incorrect, use 1 and 2 for a and b, respectively.

$$(1 + 2)^2 \neq 1^2 + 2^2, \text{ since}$$

$$9 \neq 5$$

Also,

$$(a - b) \neq a^2 - b^2$$

Consider the product $(a + b)(a - b)$. We can use the FOIL method to find this product.

$$\begin{aligned}(a + b)(a - b) &= a^2 - ab + ba - b^2 && \text{FOIL.}\\ &= a^2 - ab + ab - b^2 && ba = ab.\\ &= a^2 - b^2 && \text{Combine like terms.}\end{aligned}$$

Therefore, the resulting answer is the difference of two squares $a^2 - b^2$.

RULE

The Difference of Two Squares

$$(a + b)(a - b) = a^2 - b^2$$

$$(\text{first} + \text{last})(\text{first} - \text{last}) = (\text{first})^2 - (\text{last})^2$$

Example 3 Multiply using the Difference of Two Squares Rule.

(a) $(a + 7)(a - 7) = a^2 - 7^2$
$= a^2 - 49$

(b) $(4x + 5)(4x - 5) = (4x)^2 - 5^2$
$= 16x^2 - 25$

(c) $(2s + 3t)(2s - 3t) = (2s)^2 - (3t)^2$
$= 4s^2 - 9t^2$

NOTE *There is a "middle" term in the product $(a + b)(a - b)$. It is $0 \cdot ab$, which is zero.*

Although all of the products that we have found using our special rules could be obtained by using methods from the previous section, it is worthwhile to memorize and use these special product rules. These rules will play an important role in factoring, which is the subject of the next chapter.

We show in the following example how some of the steps in finding special products can be done mentally.

Example 4 **(a)** $(x + 5)(x - 5) = x^2 - 5^2$
$= x^2 - 25$

(b) $(4w + 3)^2 = (4w)^2 + 2(4w)(3) + 3^2$
$= 16w^2 + 24w + 9$

(c) $(3x - 2y)^2 = (3x)^2 - 2(3x)(2y) + (2y)^2$
$= 9x^2 - 12xy + 4y^2$

Example 5 (a) $(x + 2)(x - 2) - (x^2 - 3x + 1)$
(b) $(a + 1)^2 - (a - 1)^2$

Solution (a) We start by multiplying $(x + 2)(x - 2)$ which is a special product.

$$(x + 2)(x - 2) - (x^2 - 3x + 1) = x^2 - 4 - (x^2 - 3x + 1)$$
$$= x^2 - 4 - x^2 + 3x - 1$$ Change the sign of each term within the parentheses.
$$= 3x - 5$$ Combine like terms.

(b) Use the special formulas for the square of the sum and the square of the difference.

$$(a + 1)^2 - (a - 1)^2 = a^2 + 2a + 1 - (a^2 - 2a + 1)$$
$$= a^2 + 2a + 1 - a^2 + 2a - 1$$
$$= 4a$$ Combine like terms. ◀

We finish this section with two examples—one from geometry and the other a problem to find an unknown number.

Example 6 A triangle has a height of $2x - 3$ feet and a base of $2x + 3$ feet as shown in the following figure. Express the area A as a polynomial in x.

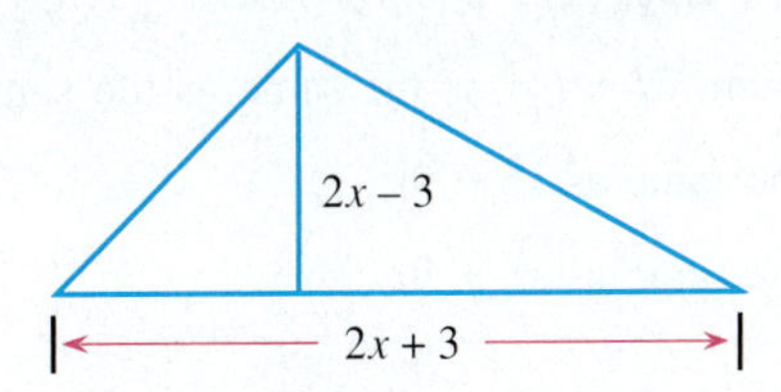

Solution The formula for the area of a triangle with base b and height h is $A = \frac{1}{2}bh$. Setting $b = 2x + 3$ and $h = 2x - 3$,

$$A = \frac{1}{2}(2x + 3)(2x - 3)$$
$$= \frac{1}{2}(4x^2 - 9)$$
$$= 2x^2 - \frac{9}{2}$$ ◀

Example 7 The square of the difference of a number and three is equal to the sum of the square of the number and ten. Find the number.

Solution Let

x = the unknown number

We next write the given sentence as an equation. The square of the difference of a number and three can be written as $(x - 3)^2$. The sum of the square of a number and ten can be written as $x^2 + 10$. These two expressions being equal give the equation

$$(x - 3)^2 = x^2 + 10$$

We now square on the left side, then use properties of equality to solve for x.

$$x^2 - 6x + 9 = x^2 + 10$$

$$x^2 - 6x + 9 - x^2 - 9 = x^2 + 10 - x^2 - 9$$ Subtract x^2 and 9 from both sides.

$$-6x = 1$$ Simplify each side.

$$x = -\frac{1}{6}$$ Divide each side by -6.

WARMING UP

Answer true or false.

1. The product of $x + 4$ and $x - 4$ is $x^2 - 16$.
2. The formula for the square of the sum is $(a + b)^2 = a^2 - 2ab + b^2$.
3. By a special products rule, $(x - 1)^2$ is $x^2 - 2x + 1$.
4. The square of the sum, $(a + b)^2$, is the same as the sum of the squares, $a^2 + b^2$.
5. $(x + 3)(x - 3)$ is the same as $x^2 - 9$.
6. $(x + 3)(x + 3)$ is the same as $x^2 + 9$.

EXERCISE SET 4.6

For Exercises 1–34, find each product.

1. $(x + 2)^2$
2. $(a + 5)^2$
3. $(z - 5)(z + 5)$
4. $(r - 7)(r + 7)$
5. $(t - 2)^2$
6. $(y - 10)^2$
7. $(2z + 4)(2z - 4)$
8. $(6t - 1)(6t + 1)$
9. $(3a - b)^2$
10. $(6s - 2t)^2$
11. $(5u - v)(5u + v)$
12. $(3r - 2s)(3r + 2s)$
13. $(a^3 + c)^2$
14. $(x^2 + 2y)^2$
15. $(3u^2 - 2c)^2$
16. $(4p - 5q^2)^2$
17. $(4z - a)^2$
18. $(3b - c)^2$
19. $(w - 10h)^2$
20. $(r - 9k)^2$
21. $(3h + a^2)^2$

22. $(5k + b^2)^2$

23. $(x^2 + 3u^2)^2$

24. $(5n^2 + m^2)^2$

25. $(2x - a)(2x + a)$

26. $(v - c)(v + c)$ $\quad v^2 - c^2$

27. $(z^2 + t)(z^2 - t)$

28. $(w + b^2)(w - b^2)$

29. $(s^3 - 2)(s^3 + 2)$ $\quad s^6 - 4$

30. $(n^4 - a)(n^4 + a)$

31. $(2.38x - 7.4)^2$

32. $(0.42a + 5.9)^2$

33. $(6.03y + 65.2)(6.03y - 65.2)$

34. $(0.38a + 0.92b)(0.38a - 0.92b)$

For Exercises 35–40, multiply and simplify.

35. $(x - 2)^2 - (x + 2)^2$

36. $(3y + 1)^2 - (3y - 1)^2$

37. $2a(a + 3) + (a - 3)^2$

38. $(2y - 3)^2 + 6(2y - 1)$

39. $(x + 1)^2 - (x + 1) - (x^2 - x)$

40. $(x + 2)^2 + (x + 2) - (x^2 + x)$

41. A rectangular driveway must be constructed as shown in the figure. Express the area A as a polynomial in x.

42. A carpenter is to make a rectangular door as shown in the figure. If the door is to have a width of $x + 1$ feet and a length of $\frac{3(x + 1)}{2}$ feet, express the area A of the door as a polynomial in x.

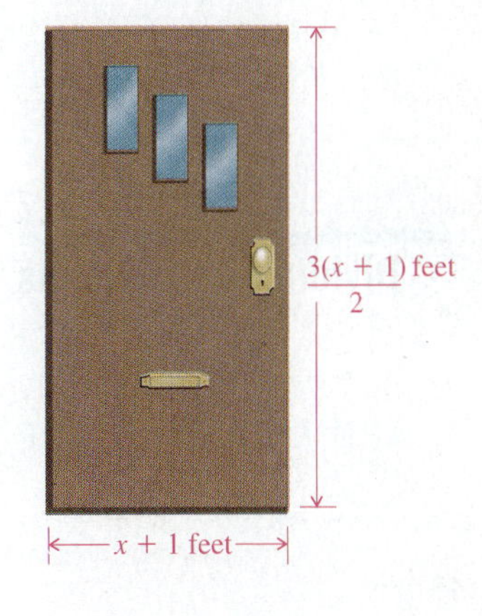

For Exercises 43–46, write each sentence as an equation using x as the unknown number. Then, solve the equation.

43. The difference of the square of a number and five is equal to the square of the sum of the number and two.

44. The sum of the square of a number and negative three is equal to the square of the sum of a number and one.

45. Find two consecutive even integers so that the difference of their squares is 196.

46. Find two consecutive odd integers so that the difference of their squares is 104.

SAY IT IN WORDS

47. In your own words, how do you find the square of a binomial?

48. Which way do you prefer to find the square of a binomial, the FOIL method or a special product rule?

REVIEW EXERCISES

The following exercises review parts of Section 4.2. Doing these exercises will help prepare you for the next section.

For Exercises 49–52, simplify. Write your answer without negative exponents.

49. $\dfrac{4x^5}{2x^2}$

50. $\dfrac{9y^4}{3y^3}$

51. $\dfrac{12a^4}{-8a^7}$

52. $\dfrac{-6z^5}{15z}$

ENRICHMENT EXERCISES

For Exercises 1–6, multiply.

1. $(a + 2)^3$

2. $(y - 1)^3$

3. $(2x + 3)^3$

4. $(s + t)^3$

5. $(u - w)^3$

6. $\left(\dfrac{1}{2}x + \dfrac{1}{3}\right)\left(\dfrac{1}{2}x - \dfrac{1}{3}\right)$

Answers to Enrichment Exercises begin on page A.1.

4.7 Dividing Polynomials

OBJECTIVES

- *To divide a polynomial by a monomial*
- *To divide a polynomial by a binomial*

LEARNING ADVANTAGE

If your instructor gives quizzes, take advantage of the situation by preparing for them; every minute you study and review now can save you two minutes of study time at the end of the chapter. Also, you may have the impression that (1) you do not know much of the mathematics that the class has studied, or (2) you can do all the problems pretty well. The results of a quiz will tell you whether your impression is right or wrong.

To divide a polynomial by a monomial, we use the property

$$\frac{a+b}{c} = \frac{a}{c} + \frac{b}{c}, \qquad c \neq 0$$

For example,

$$\frac{3x^4 + 2x^3}{x^2} = \frac{3x^4}{x^2} + \frac{2x^3}{x^2}$$
$$= 3x^2 + 2x$$

This example illustrates the following rule.

RULE

Dividing a Polynomial by a Monomial

To divide a polynomial by a monomial, divide *each term* of the polynomial by the monomial.

Example 1 Divide as indicated, then simplify.

(a) $\frac{2x^5 + 6x^4}{2x^2}$ **(b)** $\frac{z^4 - z}{z^3}$

Solution **(a)** $\frac{2x^5 + 6x^4}{2x^2} = \frac{2x^5}{2x^2} + \frac{6x^4}{2x^2}$

$$= x^{5-2} + 3x^{4-2}$$
$$= x^3 + 3x^2$$

(b) $\frac{z^4 - z}{z^3} = \frac{z^4}{z^3} - \frac{z}{z^3}$

$$= z - \frac{1}{z^2}$$

NOTE *As Example 1(b) shows, the division of polynomials need not result in a polynomial.*

We now look at a method for dividing a polynomial by a binomial. This method uses long division very similar to that of dividing two numbers.

Consider the problem of dividing $x^3 + x^2 - 5x + 5$ by $x + 3$. We first write this problem for long division and follow the instructions.

$$\begin{array}{r} x^2 \\ x + 3 \overline{) x^3 + x^2 - 5x + 5} \\ \underline{x^3 + 3x^2} \\ -2x^2 \end{array}$$

Round 1

Divide: $\frac{x^3}{x} = x^2$.

Multiply: $x^2(x + 3) = x^3 + 3x^2$.

Subtract: $x^3 - x^3 = 0$ and $x^2 - 3x^2 = -2x^2$.

Now, bring down the $-5x$ and continue the division process.

$$\begin{array}{r} x^2 - 2x \\ x + 3\overline{)x^3 + x^2 - 5x + 5} \\ \underline{x^3 + 3x^2} \\ -2x^2 - 5x \\ \underline{-2x^2 - 6x} \\ x \end{array}$$

Round 2

Divide: $\dfrac{-2x^2}{x} = -2x.$

Multiply: $-2x(x + 3) = -2x^2 - 6x.$

Subtract: $-2x^2 - (-2x^2) = 0$ and $-5x - (-6x) = x.$

Next, bring down the 5 and continue the division process.

$$\begin{array}{r} x^2 - 2x + 1 \\ x + 3\overline{)x^3 + x^2 - 5x + 5} \\ \underline{x^3 + 3x^2} \\ -2x^2 - 5x \\ \underline{-2x^2 - 6x} \\ x + 5 \\ \underline{x + 3} \\ 2 \end{array}$$

Round 3

Divide: $\dfrac{x}{x} = 1.$

Multiply: $1 \cdot (x + 3) = x + 3.$

Subtract: $x - x = 0$ and $5 - 3 = 2.$

Since the degree of 2 is less than the degree of $x + 3$, we stop the division process. The division of $x^3 + x^2 - 5x + 5$ by $x + 3$ is $x^2 - 2x + 1$ with a remainder of 2. We write this answer as

$$\frac{x^3 + x^2 - 5x + 5}{x + 3} = x^2 - 2x + 1 + \frac{2}{x + 3}$$

The general format is

$$\frac{\text{dividend}}{\text{divisor}} = \text{quotient} + \frac{\text{remainder}}{\text{divisor}}$$

Example 2 Divide $x^2 + x - 6$ by $x - 2$.

Solution

$$\begin{array}{r} x \phantom{{}+ x - 6} \\ x - 2\overline{)x^2 + x - 6} \\ \underline{x^2 - 2x} \phantom{{}- 6} \\ 3x - 6 \end{array}$$

Round 1

Divide: x goes into x^2, x times.

Multiply: $x(x - 2) = x^2 - 2x.$

Subtract: $x^2 + x - (x^2 - 2x) = 3x.$

Then, bring down the -6.

The remainder from Round 1 is $3x - 6$. Note that the degree of $3x - 6$ is *not* less than the degree of $x - 2$. We therefore go through the Divide-Multiply-Subtract sequence again.

$$\begin{array}{r} x + 3 \\ x - 2\overline{)x^2 + x - 6} \\ \underline{x^2 - 2x} \\ 3x - 6 \\ \underline{3x - 6} \\ 0 \end{array}$$

Round 2

Divide: x goes into $3x$, 3 times.

Multiply: $3(x - 2) = 3x - 6$.

Subtract: $(3x - 6) - (3x - 6) = 0$.

The remainder is 0.

Therefore, since the remainder is zero,

$$\frac{x^2 + x - 6}{x - 2} = x + 3$$

NOTE *In the long division process, notice that on each round, the* first *term is divided into the* first *term.*

Example 3 Divide $6x^2 + x - 7$ by $2x + 3$.

Solution

$$\begin{array}{r} 3x \\ 2x + 3\overline{)6x^2 + x - 7} \\ \underline{6x^2 + 9x} \\ -8x - 7 \end{array}$$

Round 1

Divide: $2x$ goes into $6x^2$, $3x$ times.

Multiply: $3x(2x + 3) = 6x^2 + 9x$.

Subtract: $(6x^2 + x) - (6x^2 + 9x) = -8x$.

Then, bring down the -7.

The remainder from Round 1 is $-8x - 7$. Note that the degree of the remainder is not less than the degree of $2x + 3$. We, therefore, go through the Divide-Multiply-Subtract sequence again.

$$\begin{array}{r} 3x - 4 \\ 2x + 3\overline{)6x^2 + x - 7} \\ \underline{6x^2 + 9x} \\ -8x - 7 \\ \underline{-8x - 12} \\ 5 \end{array}$$

Round 2

Divide: $2x$ goes into $-8x$, -4 times.

Multiply: $-4(2x + 3) = -8x - 12$.

Subtract: $(-8x - 7) - (-8x - 12) = 5$.

The remainder is 5.

The result of our long division can be written as

$$\frac{6x^2 + x - 7}{2x + 3} = 3x - 4 + \frac{5}{2x + 3}$$

NOTE *Once you are familiar with the long division method, you can use a more compact form. Example 3 would look like this.*

$$\begin{array}{r} 3x - 4 \\ 2x + 3\overline{)6x^2 + x - 7} \\ \underline{6x^2 + 9x} \\ -8x - 7 \\ \underline{-8x - 12} \\ 5 \end{array}$$

Example 4 Divide $5 - 2x + x^3$ by $x + 1$.

Solution We must first write the polynomial $5 - 2x + x^3$ in standard form.

$$5 - 2x + x^3 = x^3 - 2x + 5$$

Next, notice that there is no x^2 term in the dividend. Since all terms must appear, we insert $0 \cdot x^2$ in the proper place.

$$x^3 - 2x + 5 = x^3 + 0 \cdot x^2 - 2x + 5$$

Now we are ready to do the division.

$$\begin{array}{r} x^2 - x - 1 \\ x + 1 \overline{) x^3 + 0 \cdot x^2 - 2x + 5} \\ \underline{x^3 + x^2 } \\ -x^2 - 2x \\ \underline{-x^2 - x } \\ -x + 5 \\ \underline{-x - 1} \\ 6 \end{array}$$

Therefore,

$$\frac{5 - 2x + x^3}{x + 1} = x^2 - x - 1 + \frac{6}{x + 1}$$

WARMING UP

Answer true or false.

1. If $2x^2 + 3x$ is divided by x, the quotient is $2x + 3$.
2. The answer to the division of $35y^3 - 15y^2$ by $5y^2$ is $7y^2 - 3y$.
3. If $x^2 - x - 2$ is divided by $x + 1$, the remainder is 0.
4. If $2x^2 + 4x - 12$ is divided by $x - 2$, the remainder is -4.
5. In the expression $\dfrac{x^2 - 2x + 1}{x + 5}$, $x^2 - 2x + 1$ is the divisor.
6. If $2x^2 - x - 1$ is divided by $2x + 1$, the answer is $x - 1$.

EXERCISE SET 4.7

For Exercises 1–16, divide as indicated, then simplify your answer.

1. $\dfrac{3x^2 - 12x + 21}{3}$
2. $\dfrac{16a^2 + 120a - 64}{4}$
3. $\dfrac{7z^3 - 2z^2 - z}{z^2}$

4. $\frac{6n^4 + 2n^3 + n^2}{n^3}$

5. $\frac{12u^4 - 9u^3 + 3u^2}{3u^2}$

6. $\frac{36v^3 - 18v}{9v}$

7. $\frac{8m^3 - 4m^2 + 2m}{2m^3}$

8. $\frac{15p^4 + 12p^3 - 3}{3p^2}$

9. $\frac{2x^3 - 3x^2 + x}{-x^2}$

10. $\frac{-y^2 + 2y - 1}{-y}$

11. $\frac{t^4 - 8t^3}{-2t^3}$

12. $\frac{-z^5 + 6z^4}{-3z^2}$

13. $(4r^3 - 2r) \div 2r$

14. $(3a^2 - 6a) \div 3a$

15. $\frac{x^2y^3 - 2x^3y^4}{xy}$

16. $\frac{a^2b + 5a^3b^3}{ab}$

For Exercises 17–22, divide by using long division.

17. $\frac{x^2 + 3x - 10}{x + 5}$

18. $\frac{a^2 - a - 6}{a - 3}$

19. $\frac{z^2 - 3z - 40}{z - 8}$

20. $\frac{w^2 + 5w - 6}{w + 6}$

21. $\frac{x^2 - 8x - 15}{x - 2}$

22. $\frac{y^2 - 6y - 5}{y + 1}$

For Exercises 23–50, divide using long division. Be sure the dividends are in standard form with all terms appearing.

23. $\frac{x^2 + 4x + 7}{x + 1}$

24. $\frac{x^2 + 6x + 10}{x + 2}$

25. $\frac{x^2 - 4x - 10}{x + 3}$

26. $\frac{x^2 - 3x - 3}{x + 1}$

27. $\frac{y^2 + 2y - 8}{y - 2}$

28. $\frac{z^2 + 2z - 3}{z - 1}$

29. $\frac{x^3 - 3x^2 + x + 2}{x - 2}$

30. $\frac{x^3 + x^2 - 6x - 18}{x - 3}$

31. $\frac{x^3 + 2x^2 + 4x + 2}{x + 1}$

32. $\frac{y^3 + 4y^2 + 5y + 4}{y + 2}$

33. $\dfrac{2t^3 - 5t^2 + 2t - 12}{t - 3}$

34. $\dfrac{3x^3 + 4x^2 - 2x}{x + 1}$

35. $\dfrac{x^3 + 2x - 1}{x - 1}$

36. $\dfrac{x^3 - x + 7}{x + 2}$

37. $\dfrac{2x^2 - 12x + x^3 - 48}{x - 4}$

38. $\dfrac{3x - 3x^2 + x^3 - 2}{x - 2}$

39. $\dfrac{4 + 3x + x^3}{x + 1}$

40. $\dfrac{2 + x^3 - 3x}{x - 1}$

41. $\dfrac{4x^2 + 6x + 3}{2x + 1}$

42. $\dfrac{2x^2 - 3x + 4}{2x - 1}$

43. $\dfrac{3x^2 + x - 2}{3x - 2}$

44. $\dfrac{2x^2 - x - 6}{2x + 3}$

45. $\dfrac{4x^3 + 2x^2 - 8x + 1}{2x - 1}$

46. $\dfrac{6x^3 + 7x^2 + 5x - 1}{3x + 2}$

47. $\dfrac{6x^3 - x^2 + 1}{2x + 1}$

48. $\dfrac{9x^3 + 2x - 1}{3x - 1}$

49. $\dfrac{2x^2 + 4x^3 + 3 - 6x}{4x - 2}$

50. $\dfrac{3x - x^2 + 2 + 10x^3}{5x + 2}$

SAY IT IN WORDS

51. What is your method for dividing a polynomial by a monomial?

52. State the long division process.

REVIEW EXERCISES

The following exercises review parts of Section 4.1. Doing these exercises will help prepare you for the exercises that follow.

For Exercises 53–58, simplify.

53. a^3a^2 **54.** x^5x **55.** $2r^3r^6$

56. $5x^2(-3x^4)$ **57.** $-(-x)^3$ **58.** $y^3t^2y^5t$

ENRICHMENT EXERCISES

For Exercises 1 and 2, divide as indicated and simplify. Assume that n is a positive integer.

1. $\dfrac{x^{2n} - 2x^n}{x^n}$

2. $\dfrac{4x^{3n} + 6x^n}{2x^{2n}}$

3. $\dfrac{y^4 - 2y^3 + 2y^2 + y - 1}{y^2 - y}$

4. $\dfrac{4x^5 + 8x^4 + x^3 - x^2 + 3x + 1}{x^2 + x}$

Answers to Enrichment Exercises begin on page A.1.

CHAPTER 4 Summary and review

Definitions and rules of exponents (4.1)

Examples

If a is a number, n a positive integer, then

$$a^n = \underbrace{a \cdot a \cdot a \cdot \cdots \cdot a}_{n \text{ factors of } a}$$

Example: $2^3 = 2 \cdot 2 \cdot 2 = 8$

Definition: $a^0 = 1, \quad a \neq 0$ — Example: $\left(\frac{4}{3}\right)^0 = 1$

Definition: $a^{-n} = \frac{1}{a^n}, \quad a \neq 0$ — Example: $2^{-3} = \frac{1}{2^3} = \frac{1}{8}$

Exponent Rule 1: $a^m a^n = a^{m+n}$ — Example: $x^2x^3 = x^{2+3} = x^5$

Exponent Rule 2: $(a^m)^n = a^{mn}$ — Example: $(t^4)^2 = t^{4 \cdot 2} = t^8$

Exponent Rule 3: $(ab)^n = a^n b^n$ — Example: $(2x)^4 = 2^4x^4 = 16x^4$

Exponent Rule 4: $\frac{a^m}{a^n} = a^{m-n}, \quad a \neq 0$ — Example: $\frac{x^4}{x^3} = x^{4-3} = x$

Exponent Rule 5: $\left(\frac{a}{b}\right)^n = \frac{a^n}{b^n}, \quad b \neq 0$ — Example: $\left(\frac{3}{4}\right)^2 = \frac{3^2}{4^2} = \frac{9}{16}$

Exponent Rule 6: $\left(\frac{a}{b}\right)^{-n} = \left(\frac{b}{a}\right)^n, \quad a \neq 0 \text{ and } b \neq 0$ — Example: $\left(\frac{3}{4}\right)^{-2} = \left(\frac{4}{3}\right)^2 = \frac{16}{9}$

Scientific notation (4.2)

A number is in **scientific notation** when it is of the form $a \times 10^n$, where $1 \leq a < 10$ and n is an integer.

Examples: $3{,}450{,}000 = 3.45 \times 10^6$; $0.000081 = 8.1 \times 10^{-5}$

Multiplying and dividing monomials (4.3)

To **multiply two monomials,** multiply the numerical coefficients and add exponents.

Example: $(3x^2)(4x^5) = 12x^7$

Examples

$\frac{18x^5}{3x^2} = 6x^3$

To **divide two monomials,** divide the numerical coefficients and subtract exponents.

Addition and subtraction of polynomials (4.4)

A binomial: $x^2 - x$

A trinomial: $2a^3 + a^2 - 3$

The degree of $2x^4 - 3x^2 + 1$ is 4.

A polynomial is either a **monomial** or the sum of monomials. A **binomial** is a polynomial having two terms. A **trinomial** is a polynomial having three terms.

A polynomial is written in **standard form,** if the exponents decrease in value from left to right. The **degree of a polynomial** in one variable is the largest exponent to which the variable is raised.

$(2x + 3) + (4x - 1) = 6x + 2$

To **add two polynomials,** combine like terms.

To **find the opposite of a polynomial,** change the sign of each term in the polynomial.

$(a^3 - a^2 - 1) - (2a^3 + 3a^2 - 6)$
$= -a^3 - 4a^2 + 5$

To **subtract polynomials,** add the opposite of the second polynomial to the first.

Multiplying two polynomials (4.5)

$2x^2(x^3 - 4x) = 2x^5 - 8x^3$

$(x - 3)(x + 4)$
$= x^2 + 4x - 3x - 12$
$= x^2 + x - 12$

To **multiply two polynomials,** multiply each term of the first polynomial by each term of the second polynomial. When finding the product of a monomial and a polynomial, use the distributive property. The product of two binomials can be found using the FOIL method. Try to shorten the FOIL method by combining like terms mentally. Products of polynomials can also be found using the column format.

Special products (4.6)

Certain binomial products have special forms and can be multiplied according to the rules:

$(x + 2)^2 = x^2 + 4x + 4$

$$(a + b)^2 = a^2 + 2ab + b^2$$
$$(a - b)^2 = a^2 - 2ab + b^2$$

$(x + 3)(x - 3) = x^2 - 9$

$$(a + b)(a - b) = a^2 - b^2$$

Dividing polynomials (4.7)

$\frac{4x^3 - 3x^2}{2x^2} = 2x - \frac{3}{2}$

To **divide a polynomial by a monomial,** divide each term of the polynomial by the monomial.

$$\begin{array}{r} 3x + 1 \\ x - 1\overline{)3x^2 - 2x + 1} \\ \underline{3x^2 - 3x} \\ x + 1 \\ \underline{x - 1} \\ 2 \end{array}$$

To **divide a polynomial by a binomial,** use the long division process similar to that of dividing numbers.

CHAPTER 4 REVIEW EXERCISE SET

Section 4.1

For Exercises 1–4, simplify.

1. $x^5 \cdot x^2$

2. $(y^3)^4$

3. $(a^2b^3c)^2$

4. $(-2z^2t^3)^3$

For Exercises 5–10, simplify. Write your answer without negative exponents.

5. $2^3 \cdot 2^{-5}$

6. $\dfrac{16x^3y^2}{2x^5y^{-4}}$

7. $\dfrac{-12a^5b^{-2}}{3a^3b^2}$

8. $\dfrac{(9u^2v^3)(u^{-3}v^2)}{6(u^2v^4)^2}$

9. $\dfrac{(ab^3)^{-2}}{(a^2b)^{-1}}$

10. $\left(\dfrac{a^2}{b}\right)^{-2}\left(\dfrac{a^{-1}}{b^{-2}}\right)^{-1}$

Section 4.2

For Exercises 11 and 12, write the number in scientific notation.

11. A protein molecule consisting of the 19 basic amino acids could be built up in over 120,000,000,000,000,000 different ways.

12. One liter of ice water can dissolve 0.0005 ounce of oxygen.

13. Use scientific notation to find the value of

$$\frac{(0.000003)(160{,}000)}{0.0000006}$$

Section 4.3

For Exercises 14–17, simplify.

14. $(3x^2)(5x^3)$

15. $(2x^3y^2)(-4x^2y)$

16. $\dfrac{12x^4}{3x^3}$

17. $\dfrac{6a^3b^2c^5}{-2ab^4c^3}$

Section 4.4

For Exercises 18 and 19, write each polynomial in standard form.

18. $2 - x^2 + 4x$

19. $t^3 + t - 4 - t^4 + 5t^2$

20. Name the degree of $4x^3 - x^2 + 8x - 6x^4 + 1$.

For Exercises 21–26, simplify. Write your answer in standard form.

21. $(3x^2 - 2x + 1) + (5x^2 + 3x + 2)$

22. $(y^3 + 3y^2) + (2y^3 - 4y^2 + y)$

23. $(4x^2 + 3x - 5) - (2x^2 - 4x - 6)$

24. $(-t^2 - 4t + 2) - (-2t^2 + 2t + 5)$

25. $(3t - 4t^2 + t^3) + (4 - 2t + 2t^3)$

26. $(x^2 - 7x - 9) - (4x^2 + 5x + 6)$

Section 4.5

For Exercises 27–39, do the indicated multiplication, then simplify.

27. $z^2(-3z^4 + 6z^3)$

28. $3s(1 - s^3 + s^2)$

29. $(r - 2)(r - 1)$

30. $(x + 5)(x + 2)$

31. $(x - 3)(x + 2)$

32. $(2c - 4)(c - 2)$

33. $(4z + 3)(3z - 1)$

34. $(a + 4)(a + 4)$

35. $(x - 2t)(x + 4t)$

36. $(t + 2)(3t^2 - 3t + 1)$

37. $(2w + 3)(w^3 - w^2 - 3w)$

38. $(x^2 - 3)(x - 1)$

39. $(z - 3)(2z - 1) - (2z + 4)(z + 2)$

Section 4.6

For Exercises 40–46, find each product.

40. $(t - 3)(t + 3)$

41. $(a + 5)^2$

42. $(x - 6)^2$

43. $(3v - 2r)^2$

44. $(3t + 4)(3t - 4)$

45. $(4z^2 - 1)^2$

46. $(x^2 - 3y^2)^2$

Section 4.7

For Exercises 47–49, divide as indicated, then simplify your answer.

47. $\dfrac{4y^2 - 6y - 12}{2}$

48. $\dfrac{5x^4 + 8x^3}{x^2}$

49. $\dfrac{10c^5 - c^4}{c^5}$

For Exercises 50–54, divide using long division.

50. $\dfrac{x^2 + 7x + 12}{x + 3}$

51. $\dfrac{12x^2 + 5x + 3}{3x - 1}$

52. $\dfrac{y^3 - 27}{y - 3}$

53. $\dfrac{x^3 - 2x + 6}{x - 1}$

54. $\dfrac{5x + 6x^3 + 6 - 2x^2}{3x + 2}$

CHAPTER 4 TEST

1. Find the value of $\left(-\dfrac{2}{3}\right)^3$.

For Problems 2 and 3, simplify. Assume that the variables are nonzero.

2. $(x^3)^4x^5$

3. $(-4xy)^0$

For Problems 4 and 5, simplify. Write your answer without negative exponents.

4. $\left(\frac{4x}{y}\right)^{-2}$

5. $(a^2b^{-3}c)^{-2}(a^{-1}b^{-2})^{-1}$

6. Write 0.000301 in scientific notation.

7. Use scientific notation to find the value of

$$\frac{(3{,}000)(0.00008)}{(2{,}000{,}000)(0.006)}$$

8. Divide $16a^2bc^4$ by $-8ab^4c^4$.

For Problems 9 and 10, simplify. Write your answer without negative exponents.

9. $(2r^2st^4)(-9r^2s^4)(st^3)$

10. $\frac{(3x^2)(x^2yz^4)^3}{6x^2y^5}$

11. Find the degree of the polynomial: $2x^6 - x^3 + 3x^5 + x^8 + 5$

For Problems 12 and 13, simplify. Write your answer in standard form.

12. $(9r^5 - 2r^4 + r^3 + 2r^2 - 1) + (10r^4 - 3r^3 + r - 6)$

13. $(m^4 - 2m^3 + m) - (m^4 + 3m^3) - (m^3 + m + 2)$

For Problems 14–17, multiply, and then simplify.

14. $-4(c^3 - 2c - 5)$

15. $5x^3(x^5 - 2x^3 + 6)$

16. $(y - 2)(3y^2 - 2y + 5)$

17. $(a^4 - 3b)(a^4 - 4b)$

For Problems 18 and 19, find each product.

18. $(3x - 2)^2$

19. $(x + 2y)(x - 2y)$

20. Divide as indicated, then simplify your answer:

$$\frac{16x^6 - 12x^3 + 8x^2}{4x^2}$$

For Problems 21 and 22, perform the indicated division.

21. $\frac{x^3 - 3x + 2}{x + 2}$

22. $\frac{2x^2 + 5 - 4x + 3x^3}{3x - 1}$

CHAPTER 5 Factoring

CONNECTIONS

A photography student must reduce an 8- by 10-inch photograph by decreasing the length and width by the same amount. What will be the dimensions of the reduced print if it must have 30 percent the area of the original? See Enrichment Exercise 2 in Section 5.7.

Overview

Factoring a polynomial is the reverse of multiplying two or more polynomials. In this chapter, we will learn several methods of factoring. By developing these factoring techniques, we will be able to solve certain types of quadratic equations.

5.1 Greatest Common Factor and Factoring by Grouping

OBJECTIVES

- *To determine if a number is prime or composite*
- *To factor a number into primes*
- *To factor a monomial*
- *To factor out the greatest common monomial from a polynomial*

LEARNING ADVANTAGE

Each chapter ends with a Summary and Review that highlights the major points of the chapter. Also, there is a Review Exercise Set. Make these part of your assignment at least two nights before the test; then you will still have time to see your instructor if you have any questions.

In the last chapter, we developed methods of multiplying polynomials. The reverse of multiplication is called *factoring.* In this chapter, we will study methods to factor polynomials.

We start our discussion with factoring real numbers. If the number 84 is written as a product of other numbers, then these numbers are called **factors** of 84. For example, since $84 = 2 \cdot 42$, the numbers 2 and 42 are each factors of 84. Also,

$$84 = 4 \cdot 21 \qquad 84 = 2 \cdot 6 \cdot 7 \qquad 84 = 1 \cdot 14 \cdot 6$$

So there are many ways of factoring a number. We will be interested in factors that are *prime numbers.*

DEFINITION

A positive integer p, greater than one, is a **prime number** if the only positive integer factors are 1 and p. Otherwise p is called a **composite number.**

NOTE *The number 1 is considered neither prime nor composite.*

Examples of prime numbers:

$$2, 3, 5, 7, 11, 13, 17, 19, \ldots$$

Examples of composite numbers:

$$4, 6, 8, 9, 10, 12, 14, 15, \ldots$$

Both lists are infinite.

Example 1 Determine if the number is composite or prime.

(a) 27 **(b)** 1074 **(c)** 23

Solution **(a)** 27 is a composite number, since $27 = 3 \cdot 9$.

(b) 1074 is a composite number, since it is even and any even number has 2 as a factor.

(c) 23 is a prime number, since no positive integer greater than 1 but less than 23 is a factor. ◀

We say that a composite number is **factored into primes** if it is written as a product of prime numbers. To factor into primes, we begin by using any two factors of the number and continue factoring where possible.

Example 2 Factor 84 into primes.

Solution We may start with any two factors of 84. Since 84 is even, we divide by 2.

$$\begin{aligned} 84 &= 2 \cdot 42 \\ &= 2 \cdot 2 \cdot 21 && 42 = 2 \cdot 21. \\ &= 2 \cdot 2 \cdot 3 \cdot 7 && 21 = 3 \cdot 7. \\ &= 2^2 \cdot 3 \cdot 7 && 2 \cdot 2 = 2^2. \end{aligned}$$

◀

We now consider the problem of factoring a monomial from a polynomial. **Factoring out common factors** will be done using the distributive property. In the last chapter we used this property to multiply.

$$\text{To multiply: } 2(3x + 4) = 6x + 8$$

We will use the same property in the reverse direction to factor.

$$\text{To factor: } 6x + 8 = 2(3x + 4)$$

Multiplying changes a product of factors into a sum of terms and factoring changes a sum of terms into a product of factors.

To help us factor, we must find the **greatest common factor** of two or more integers or terms. For example, the greatest common factor of 4 and 6 is 2, since 2 is the largest factor common to both numbers.

The greatest common factor of two numbers such as 60 and 24 may not be seen immediately. If this happens, factor each number in primes and this will show you the greatest common factor.

$$\left.\begin{aligned}60 &= 2 \cdot 2 \cdot 3 \cdot 5\\ 24 &= 2 \cdot 2 \cdot 2 \cdot 3\end{aligned}\right\} \text{ } 2 \cdot 2 \cdot 3 \text{ or } 12 \text{ is the largest factor common to both numbers}$$

Therefore, the greatest common factor of 60 and 24 is 12. In a similar way, we can find the greatest common factor of two terms such as $8x^3$ and $12x^2y$.

$$8x^3 = 2 \cdot 2 \cdot 2 \cdot x \cdot x \cdot x$$
$$12x^2y = 2 \cdot 2 \cdot 3 \cdot x \cdot x \cdot y$$

The greatest common factor of $8x^3$ and $12x^2y$ is $2 \cdot 2 \cdot x \cdot x$ or $4x^2$.

$$8x^3 = 2 \cdot 2 \cdot 2 \cdot x \cdot x \cdot x = 4x^2(2x)$$
$$12x^2y = 2 \cdot 2 \cdot 3 \cdot x \cdot x \cdot y = 4x^2(3y)$$

Example 3 Find the greatest common factor of $20x^3y^2$, $12x^2y^3$, and $36x^4y^2$.

Solution By factoring the three monomials,

$$20x^3y^2 = 2 \cdot 2 \cdot 5 \cdot x \cdot x \cdot x \cdot y \cdot y = 4x^2y^2(5x)$$
$$12x^2y^3 = 2 \cdot 2 \cdot 3 \cdot x \cdot x \cdot y \cdot y \cdot y = 4x^2y^2(3y)$$
$$36x^4y^2 = 2 \cdot 2 \cdot 3 \cdot 3 \cdot x \cdot x \cdot x \cdot x \cdot y \cdot y = 4x^2y^2(9x^2)$$

we see that the greatest common factor is $4x^2y^2$.

Suppose the three monomials in Example 3 are the terms of a polynomial:

$$20x^3y^2 + 12x^2y^3 + 36yx^4y^2$$

The greatest common factor, $4x^2y^2$, of these terms is called the greatest common factor of the polynomial. We can factor out $4x^2y^2$ from this polynomial. Here are the details:

$$\begin{aligned}&20x^3y^2 + 12x^2y^3 + 36x^4y^2\\ &\quad = 4x^2y^2(5x) + 4x^2y^2(3y) + 4x^2y^2(9x^2) &&\text{Factor each term.}\\ &\quad = 4x^2y^2(5x + 3y + 9x^2) &&\text{Distributive property.}\end{aligned}$$

Example 4 Factor out the greatest common factor.

(a) $9x - 6$ **(b)** $12y^3 - 6y^2 + 18y$

Solution **(a)** The greatest common factor of $9x$ and -6 is 3. Therefore,

$$\begin{aligned}9x - 6 &= 3(3x) - 3(2)\\ &= 3(3x - 2)\end{aligned}$$

(b) The greatest common factor of the three terms, $12y^3$, $-6y^2$, and $18y$, is $6y$. Therefore,

$$\begin{aligned} 12y^3 - 6y^2 + 18y &= 6y(2y^2) - 6y(y) + 6y(3) \\ &= 6y(2y^2 - y + 3) \end{aligned}$$

NOTE *All polynomials have a greatest common factor, but it could just be 1. For example, the greatest common factor of $2x + 3$ is 1.*

With this strategy, we summarize the steps in factoring out the greatest common factor:

STRATEGY

Factoring Out the Greatest Common Factor from a Polynomial

Step 1 Factor out the greatest common number factor, if other than one. When necessary, first factor each coefficient into primes.

Step 2 For each variable, factor out the greatest common variable factor, if any. This factor is the variable raised to the smallest exponent appearing in the polynomial.

Suppose a polynomial that is in standard form has a negative coefficient on the first term. Then the negative sign (actually (-1)) is usually factored out with the greatest common factor.

Example 5 Factor out the greatest common factor along with -1 from $-2x^3 + 4x^2 - 6x$.

Solution The greatest common factor is $2x$. Therefore, we factor out $-2x$.

$$\begin{aligned} -2x^3 + 4x^2 - 6x &= -2x(x^2) - 2x(-2x) - 2x(3) \\ &= -2x(x^2 - 2x + 3) \end{aligned}$$

Be careful of the signs.

Consider factoring the polynomial

$$a(x + 2y) + b(x + 2y).$$

This polynomial has two terms with a common factor $(x + 2y)$. This common factor can be factored out

$$a(x + 2y) + b(x + 2y) = (x + 2y)(a + b)$$

The polynomial is now factored.

NOTE *The expression $a(x + 2y) + b(x + 2y)$ is a* sum *of two terms and $(x + 2y)(a + b)$ is a* product *of two terms.*

Example 6 Factor each polynomial.

(a) $x(u - v) + y(u - v)$
(b) $a^2(x^2 + y^2) + b^2(x^2 + y^2)$
(c) $r(a + b^2) + (a + b^2)$

Solution **(a)** $(u - v)$ is a common factor of the two terms. Therefore,

$$x(u - v) + y(u - v) = (u - v)(x + y)$$

(b) $(x^2 + y^2)$ is a factor common to the two terms.

$$a^2(x^2 + y^2) + b^2(x^2 + y^2) = (x^2 + y^2)(a^2 + b^2)$$

(c) Think of $(a + b^2)$ as $1 \cdot (a + b^2)$. Therefore,

$$\begin{aligned} r(a + b^2) + (a + b^2) &= r(a + b^2) + 1 \cdot (a + b^2) \\ &= (a + b^2)(r + 1) \end{aligned}$$

Consider factoring the polynomial

$$ax + ay + bx + by$$

Notice that a is a factor common to the first two terms and b is a factor common to the last two terms. If we group the first two terms and last two terms together,

$$\begin{aligned} ax + ay + bx + by &= (ax + ay) + (bx + by) \\ &= a(x + y) + b(x + y) && \text{Factor } a \text{ from the first grouping and factor } b \text{ from the second.} \\ &= (x + y)(a + b) && \text{Factor out the common factor } (x + y). \end{aligned}$$

The polynomial is now factored, and the technique used is called **factoring by grouping.**

Example 7 Factor by grouping.

(a) $bu + bv^2 + cu + cv^2$
(b) $x^2 + xd + ax + ad$
(c) $ax^2 + 2ay - x^2 - 2y$
(d) $ax - ay + b^2y - b^2x$

Solution **(a)**

$$\begin{aligned} bu + bv^2 + cu + cv^2 &= (bu + bv^2) + (cu + cv^2) \\ &= b(u + v^2) + c(u + v^2) \\ &= (u + v^2)(b + c) \end{aligned}$$

(b)

$$\begin{aligned} x^2 + xd + ax + ad &= (x^2 + xd) + (ax + ad) \\ &= x(x + d) + a(x + d) \\ &= (x + d)(x + a) \end{aligned}$$

(c) Start by grouping the first two terms together, but for the last two, factor out -1.

$$\begin{aligned} ax^2 + 2ay - x^2 - 2y &= (ax^2 + 2ay) - (x^2 + 2y) \\ &= a(x^2 + 2y) - 1 \cdot (x^2 + 2y) \\ &= (x^2 + 2y)(a - 1) \end{aligned}$$

(d) $ax - ay + b^2y - b^2x = (ax - ay) + (b^2y - b^2x)$

$$= a(x - y) + b^2(y - x)$$

Notice that $y - x = -(x - y)$. Therefore, the next step is

$$\begin{aligned} &= a(x - y) - b^2(x - y) \\ &= (x - y)(a - b^2) \end{aligned}$$

Sometimes there is more than one way to group the four terms. For instance, the polynomial in Example 7(c) can be factored by grouping the first and third terms together along with the second and fourth terms.

$$\begin{aligned} ax^2 + 2ay - x^2 - 2y &= (ax^2 - x^2) + (2ay - 2y) \\ &= x^2(a - 1) + 2y(a - 1) \\ &= (a - 1)(x^2 + 2y) \end{aligned}$$

Compare this result to the answer in Part (c). They are the same by the commutative property of multiplication.

WARMING UP

Answer true or false.

1. 18 is a composite number.

2. 31 is a prime number.

3. $32 = 2^4 \cdot 3$

4. $2x^2$ is the greatest common factor of $16x^3$ and $18x^2$.

5. 7 is the greatest common factor of 14 and 28.

6. The greatest common factor of 8, 12, and 16 is 4.

7. The greatest common factor of $4a^2b$ and $12ab^2$ is $12a^2b^2$.

8. The monomial $3x^2$ is the greatest common factor of $6x^3$, $9x^2$, and $12x^4$.

EXERCISE SET 5.1

For Exercises 1–6, determine if the number is composite or prime.

1. 9 **2.** 16 **3.** 29

4. 31 **5.** 1674 **6.** 2985

For Exercises 7–20, factor each composite number into primes. If the number is already prime, write "prime."

7. 42 **8.** 90 **9.** 36 **10.** 180

11. 525 **12.** 315 **13.** 37 **14.** 41

15. 70 **16.** 105 **17.** 231 **18.** 154

19. 43 **20.** 47

For Exercises 21–32, find the greatest common factor.

21. 4 and 16

22. 12 and 3

23. $6x^2$ and $9x$

24. $7a^3$ and $21a^2$

25. 6, 15, and 24

26. 12, 24, and 18

27. $3x^4$, $9x^5$, and $8x^3$

28. $6z^2$, $18z^3$, and $4z$

29. $12x^3y^4$ and $8x^2y^3$

30. $6a^2b^3$ and $15a^3b^2$

31. $18r^3t^4$, $54r^3t^3$, and $36r^4t^3$

32. $24x^5y^4$, $16x^4y^6$, and $40x^3y^4$

For Exercises 33–52, factor out the greatest common factor, if other than 1, from each polynomial.

33. $6x + 3$

34. $4y + 6$

35. $x^3 - x^2$

36. $y^4 + 2y^3$

37. $8a^2 + 12a^4$

38. $27m^3 - 9m$

39. $2x^2 + 8x + 4$

40. $3y^2 + 21y + 9$

41. $5t^2 - 25t + 10$

42. $4v^2 - 16v - 20$

43. $12b^2 + 28b + 72$

44. $9h^2 - 15h + 33$

45. $3s^4 - 5s + 7$

46. $14t^3 - 9t^2 + 23$

47. $36p^3 - 12p^2 + 30$

48. $45x^4 + 75x^2 - 30$

49. $2a^3 + 3a^2 - 11a$

50. $5z^6 - 3z^5 + 2z^3$

51. $14y^4 - 17y^3 + 15y^2$

52. $12b^7 - 13b^6 + 9b^5$

For Exercises 53–62, factor out the greatest common factor along with -1.

53. $-2x^2 + 10$

54. $-3y^2 + 21$

55. $-y^2 + 1$

56. $-a^3 + 8$

57. $-4x^2 + 8x - 4$

58. $-2t^2 - 6t + 4$

59. $-x^5 - 2x^4 + x^3$

60. $-3b^4 + 6b^3 - 3b^2$

61. $-4x^3 - 12x^2 - 18x$

62. $-9c^6 - 15c^5 + 18c^2$

For Exercises 63–68, factor out the greatest common factor.

63. $x(c + d) + z(c + d)$

64. $y(r^2 + t) + x(r^2 + t)$

65. $a(x^2 - 2) + b^2(x^2 - 2)$

66. $y^2(x + 5) + z(x + 5)$

67. $z(y + x) + (y + x)$

68. $m^2(a - b) + (a - b)$

For Exercises 69–80, factor by grouping.

69. $ax + ay^2 + bx + by^2$

70. $ya + yz + ba + bz$

71. $y^2 + yz + by + bz$

72. $x^2 + xt + ax + at$

73. $ct^2 + 3cx - t^2 - 3x$

74. $dx^3 + 2dy - x^3 - 2y$

75. $bx - by + cy - cx$

76. $az^2 - at + dt - dz^2$

77. $3a - 3b + a^2 - ab$

78. $2x^2 + 4y + x^4 + 2x^2y$

79. $a^3 - a^2 - xa + x$

80. $4u - 8v - au + 2av$

SAY IT IN WORDS

81. Explain how you factor out the greatest common factor from a polynomial.

82. Explain factoring by grouping.

83. What is the difference between multiplying and factoring?

REVIEW EXERCISES

The following exercises review parts of Sections 4.5 and 4.6. Doing these exercises will help prepare you for the next section.

For Exercises 84–89, multiply. Concentrate on combining like terms mentally.

84. $(x + 3)(x + 1)$

85. $(x + 4)(x - 3)$

86. $(y - 3)(y - 2)$

87. $(3x - 2)(x - 5)$

88. $(x + y)(x - 3y)$

89. $(a + 2b)(a - 2b)$

ENRICHMENT EXERCISES

Factor out the greatest common factor from each polynomial.

1. $27x^5 - 18x^4 + 36x^3 + 63x^2$

2. $12y^6 + 36y^4 - 84y^5 - 48y^3$

Factor out the greatest common factor, then factor by grouping.

3. $2ax^2 + 2ay^2 - 2bx^2 - 2by^2$

4. $x^3 - ax^2 - 2x^2 + 2ax$

Answers to Enrichment Exercises begin on page A.1.

5.2 Factoring Special Trinomials

OBJECTIVES

- *To factor a trinomial of the form $ax^2 + bx + c$, where $a = 1$*
- *To factor out the greatest common factor, then factor the resulting trinomial*

LEARNING ADVANTAGE

Your instructor is more willing to help you one week before a test than one day before a test. Ask for extra help early so that you have more time to review methods or materials.

In this section, we factor special trinomials of the form $ax^2 + bx + c$, where the leading coefficient a is 1. In Chapter 4, we multiplied two binomials, such as $x + 2$ and $x + 4$, to obtain a trinomial.

$$(x + 2)(x + 4) = (x + 2)x + (x + 2)4 \quad \text{Distributive property.}$$
$$= x^2 + 2x + 4x + 8 \quad \text{Distributive property used again (twice).}$$
$$= x^2 + 6x + 8 \quad \text{Combine like terms.}$$

Now to factor $x^2 + 6x + 8$, we will reverse our steps. Notice that the coefficient 6 is the sum of two numbers that have a product of 8. Namely, 2 and 4. Therefore, write $6x$ as $2x + 4x$.

$$x^2 + 6x + 8 = x^2 + 2x + 4x + 8$$

Next, we factor the right side by grouping.

$$x^2 + 6x + 8 = x^2 + 2x + 4x + 8$$
$$= (x^2 + 2x) + (4x + 8)$$
$$= x(x + 2) + 4(x + 2)$$
$$= (x + 2)(x + 4) \quad \text{Factor out } x + 2.$$

In summary, the numbers 2 and 4 have two properties:

1. The product of 2 and 4 is 8, the constant term of $x^2 + 6x + 8$.
2. The sum of 2 and 4 is 6, the coefficient of x in $x^2 + 6x + 8$.

From these observations, we can suggest a strategy for factoring a trinomial of the form $x^2 + bx + c$.

STRATEGY

Factoring the Trinomial $ax^2 + bx + c$, where $a = 1$

1. Find two integers whose product is c and whose sum is b.
2. Then,

$$x^2 + bx + c = (x + \text{the first integer})(x + \text{the second integer})$$

We illustrate this strategy in the next example.

Example 1 Factor.

(a) $x^2 + 5x + 6$ (b) $x^2 + 4x - 12$ (c) $a^2 - 9a + 20$

Solution (a) To have a product of 6 and a sum of 5, use 2 and 3. Therefore,

$$x^2 + 5x + 6 = (x + 2)(x + 3)$$

Check by multiplying.

(b) To have a product of -12 and a sum of 4, use 6 and -2. Therefore,

$$x^2 + 4x - 12 = (x + 6)(x - 2)$$

Check by multiplying.

(c) To have a product of 20 and a sum of -9, use -4 and -5. Therefore,

$$a^2 - 9a + 20 = (a - 4)(a - 5)$$

Check the answer by multiplying. ◀

NOTE *The order in which you write the two factors makes no difference. For example,* $x^2 + 4x - 12 = (x + 6)(x - 2)$*, which is the same as* $(x - 2)(x + 6)$*.*

Example 2 Factor $x^2 + 4x + 6$.

Solution We are looking for two positive integers having a product of 6 with a sum of 4. There are two possibilities, 1 and 6 or 2 and 3. But,

$$1 + 6 = 7 \qquad \text{and} \qquad 2 + 3 = 5$$

Since neither gave the correct sum of 4, we conclude that $x^2 + 2x + 6$ cannot be factored using integers. ◀

A polynomial such as $x^2 + 4x + 6$ of Example 2 that cannot be factored *using integers* is called a **prime polynomial.** A prime polynomial may be factorable using numbers that are not integers. This is a topic that is covered in later chapters.

We show in the next example how to factor a special trinomial containing more than one variable.

Example 3 Factor $x^2 + 2xy - 3y^2$.

Solution We apply the same technique as in the previous examples. Now, we want two numbers whose product is -3 and sum is 2. They are 3 and -1. Therefore,

$$x^2 + 2xy - 3y^2 = (x + 3y)(x - y)$$ ◀

A polynomial is **factored completely** when each factor is a prime polynomial. We show in the next example how to combine factoring methods to completely factor a polynomial.

Example 4 Factor completely: $2z^3 + 8z^2 - 42z$.

Solution For this polynomial, first factor out the greatest common factor. The greatest common factor is $2z$.

$$2z^3 + 8z^2 - 42z = 2z(z^2 + 4z - 21)$$

Next, we want to factor $z^2 + 4z - 21$, if possible. Using the factoring strategy of this section, we look for two integers whose product is -21 and sum is 4. The numbers -3 and 7 work. Therefore,

$$\begin{aligned} 2z^3 + 8z^2 - 42z &= 2z(z^2 + 4z - 21) \\ &= 2z(z + 7)(z - 3) \end{aligned}$$

As illustrated in the last example, we should first determine if the polynomial to be factored has a greatest common factor (other than 1).

STRATEGY

When factoring a polynomial, always check first for common factors.

WARMING UP

Decide if the factoring is correct (true) or incorrect (false).

1. $x^2 + 7x + 6 = (x + 2)(x + 3)$
2. $y^2 + 6y + 5 = (y + 5)(y + 1)$
3. $x^2 + 2x - 3 = (x + 3)(x - 1)$
4. $m^2 + 6m - 8 = (m - 4)(m - 2)$
5. $x^2 - 3x - 10 = (x - 5)(x + 2)$
6. $x^3 + 6x^2 + 8 = x(x + 2)(x + 4)$
7. $t^3 + 8t^2 + 7t = t(t + 1)(t + 7)$
8. $2m^4 - 2m^3 - 12m^2 = 2m^2(m - 3)(m + 2)$

EXERCISE SET 5.2

For Exercises 1–38, factor the trinomial. If it is prime, write "prime."

1. $x^2 + 6x + 8$
2. $y^2 + 10y + 21$
3. $t^2 + 9t + 14$
4. $k^2 + 3k + 2$
5. $a^2 + 10a + 16$
6. $x^2 + 11x + 24$
7. $y^2 + 9y + 20$
8. $u^2 + 13u + 36$
9. $c^2 + 13c + 30$
10. $x^2 + 7x - 8$
11. $t^2 + t - 2$
12. $n^2 + n - 6$

13. $x^2 + 7x + 4$

14. $z^2 + 6z + 4$

15. $p^2 + 2p - 15$

16. $x^2 - 7x - 8$

17. $s^2 - 4s + 3$

18. $x^2 - 7x + 12$

19. $y^2 - 13y + 30$

20. $y^2 - 11y + 24$

21. $z^2 + 4x + 5$

22. $x^2 + 6x + 6$

23. $x^2 - 12x + 35$

24. $y^2 + 10y + 16$

25. $k^2 - 3k - 4$

26. $m^2 + 4m - 5$

27. $d^2 + 7d + 12$

28. $x^2 + 3x - 5$

29. $y^2 - 2y - 8$

30. $x^2 - x - 20$

31. $x^2 + 7xa + 12a^2$

32. $y^2 + 5yb + 6b^2$

33. $m^2 - 7mn + 6n^2$

34. $s^2 + st - 12t^2$

35. $x^2 - 4xy + 3y^2$

36. $x^2 + xy - 6y^2$

37. $r^2 + 7rs - 8s^2$

38. $y^2 + yz - 2z^2$

For Exercises 39–50, factor completely. See Example 4.

39. $x^3 + x^2 - 6x$

40. $y^3 + 3y^2 + 2y$

41. $z^3 + 3z^2 - 10z$

42. $x^3 - 2x^2 - 8x$

43. $2x^4 + 6x^3 + 4x^2$

44. $3y^4 + 12y^3 + 9y^2$

45. $4t^5 + 4t^4 - 24t^3$

46. $2x^5 - 10x^4 + 12x^3$

47. $5a^6 - 5a^5 - 10a^4$

48. $3z^5 + 15z^4 + 18z^3$

49. $2m^5 - 6m^4 - 8m^3$

50. $2x^6 + 2x^5 - 24x^4$

51. Is $(x - 2)(2x + 8)$ a complete factoring of $2x^2 + 4x - 16$? Why?

52. Is $(3a + 1)(3a - 6)$ a complete factoring of $9a^2 - 15a - 6$? Why?

SAY IT IN WORDS

53. What is your method when factoring a trinomial $ax^2 + bx + c$, where $a = 1$?

REVIEW EXERCISES

The following exercises review parts of Section 4.5. Doing these exercises will help prepare you for the next section.

For Exercises 54–63, multiply. Concentrate on combining like terms mentally.

54. $(3x - 4)(x - 1)$

55. $(a + 4)(2a - 3)$

56. $(2z - 1)(3z + 1)$

57. $(4y + 3)(y - 2)$

58. $(2x + 3)(2x - 2)$

59. $(4x + 1)(2x + 1)$

60. $(5a - 2)(a - 3)$

61. $(2x + 6)(4x + 3)$

62. $(2a + 3b)(3a - 2b)$

63. $(x - 5y)(3x - y)$

ENRICHMENT EXERCISES

Factor.

1. $x^4 + 10x^2 + 21$

2. $y^4 + 3y^2 + 2$

3. $t^4 + 2t^2 - 15$

4. $a^5 + 11a^3 + 24a$

Answers to Enrichment Exercises begin on page A.1.

5.3 Factoring General Trinomials

OBJECTIVES

- *To factor the trinomial $ax^2 + bx + c$, where $a \neq 1$*
- *To factor out the greatest common factor, then factor the resulting trinomial*

LEARNING ADVANTAGE

If your instructor provides review activities in class, take them seriously. The instructor may have just finished writing the test and may be creating review problems that are similar to those that will be on the test. However, do not assume that a topic will not *be on your test just because it was not covered in the review lesson.*

In this section, we factor general trinomials of the form $ax^2 + bx + c$, where the leading coefficient is not equal to 1. We can use a method similar to the one developed in the last section.

Consider $2x^2 + 7x + 3$. Find the product of the leading coefficient and the constant term. For this trinomial, it is $2 \cdot 3 = 6$. Next, find two numbers whose product is 6 with a sum of 7. The pairs of numbers of numbers whose product is 6 are

1 and 6 and 2 and 3

Only the pair 1 and 6 has a sum of 7. Therefore,

$$\begin{aligned}2x^2 + 7x + 3 &= 2x^2 + x + 6x + 3 \\ &= (2x^2 + x) + (6x + 3) \\ &= x(2x + 1) + 3(2x + 1) \\ &= (2x + 1)(x + 3)\end{aligned}$$

STRATEGY

Factoring the trinomial $ax^2 + bx + c$, where $a \neq 1$.

1. Find two integers whose product is ac and whose sum is b.
2. Write bx as the sum of two terms using the two integers found in Step 1 as coefficients.
3. Factor by grouping.

Example 1 Factor.

(a) $2x^2 - x - 6$ (b) $6x^2 - 11x + 3$ (c) $2x^2 + xy - 10y^2$

Solution (a) Since $2(-6) = -12$, we want two numbers with a product of -12 and with a sum of -1. The numbers are 3 and -4. Therefore,

$$\begin{aligned} 2x^2 - x - 6 &= 2x^2 + 3x - 4x - 6 \\ &= x(2x + 3) - 2(2x + 3) \\ &= (2x + 3)(x - 2) \end{aligned}$$

(b) Since $6 \cdot 3 = 18$, we want two numbers with a product of 18 and a sum of -11. The numbers are -9 and -2. Therefore,

$$\begin{aligned} 6x^2 - 11x + 3 &= 6x^2 - 9x - 2x + 3 \\ &= 3x(2x - 3) - (2x - 3) \\ &= (2x - 3)(3x - 1) \end{aligned}$$

(c) Since $2(-10) = -20$, we want two numbers with a product of -20 and a sum of 1. The two numbers are 5 and -4. Therefore,

$$\begin{aligned} 2x^2 + xy - 10y^2 &= 2x^2 + 5xy - 4xy - 10y^2 \\ &= x(2x + 5y) - 2y(2x + 5y) \\ &= (2x + 5y)(x - 2y) \end{aligned}$$

When factoring a trinomial, we should always first think of factoring out the greatest common factor.

Example 2 Factor completely by first factoring out the greatest common factor.

(a) $6x^4 + 20x^3 + 6x^2$ (b) $6cx^2 - 18cx + 9c$

Solution (a) The greatest common factor is $2x^2$. Therefore,

$$6x^4 + 20x^3 + 6x^2 = 2x^2(3x^2 + 10x + 3)$$

To factor the resulting trinomial, we want two numbers with a product of $3 \cdot 3 = 9$ and a sum of 10. They are 9 and 1. Continuing the factoring,

$$\begin{aligned} 2x^2(3x^2 + 10x + 3) &= 2x^2(3x^2 + 9x + x + 3) \\ &= 2x^2[3x(x + 3) + (x + 3)] \\ &= 2x^2(x + 3)(3x + 1) \end{aligned}$$

(b) The greatest common factor is $3c$.

$$6cx^2 - 18cx + 9c = 3c(2x^2 - 6x + 3)$$

To factor the resulting trinomial, we need two integers with a product of $2 \cdot 3 = 6$ and a sum of -6. No two integers work. Therefore, $2x^2 - 6x + 3$ is prime and the factoring is complete. ◀

WARMING UP

Decide if the factoring is correct (true) or incorrect (false).

1. $3x^2 + 5x + 2 = (3x + 2)(x + 1)$

2. $2y^2 + 4x + 3 = (2y + 3)(y + 1)$

3. $2y^2 - 11y - 21 = (2y + 3)(y - 7)$

4. $9x^2 - 4x - 5 = (9x + 5)(x - 1)$

5. $18t^2 - 41t + 10 = (18t - 5)(t - 2)$

6. $2x^2 + 5xy + 2y^2 = (2x + y)(x + 2y)$

7. $2x^3 + 5x^2 + 3 = x(2x + 3)(x + 1)$

8. $ax^2 - 2ax - 8a = a(x - 4)(x + 2)$

EXERCISE SET 5.3

For Exercises 1–38, factor the trinomial. If it is prime, write "prime."

1. $2x^2 + 5x + 3$

2. $3x^2 + 7x + 2$

3. $4x^2 + 13x + 3$

4. $2x^2 + 9x + 4$

5. $2x^2 - 5x - 3$

6. $3x^2 - 5x - 2$

7. $6y^2 + 5y + 1$

8. $8y^2 + 6y + 1$

9. $3x^2 + 6x + 2$

10. $2x^2 + 4x + 3$

11. $6t^2 - 5t + 1$

12. $2s^2 - 7s + 6$

13. $8z^2 - 10z - 3$

14. $5x^2 - 6x + 1$

15. $6x^2 - 5x - 1$

16. $7y^2 - 9y + 2$

17. $5k^2 + 7k + 2$

18. $6x^2 + 11x + 5$

19. $7v^2 - 10v + 3$

20. $3z^2 + 7z - 6$

21. $y^2 - 6y + 6$

22. $6x^2 - 4x + 1$

23. $3x^2 + 11x - 4$

24. $5u^2 - 14u - 3$

25. $6n^2 + n - 5$

26. $4z^2 - 11z + 6$

27. $10x^2 + x - 3$

28. $4y^2 - 17y + 4$

29. $3z^2 - 10z + 3$

30. $12t^2 + 4t - 1$

31. $2x^2 + 3xy + y^2$

32. $2x^2 + 7xy + 3y^2$

33. $2x^2 + xy - y^2$

34. $3x^2 - 5xy + 2y^2$

35. $6s^2 + st - 2t^2$

36. $3s^2 + 7st + 2t^2$

37. $6u^2 - uv - 2v^2$

38. $8m^2 - 2mn - 3n^2$

For Exercises 39–48, factor completely by first factoring out the greatest common factor.

39. $2x^4 + 7x^3 + 6x^2$

40. $3x^3 + 11x^2 - 4x$

41. $4x^3 - 14x^2 + 6x$

42. $6x^3 + 3x^2 - 9x$

43. $6y^5 + 9y^4 - 15y^3$

44. $12y^4 - 26y^3 + 12y^2$

45. $6tx^2 - tx - 15t$

46. $10ax^2 - 21ax - 10a$

47. $6m^2p^2 - 27m^2p - 54m^2$

48. $12c^2x^2 - 22c^2x + 6c^2$

49. Which pair of factors, $(3x + 2)$ and $(x + 4)$ or $(3x - 2)$ and $(x - 4)$, does *not* give the product as $3x^2 - 14x + 8$?

50. What polynomial can be factored as $(5x - 4)(2x + 3)$?

51. The area of a rectangle is $6x^2 + 13x + 6$ square inches. If one side measures $2x + 3$ inches, use factoring to find a binomial that represents the length of the other side.

52. The area of a triangle is $3x^2 + 13x + 4$ square feet. If the height is $x + 4$ feet, use factoring to find a binomial that represents the length of the base.

SAY IT IN WORDS

53. Explain your method for factoring the trinomial $ax^2 + bx + c$, where $a \neq 1$.

REVIEW EXERCISES

The following exercises review parts of Sections 4.5 and 4.6. Doing these exercises will help prepare you for the next section.

For Exercises 54–66, multiply.

54. $(x + 3)(x - 3)$

55. $(3t + 10)(3t - 10)$

56. $(2z - 1)(2z + 1)$

57. $(a + 3)^2$

58. $(b - 2)^2$

59. $(3x - 1)^2$

60. $(4y + 3)^2$

61. $(x - 2y)^2$

62. $(3r + t)^2$

63. $(x - 2)(x^2 + 2x + 4)$

64. $(s + 1)(s^2 - s + 1)$

65. $(2a - 3)(4a^2 + 6a + 9)$

66. $(x - y)(x^2 + xy + y^2)$

ENRICHMENT EXERCISES

For Exercises 1–4, factor completely.

1. $6x^4 - 7x^2 - 5$

2. $3ax^6 - 5ax^3 - 2a$

3. $3xy^5 + 5xy^3 - 2xy$

4. $3x^3y + 5x^2y^2 + 2xy^3$

Answers to Enrichment Exercises begin on page A.1.

5.4 Factoring Special Polynomials

OBJECTIVES

- *To factor the difference of two squares,* $a^2 - b^2$
- *To recognize and factor perfect square trinomials*
- *To factor the sum of two cubes,* $a^3 + b^3$
- *To factor the difference of two cubes,* $a^3 - b^3$

LEARNING ADVANTAGE *Some instructors allow time for your requests during a review lesson. If so, bring up topics you would like the instructor to review. Above all, remember that a review lesson can be a costly lesson for you to miss. If you are absent for a review, be sure to contact someone in your class so that you can find out exactly what work was covered.*

In this section, we use the special product rules of Section 4.6. Recall the Difference of Two Squares Rule:

$$(a + b)(a - b) = a^2 - b^2$$

We can reverse this equation to obtain the factoring pattern:

RULE

The Difference of Two Squares

$$a^2 - b^2 = (a + b)(a - b)$$

That is,

$$(\textbf{first})^2 - (\textbf{second})^2 = (\textbf{first} + \textbf{second})(\textbf{first} - \textbf{second})$$

Example 1 Factor each polynomial.

(a) $x^2 - 4$ **(b)** $4a^2 - 9$ **(c)** $k^2 - 16b^2$ **(d)** $36z^2 - 25n^2$

Solution We write each one as a difference of squares.

(a) $x^2 - 4 = x^2 - 2^2 = (x + 2)(x - 2)$
(b) $4a^2 - 9 = (2a)^2 - 3^2 = (2a + 3)(2a - 3)$
(c) $k^2 - 16b^2 = k^2 - (4b)^2 = (k + 4b)(k - 4b)$
(d) $36z^2 - 25n^2 = (6z)^2 - (5n)^2 = (6z + 5n)(6z - 5n)$ ◀

The difference of two squares, such as $x^2 - 4$, can be factored, but the *sum* of two squares, such as $x^2 + 4$, cannot be factored using integers; it is prime.

RULE

The Sum of Two Squares

$$a^2 + b^2 \text{ is prime}$$

Even though the sum of two squares, such as $x^2 + 9$, cannot be factored, the sum of two *cubes* can be factored. To see this, let us multiply $a + b$ and $a^2 - ab + b^2$.

$$\begin{aligned}(a + b)(a^2 - ab + b^2) &= a^3 - a^2b + ab^2 + a^2b - ab^2 + b^3 \\ &= a^3 + b^3\end{aligned}$$

Likewise,

$$\begin{aligned}(a - b)(a^2 + ab + b^2) &= a^3 + a^2b + ab^2 - a^2b - ab^2 - b^3 \\ &= a^3 - b^3\end{aligned}$$

RULE

The Sum and Difference of Cubes

$$a^3 + b^3 = (a + b)(a^2 - ab + b^2)$$
$$a^3 - b^3 = (a - b)(a^2 + ab + b^2)$$

Example 2 Factor each binomial.

(a) $x^3 + 8$ **(b)** $8s^3 + 27$ **(c)** $27x^3 - 8y^3$

Solution **(a)** Since 8 is 2^3, we can write $x^3 + 8$ as the sum of two cubes.

$$\begin{aligned}x^3 + 8 = x^3 + 2^3 &= (x + 2)(x^2 - x \cdot 2 + 2^2) \\ &= (x + 2)(x^2 - 2x + 4)\end{aligned}$$

(b) Write $8s^3 + 27$ as the sum of two cubes.

$$\begin{aligned}8s^3 + 27 &= (2s)^3 + 3^3 \\ &= (2s + 3)[(2s)^2 - (2s)(3) + 3^2] \\ &= (2s + 3)(4s^2 - 6s + 9)\end{aligned}$$

(c) $$\begin{aligned}27x^3 - 8y^3 &= (3x)^3 - (2y)^3 \\ &= (3x - 2y)(9x^2 + 6xy + 4y^2)\end{aligned}$$

If a trinomial is the square of a binomial, it is called a **perfect square trinomial.** For example, $x^2 + 2x + 1 = (x + 1)^2$, so $x^2 + 2x + 1$ is a perfect square trinomial. Perfect square trinomials can be factored just like any other trinomial, but using the following rule may be faster.

RULE

Perfect Square Trinomials

$$a^2 + 2ab + b^2 = (a + b)^2$$
$$a^2 - 2ab + b^2 = (a - b)^2$$

Example 3 Factor the perfect square trinomials.

(a) $x^2 + 8x + 16$ **(b)** $x^2 - 4xy + 4y^2$

Solution **(a)** Since $x^2 + 8x + 16 = x^2 + 2 \cdot 4x + 4^2$, it is a perfect square trinomial and

$$x^2 + 8x + 16 = (x + 4)^2$$

(b) We can write this trinomial in the perfect square form.

$$\begin{aligned} x^2 - 4xy + 4y^2 &= x^2 - 2x(2y) + (2y)^2 \\ &= (x - 2y)^2 \end{aligned}$$

We summarize the results of this section.

Special Factoring Forms

The difference of two squares

$$a^2 - b^2 = (a + b)(a - b)$$

$$a^2 + b^2 \text{ is prime}$$

The sum of two cubes

$$a^3 + b^3 = (a + b)(a^2 - ab + b^2)$$

The difference of two cubes

$$a^3 - b^3 = (a - b)(a^2 + ab + b^2)$$

Perfect squares

$$a^2 + 2ab + b^2 = (a + b)^2$$

$$a^2 - 2ab + b^2 = (a - b)^2$$

In the next example we combine factoring methods.

Example 4 Factor each polynomial completely.

(a) $3x^2 - 12$ **(b)** $2v^7 - 16v^4$ **(c)** $3s^3 - 6s^2t + 3st^2$

Solution **(a)** We factor out the greatest common factor, 3, from $3x^2 - 12$, then apply the difference of two squares form.

$$3x^2 - 12 = 3(x^2 - 4) = 3(x + 2)(x - 2)$$

(b) We factor out the greatest common factor, $2v^4$, from the polynomial, then use the difference of two cubes form.

$$\begin{aligned} 2v^7 - 16v^4 &= (2v^4)(v^3 - 8) \\ &= (2v^4)(v - 2)(v^2 + 2v + 4) \end{aligned}$$

(c) We factor out the greatest common factor, $3s$, from the polynomial, then use a perfect square trinomial form.

$$3s^3 - 6s^2t + 3st^2 = (3s)(s^2 - 2st + t^2)$$
$$= (3s)(s - t)^2$$

WARMING UP

Answer true or false. Decide if the factoring is correct (true) or incorrect (false).

1. $x^2 + 4 = (x + 2)(x - 2)$

2. $x^2 - 9 = (x + 3)(x - 3)$

3. $y^3 - 27 = (y + 3)(y - 3)$

4. $z^2 - 2z + 4 = (z - 2)^2$

5. $a^2 + 4a + 4 = (a + 2)^2$

6. $8m^3 + 1 = (2m - 1)(4m^2 + 2m + 1)$

7. $x^3 - 16x = x(x + 4)(x - 4)$

8. $c^2 + 49$ is prime.

EXERCISE SET 5.4

For Exercises 1–30, factor each binomial. If a binomial is prime, write "prime."

1. $x^2 - 25$

2. $a^2 - 9$

3. $t^2 - 100$

4. $u^2 - 49$

5. $n^2 - p^2$

6. $m^2 - q^2$

7. $s^2 - t^2$

8. $u^2 - v^2$

9. $4y^2 - 1$

10. $16z^2 - 9$

11. $k^2 - 25t^2$

12. $q^2 - 100r^2$

13. $x^2 + 1$

14. $y^2 + 9$

15. $81k^2 - h^2$

16. $64w^2 - b^2$

17. $16s^2 - 9y^2$

18. $25r^2 - 49s^2$

19. $z^3 + 27$

20. $a^3 + 8$

21. $64t^3 + 1$

22. $8p^3 + 27$

23. $x^3 - 8$

24. $z^3 - 27$

25. $b^3 - 1$

26. $y^3 - 216$

27. $8t^3 - v^3$

28. $u^3 - 27x^3$

29. $27m^3 + 1{,}000n^3$

30. $z^3 - 8c^3$

For Exercises 31–42, factor each perfect square trinomial.

31. $x^2 + 10x + 25$

32. $y^2 + 2y + 1$

33. $x^2 + 4x + 4$

34. $x^2 + 12x + 36$

35. $t^2 - 6t + 9$

36. $z^2 - 10z + 25$

37. $c^2 - 12c + 36$

38. $x^2 - 14x + 49$

39. $x^2 - 2xy + y^2$

40. $x^2 - 4xt + 4t^2$

41. $4x^2 + 4xz + z^2$

42. $a^2 + 6ab + 9b^2$

For Exercises 43–60, factor each polynomial completely. Remember to first check for a greatest common factor.

43. $7y^2 - 63$

44. $3a^2 - 48$

45. $-2w^2 + 50$

46. $-4n^2 + 36$

47. $x^5 - 81x^3$

48. $w^6 - 9w^4$

49. $4y^7 - 16y^5$

50. $5p^5 - 125p^3$

51. $6m^2 + 36m + 54$

52. $4c^2 + 16c + 16$

53. $10k^2 - 80k + 160$

54. $7h^2 - 14h + 7$

55. $3z^4 + 12z^3 + 12z^2$

56. $4n^5 + 8n^4 + 4n^3$

57. $25x^8 - 50x^7 + 25x^6$

58. $3z^7 - 36z^6 + 108z^5$

59. $a^3 - 2ba^2 + b^2a$

60. $st^2 - 2s^2t + s^3$

SAY IT IN WORDS

61. Explain the rule for the sum or difference of squares.

62. How can you tell if a trinomial is a perfect square?

REVIEW EXERCISES

The following exercises review parts of Section 5.1. Doing these exercises will help prepare you for the next section.

For Exercises 63–68, factor out the greatest common factor, if other than 1, from each polynomial.

63. $4x - 12y$

64. $15a + 21b$

65. $4z^3 - 2z + 1$

66. $2c^3 - 2c^4 + 6c$

67. $ax - ay$

68. $nz + nt$

ENRICHMENT EXERCISES

Factor completely.

1. $z^4 - 16$

2. $a^6 + b^9$

3. $u^{12} - v^6$

4. $2s^3r^4 - 12s^4r^3 + 18s^5r^2$

5. $(8x + 7)^2 - (8x - 7)^2$

6. $y(3x + 2)^2 - 2y(3x + 2) + y$

Answers to Enrichment Exercises begin on page A.1.

5.5 A Strategy for General Factoring

OBJECTIVE

▶ *To use the strategy for general factoring*

We have finished the discussion on factoring. Two or more methods might be needed to factor a polynomial completely. Therefore, we list steps that can be used as a strategy when factoring.

LEARNING ADVANTAGE

If you save all of your assignments, correct and write the details of every problem, you will be amazed at how quickly you can review and remember all the things that seemed so difficult a few days ago. You can cut the time spent reviewing homework in half if you spend enough time studying your assignments each day.

STRATEGY

General Factoring

1. **Does the polynomial have a greatest common factor?** Factor out the greatest common factor, if other than one.
2. **How many terms are in the polynomial?** If the polynomial has
   (a) **Two terms:** Check for the difference of two squares or the sum or difference of two cubes. Remember $a^2 + b^2$ is prime.
   (b) **Three terms:** Check for a perfect square trinomial. Otherwise, use the factoring methods of Sections 5.2 or 5.3.
   (c) **Four terms:** Check for factoring by grouping.
3. **Has the polynomial been factored completely?** Check to see if the factors themselves can be factored further.
4. **Check by multiplying.** Multiply the factors of your answer to see if you get the original polynomial.

We use the strategy in the next examples.

Example 1 Factor completely: $2x^2 - 8$

Solution We start by factoring out the greatest common factor.

$$2x^2 - 8 = 2(x^2 - 4)$$

There are two terms in $x^2 - 4$, which is a difference of squares.

$$\begin{aligned} 2x^2 - 8 &= 2(x^2 - 4) \\ &= 2(x + 2)(x - 2) \end{aligned}$$

Check by multiplying. ◀

Example 2 Factor completely: $y^4 + y$

Solution The greatest common factor is y.

$$y^4 + y = y(y^3 + 1)$$

The resulting polynomial has two terms. It is the sum of cubes.

$$\begin{aligned} y^4 + y &= y(y^3 + 1) \\ &= y(y + 1)(y^2 - y + 1) \end{aligned}$$

Check by multiplying.

Example 3 Factor completely:

$$6t^3 + 27t^2 - 15t$$

Solution The greatest common factor is $3t$.

$$\begin{aligned} 6t^3 + 27t^2 - 15t &= 3t(2t^2 + 9t - 5) \\ &= 3t(2t - 1)(t + 5) \end{aligned}$$

Check by multiplying.

Example 4 Factor completely:

$$2ax - 2ay + 2t^2x - 2t^2y$$

Solution This polynomial of four terms has a greatest common factor of 2.

$$2ax - 2ay + 2t^2x - 2t^2y = 2(ax - ay + t^2x - t^2y)$$

Next, factor the resulting polynomial by grouping.

$$\begin{aligned} &= 2[a(x - y) + t^2(x - y)] \\ &= 2(x - y)(a + t^2) \end{aligned}$$

Example 5 Factor completely:

$$3xr^2 - 3xs^2 - 2r^2 + 2s^2$$

Solution There is no greatest common factor other than 1. Since there are four terms, we try factoring by grouping.

$$\begin{aligned} 3xr^2 - 3xs^2 - 2r^2 + 2s^2 &= 3x(r^2 - s^2) - 2(r^2 - s^2) \\ &= (r^2 - s^2)(3x - 2) \end{aligned}$$

The factor $r^2 - s^2$ is a difference of squares.

$$= (r + s)(r - s)(3x - 2)$$

TEAM PROJECT

(3 or 4 Students)

FACTORING POLYNOMIALS

Course of Action: Each person on the team selects three problems from Exercise Set 5.5. Team members factor their own selected problems completely.

Team members then give their problem numbers to the person on their left. Each member then solves the problems given to him or her and returns the solutions to the person on the right. All team members then compare their neighbors' solutions to their own.

Next, team members select another three problems from Exercise Set 5.5, solve the problems, and give the problem numbers to the person on their right. Each member then solves the problems given to him or her and returns the solutions to the person on the left. Team members then compare their neighbors' solutions to their own.

Group Report:

1. How successful was the team in factoring polynomials?
2. What difficulties, if any, did team members encounter?
3. Write a paragraph explaining how to factor a polynomial.

WARMING UP

Decide if the polynomial is factored completely (true) or not (false).

1. $2x^2 + 2 = 2(x^2 + 1)$

2. $4x^3 + 4 = 4(x^3 + 1)$

3. $y^3 - 16y = y(y + 4)(y - 4)$

4. $x^2 + x + 2$

5. $t^2 - 4ta + 4a^2 = (t - 2a)^2$

6. $4x^2 - 16y^2 = (2x + 4y)(2x - 4y)$

7. $2x^2 + 2y^2 + 3tx^2 + 3ty^2 = (x^2 + y^2)(2 + 3t)$

8. $x^3 - 5x^2 - 6x = x(x + 1)(x - 6)$

EXERCISE SET 5.5

Use the strategy for general factoring to factor each polynomial completely. If a polynomial is prime, write "prime." The even-numbered exercises are not necessarily matched to the preceding odd-numbered exercise.

1. $x^5 + 4x^3$

2. $x^4 - x^2$

3. $2t^2 + t - 3$

4. $a^2 - 6a + 9$

5. $2y^2 + 4y - 6$

6. $2x^2 + 11x + 12$

7. $m^3 + 8$

8. $2z^3 + 2z^2 - 60z$

9. $9x^2 - 16y^2$

10. $12 - 6a + 10b - 5ab$

11. $3x^5 - 81x^2$

12. $2y^3 - y^2 - y$

13. $z^2 + 49$

14. $a^2(z + 2) - 9(z + 2)$

15. $a^2 + 8a + 16$

16. $tx + ty + x + y$

17. $y^2 - 18y + 81$

18. $12x^2y + 30xy^2 + 18xy$

19. $3h^2 - 9h + 6$

20. $x^2 + 4x + 5$

21. $9x^2 + 21x - 8$

22. $z^3 + 2z^2 - 24z$

23. $10b^3c^2 + 20b^2c^3 - 5b^2c^2$

24. $8y^2 - 18y + 7$

25. $6x^2 + 30x - 36$

26. $2m^4 - 16m$

27. $12t^3 - 3t$

28. $x^2 - 4xy + 4y^2$

29. $2h^4 + 54h$

30. $6x^2 - xt - 2t^2$

31. $m^4 + 25m^2$

32. $6cd^2 + cd - 12c$

33. $x^3 - 3x^2 + 5x$

34. $3 - 3a^2$

35. $z^2 - 12zc + 27c^2$

36. $p^4q - 8pq^4$

37. $6x^3y - 19x^2y + 10xy$

38. $a^4b^5 - 9a^2b^3$

39. $3abx^2 - 75ab$

40. $2st^3 - 12s^2t^2 + 18s^3t$

41. $2bc^3 - 5bc^2d + 2bcd^2$

42. $9x^2 + 12xa + 4a^2$

43. $4x^2 + 25$

44. $x^3 - 7x^2 + 12x$

45. $7 - 14m + 2k - 4km$

46. $10u^4 - 10uv^3$

47. $x^2a^3 + x^2 - y^2a^3 - y^2$

48. $x(a^2 - 49) - y(a^2 - 49)$

49. $y^5 - y^3 - y^2 + 1$

50. $2a^4b + 2ab^4$

5.6 Solving Quadratic Equations by Factoring

OBJECTIVES

- *To solve quadratic equations by factoring*
- *To solve other equations by factoring*

LEARNING ADVANTAGE

Quizzes you took during the study of a chapter can tell you some of the things your instructor will expect you to know at the end of a chapter. Save the quizzes and use them to review for the chapter test.

In Chapter 1, we studied the multiplication property of zero.

$$5 \cdot 0 = 0, \qquad (-4) \cdot 0 = 0, \qquad 0 \cdot \left(\frac{1}{2}\right) = 0, \qquad 0 \cdot 0 = 0$$

Notice a common feature in all of these examples. When the product of two numbers is zero, at least one of the factors in the product is zero. The statement is true for variable expressions.

PROPERTY

The Zero-Product Property

Suppose A and B are two expressions such that $A \cdot B = 0$, then either $A = 0$, or $B = 0$, or both A and B are zero.

That is, if the product of two factors is zero, at least one of them is zero.

We can use this property to solve an equation like

$$(x - 6)(x - 2) = 0$$

In the product on the left side, think of $x - 6$ as A and $x - 2$ as B. By the Zero-Product Property, $A \cdot B = 0$ means that either A is zero or B is zero. That is, either

$$x - 6 = 0 \qquad \text{or} \qquad x - 2 = 0$$

Solving each of these equations, either $x = 6$ or $x = 2$.

We can check our two answers by seeing if they each make the original equation true.

$$(x - 6)(x - 2) = 0$$

Check $x = 6$:

$$\begin{aligned} (x - 6)(x - 2) &= 0 \\ (6 - 6)(6 - 2) &\stackrel{?}{=} 0 \\ (0)(4) &\stackrel{?}{=} 0 \\ 0 &= 0 \end{aligned}$$

$x = 2$:

$$\begin{aligned} (x - 6)(x - 2) &= 0 \\ (2 - 6)(2 - 2) &\stackrel{?}{=} 0 \\ (-4)(0) &\stackrel{?}{=} 0 \\ 0 &= 0 \end{aligned}$$

Since they each make the original equation true, both 6 and 2 are solutions.

Example 1 Solve the equation

$$(2y - 1)(5y + 3) = 0$$

Then, check your answers.

Solution By the Zero-Product Property, either

$$\begin{aligned} 2y - 1 &= 0 & \text{or} \qquad 5y + 3 &= 0 \\ 2y &= 1 & 5y &= -3 \\ y &= \frac{1}{2} & y &= -\frac{3}{5} \end{aligned}$$

We now check our answers in the original equation.

Check $\frac{1}{2}$:

$$(2y - 1)(5y + 3) = 0$$

$$\left[2\left(\frac{1}{2}\right) - 1\right]\left[5\left(\frac{1}{2}\right) + 3\right] \stackrel{?}{=} 0$$

$$[1 - 1]\left[\frac{11}{2}\right] \stackrel{?}{=} 0$$

$$[0]\left[\frac{11}{2}\right] \stackrel{?}{=} 0$$

$$0 = 0$$

Check $-\frac{3}{5}$:

$$(2y - 1)(5y + 3) = 0$$

$$\left[2\left(-\frac{3}{5}\right) - 1\right]\left[5\left(-\frac{3}{5}\right) + 3\right] \stackrel{?}{=} 0$$

$$\left[-\frac{11}{5}\right][0] \stackrel{?}{=} 0$$

$$0 = 0$$

The solution set is $\left\{\frac{1}{2}, -\frac{3}{5}\right\}$

The Zero-Product Property is true for any number of factors.

If the product of two or more expressions is zero, then at least one of them is zero.

Example 2 Solve the equation

$$t(t - 5)(t + 11) = 0$$

Solution Since the product of three factors is equal to zero, then at least one of them is zero. That is,

$$t = 0 \quad \text{or} \quad t - 5 = 0 \quad \text{or} \quad t + 11 = 0$$

Therefore, either $t = 0$, $t = 5$, or $t = -11$.

We check these three answers in the original equation.

Check 0:

$$t(t - 5)(t + 11) = 0$$

$$0(0 - 5)(0 + 11) \stackrel{?}{=} 0$$

$$0(-5)(11) \stackrel{?}{=} 0$$

$$0 = 0$$

Check 5:

$$t(t - 5)(t + 11) = 0$$

$$5(5 - 5)(5 + 11) \stackrel{?}{=} 0$$

$$5(0)(16) \stackrel{?}{=} 0$$

$$0 = 0$$

Check -11:

$$\begin{aligned} t(t-5)(t+11) &= 0 \\ (-11)(-11-5)(-11+11) &\stackrel{?}{=} 0 \\ (-11)(-16)(0) &\stackrel{?}{=} 0 \\ 0 &= 0 \end{aligned}$$

Therefore, the solution set to the original equation is $\{0, 5, -11\}$.

We will use the Zero-Product Property to solve certain *quadratic equations.*

DEFINITION

A **quadratic equation** is an equation that can be put in the form

$$ax^2 + bx + c = 0$$

where a, b, and c are numbers with $a \neq 0$

The form $ax^2 + bx + c = 0$ is called the **standard form** of a quadratic equation. Examples of quadratic equations in standard form are

$$x^2 - 3x + 1 = 0 \qquad \text{and} \qquad 4t^2 - 5t + 1 = 0$$

Examples of quadratic equations that are *not* in standard form are

$$x^2 - 6x = 5 \qquad \text{and} \qquad 1 + v - v^2 = 0$$

If the polynomial on the left side of a quadratic equation in standard form can be factored, then we can solve the equation using the Zero-Product Property.

Example 3 Solve the quadratic equation $z^2 + 3z - 10 = 0$.

Solution We start by attempting to factor the left side.

$$\begin{aligned} z^2 + 3z - 10 &= 0 \\ (z+5)(z-2) &= 0 \qquad \text{Factor } z^2 + 3z - 10. \end{aligned}$$

Using the Zero-Product Property, either $z = -5$ or $z = 2$. We check our solutions in the original equation.

Check -5:

$$\begin{aligned} z^2 + 3z - 10 &= 0 \\ (-5)^2 + 3(-5) - 10 &\stackrel{?}{=} 0 \\ 25 - 15 - 10 &\stackrel{?}{=} 0 \\ 0 &= 0 \end{aligned}$$

Check 2:

$$\begin{aligned} z^2 + 3z - 10 &= 0 \\ 2^2 + 3(2) - 10 &\stackrel{?}{=} 0 \\ 4 + 6 - 10 &\stackrel{?}{=} 0 \\ 0 &= 0 \end{aligned}$$

Therefore, the solution set is $\{-5, 2\}$.

Example 4 Solve for b: $6b^2 + 13b = 5$.

Solution This equation is not in standard form. We start by subtracting 5 from both sides, then we factor the trinomial on the left side.

$$6b^2 + 13b = 5$$

$$6b^2 + 13b - 5 = 5 - 5 \quad \text{Subtract 5.}$$

$$6b^2 + 13b - 5 = 0$$

$$(3b - 1)(2b + 5) = 0 \quad \text{Factor } 6b^2 + 13b - 5.$$

By the Zero-Product Property, either

$$3b - 1 = 0 \quad \text{or} \quad 2b + 5 = 0$$

$$b = \frac{1}{3} \qquad\qquad b = -\frac{5}{2}$$

It is up to you to check these two solutions in the original equation. The solution set is $\left\{\frac{1}{3}, -\frac{5}{2}\right\}$. ◀

Example 5 Solve for n: $9n^2 + 12n + 4 = 0$.

Solution This equation is in standard form. To factor the left side, notice it is a perfect square trinomial.

$$9n^2 + 12n + 4 = 0$$

$$(3n + 2)^2 = 0 \quad \text{Factor } 9n^2 + 12n + 4.$$

We can think of $(3n + 2)^2 = 0$ as

$$(3n + 2)(3n + 2) = 0$$

Using the Zero-Product Property, either

$$3n + 2 = 0 \quad \text{or} \quad 3n + 2 = 0$$

Notice, however, that there is really only one equation, $3n + 2 = 0$. Solving this equation, we have $n = -\frac{2}{3}$.

We check this solution in the original equation.

Check $-\frac{2}{3}$:

$$9n^2 + 12n + 4 = 0$$

$$9\left(-\frac{2}{3}\right)^2 + 12\left(-\frac{2}{3}\right) + 4 \stackrel{?}{=} 0$$

$$9\left(\frac{4}{9}\right) - 8 + 4 \stackrel{?}{=} 0$$

$$4 - 8 + 4 \stackrel{?}{=} 0$$

$$0 = 0$$

Therefore, the solution set is $\left\{-\frac{2}{3}\right\}$. ◀

It should be noted that not all quadratic equations can be solved by factoring. This technique only works when the trinomial $ax^2 + bx + c$ can be factored. We will study other methods for solving quadratic equations in Chapter 9.

We now give a strategy for solving quadratic equations when factoring can be used.

STRATEGY

Solving Quadratic Equations by Factoring

Step 1 Put in standard form, if necessary.

Step 2 Factor the polynomial completely.

Step 3 Use the Zero-Product Property and set each factor equal to zero.

Step 4 Solve the resulting equations.

Step 5 Check each answer from Step 4 in the original equation.

Example 6 Solve $2x^2 + 1 = 3(12 + x)$.

Solution We follow the strategy for solving quadratic equations. First, we put it in standard form.

$$2x^2 + 1 = 36 + 3x \qquad \text{Distributive property.}$$

$$2x^2 + 1 - 36 - 3x = 0 \qquad \text{Subtract 36 and subtract } 3x.$$

$$2x^2 - 35 - 3x = 0$$

$$2x^2 - 3x - 35 = 0 \qquad \text{Rearrange terms.}$$

Next, we factor the polynomial.

$$(2x + 7)(x - 5) = 0$$

By the Zero-Product Property, either

$$2x + 7 = 0 \quad \text{or} \quad x - 5 = 0$$

Solving each equation, either

$$x = -\frac{7}{2} \quad \text{or} \quad x = 5$$

We check our answers in the original equation.

Check $-\dfrac{7}{2}$: $\quad 2x^2 + 1 = 3(12 + x)$

$$2\left(-\frac{7}{2}\right)^2 + 1 \stackrel{?}{=} 3\left(12 - \frac{7}{2}\right)$$

$$2\left(\frac{49}{4}\right) + 1 \stackrel{?}{=} 3\left(\frac{24}{2} - \frac{7}{2}\right)$$

$$\frac{49}{2} + \frac{2}{2} \stackrel{?}{=} 3\left(\frac{17}{2}\right)$$

$$\frac{51}{2} = \frac{51}{2}$$

Check 5:

$$2x^2 + 1 = 3(12 + x)$$

$$2(5)^2 + 1 \stackrel{?}{=} 3(12 + 5)$$

$$2(25) + 1 \stackrel{?}{=} 3(17)$$

$$51 = 51$$

Therefore, the solution set is $\left\{-\frac{7}{2}, 5\right\}$.

COMMON ERROR

Consider the equation

$$(x + 6)(x + 5) = 2$$

Note that the right-hand side is 2, *not* 0. Therefore, the Zero-Product Property does *not* apply. It is incorrect to state that either $x + 6 = 2$ or $x + 5 = 2$. Instead, the first step is to put the equation in standard form.

Example 7 Solve the equation

$$(x + 6)(x + 5) = 2$$

Solution To put this equation into standard form, we first do the multiplication on the left side.

$$x^2 + 11x + 30 = 2$$

$$x^2 + 11x + 30 - 2 = 2 - 2 \qquad \text{Subtract 2.}$$

$$x^2 + 11x + 28 = 0$$

$$(x + 4)(x + 7) = 0 \qquad \text{Factor the left side.}$$

Therefore, either

$$x + 4 = 0 \quad \text{or} \quad x + 7 = 0$$

That is,

$$x = -4 \quad \text{or} \quad x = -7$$

Check these two solutions in the original equation. The solution set is $\{-4, -7\}$.

We show in the next two examples how some other equations can be solved by factoring and then using the Zero-Product Property.

Example 8 Solve $2x^3 - 2x = 0$.

Solution We first factor the common factor $2x$ from $2x^3 - 2x$.

$$2x^3 - 2x = 0$$
$$2x(x^2 - 1) = 0$$
$$2x(x + 1)(x - 1) = 0 \quad \text{Factor } x^2 - 1.$$

By the Zero-Product Property, either

$$2x = 0, \quad x + 1 = 0, \quad \text{or} \quad x - 1 = 0$$

Therefore, the solution set is $\{0, \pm 1\}$

WARMING UP

For each equation, decide if the given set is the solution set (true) or is *not* the solution set (false).

1. $(2x - 1)(x - 1) = 0; \left\{-\frac{1}{2}, -1\right\}$

2. $x(x - 2)(x + 3) = 0; \{2, -3\}$

3. $x^2 - 1 = 0; \{\pm 1\}$

4. $y^2 + 2y - 8 = 0; \{2, -4\}$

5. $y^2 - 3y = 0; \{3\}$

6. $4x^2 - 10x + 6 = 0; \left\{1, \frac{3}{2}\right\}$

7. $m^3 - 9m = 0; \{0, \pm 3\}$

8. $x^3 + 5x^2 - 14x = 0; \{0, 2, -7\}$

EXERCISE SET 5.6

For Exercises 1–12, write the solution set.

1. $(x - 2)(x - 3) = 0$

2. $(y + 1)(y - 5) = 0$

3. $(a + 3)(2a - 1) = 0$

4. $(x - 2)(3x + 2) = 0$

5. $x(x + 3) = 0$

6. $m(m - 2) = 0$

7. $2y(y + 1) = 0$

8. $3z(z - 8) = 0$

9. $(x - 1)(x - 2)(x + 1) = 0$

10. $(c + 3)(c - 4)(c + 5) = 0$

11. $(2y - 3)(y + 2)(y - 2) = 0$

12. $(x + 4)(3x - 2)(x + 7) = 0$

For Exercises 13–60, solve each equation by using the strategy for solving quadratic equations by factoring.

13. $x^2 - 3x + 2 = 0$

14. $x^2 - 2x - 3 = 0$

15. $x^2 + x - 6 = 0$

16. $y^2 - 3y - 4 = 0$

17. $2x^2 + 3x + 1 = 0$

18. $3z^2 + z - 2 = 0$

19. $3t^2 + 7t - 6 = 0$

20. $2y^2 - 7y + 6 = 0$

21. $4x^2 - 4x - 3 = 0$

22. $6k^2 - 5k - 6 = 0$

23. $6n^2 + 7n - 3 = 0$

24. $8x^2 - 10x + 3 = 0$

25. $x^2 - 4 = 0$

26. $y^2 - 9 = 0$

27. $x^2 = 16$

28. $s^2 = 25$

29. $2x^2 - 8x + 6 = 0$

30. $3x^2 - 6x - 24 = 0$

31. $2x^2 - 2 = 0$

32. $3t^2 = 12$

33. $t^3 = 100t$

34. $x^3 = x$

35. $3a^2 - 14a = 5$

36. $5x^2 - 12 = 17x$

37. $10(t + 5) - 3t(t + 5) = 0$

38. $2y(y - 2) + 5(y - 2) = 0$

39. $(2k + 3)(6k - 5) = -11$

40. $(x - 2)(6x - 1) = -4$

41. $(x + 1)^2 = 4$

42. $(r + 1)^2 - r = 7$

43. $x^3 - x^2 - 4x + 4 = 0$

44. $2y^3 - 2y^2 - 32y + 32 = 0$

45. $3y^2 - 5 = -14y$

46. $4z^2 - 15 = 17z$

47. $5a^2 - 17a = 12$

48. $b^2 = 16b - 48$

49. $2n(n - 2) = 11(n - 2)$

50. $3z(z + 5) = 10(z + 5)$

51. $(x + 1)^2 = x + 7$

52. $(r - 3)^2 = 4$

53. $(k + 1)(k - 2) = 28$

54. $(t - 2)(t + 5) = 18$

55. $x^3 - 9x = 0$

56. $y^3 - 16y = 0$

57. $4z^3 - 25z = 0$

58. $16a^3 - 9a = 0$

59. $(3t - 2)(t^2 - t - 2) = 0$

60. $(s + 1)(s^2 - 4s + 3) = 0$

SAY IT IN WORDS

61. Explain the Zero-Product Property.

62. Explain the standard form of a quadratic equation.

REVIEW EXERCISES

The following exercises review parts of Section 2.7. Doing these problems will help prepare you for the next section.

63. If $y = 2x - 5$, find y when $x = 6$.

64. If $P = 2w + 2l$, find P when $w = 4$ and $l = 7$.

65. If $3y - 2x = -4$, find x if $y = -2$.

66. Solve $y = -\frac{3}{4}x + 8$ for x.

67. Solve $\frac{y}{3} - \frac{x}{2} = 1$ for y.

ENRICHMENT EXERCISES

For Exercises 1–5, solve the equation.

1. $\frac{1}{3}x^2 - 3 = 0$

2. $x^2 - \frac{1}{4} = 0$

3. $y^2 - \frac{8}{50} = 0$

4. $6 - x^2 - x = 0$

5. $x^2 + 25 = 0$

Answers to Enrichment Exercises begin on p. A.1.

5.7 Applications of Quadratic Equations

OBJECTIVES

- ▶ *To solve word problems using quadratic equations*
- ▶ *To use the Pythagorean Theorem to solve problems*

LEARNING ADVANTAGE *Some instructors will not cover the same topics on a chapter test that appeared on an earlier quiz. Look for important ideas you have studied that have not appeared on tests or quizzes before. You may find them on the chapter test.*

In this section, we solve word problems that involve quadratic equations.

Example 1 Find two consecutive positive integers whose product is 8 more than twice the sum of the two integers.

Solution Let x and $x + 1$ be the two consecutive positive integers. From the information,

the product	is	twice the sum plus 8
$x(x + 1)$	$=$	$2(x + x + 1) + 8$

Next, we solve this equation for x.

$$x^2 + x = 2(2x + 1) + 8 \quad \text{Simplify.}$$
$$x^2 + x = 4x + 2 + 8 \quad \text{Distributive property.}$$
$$x^2 + x - 4x - 10 = 0 \quad \text{Subtract } 4x + 10 \text{ from both sides.}$$
$$x^2 - 3x - 10 = 0 \quad \text{Simplify.}$$
$$(x - 5)(x + 2) = 0$$

Therefore, either $x = 5$ or $x = -2$. Since the two numbers must be positive, we discard the negative solution.

The two consecutive positive integers are 5 and 6. Next, we check our answers. The product of 5 and 6 is 30. Is this product 8 more than twice the sum?

$$2(5 + 6) + 8 = 22 + 8 = 30$$

Our answer checks. ◀

Example 2 A carpenter needs a rectangular piece of plywood with a length that is 2 feet less than twice the width. If the area must be 40 square feet, find the dimensions of the plywood.

Solution Let x be the width of the plywood in feet, then the length is $2x - 2$ feet, as shown in the figure. Since width times length is the area,

$2x - 2$ feet

x feet

$$(\text{width})(\text{length}) = \text{area}$$
$$x(2x - 2) = 40$$
$$2x^2 - 2x = 40 \qquad \text{Distributive property.}$$
$$2x^2 - 2x - 40 = 0$$
$$2(x^2 - x - 20) = 0 \qquad \text{Factor out 2.}$$
$$2(x + 4)(x - 5) = 0 \qquad \text{Factor the resulting trinomial.}$$

Therefore, either $x = -4$ or $x = 5$. Since x represents a width, x must be positive. We discard -4, and so the width is 5 feet. The length is $2x - 2 = 2(5) - 2$, or 8 feet. ◀

COMMON ERROR

Always check your answers in a word problem to see if they make sense. In Example 2, we obtained a negative value for x, but x represented a width.

An important theorem from geometry is the Pythagorean Theorem, which gives a relationship among the sides of a right triangle. A right triangle is a triangle with a 90-degree angle.

Pythagorean Theorem. Consider the right triangle with the longest side (hypotenuse) having length c and the other two sides (legs) having lengths a and b. Then
$a^2 + b^2 = c^2$

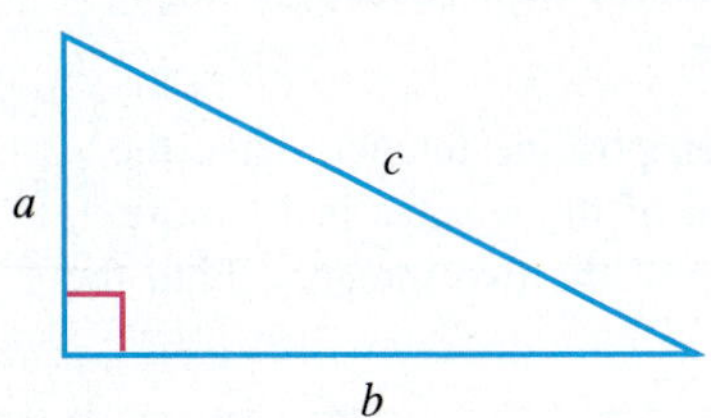

Example 3 The front of a box has a diagonal measurement of 10 inches. Find the dimensions of the front if the height is 2 inches more than the width. See the figure.

Solution Let x be the width, then $x + 2$ is the height. The diagonal is 10 inches, and it is the length of the hypotenuse of a triangle whose legs are x and $x + 2$ inches. Using the Pythagorean Theorem,

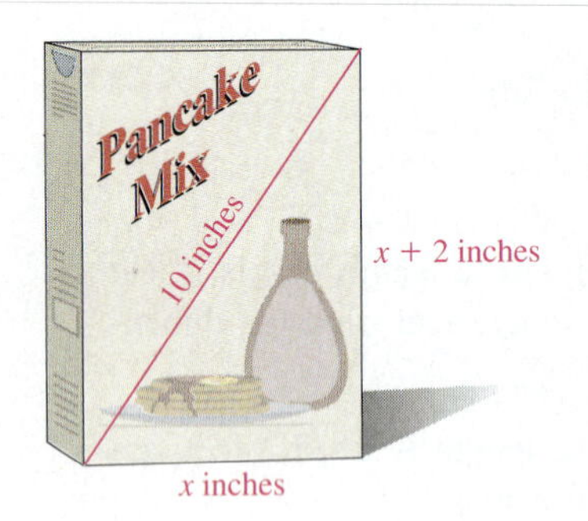

$$x^2 + (x + 2)^2 = 10^2$$

Next, we solve this equation.

$$x^2 + x^2 + 4x + 4 = 100$$

$$2x^2 + 4x - 96 = 0 \quad \text{Put in standard form.}$$

$$2(x^2 + 2x - 48) = 0 \quad \text{Factor out 2.}$$

$$2(x + 8)(x - 6) = 0 \quad \text{Factor the resulting trinomial.}$$

Therefore, either $x = -8$ or $x = 6$. We discard the negative answer. (Why?) We use 6 inches as the width and $6 + 2 = 8$ inches as the height.

WARMING UP

Answer true or false.

1. Two consecutive integers can be represented by x and $x + 1$.
2. Two consecutive even integers can be represented by x and $2x$.
3. Two consecutive odd integers can be represented by x and $x + 2$.
4. The area of a rectangle that measures x feet by $2x - 1$ feet is $2x^2 - x$ square feet.
5. The area of a triangle with a height of x inches and a base of $2x + 2$ inches is $2x^2 + 2x$ square inches.
6. If c is the length of the hypotenuse and a and b are the lengths of the two legs, respectively, of a right triangle, then $a^2 + b^2 = c^2$.

EXERCISE SET 5.7

For Exercises 1–8, solve. See Example 1.

1. Find two consecutive positive integers whose product is 3 more than 3 times the sum of the two integers.
2. Find two consecutive positive integers whose product is 4 more than 4 times the sum of the two numbers.
3. Two consecutive positive integers have the property that twice the sum of their squares is 50. Find the two integers.
4. Two consecutive positive integers have the property that 2 times the sum of their squares is 82. Find the two numbers.
5. Two consecutive even positive integers have the property that the sum of the squares is 12 more than 4 times the sum of the two integers. Find the two numbers.
6. Find two consecutive odd positive integers whose product is 11 more than twice the sum of the two integers.

7. Find two consecutive negative integers whose product is 11 more than -3 times the sum of the two integers.

8. Two consecutive negative integers have the property that the sum of their squares is 5 more than -4 times the sum of the two integers. Find the two integers.

For Exercises 9–14, solve the problem. See Example 2.

9. A steel company has an order to fill that requires making a rectangular sheet of steel. The length of the sheet must be 1 yard longer than four times the width. If the area is to be 18 square yards, what are the dimensions of the sheet?

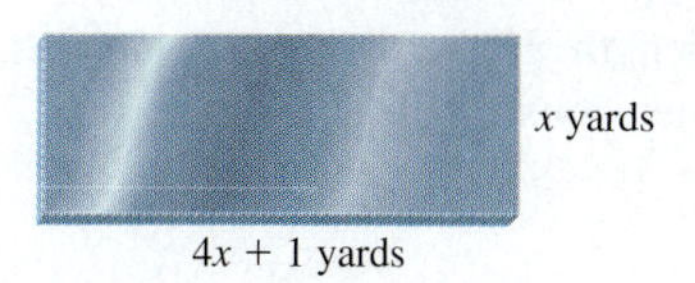

10. Acme Trophies plans to make a rectangular plaque with a length that is 3 inches shorter than twice the width. If the area is to be 54 square inches, what are dimensions of the plaque?

11. The Kislers are planning to build a rectangular work station in their kitchen. They want the surface area to be 15 square feet. If the length needs to be 2 feet more than the width, what are the dimensions of this rectangular surface?

12. The NO OUTLET sign shown in the photograph is in the shape of a square set on its corner. If the area is 9 square feet, find the length of its side.

13. A triangular reflection pool has a base of 6 feet and a height of 4 feet. It is decided to increase the area of the pool by extending the base and the height each by x feet. Find x so that the area becomes 24 square feet.

14. The triangular entrance of a tent has a base that is 2 feet less than twice the height. If the area is 30 square feet, find the base and the height of the entrance.

For Exercises 15–20, solve. See Example 3.

15. A 13-foot loading ramp is connected to a dock 5 feet high. How far is the foot of the ramp from the loading dock?

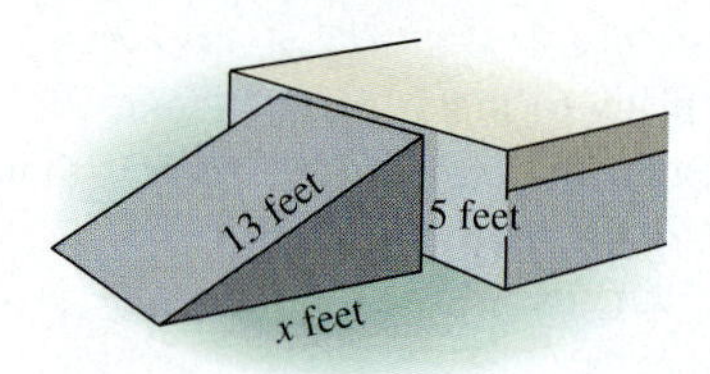

16. Kim hikes 12 miles east from camp, then 5 miles north. How far is she from camp?

17. A rectangular field has a length that is 20 yards longer than twice the width. A path that connects two opposite corners of the field is 130 yards long. What are the dimensions of the field?

18. A rectangular lot has a width that is 10 feet shorter than its length. If the diagonal is 50 feet, what are the dimensions of the lot?

19. The Barkleys are having their family room paneled. What is the widest piece of paneling that the workers can carry through the door? See the figure.

20. Two motorboats started at the same spot on a lake, one going south and the other going east. When the boat going east has traveled x miles, the other boat has traveled $x + 1$ miles and the distance between them is $x + 2$. How far apart are the two boats?

SAY IT IN WORDS

21. Explain the Pythagorean Theorem.

REVIEW EXERCISES

The following exercises review parts of Section 1.1. Doing these exercises will help prepare you for the next section.

For Exercises 22–27, write the answer as a single fraction in lowest terms.

22. $\left(\frac{3}{4}\right)\left(\frac{2}{15}\right)$

23. $\frac{3}{5} \cdot \frac{5}{9}$

24. $\frac{4}{3} + \frac{5}{3}$

25. $\frac{3}{2} + \frac{5}{6}$

26. $\frac{3}{5} \div \frac{9}{10}$

27. $\frac{2}{3} \div \frac{2}{3}$

ENRICHMENT EXERCISES

1. Sam wants to double the area of his rectangular vegetable garden by adding a strip of equal width to each of the four sides (see the figure). How wide a strip must be added?

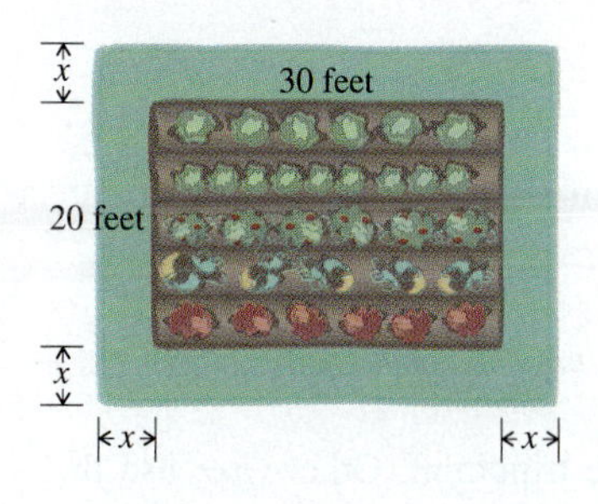

2. David, a photography student, must reduce an 8- by 10-inch photograph by decreasing the length and width by the same amount. What will be the dimensions of the reduced print if it must have only 30 percent the area of the original?

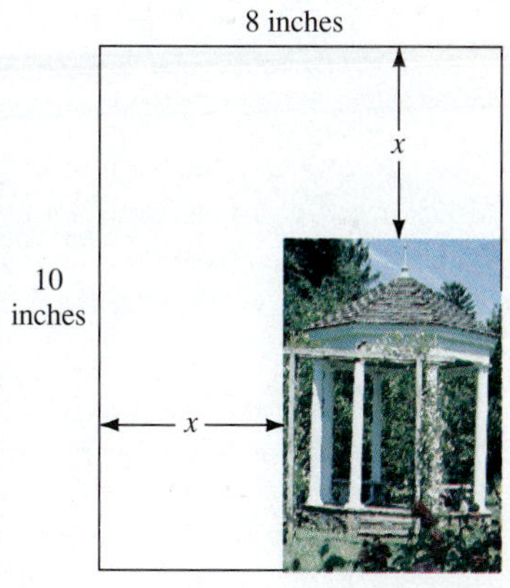

Answers to Enrichment Exercises begin on page A.1.

CHAPTER 5 Summary and review

Greatest common factor and factoring by grouping (5.1)

Examples

Factor 20 into prime numbers:
$20 = 4 \cdot 5 = 2^2 \cdot 5$

$8x^3 - 6x^2 = 2x^2(4x - 3)$

$2x^2 + 2x - ax - a = 2x(x + 1) - a(x + 1) = (x + 1)(2x - a)$

Factoring is the reverse of multiplication. $6 = 2 \cdot 3$ means 6 has been factored into $2 \cdot 3$.

The **greatest common factor** of a polynomial is obtained by factoring out the greatest common number factor, then factoring out the greatest common variable factor for each variable. **Factoring by grouping** is used when the polynomial has four terms.

Factoring trinomials (5.2 and 5.3)

$x^2 + 5x + 4 = (x + 1)(x + 4)$

$x^2 + 3x - 10 = (x + 5)(x - 2)$

$8t^2 + 6t + 1 = (2t + 1)(4t + 1)$

$6y^2 - 7y + 2 = (3y - 2)(2y - 1)$

To **factor a trinomial, $ax^2 + bx + c$,** make a list of pairs of binomials so that the product of the first terms give the first term of the trinomial and the product of the last terms give the last term of the trinomial. Then, pick out the pair that gives the correct middle term of the trinomial. With practice you will become better at finding the right pair without having to go through all possibilities.

Factoring special polynomials (5.4)

Examples

$x^2 - 4y^2 = (x + 2y)(x - 2y)$

$x^2 + 6x + 9 = (x + 3)^2$

$y^3 - 8 = (y - 2)(y^2 + 2y + 4)$

$$a^2 - b^2 = (a + b)(a - b)$$
$$a^2 + 2ab + b^2 = (a + b)^2$$
$$a^2 - 2ab + b^2 = (a - b)^2$$
$$a^3 + b^3 = (a + b)(a^2 - ab + b^2)$$
$$a^3 - b^3 = (a - b)(a^2 + ab + b^2)$$

A strategy for general factoring (5.5)

1. **Does the polynomial have a greatest common factor?** Factor out the greatest common factor, if other than one.
2. **How many terms are in the polynomial?** If the polynomial has
   (a) **Two terms:** Check for the difference of two squares or the sum or difference of two cubes.
   (b) **Three terms:** Check for a perfect square trinomial. Otherwise, use the factoring methods of Sections 5.2 or 5.3.
   (c) **Four terms:** Check for factoring by grouping.
3. **Has the polynomial been factored completely?** Check to see if the factors themselves can be factored further.

Solving quadratic equations by factoring (5.6)

Step 1 Put in the standard form $ax^2 + bx + c = 0$, if necessary.

Step 2 Factor the polynomial completely.

Step 3 Use the Zero-Product Property and set each factor equal to zero.

Step 4 Solve the resulting equations.

Step 5 Check each answer from Step 4 in the original equation.

Solve $x(2x + 5) = 3$.

$2x^2 + 5x = 3$

$2x^2 + 5x - 3 = 0$

$(x + 3)(2x - 1) = 0$

Either $x + 3 = 0$

$x = -3$

or $2x - 1 = 0$

$x = \frac{1}{2}$

Applications of quadratic equations (5.7)

An important theorem from geometry is the Pythagorean Theorem.

Consider the first triangle with the longest side (hypotenuse) having length c and the other two sides (legs) having lengths a and b.

The lengths, in meters, of three sides of a right triangle are three consecutive integers. Find these three lengths.

Let n = the smallest integer. Then the three lengths are n, $n + 1$, and $n + 2$, with $n + 2$ being the length of the hypotenuse. By the Pythagorean Theorem,

Examples

$n^2 + (n+1)^2 = (n+2)^2$

$n^2 + n^2 + 2n + 1 = n^2 + 4n + 4$

$n^2 - 2n - 3 = 0$

$(n-3)(n+1) = 0$

Either $n = 3$ or $n = -1$

The triangle has lengths of 3 meters, 4 meters, and 5 meters.

Check $3^2 + 4^2 = 9 + 16 = 25$ or 5^2

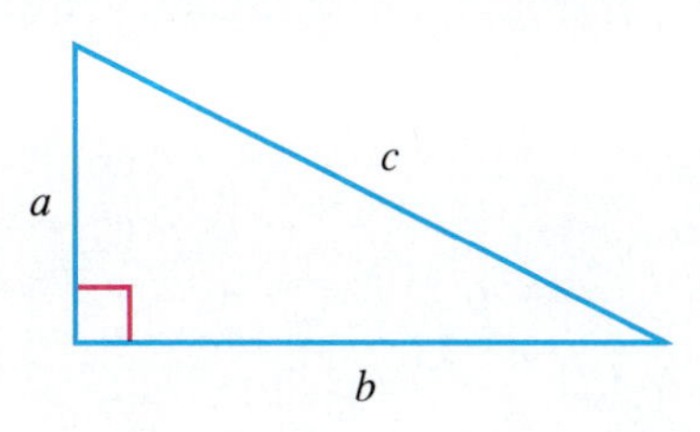

Then,

$$a^2 + b^2 = c^2$$

CHAPTER 5 REVIEW EXERCISE SET

Section 5.1

For Exercises 1 and 2, determine if the number is composite or prime.

1. 35

2. 41

3. Factor 72 into primes.

For Exercises 4–6, factor out the greatest common factor.

4. $2x^2 - 6x$

5. $y^4 - y^3 + y^2$

6. $9t^2 - 12t + 30$

For Exercises 7–10, factor by grouping.

7. $ax + 2a + b^2x + 2b^2$

8. $by - bx - 4y + 4x$

9. $z + 1 - az - a$

10. $4cx^2 + 2cz^2 - 2bx^2 - bz^2$

Section 5.2

For Exercises 11–15, factor.

11. $x^2 + 7x + 10$

12. $y^2 - y - 12$

13. $m^2 - m - 30$

14. $x^2 - 9x + 20$

15. $x^2 + 3x + 4$

Section 5.3

For Exercises 16–20, factor completely.

16. $6x^2 + 11x + 3$

17. $6y^2 - 7y - 20$

18. $4a^2 - 16a + 15$

19. $2x^2 + xt - 3t^2$

20. $8k^3 - 8k^2 - 6k$

Section 5.4

For Exercises 21–26, factor completely.

21. $x^2 - 100$

22. $9y^2 + 12y + 4$

23. $4t^2 - 9s^2$

24. $m^3 + 8$

25. $8a^3 - 1$

26. $x^5 - x^3$

Section 5.5

For Exercises 27–36, factor completely.

27. $tx^2 - ty^2 + sx^2 - sy^2$

28. $4c^5 - 12c^4 + 5c^3$

29. $2t^4 - 8t^3 + 8t^2$

30. $5x^5 - 20x^3$

31. $2at^2 + 8t^2 - 18a - 72$

32. $16d^3 + 4d^2 - 30d$

33. $18x^2 + 6x - 4$

34. $2y^2 + 50$

35. $2a^2 - 5ab + 2b^2$

36. $6x^3y + 11x^2y^2 + 3xy^3$

Section 5.6

For Exercises 37–42, solve the equation.

37. $(x - 4)(2x - 1) = 0$

38. $x^2 - 4x - 21 = 0$

39. $y^3 - 10y^2 + 25y = 0$

40. $(x - 4)(x - 1) = -2$

41. $x(2x + 3) = (x + 1)^2 + 5$

42. $x^2 + (x + 2)^2 = 8(x + 1) + 2$

Section 5.7

43. Find two consecutive positive integers so that the sum of the squares is 5 more than 4 times the sum of the two integers.

44. John is planning to build a rectangular storage shed in his back yard. He estimates that he will need an area of 60 square feet to store his lawnmower and garden equipment. If the width of the shed must be 4 feet shorter than the length, what should be the width and length of this shed?

45. A rectangular brass plate is being made so that the area is 35 square inches. If the length must be 2 inches longer than the width, what are the dimensions of the plate?

46. Sally starts at home and walks 3 miles south, then 4 miles west. How far from home is she?

CHAPTER 5 TEST

1. Factor 420 into primes

For Problems 2–4, factor out the greatest common factor, if other than 1.

2. $x^8 - x^6 + x^4$

3. $4a^4b^3 - 16a^2b^5 + 20ab^8$

4. $20x^2 - 16x + 12$

For Problems 5–19, factor completely.

5. $r^2 - 12r + 35$

6. $x^2 + 9xy - 18y^2$

7. $a^2 - 8a - 20$

8. $18b^2 + 5b - 2$

9. $7x^2 - 29x + 4$

10. $3x^4y + 13x^3y^2 - 10x^2y^3$

11. $100x^2 - 81$

12. $16a^2 - 40a + 25$

13. $9x^2 + 21x + 49$

14. $8x^3 - 125$

15. $4r^2 + 16$

16. $s^2 - 16st + 64t^2$

17. $z^3 + 64$

18. $4m^2 + m - 8mn - 2n$

19. $a^3 + 5a^2 - 16a - 80$

For Problems 20–22, solve each equation.

20. $x^2 - 7x + 12 = 0$

21. $14a^3 - 19a^2 - 3a = 0$

22. $(x - 4)(x + 6) = -21$

23. Two consecutive negative odd integers have the property that the sum of their squares is 74. Find the two integers.

24. A rectangle has a length that is 2 feet longer than its width. Find the dimensions of the rectangle if its area is 24 square feet.

CHAPTER 6

Rational Expressions

CONNECTIONS

One pipe can fill a swimming pool in 6 hours. A second pipe can do it in $4\frac{1}{2}$ hours. How long will it take to fill the pool using both sources of water? See Exercise 30 in Section 6.7.

Overview

In this chapter, we cover the algebra of rational expressions. Just as rational numbers are numerical fractions, rational expressions are algebraic fractions. We will combine algebraic fractions under the rules of addition, subtraction, multiplication, and division. In the last three sections, we solve equations with rational expressions and then look at some applications.

6.1 Rational Expressions

OBJECTIVES

- *To evaluate a rational expression*
- *To determine where a rational expression is undefined*
- *To simplify a rational expression*

LEARNING ADVANTAGE *When taking a test, remember, you will not earn points for blank paper. Always try to record as much information as possible in any problem that demands the use of more than one procedure or concept. Never hesitate to fill in any gaps in a multipart problem. However, to avoid causing confusion to the person grading your paper, it is very important that you explain what you are doing.*

Recall from Chapter 1 that rational numbers are fractions such as

$$\frac{3}{2}, \quad \frac{-21}{7}, \quad \frac{2}{1}, \quad \text{and} \quad \frac{0}{9}$$

In this chapter, we deal with fractions where the numerator and denominator are polynomials. Such algebraic fractions are called *rational expressions.*

DEFINITION

A **rational expression** is a fraction $\frac{A}{B}$, where A and B are polynomials with $B \neq 0$.

Examples of rational expressions are

$\frac{7}{y}$, $\frac{6x^2y}{9xy^2}$, and $\frac{n^2 + n}{n^2 - n + 6}$ Rational expressions can be evaluated when the variable is replaced by a number.

Example 1 Evaluate $\frac{x+5}{3x^2+1}$ when

(a) x is 1 **(b)** x is -2

Solution **(a)** Replace x by 1.

$$\frac{x+5}{3x^2+1} = \frac{1+5}{3(1)^2+1}$$
$$= \frac{6}{3+1}$$
$$= \frac{6}{4}$$
$$= \frac{3}{2} \quad \text{Divide numerator and denominator by 2.}$$

Therefore, the value of $\frac{x+5}{3x^2+1}$ is $\frac{3}{2}$ when x is 1.

(b) Replace x with -2.

$$\frac{x+5}{3x^2+1} = \frac{(-2)+5}{3(-2)^2+1}$$
$$= \frac{3}{3 \cdot 4 + 1}$$
$$= \frac{3}{13}$$

The value of $\frac{x+5}{3x^2+1}$ is $\frac{3}{13}$ when x is -2.

Given an algebraic fraction, we want to know the values of the variable for which it is undefined. To find these values, we set the denominator equal to zero and solve the resulting equation. This method is shown in the next example.

Example 2 Determine any values of x for which $\frac{x+2}{x-3}$ is undefined.

Solution The rational expression is undefined when the denominator is zero. Therefore, we set the denominator, $x - 3$, equal to zero and solve the resulting equation for x.

$$x - 3 = 0$$
$$x = 3$$

Therefore, the rational expression is undefined when x is 3.

Example 3 For what values of x, if any, is the rational expression undefined?

(a) $\frac{4x+3}{x^2+1}$ **(b)** $\frac{3}{2x^2+9x-5}$

Solution **(a)** Any value of x for which the rational expression is undefined would occur where the denominator, $x^2 + 1$, is zero. However, $x^2 + 1$ is positive no matter what number is substituted for x. Therefore, the rational expression is defined for all real numbers.

(b) We set the denominator, $2x^2 + 9x - 5$, equal to zero and solve for x.

$$2x^2 + 9x - 5 = 0$$
$$(2x - 1)(x + 5) = 0 \quad \text{Factor.}$$

Therefore, either $x = \frac{1}{2}$ or $x = -5$. Thus, the rational expression is undefined at $\frac{1}{2}$ and at -5.

Next, we turn to simplifying rational expressions. Recall from Section 1.1, a fraction like $\frac{14}{21}$ is reduced to lowest terms by dividing out factors common to both the numerator and denominator.

$$\frac{14}{21} = \frac{2 \cdot 7}{3 \cdot 7} = \frac{2 \cdot \cancel{7}}{3 \cdot \cancel{7}} = \frac{2}{3}$$

This same method can be used on rational expressions.

RULE

The Principle of Rational Expressions

If $\frac{A}{B}$ is a rational expression and K is a nonzero polynomial, then

$$\frac{AK}{BK} = \frac{A}{B}$$

That is, if a rational expression has a factor K, common to both numerator and denominator, then K can be divided out of the fraction.

$$\frac{AK}{BK} = \frac{A\cancel{K}}{B\cancel{K}} = \frac{A}{B}$$

We can use this principle to simplify rational expressions. A rational expression is in **lowest terms** if the numerator and denominator have no common factors. In Section 3.3, we learned how to divide monomials. This same technique is used to reduce a rational expression to lowest terms when the numerator and denominator are both monomials. We illustrate this process in the next example.

Example 4 Reduce $\dfrac{12x^3y^2}{15x^2y^4}$ to lowest terms.

Solution For the numerical coefficients, divide the numerator and the denominator by 3. For the variable parts, divide the powers of x by x^2 and the powers of y by y^2.

$$\frac{12x^3y^2}{15x^2y^4} = \frac{\overset{4x}{\cancel{12x^3y^2}}}{\underset{5\ \ y^2}{\cancel{15x^2y^4}}} = \frac{4x}{5y^2}$$

We can reduce more complicated rational expressions to lowest terms by *factoring* both the numerator and the denominator completely, then dividing out common factors. For example, consider the rational expression $\dfrac{x^2-4}{x-2}$. The numerator factors as $(x+2)(x-2)$, and, therefore,

$$\frac{x^2-4}{x-2} = \frac{(x+2)(x-2)}{x-2} = \frac{(x+2)\cancel{(x-2)}}{\cancel{x-2}} = \frac{x+2}{1} \quad \text{or simply} \quad x+2$$

NOTE *Notice that when $x = 2$, $\dfrac{x^2-4}{x-2}$ is undefined since the denominator is then zero, but $x+2$ has a value of $2+2$ or 4. Therefore, $\dfrac{x^2-4}{x-2}$ is not the same as $x+2$ unless it is said that x cannot equal 2.*

Example 5 Reduce each rational expression to lowest terms.

(a) $\dfrac{x^2+x-6}{x^2-9}$ **(b)** $\dfrac{2x^2-2x-12}{6x^2+15x+6}$ **(c)** $\dfrac{2t-10}{3t-15}$

Solution **(a)** We factor both the numerator and denominator completely, then divide out common factors.

$$\frac{x^2 + x - 6}{x^2 - 9} = \frac{(x + 3)(x - 2)}{(x + 3)(x - 3)}$$

$$= \frac{\cancel{(x + 3)}(x - 2)}{\cancel{(x + 3)}(x - 3)} \quad \text{Divide out the common factor, } x + 3.$$

$$= \frac{x - 2}{x - 3}$$

(b)
$$\frac{2x^2 - 2x - 12}{6x^2 + 15x + 6} = \frac{2(x^2 - x - 6)}{3(2x^2 + 5x + 2)}$$

$$= \frac{2(x + 2)(x - 3)}{3(x + 2)(2x + 1)}$$

$$= \frac{2(x - 3)}{3(2x + 1)} \quad \text{Divide out the common factor } x + 2.$$

(c) The numerator has a common factor of 2 and the denominator has a common factor of 3.

$$\frac{2t - 10}{3t - 15} = \frac{2(t - 5)}{3(t - 5)}$$

$$= \frac{2\cancel{(t - 5)}}{3\cancel{(t - 5)}} \quad \text{Divide out the common factor, } t - 5.$$

$$= \frac{2}{3}$$

COMMON ERROR

Be sure not to use incorrect "shortcuts" when dividing out common factors.

Incorrect:
$$\frac{x + 4}{x^2 - 16} = \frac{\overset{1}{\cancel{x}} + \overset{1}{\cancel{4}}}{\underset{x}{\cancel{x^2}} - \underset{4}{\cancel{16}}} = \frac{2}{x - 4}$$

Correct:
$$\frac{x + 4}{x^2 - 16} = \frac{(x + 4)}{(x + 4)(x - 4)}$$

$$= \frac{\cancel{x + 4}}{\cancel{(x + 4)}(x - 4)}$$

$$= \frac{1}{x - 4}$$

There is an important property concerning the sign of a fraction. For example,

$$\frac{2}{-3} = \frac{-2}{3} = -\frac{2}{3}$$

In general,

PROPERTY

The sign property of rational expressions of $\frac{A}{B}$ is an algebraic fraction,

$$\frac{A}{-B} = \frac{-A}{B} = -\frac{A}{B}$$

Example 6 Reduce to lowest terms:

(a) $\frac{2 - x}{x - 2}$ **(b)** $\frac{y + 7}{49 - y^2}$

Solution **(a)** First, we factor out -1 from the numerator:

$$\frac{2 - x}{x - 2} = \frac{-1(-2 + x)}{x - 2}$$

$$= \frac{-(x - 2)}{x - 2} \quad a + b = b + a.$$

$$= -\frac{x - 2}{x - 2} \quad \text{Sign property.}$$

$$= -\frac{\cancel{x - 2}}{\cancel{x - 2}} \quad \text{Divide numerator and denominator by } x - 2.$$

$$= -1$$

(b) Since the denominator, $49 - y^2$, has a negative coefficient on the highest power, we start by factoring out -1.

$$\frac{y + 7}{49 - y^2} = \frac{y + 7}{-(-49 + y^2)}$$

$$= \frac{y + 7}{-(y^2 - 49)} \quad a + b = b + a.$$

$$= -\frac{y + 7}{y^2 - 49} \quad \text{Sign property.}$$

$$= -\frac{y + 7}{(y + 7)(y - 7)} \quad \text{Factor } y^2 - 49.$$

$$= -\frac{\cancel{y+7}}{\cancel{(y+7)}(y-7)}$$

$$= -\frac{1}{y-7}$$

WARMING UP

Answer true or false.

1. $\frac{x+2}{x^2+3}$ is in lowest terms.
2. $\frac{4}{x}$ is a rational expression.
3. $\frac{2-x}{x^2+1}$ is not defined when x is 2.
4. The value of $\frac{x^2}{x+3}$ when x is 1 is $\frac{1}{4}$.
5. When x is -1, the value of $\frac{2}{x}$ is 2.
6. $\frac{3}{-4} = \frac{-3}{4}$
7. $\frac{x^2-4}{x-2}$ reduces to $\frac{1}{x+2}$
8. $\frac{y-3}{9-y^2}$ reduces to $-\frac{1}{y+3}$

EXERCISE SET 6.1

1. Evaluate $\frac{3}{x+1}$ when
   (a) x is 0 (b) x is 2 (c) x is -5
2. Evaluate $\frac{2y}{3y^2+1}$ when
   (a) y is 0 (b) y is 1 (c) y is -1

For Exercises 3–16, find the values of x, if any, for which the rational expression undefined.

3. $\frac{2}{x-5}$
4. $\frac{3}{x+4}$
5. $\frac{7}{x}$
6. $\frac{-1}{x}$
7. $\frac{8}{2x-1}$
8. $\frac{5}{3x+2}$
9. $\frac{x}{3}$
10. $\frac{x-2}{5}$
11. $\frac{3}{x^2-9}$
12. $\frac{2x}{(x-1)(x-3)}$
13. $\frac{3x+2}{x^2-8x+15}$
14. $\frac{4x-1}{x^2+4x-21}$
15. $\frac{1}{2x^2-5x+3}$
16. $\frac{3x^2+2}{6x^2+x-1}$

For Exercises 17–52, reduce to lowest terms.

17. $\dfrac{3x^4}{x^6}$

18. $\dfrac{x^5}{2x^8}$

19. $\dfrac{12x^4}{4x^7}$

20. $\dfrac{15y^2}{3y^6}$

21. $\dfrac{a^3b^2}{a^2b^5}$

22. $\dfrac{x^4y^3}{x^2y^5}$

23. $\dfrac{3u^5v^2}{6u^2v^2}$

24. $\dfrac{4y^4z^3}{16y^4z^2}$

25. $\dfrac{5x^4t^5}{10x^6t^4}$

26. $\dfrac{6x^3y^4}{8x^5y^6}$

27. $\dfrac{x^2 - 2x - 3}{x^2 - x - 6}$

28. $\dfrac{x^2 - 5x + 6}{x^2 - 3x + 2}$

29. $\dfrac{3y^2 - y - 2}{3y^2 + 11y + 6}$

30. $\dfrac{2x^2 + x - 6}{2x^2 - 5x + 3}$

31. $\dfrac{x^2 - xy - 2y^2}{xz - xa - 2yz + 2ya}$

32. $\dfrac{x^2 - 4}{xb + xc - 2b - 2c}$

33. $\dfrac{7a + 14}{6a + 12}$

34. $\dfrac{2y + 2}{3y + 3}$

35. $\dfrac{3x - 6}{9x - 18}$

36. $\dfrac{5m + 20}{m + 4}$

37. $\dfrac{3y + 9}{y^2 + 6y + 9}$

38. $\dfrac{2x - 4}{x^2 - 4x + 4}$

39. $\dfrac{h - 5}{-h + 5}$

40. $\dfrac{a^2 - 3}{3 - a^2}$

41. $\dfrac{x - 3}{9 - x^2}$

42. $\dfrac{y + 6}{36 - y^2}$

43. $\dfrac{8x - 10}{5 - 4x}$

44. $\dfrac{2y - 3}{6 - 4y}$

45. $\dfrac{2k^2 - 18}{9 - 3k}$

46. $\dfrac{2 - 2x}{4x^2 - 4}$

47. $\dfrac{xa - xb - ay + by}{xa + xb - ay - by}$

48. $\dfrac{2xt - xv + 2y^2t - y^2v}{2xt + xv + 2y^2t + y^2v}$

49. $\dfrac{x^2 - xy - 2y^2}{x^2 + 2xy + y^2}$

50. $\dfrac{2x^2 + xy - 3y^2}{4x^2 + 8xy + 3y^2}$

51. $\dfrac{3x^2 - 3x - 6}{8x^2 - 12x - 8}$

52. $\dfrac{4y^2 + 2y - 6}{6y^2 + 21y + 18}$

53. What is wrong with this method of reducing a rational expression:

$$\frac{x^2 - 3}{x} = \frac{\overset{x}{\cancel{x^2}} - 3}{\cancel{x}} = x - 3?$$

54. What is wrong with this method of reducing a rational expression:

$$\frac{a^3 + b^3}{a + b} = \frac{\overset{a^2}{\cancel{a^3}} + \overset{b^2}{\cancel{b^3}}}{\cancel{a} + \cancel{b}} = \frac{a^2 + b^2}{2}$$

55. Evaluate $\frac{x^3 + 1}{x^2 - x + 1}$ and $x + 1$ when $x = 2$. Why do you get the same answer?

56. Evaluate $\frac{y^3 - 8}{y - 2}$ and $y^2 + 2y + 4$ when $y = -2$. Why do you get the same answer?

57. Evaluate $\frac{4n + 5}{5n^2 - 14n - 24}$ for

(a) $n = 1.6$
(b) $n = -2$
(c) What happens when you attempt to evaluate the rational expression for $n = -1.2$?

SAY IT IN WORDS

58. What is a rational expression?

59. How do you write an algebraic fraction in lowest terms?

REVIEW EXERCISES

The following exercises review parts of Section 1.1. Doing these exercises will help prepare you for the next section.

For Exercises 60–64, find the product or quotient and write the answer in lowest terms.

60. $\frac{4}{5} \cdot \frac{15}{8}$

61. $\frac{3}{2} \cdot \frac{8}{9}$

62. $\frac{1}{2}\left(-\frac{4}{3}\right)$

63. $\frac{6}{7} \div \frac{9}{14}$

64. $\left(-\frac{4}{5}\right) \div \left(-\frac{8}{15}\right)$

ENRICHMENT EXERCISES

Reduce to lowest terms.

1. $\frac{x^4 - 4}{2x^2 + 4}$

2. $\frac{y^4 - 1}{-y^2 + y + 2}$

3. $\frac{x^3 + 1}{x^2 - 2x - 3}$

4. $\frac{y^3 - 8}{y^3 + 2y^2 + 4y}$

Answers to Enrichment Exercises begin on page A.1.

6.2 Multiplication and Division of Rational Expressions

OBJECTIVES

- *To multiply rational expressions and write in lowest terms*
- *To divide rational expressions and write in lowest terms*

LEARNING ADVANTAGE

The solution to a problem is not always immediate. When working on a problem where the answer is not obvious on the first effort, keep trying. Continue to work until you have the answer or until you are frustrated; then put it aside. During the time you do not work on the problem, your subconscious continues to work. Sometimes the solution will appear to you when you least expect it. But be sure to return to the problem.

To multiply two algebraic fractions, multiply the numerators and multiply the denominators.

RULE

Multiplication of Two Rational Expressions

If $\frac{A}{B}$ and $\frac{C}{D}$ are two rational expressions, then

$$\frac{A}{B} \cdot \frac{C}{D} = \frac{AC}{BD}$$

For example,

$$\begin{aligned}\frac{3}{5} \cdot \frac{25}{36} &= \frac{3 \cdot 25}{5 \cdot 36} \\ &= \frac{75}{180} \\ &= \frac{\cancel{15} \cdot 5}{\cancel{15} \cdot 12} \\ &= \frac{5}{12}\end{aligned}$$

The reducing we did after multiplying is usually done before multiplying. First reduce, then multiply.

$$\frac{3}{5} \cdot \frac{25}{36} = \frac{\cancel{3}}{\cancel{5}} \cdot \frac{\cancel{5} \cdot 5}{\cancel{3} \cdot 12} = \frac{5}{12}$$

Example 1 Multiply.

(a) $\dfrac{3x^2}{14} \cdot \dfrac{7}{x^3}$ (b) $\dfrac{(x-2)^5}{x+3} \cdot \dfrac{x+3}{(x-2)^3}$

Solution (a) Divide out common factors, then multiply.

$$\frac{3x^2}{14} \cdot \frac{7}{x^3} = \frac{3\cancel{x^2}}{\underset{2}{\cancel{14}}} \cdot \frac{\cancel{7}}{\underset{x}{\cancel{x^3}}}$$

$$= \frac{3 \cdot 1}{2} \cdot \frac{1}{x}$$

$$= \frac{3}{2x}$$

(b) $$\frac{(x-2)^5}{x+3} \cdot \frac{x+3}{(x-2)^3} = \frac{\overset{(x-2)^2}{\cancel{(x-2)^5}}}{\cancel{x+3}} \cdot \frac{\cancel{x+3}}{\cancel{(x-2)^3}}$$

$$= \frac{(x-2)^2}{1} \cdot \frac{1}{1}$$

$$= (x-2)^2$$

In the next example, we find the product of more complicated rational expressions. Be sure to factor the numerators and denominators completely, then divide out common factors.

Example 2 Find the indicated product.

$$\frac{4t+2}{3t-4} \cdot \frac{3t-4}{2t^2-9t-5}$$

Solution First, we factor the numerator of the first fraction and the denominator of the second fraction, then multiply. The final step is to divide out all common factors.

$$\frac{4t+2}{3t-4} \cdot \frac{3t-4}{2t^2-9t-5} = \frac{2(2t+1)}{3t-4} \cdot \frac{3t-4}{(2t+1)(t-5)}$$ Factor completely.

$$= \frac{2\cancel{(2t+1)}}{\cancel{3t-4}} \cdot \frac{\cancel{3t-4}}{\cancel{(2t+1)}(t-5)}$$ Divide out common factors.

$$= \frac{2}{t-5}$$ Multiply numerators and denominators.

The next example shows how to multiply a polynomial and a rational expression.

Example 3 Multiply: $(2x+1)\cdot\dfrac{5}{x^2-7x-4}$

Solution First, write $2x + 1$ as $\dfrac{2x+1}{1}$.

$$(2x+1)\cdot\frac{5}{x^2-7x-4} = \frac{2x+1}{1}\cdot\frac{5}{x^2-7x-4}$$

$$= \frac{2x+1}{1}\cdot\frac{5}{(2x+1)(x-4)} \qquad \text{Factor.}$$

$$= \frac{\cancel{2x+1}}{1}\cdot\frac{5}{\cancel{(2x+1)}(x-4)} \qquad 2x+1 \text{ is a common factor.}$$

$$= \frac{5}{x-4} \qquad \text{Multiply.}$$

Example 4 Multiply: $\dfrac{x^2-y^2}{x+2y}\cdot\dfrac{2x+4y}{3x^2+2xy-y^2}$

Solution

$$\frac{x^2-y^2}{x+2y}\cdot\frac{2x+4y}{3x^2+2xy-y^2} = \frac{(x+y)(x-y)}{x+2y}\cdot\frac{2(x+2y)}{(x+y)(3x-y)} \qquad \text{Factor.}$$

$$= \frac{\cancel{(x+y)}(x-y)}{\cancel{x+2y}}\cdot\frac{2\cancel{(x+2y)}}{\cancel{(x+y)}(3x-y)} \qquad \text{Divide out common factors.}$$

$$= \frac{2(x-y)}{3x-y} \qquad \text{Multiply.}$$

We now summarize the steps in multiplying rational expressions.

PROPERTY

Multiplying Rational Expressions

Step 1 Factor the numerators and denominators.

Step 2 Divide out any common factors.

Step 3 Write the product of the factors that remain in the numerators over the product of the factors that remain in the denominators.

From Chapter 1, recall the method for dividing fractions.

$$\frac{3}{2}\div\frac{9}{8}=\frac{3}{2}\cdot\frac{8}{9}$$

Division was performed by multiplying the first fraction by the reciprocal of the second fraction. Dividing two algebraic fractions follows the same method.

RULE

Division of Rational Expressions

If $\frac{A}{B}$ and $\frac{C}{D}$ are two rational expressions, where $\frac{C}{D} \neq 0$,

$$\frac{A}{B} \div \frac{C}{D} = \frac{A}{B} \cdot \frac{D}{C}$$

That is, invert the divisor $\frac{C}{D}$ and multiply.

Example 5 Divide:

(a) $\frac{6x^3}{y^2} \div \frac{3x^5}{y^4}$ **(b)** $\frac{10t - 20}{(t + 3)^3} \div \frac{5}{2t + 6}$

Solution **(a)** $\frac{6x^3}{y^2} \div \frac{3x^5}{y^4} = \frac{6x^3}{y^2} \cdot \frac{y^4}{3x^5}$ Invert and multiply.

$= \frac{\overset{2}{\cancel{6}}\cancel{x^3}}{\cancel{y^2}} \cdot \frac{\overset{y^2}{\cancel{y^4}}}{\cancel{3}\underset{x^2}{\cancel{x^5}}}$ Divide out common factors.

$= \frac{2y^2}{x^2}$ Multiply.

(b) $\frac{10t - 20}{(t + 3)^3} \div \frac{5}{2t + 6} = \frac{10t - 20}{(t + 3)^3} \cdot \frac{2t + 6}{5}$ Invert and multiply.

$= \frac{10(t - 2)}{(t + 3)^3} \cdot \frac{2(t + 3)}{5}$ Factor.

$= \frac{\overset{2}{\cancel{10}}(t - 2)}{\underset{(t+3)^2}{\cancel{(t + 3)^3}}} \cdot \frac{2\cancel{(t + 3)}}{\cancel{5}}$ Divide out common factors.

$= \frac{4(t - 2)}{(t + 3)^2}$ Multiply. ◀

Example 6 Divide:

(a) $\frac{x^2 - x - 2}{x + 5} \div (x - 2)$ **(b)** $\frac{3 - y}{6y + 12} \div \frac{y^2 - 9}{4y + 8}$

Solution (a) The reciprocal of $x - 2$ is $\frac{1}{x-2}$, therefore

$$\frac{x^2 - x - 2}{x + 5} \div (x - 2) = \frac{x^2 - x - 2}{x + 5} \cdot \frac{1}{x - 2}$$

$$= \frac{(x+1)(x-2)}{x+5} \cdot \frac{1}{x-2} \quad \text{Factor.}$$

$$= \frac{x+1}{x+5} \quad \text{Reduce.}$$

(b)
$$\frac{3 - y}{6y + 12} \div \frac{y^2 - 9}{4y + 8} = \frac{3 - y}{6y + 12} \cdot \frac{4y + 8}{y^2 - 9}$$

$$= \frac{-(y-3)}{6(y+2)} \cdot \frac{4(y+2)}{(y+3)(y-3)} \quad \text{Factor.}$$

$$= \frac{-2}{3(y+3)} \quad \text{Reduce.}$$

$$= -\frac{2}{3(y+3)} \quad \frac{-A}{B} = -\frac{A}{B}.$$

WARMING UP

Decide if the multiplication or division is correct (true) or incorrect (false).

1. $\frac{2}{3} \cdot \frac{9}{4} = \frac{3}{2}$

2. $\frac{4}{15} \div \frac{16}{5} = 12$

3. $\frac{x^2}{y^3} \cdot \frac{y^4}{x^5} = \frac{x^3}{y}$

4. $\frac{4x^2}{7y^4} \div \frac{8x^3}{7y^4} = \frac{1}{2x}$

5. $\frac{(x-2)^2}{(x-3)^4} \cdot \frac{x-3}{x-2} = \frac{x-2}{(x-3)^5}$

6. $\frac{x^2 - 3xy + 2y^2}{x + y} \div (x - y) = \frac{x - 2y}{x + y}$

EXERCISE SET 6.2

For Exercises 1–18, multiply.

1. $\frac{2}{5x^3} \cdot \frac{15x^5}{4}$

2. $\frac{2x^4}{3} \cdot \frac{9}{x^6}$

3. $\frac{5y^3}{2} \cdot \frac{8}{4y^2}$

4. $\frac{4}{3t^2} \cdot \frac{9t^2}{2}$

5. $\frac{3u^2v^4}{8} \cdot \frac{8}{u^3v^5}$

6. $\frac{x^3y^2}{14} \cdot \frac{2}{x^2y^3}$

7. $\frac{(x+4)^2}{(x+3)^3} \cdot \frac{(x+3)^4}{(x+4)^4}$

8. $\frac{(a-2)}{(a+5)^2} \cdot \frac{(a+5)^3}{(a-2)^2}$

9. $\frac{3(t+6)^3}{2(t-1)} \cdot \frac{4(t-1)}{9(t+6)}$

10. $\frac{7(x+3)^4}{8(x-3)^5} \cdot \frac{16(x-3)^2}{21(x+3)^4}$

11. $\frac{x+1}{x-2} \cdot \frac{x-2}{x^2-2x-3}$

12. $\frac{y^2-5y+4}{(y+5)^2} \cdot \frac{y+5}{y-4}$

13. $\frac{x^2-4}{12} \cdot \frac{18}{x^2-x-2}$

14. $\frac{t+1}{t-3} \cdot \frac{t^2-9}{t^2-1}$

15. $(3x-1)\frac{x^2-4x-5}{3x^2-x}$

16. $(2y+6)\frac{y^2-3y-4}{y+3}$

17. $\frac{s^2-t^2}{3s^2+6st+3t^2} \cdot \frac{t+s}{t-s}$

18. $\frac{a-2ab+b^2}{a^2-b^2} \cdot \frac{a+b}{2a^2-ab-b^2}$

For Exercises 19–30, divide.

19. $\frac{2x^4}{3y^5} \div \frac{4x^2}{9y^4}$

20. $\frac{4a^3}{5b^4} \div \frac{16a^6}{25b^5}$

21. $\frac{x^3+x^2}{x-4} \div \frac{x^2-1}{x^2-16}$

22. $\frac{2y-1}{3y^8} \div \frac{2y-1}{4y}$

23. $\frac{x-10}{15} \div \frac{(x-10)^2}{5}$

24. $\frac{5y-8}{10y-1} \div \frac{(5y-8)^3}{(10y-1)^2}$

25. $\frac{3-b}{12} \div \frac{b-3}{4}$

26. $\frac{x^2-25}{x-2} \div \frac{x+5}{4-x^2}$

27. $\frac{4z^2-13z+3}{2z+1} \div (4z-1)$

28. $\frac{2x^2-x-3}{x-1} \div (2x-3)$

29. $\frac{x^2+2xy+y^2}{2x^2-3xy+y^2} \div \frac{x+y}{x-y}$

30. $\frac{2a^2+3ab+b^2}{a^2-b^2} \div \frac{2a^2-ab-b^2}{a-b}$

SAY IT IN WORDS

31. How do you multiply rational expressions?

32. Explain the steps you take to divide rational expressions.

REVIEW EXERCISES

The following exercises review parts of Section 6.1. Doing these exercises will help prepare you for the next section.

For Exercises 33–38, write each rational expression in lowest terms.

33. $\frac{x+1}{2(x+1)^2}$

34. $\frac{4t}{10t^3}$

35. $\frac{6x-3}{2x^2+5x-3}$

36. $\frac{12w-12}{w^2+w-2}$

37. $\frac{2y^2-5y+2}{4-2y}$

38. $\frac{a^2-b^2}{a^2-2ab+b^2}$

ENRICHMENT EXERCISES

Perform the operations.

1. $24x^3 \div 12x^5 \div 6x$

2. $\dfrac{x^4 - y^4}{x^2 - 2xy + y^2} \cdot \dfrac{x + y}{x^2 + y^2}$

3. $\dfrac{xt - at - xv + av}{t - v} \div (x - a)$

Answers to Enrichment Exercises begin on page A.1.

6.3 Addition and Subtraction of Rational Expressions

OBJECTIVES

- *To add and subtract rational expressions with the same denominators*
- *To add and subtract rational expressions with different denominators*
- *To simplify the result of combining rational expressions*

LEARNING ADVANTAGE

Is your answer reasonable? *When working on a word problem, make sure that the answer makes sense. If your answer is that 18.4 people bought tickets for the concert, or that the toolbox measured negative 5 feet on each side, stop and review your calculations.*

We add and subtract rational expressions by the same methods as we add or subtract fractions. If they have the same denominators, we have the following rule.

RULE

Addition and Subtraction of Rational Expressions with Common Denominators

If $\dfrac{A}{B}$ and $\dfrac{C}{B}$ are rational expressions with common denominator B, then

$$\frac{A}{B} + \frac{C}{B} = \frac{A + C}{B}$$

and

$$\frac{A}{B} - \frac{C}{B} = \frac{A - C}{B}$$

That is, to add or subtract two rational expressions with common denominators, add or subtract the numerators, keeping the common denominator.

Example 1 Add or subtract as indicated.

(a) $\frac{5}{a} + \frac{7}{a} = \frac{5+7}{a}$ Add numerators.

$= \frac{12}{a}$

(b) $\frac{14}{t} - \frac{18}{t} = \frac{14-18}{t}$ Subtract numerators.

$= -\frac{4}{t}$

(c) $\frac{3y-2}{y-5} + \frac{8y+7}{y-5} = \frac{3y-2+8y+7}{y-5}$ Add numerators.

$= \frac{11y+5}{y-5}$ Combine terms in the numerator.

Sometimes the result from adding or subtracting two rational expressions can be simplified by reducing it to lowest terms.

Example 2 Subtract, then simplify by reducing the answer to lowest terms.

$$\frac{3t}{(t-1)(t-2)} - \frac{6}{(t-1)(t-2)}$$

Solution We subtract numerators, keeping the same denominator.

$$\frac{3t}{(t-1)(t-2)} - \frac{6}{(t-1)(t-2)} = \frac{3t-6}{(t-1)(t-2)}$$

$$= \frac{3(t-2)}{(t-1)(t-2)} \quad \text{Factor the numerator.}$$

$$= \frac{3\cancel{(t-2)}}{(t-1)\cancel{(t-2)}} \quad \text{Divide out } t-2.$$

$$= \frac{3}{t-1}$$

When adding or subtracting two rational expressions with unlike denominators, first find the *least common denominator.*

DEFINITION

The **least common denominator (LCD)** of a collection of rational expressions is the polynomial of lowest degree that all denominators divide into evenly.

For example, a common denominator of $\frac{5}{4}$ and $\frac{1}{3}$ is 12, since 4 and 3 both divide into 12 evenly. Furthermore, 12 is the *smallest* positive number that both 4 and 3 divide. Therefore, 12 is the *least* common denominator of $\frac{5}{4}$ and $\frac{1}{3}$. We write LCD = 12.

To find the least common denominator of two rational expressions, use the following guide.

STRATEGY

To Find the Least Common Denominator of Two Rational Expressions

Step 1 Factor each denominator completely.

Step 2 Write the product of each different factor taken the highest number of times that it occurs in any denominator.

The LCD is this product.

Example 3 Find the least common denominator of $\frac{5}{12}$ and $\frac{7}{30}$.

Solution We use the two-step method for finding the least common denominator. First, factor the two denominators 12 and 30 completely.

Step 1 $12 = 2 \cdot 2 \cdot 3$ and $30 = 2 \cdot 3 \cdot 5$.

Step 2 The least common denominator is formed by taking each factor the highest number of times that it occurs in either denominator. The factor 2 occurs twice in one product and once in the other, so we use $2 \cdot 2$; the numbers 3 and 5 each occur once. Therefore,

$$\text{LCD} = 2 \cdot 2 \cdot 3 \cdot 5 = 60$$

Example 4 Find the least common denominator of $\frac{9}{8a^5}$ and $\frac{a+2}{12a^2}$.

Solution The two denominators are $8a^5$ and $12a^2$.

Step 1 We factor both denominators writing the numerical coefficients as products of prime numbers.

$$8a^5 = 2 \cdot 2 \cdot 2a^5 \qquad \text{and} \qquad 12a^2 = 2 \cdot 2 \cdot 3a^2$$

Step 2 The least common denominator is obtained by forming the product of 2 used three times, 3 used once, and a used five times:

$$\text{LCD} = 2^3 \cdot 3 \cdot a^5 = 24a^5$$

Example 5 Find the least common denominator.

(a) $\dfrac{5}{6q^2}$ and $\dfrac{8}{3q^2 - 3q}$ (b) $\dfrac{2z-3}{z-1}$ and $\dfrac{6}{z^2-1}$

(c) $\dfrac{2x}{x^2+x-2}$ and $\dfrac{3}{2x^2+x-6}$ (d) $\dfrac{1}{m-3}$ and $\dfrac{2}{3-m}$

(e) $r - 1$ and $\dfrac{5}{2r}$

Solution (a) The denominators are $6q^2$ and $3q^2 - 3q$.

Step 1 We factor the two denominators, $6q^2$ and $3q^2 - 3q$.

$$6q^2 = 2 \cdot 3q^2 \qquad \text{and} \qquad 3q^2 - 3q = 3q(q-1)$$

Step 2 The least common denominator is

$$\text{LCD} = 2 \cdot 3q^2(q-1) = 6q^2(q-1)$$

(b) The two denominators are $z - 1$ and $z^2 - 1$.

Step 1 The denominator $z - 1$ cannot be factored. The other denominator, $z^2 - 1$, factors into

$$(z+1)(z-1)$$

Step 2 Since $z + 1$ and $z - 1$ each occur once in either denominator, the least common denominator is

$$\text{LCD} = (z+1)(z-1) = z^2 - 1$$

(c) Start by factoring.

Step 1 Factor each denominator.

$$x^2 + x - 2 = (x+2)(x-1)$$
$$2x^2 + x - 6 = (x+2)(2x-3)$$

Step 2 The least common denominator is

$$(x+2)(x-1)(2x-3)$$

(d) Since

$$m - 3 = m - 3$$

and

$$3 - m = (-1)(m-3)$$

one denominator is the negative of the other. Therefore, either $m - 3$ or $3 - m$ can be used as the LCD.

(e) The expression $r - 1$ can be written as $\frac{r-1}{1}$. Therefore, the two denominators are 1 and $2r$. The least common denominator is $1 \cdot 2r = 2r$.

We use the same method to find the least common denominator of three or more fractions.

Example 6 Find the least common denominator.

(a) $\frac{5}{6}, \frac{1}{12}$, and $\frac{7}{15}$ **(b)** $\frac{4}{3y}, \frac{5}{2y^2}$, and $\frac{8y}{y-1}$

Solution **(a)** We factor the three denominators 6, 12, and 15.

$$6 = 2 \cdot 3, \qquad 12 = 2^2 \cdot 3, \qquad 15 = 3 \cdot 5$$

Therefore,

$$\text{LCD} = 2^2 \cdot 3 \cdot 5 = 60$$

(b) The three denominators are $3y$, $2y^2$, and $y - 1$. The least common denominator is

$$3 \cdot 2 \cdot y^2(y-1) = 6y^2(y-1)$$

Using the principle of rational expressions, we are allowed to reduce an expression to lowest terms. We can use the same principle to "build up" fractions. Namely, for nonzero K,

$$\frac{A}{B} = \frac{AK}{BK}$$

We will use this principle to add two rational expressions with unlike denominators.

Example 7 Add: $\frac{3}{2} + \frac{5}{4}$

Solution The least common denominator of $\frac{3}{2}$ and $\frac{5}{4}$ is 4. So, we "buildup" $\frac{3}{2}$ as $\frac{3 \cdot 2}{2 \cdot 2}$ or $\frac{6}{4}$.

$$\begin{aligned} \frac{3}{2} + \frac{5}{4} &= \frac{3 \cdot 2}{2 \cdot 2} + \frac{5}{4} \\ &= \frac{6}{4} + \frac{5}{4} \\ &= \frac{6+5}{4} \\ &= \frac{11}{4} \end{aligned}$$

STRATEGY

Addition and Subtraction of Rational Expressions with Unlike Denominators

Step 1 Find the least common denominator.

Step 2 Rewrite each fraction using the least common denominator.

Step 3 Add or subtract the numerators, keeping the same denominator.

Step 4 Simplify the numerator, then write the fraction in lowest terms.

Example 8 Add or subtract as indicated. Write the answer in lowest terms.

(a) $\dfrac{5}{3y} - \dfrac{7}{6y}$ **(b)** $\dfrac{12v}{v^2 - 4} + \dfrac{4v}{v - 2}$

Solution **(a)** The least common denominator of $\dfrac{5}{3y}$ and $\dfrac{7}{6y}$ is $6y$. Therefore,

$$\begin{aligned}\frac{5}{3y} - \frac{7}{6y} &= \frac{5 \cdot 2}{3y \cdot 2} - \frac{7}{6y}\\ &= \frac{10}{6y} - \frac{7}{6y}\\ &= \frac{10 - 7}{6y}\\ &= \frac{3}{6y}\\ &= \frac{1}{2y} \qquad \text{3 is a common factor.}\end{aligned}$$

(b) Write the two denominators in factored form.

$$v^2 - 4 = (v + 2)(v - 2)$$
$$v - 2 = v - 2$$

Therefore, the least common denominator is

$$(v + 2)(v - 2)$$

Next, write both fractions with this common denominator, then add.

$$\begin{aligned}\frac{12v}{v^2 - 4} + \frac{4v}{v - 2} &= \frac{12v}{(v + 2)(v - 2)} + \frac{4v}{v - 2}\\ &= \frac{12v}{(v + 2)(v - 2)} + \frac{4v(v + 2)}{(v + 2)(v - 2)}\end{aligned}$$

$$= \frac{12v + 4v(v + 2)}{(v + 2)(v - 2)}$$

$$= \frac{12v + 4v^2 + 8v}{(v + 2)(v - 2)} \quad \text{Distributive property.}$$

$$= \frac{4v^2 + 20v}{(v + 2)(v - 2)} \quad \text{Combine terms in the numerator.}$$

Next, check to see if the answer is in lowest terms. We factor the numerator.

$$\frac{4v^2 + 20v}{(v + 2)(v - 2)} = \frac{4v(v + 5)}{(v + 2)(v - 2)}$$

In this factored form, we see that there are no common factors to be divided out. Our answer is already in lowest terms. It is acceptable to keep the answer in this factored form. ◀

Example 9 Add or subtract as indicated, then simplify.

(a) $\frac{v^2 - 11}{v^2 - 5v + 6} - \frac{v - 5}{v - 3}$ **(b)** $(2w - 3) + \frac{w}{w + 5}$

Solution **(a)** $\frac{v^2 - 11}{v^2 - 5v + 6} - \frac{v - 5}{v - 3} = \frac{v^2 - 11}{(v - 3)(v - 2)} - \frac{v - 5}{v - 3}$

$$= \frac{v^2 - 11}{(v - 3)(v - 2)} - \frac{(v - 5)(v - 2)}{(v - 3)(v - 2)}$$

The least common denominator is $(v - 3)(v - 2)$.

$$= \frac{(v^2 - 11) - (v - 5)(v - 2)}{(v - 3)(v - 2)} \quad \text{Subtract numerators.}$$

$$= \frac{v^2 - 11 - (v^2 - 7v + 10)}{(v - 3)(v - 2)} \quad \text{Simplify the numerator.}$$

$$= \frac{v^2 - 11 - v^2 + 7v - 10}{(v - 3)(v - 2)}$$

$$= \frac{7v - 21}{(v - 3)(v - 2)}$$

$$= \frac{7(v - 3)}{(v - 3)(v - 2)} \quad \text{Factor } 7v - 21.$$

$$= \frac{7(v - 3)}{(v - 3)(v - 2)} \quad \text{Divide out } v - 3.$$

$$= \frac{7}{v - 2}$$

(b) $(2w - 3) + \dfrac{w}{w+5} = \dfrac{2w-3}{1} + \dfrac{w}{w+5}$

The least common denominator is $1 \cdot (w + 5)$.

$$= \frac{(2w-3)(w+5)}{w+5} + \frac{w}{w+5}$$

$$= \frac{(2w-3)(w+5) + w}{w+5}$$

$$= \frac{2w^2 + 7w - 15 + w}{w+5}$$

$$= \frac{2w^2 + 8w - 15}{w+5}$$

This fraction is in lowest terms. Check.

WARMING UP

Decide if the addition or subtraction is correct (true) or incorrect (false).

1. $\dfrac{2}{3} + \dfrac{4}{3} = 2$

2. $\dfrac{8}{7} - \dfrac{1}{7} = 7$

3. $\dfrac{3}{4} + \dfrac{1}{2} = \dfrac{5}{4}$

4. $\dfrac{4}{7} - \dfrac{3}{2} = -\dfrac{13}{14}$

5. $\dfrac{2}{x} + \dfrac{3}{x} = \dfrac{6}{x}$

6. $\dfrac{2x}{x+1} + \dfrac{2}{x+1} = 2$

7. $\dfrac{3}{x-1} + \dfrac{4}{x+1} = \dfrac{7}{(x-1)(x+1)}$

8. $\dfrac{1}{x^2 - 3x + 2} + \dfrac{2}{x-1} = \dfrac{2x-3}{(x-2)(x-1)}$

EXERCISE SET 6.3

For Exercises 1–14, add or subtract as indicated. Write the answer in lowest terms.

1. $\dfrac{6}{z} + \dfrac{11}{z}$

2. $\dfrac{14}{a} - \dfrac{12}{a}$

3. $\dfrac{40}{z^2} - \dfrac{20}{z^2}$

4. $\dfrac{18}{r^3} - \dfrac{13}{r^3}$

5. $\dfrac{3}{b+1} + \dfrac{7}{b+1}$

6. $\dfrac{18}{c-3} + \dfrac{5}{c-3}$

7. $\dfrac{17}{a^2+7} - \dfrac{8}{a^2+7}$

8. $\dfrac{8}{2+w^2} - \dfrac{5}{2+w^2}$

9. $\dfrac{2y}{(y-3)(y+1)} + \dfrac{2}{(y-3)(y+1)}$

10. $\dfrac{3r}{(r-1)(r-2)} - \dfrac{6}{(r-1)(r-2)}$

11. $\dfrac{2t}{(2t-3)(t+1)} - \dfrac{3}{(2t-3)(t+1)}$

12. $\dfrac{3v}{(3v-1)(v+2)} - \dfrac{1}{(3v-1)(v+2)}$

13. $\dfrac{u}{u^2-9} - \dfrac{3}{u^2-9}$

14. $\dfrac{x}{x^2-100} + \dfrac{10}{x^2-100}$

For Exercises 15–34, find the least common denominator.

15. $\dfrac{3}{16}$ and $\dfrac{1}{8}$

16. $\dfrac{5}{24}$ and $\dfrac{3}{8}$

17. $\dfrac{-1}{12}$ and $\dfrac{10}{9}$

18. $\dfrac{3}{10}$ and $\dfrac{-2}{15}$

19. $\dfrac{13}{3y}$ and $\dfrac{9}{4y}$

20. $\dfrac{8}{5x}$ and $\dfrac{2}{3x}$

21. $\dfrac{1}{3a^2}$ and $\dfrac{-5}{7a^5}$

22. $\dfrac{25}{4b^3}$ and $\dfrac{-16}{3b^2}$

23. $\dfrac{3}{4t^3}$ and $\dfrac{7}{15t^4}$

24. $\dfrac{19}{9u^4}$ and $\dfrac{11}{12u^2}$

25. $\dfrac{2w}{w^2+3w}$ and $\dfrac{w-4}{3w^2+9w}$

26. $\dfrac{-16}{4x^2-4x}$ and $\dfrac{1-x}{2x^2-2x}$

27. $\dfrac{1+a^2}{a^2-1}$ and $\dfrac{a}{a+1}$

28. $\dfrac{x+1}{x^2-9}$ and $\dfrac{18}{x-3}$

29. $\dfrac{n+1}{n^2+n-2}$ and $\dfrac{3+n}{n^2-n-6}$

30. $\dfrac{y-1}{y^2+y-12}$ and $\dfrac{1}{y^2-2y-3}$

31. $\dfrac{5}{1-z}$ and $\dfrac{2}{z-1}$

32. $\dfrac{-34}{4-s}$ and $\dfrac{21}{s-4}$

33. $2c+1$ and $\dfrac{1}{c-1}$

34. $\dfrac{1}{2h-7}$ and $h+5$

For Exercises 35–60, add or subtract as indicated. Write the answer in lowest terms.

35. $\dfrac{3}{5} + \dfrac{7}{10}$

36. $\dfrac{2}{3} + \dfrac{1}{6}$

37. $\dfrac{11}{12} - \dfrac{5}{9}$

38. $\dfrac{13}{15} - \dfrac{1}{6}$

39. $\dfrac{4}{5y} + \dfrac{6}{15y}$

40. $\dfrac{5}{12s} + \dfrac{5}{3s}$

41. $\dfrac{17}{18h} - \dfrac{7}{9h}$

42. $\dfrac{12}{7c} - \dfrac{9}{14c}$

43. $\dfrac{1}{r^2-4} + \dfrac{r}{r+2}$

44. $\dfrac{2s}{s-5} + \dfrac{12}{s^2-25}$

45. $\dfrac{4v+3}{v^2-9} - \dfrac{v+1}{v-3}$

46. $\dfrac{2n-1}{n+6} - \dfrac{6-13n}{n^2-36}$

47. $\dfrac{3}{m^2-3m+2} + \dfrac{2}{m^2-m-2}$

48. $\dfrac{4}{p^2+5p+6} + \dfrac{1}{p^2+2p-3}$

49. $\dfrac{z}{2z^2-5z+3} - \dfrac{2z}{2z^2-z-3}$

50. $\dfrac{3c}{2c^2 + 7c - 4} - \dfrac{4c}{c^2 + 2c - 8}$

51. $\dfrac{r + 2}{r(r + 1)} + \dfrac{r - 1}{r(r + 2)}$

52. $\dfrac{2h}{h^2(h + 2)} + \dfrac{1}{h^2(h + 3)}$

53. $\dfrac{3(t - 1)}{t(t - 1)} - \dfrac{t^2 - 3t}{t^2(t - 1)}$

54. $\dfrac{z + 1}{z(z + 1)} - \dfrac{z - z^2}{z^2(z + 1)}$

55. $\dfrac{y^2 - 19}{y^2 + 4y - 5} - \dfrac{y - 4}{y - 1}$

56. $\dfrac{z^2 + 14}{z^2 + z - 2} - \dfrac{z - 4}{z + 2}$

57. $(z + 5) + \dfrac{2}{z + 3}$

58. $(2t - 1) - \dfrac{3t}{t + 4}$

59. $3 - \dfrac{v}{2v + 12}$

60. $-2 + \dfrac{3a}{a + 1}$

61. Orlene plants $\dfrac{1}{n}$ of her vegetable garden in the morning and after lunch plants $\dfrac{1}{2n}$ of the garden, where n is an integer greater than one. Write as a rational expression the amount of garden that is left to plant.

62. A triangle has sides of lengths $\dfrac{x + 3}{2x}$, $\dfrac{x + 3}{x}$, and $\dfrac{x + 4}{x}$ inches. Express the perimeter as a rational expression.

For Exercises 63–66, let x represent a number. Write an algebraic expression from the information given, then simplify it into a rational expression.

63. The sum of a number and three times its reciprocal.

64. Five times a number reduced by the reciprocal of the number.

65. The reciprocal of the sum of a number and four, reduced by twice the number.

66. The sum of the reciprocal of a number and the reciprocal of twice the number.

SAY IT IN WORDS

67. How do you find the least common denominator of two rational expressions?

68. Explain the steps you take when adding two algebraic fractions with unlike denominators.

REVIEW EXERCISES

The following exercises review parts of Section 1.1. Doing these exercises will help prepare you for the next section.

For Exercises 69–74, write the answer as a single fraction in lowest terms.

69. $\frac{2}{3} \div \frac{4}{5}$

70. $\frac{4}{3} \div \frac{12}{7}$

71. $\frac{1}{5} \div \frac{3}{10}$

72. $\frac{8}{3} \div \frac{24}{15}$

73. $\frac{5}{6} \cdot \frac{2}{3} \div \frac{5}{36}$

74. $3 \cdot \frac{14}{5} \div \frac{24}{5}$

ENRICHMENT EXERCISES

For Exercises 1–4, add or subtract as indicated.

1. $\frac{3x}{5} - \frac{7x}{10} + \frac{6x}{15}$

2. $\frac{3}{x} - \frac{5}{2x} + \frac{7}{5x^2}$

3. $\frac{4}{3x} + \frac{5}{9x^2} - \frac{7}{18x}$

4. $\frac{1}{x-1} + \frac{1}{3x} - \frac{1}{x^2(x-1)}$

Answers to Enrichment Exercises begin on page A.1.

6.4 Complex Fractions

OBJECTIVE

▶ *To simplify complex fractions*

LEARNING ADVANTAGE

Is your answer reasonable? *If your answer to a word problem is unreasonable because the answer is negative when it should be positive, check each step in the solution where negative or positive numbers were combined through addition, subtraction, multiplication, or division.*

A rational expression that has fractions in its numerator or denominator is called a **complex fraction.** Examples of complex fractions are the following:

$$\frac{\frac{3}{2}}{\frac{5}{8}}, \quad \frac{7 - \frac{1}{x}}{5 + 3x}, \quad \frac{8 + 9a}{6 - \frac{5}{a}}$$

A complex fraction can be simplified to an ordinary fraction.

Example 1 Simplify each complex fraction.

(a) $\frac{\frac{3}{2}}{\frac{5}{8}}$ **(b)** $\frac{\frac{6}{x^2}}{\frac{9}{x}}$

Solution **(a)** $\dfrac{\frac{3}{2}}{\frac{5}{8}}$ means divide $\frac{3}{2}$ by $\frac{5}{8}$. Therefore,

$$\frac{\frac{3}{2}}{\frac{5}{8}} = \frac{3}{2} \div \frac{5}{8}$$

$$= \frac{3}{2} \cdot \frac{8}{5}$$

$$= \frac{3 \cdot 8}{2 \cdot 5}$$

$$= \frac{3 \cdot \overset{4}{\cancel{8}}}{\cancel{2} \cdot 5}$$

$$= \frac{3 \cdot 4}{1 \cdot 5}$$

$$= \frac{12}{5}$$

(b) $\dfrac{\frac{6}{x^2}}{\frac{9}{x}}$ means divide $\frac{6}{x^2}$ by $\frac{9}{x}$. Therefore,

$$\frac{\frac{6}{x^2}}{\frac{9}{x}} = \frac{6}{x^2} \div \frac{9}{x}$$

$$= \frac{6}{x^2} \cdot \frac{x}{9}$$

$$= \frac{6x}{9x^2}$$

$$= \frac{\overset{2}{\cancel{6}}\cancel{x}}{\underset{3}{\cancel{9}}\underset{x}{\cancel{x^2}}}$$

$$= \frac{2}{3x}$$

When the numerator or denominator of a complex fraction contains two or more terms, there are two methods available to simplify them.

STRATEGY

Two Methods for Simplifying a Complex Fraction

Method 1 Add and subtract the fractions in the numerator, add and subtract the fractions in the denominator, then divide.

Method 2 Find the least common denominator (LCD) of all the fractions appearing in the numerator and denominator, then multiply both the numerator and denominator by the LCD to clear fractions.

Example 2 Use Method 2 to simplify $\dfrac{\dfrac{2}{3}-\dfrac{1}{x^2}}{\dfrac{5}{x}+\dfrac{1}{3x^2}}$.

Solution The least common denominator of $\dfrac{2}{3}$, $\dfrac{1}{x^2}$, $\dfrac{5}{x}$, and $\dfrac{1}{3x^2}$ is $3x^2$. Therefore, we multiply numerator and denominator by $3x^2$.

$$\frac{\dfrac{2}{3}-\dfrac{1}{x^2}}{\dfrac{5}{x}+\dfrac{1}{3x^2}}=\frac{3x^2\left(\dfrac{2}{3}-\dfrac{1}{x^2}\right)}{3x^2\left(\dfrac{5}{x}+\dfrac{1}{3x^2}\right)}$$

$$=\frac{3x^2\left(\dfrac{2}{3}\right)+3x^2\left(-\dfrac{1}{x^2}\right)}{3x^2\left(\dfrac{5}{x}\right)+3x^2\left(\dfrac{1}{3x^2}\right)}$$ Distributive property.

$$=\frac{2x^2-3}{15x+1}$$ Simplify each term of the numerator and denominator.

Example 3 Use Method 2 to simplify $\dfrac{\dfrac{5}{n+1}-\dfrac{1}{n-1}}{\dfrac{1}{n+1}-\dfrac{2}{n-1}}$.

Solution The LCD of the four fractions

$$\frac{5}{n+1}, \quad \frac{1}{n-1}, \quad \frac{1}{n+1}, \quad \text{and} \quad \frac{2}{n-1}$$

is $(n+1)(n-1)$. Therefore,

$$\frac{\frac{5}{n+1}-\frac{1}{n-1}}{\frac{1}{n+1}-\frac{2}{n-1}} = \frac{(n+1)(n-1)\left(\frac{5}{n+1}-\frac{1}{n-1}\right)}{(n+1)(n-1)\left(\frac{1}{n+1}-\frac{2}{n-1}\right)}$$

$$= \frac{(n+1)(n-1)\frac{5}{n+1}-(n+1)(n-1)\frac{1}{n-1}}{(n+1)(n-1)\frac{1}{n+1}-(n+1)(n-1)\frac{2}{n-1}}$$

$$= \frac{(n-1)(5)-(n+1)}{(n-1)-(n+1)(2)}$$

$$= \frac{5n-5-n-1}{n-1-2n-2}$$

$$= \frac{4n-6}{-n-3} \qquad \text{Simplify numerator and denominator.}$$

$$= -\frac{4n-6}{n+3} \qquad \frac{A}{-B} = -\frac{A}{B}$$

WARMING UP

Decide if the simplification of the complex fraction is correct (true) or incorrect (false).

1. $\dfrac{\frac{2}{3}}{\frac{5}{6}} = \dfrac{4}{5}$

2. $\dfrac{\frac{5}{x^3}}{\frac{15}{x^7}} = \dfrac{x^4}{3}$

3. $\dfrac{\frac{1}{5}-\frac{3}{2}}{\frac{7}{10}+\frac{2}{5}} = -\dfrac{11}{13}$

4. $\dfrac{\frac{1}{y}+\frac{3}{2y^2}}{\frac{2}{y}-\frac{1}{2}} = \dfrac{2y+3}{4y-y^2}$

5. $\dfrac{12+\frac{2}{x^2}}{\frac{3}{x}-\frac{2}{x^2}} = \dfrac{2+12x^2}{3x-2}$

EXERCISE SET 6.4

For Exercises 1–20, simplify each complex fraction by using either Method 1 or Method 2.

1. $\dfrac{\frac{1}{2}}{\frac{3}{4}}$

2. $\dfrac{\frac{5}{8}}{\frac{1}{4}}$

3. $\dfrac{\frac{1}{a}}{\frac{2}{b}}$

4. $\dfrac{\frac{3}{x}}{\frac{1}{r}}$

5. $\dfrac{\frac{15}{s^2}}{\frac{12}{s}}$

6. $\dfrac{\frac{46}{w^3}}{\frac{36}{w^2}}$

7. $\dfrac{\frac{5}{a}+\frac{3}{2}}{\frac{1}{a}+\frac{1}{2}}$

8. $\dfrac{\frac{7}{t}+\frac{1}{3}}{\frac{2}{3}+\frac{5}{t}}$

9. $\dfrac{\frac{12}{w^2}-\frac{3}{w}}{\frac{15}{w}-\frac{9}{w^2}}$

10. $\dfrac{\frac{4}{h^3}+\frac{16}{h^2}}{\frac{20}{h^2}-\frac{36}{h^3}}$

11. $\dfrac{\frac{1}{z}+\frac{2}{z^2}}{\frac{2}{z}+1}$

12. $\dfrac{\frac{2}{r}-\frac{3}{r^2}}{2-\frac{3}{r}}$

13. $\dfrac{\frac{2}{3q}-\frac{5}{3}}{\frac{2}{q^2}-\frac{5}{q}}$

14. $\dfrac{\frac{4}{5h}+\frac{9}{5}}{\frac{4}{h^2}+\frac{9}{h}}$

15. $\dfrac{\frac{7}{c^2}-\frac{3}{c}}{\frac{7}{c^3}-\frac{3}{c^2}}$

16. $\dfrac{\frac{6}{b^3}+\frac{11}{b^2}}{\frac{6}{b^4}+\frac{11}{b^3}}$

17. $\dfrac{\frac{10}{a^2-9}}{\frac{5}{a+3}}$

18. $\dfrac{\frac{3}{z+4}}{\frac{15}{z^2-16}}$

19. $\dfrac{\frac{3}{x-2}-\frac{4}{x+2}}{\frac{7}{x^2-4}}$

20. $\dfrac{\frac{1}{t+5}}{\frac{12}{t^2-25}+\frac{2}{t-5}}$

SAY IT IN WORDS

21. Explain the two methods for simplifying a complex fraction.

22. In reference to Exercise 21, which method do you prefer? Why?

REVIEW EXERCISES

The following exercises review parts of Sections 2.4 and 5.6. Doing these exercises will help prepare you for the next section.

Section 2.4

For Exercises 23–26, solve each equation.

23. $3x - 8x = 35$

24. $3(2t - 3) = -4$

25. $z^2 - 3z + 2 = z^2 - 5z + 8$

26. $\frac{w}{5} + \frac{3w + 8}{15} = 0$

Section 5.6

For Exercises 27–30, solve each equation by factoring.

27. $2x^2 - 3x - 2 = 0$

28. $3z^2 - z - 2 = 0$

29. $21y^2 + 2y - 3 = 0$

30. $6a^3 - 5a^2 - a = 0$

ENRICHMENT EXERCISES

For Exercises 1–4, simplify.

1. $$\frac{1 - \dfrac{1}{1 - \dfrac{1}{5}}}{\dfrac{3}{4} - \dfrac{1}{2}}$$

2. $$\frac{\dfrac{1}{3} - \dfrac{4}{9}}{\dfrac{1}{1 - \dfrac{5}{2}} + \dfrac{1}{9}}$$

3. $$\frac{\dfrac{1}{x + h} - \dfrac{1}{x}}{h}$$

4. $$\frac{\dfrac{1}{(x + h)^2} - \dfrac{1}{x^2}}{h}$$

Answers to Enrichment Exercises begin on page A.1.

6.5 Equations with Rational Expressions

OBJECTIVES

- *To solve equations containing rational expressions*
- *To use formulas containing rational expressions*

LEARNING ADVANTAGE

Is your answer reasonable? *When solving a word problem, estimating and intelligent guessing are not improper and have their place. For example, intuitive guessing about the size or type of numbers that are reasonable answers is something to keep in mind as you solve the problem.*

An equation such as $\frac{x}{x-2} - \frac{3x}{x-5} = \frac{1}{10}$ contains rational expressions. The best way to solve this kind of equation is to "clear fractions" by multiplying both sides by the least common denominator of all fractions appearing. A solution of the resulting equation will usually be a solution of the original equation. However, always check your answer in the original equation.

Example 1 Solve $\frac{2x}{x+2} + \frac{3}{x+2} = 1$

Solution To clear fractions, multiply both sides by $x + 2$, then simplify.

$$\frac{2x}{x+2} + \frac{3}{x+2} = 1$$

$$(x+2)\left(\frac{2x}{x+2} + \frac{3}{x+2}\right) = (x+2) \cdot 1$$

$$(x+2) \cdot \frac{2x}{x+2} + (x+2) \cdot \frac{3}{x+2} = x + 2 \qquad \text{Distributive property.}$$

$$2x + 3 = x + 2 \qquad \text{Simplify.}$$

$$2x + 3 - x - 3 = x + 2 - x - 3$$

$$x = -1$$

Next, we check to see if replacing x by -1 makes a denominator 0 in the original equation. When x is -1, the denominator $x + 2 = -1 + 2 = 1 \neq 0$. The solution set is $\{-1\}$.

STRATEGY

Solving Equations Containing Fractions

Step 1 Clear the equation of fractions by multiplying both sides by the LCD of all fractions.

Step 2 Solve the resulting equation.

Step 3 Check each solution from Step 2 to see if it makes any denominator of the original equation have a value of 0. If it does, discard the number from the solution set.

Example 2 Solve $\frac{1}{x} + \frac{3}{2} = \frac{1}{3}$.

Solution The LCD of the fractions is $6x$.

$$6x\left(\frac{1}{x} + \frac{3}{2}\right) = 6x \cdot \frac{1}{3}$$

$$6 + 9x = 2x \quad \text{Simplify.}$$
$$7x = -6$$
$$x = -\frac{6}{7}$$

The solution set is $\left\{-\frac{6}{7}\right\}$.

Example 3 Solve $\frac{t}{t-3} - \frac{2}{t+3} = 1$.

Solution Multiply both sides by the least common denominator, $(t-3)(t+3)$, and simplify the result.

$$(t-3)(t+3)\left(\frac{t}{t-3} - \frac{2}{t+3}\right) = (t-3)(t+3)(1)$$
$$(t-3)(t+3)\left(\frac{t}{t-3}\right) - (t-3)(t+3)\left(\frac{2}{t+3}\right) = (t-3)(t+3)$$
$$\cancel{(t-3)}(t+3)\left(\frac{t}{\cancel{t-3}}\right) - (t-3)\cancel{(t+3)}\left(\frac{2}{\cancel{t+3}}\right) = (t-3)(t+3)$$
$$(t+3)(t) - (t-3)(2) = (t-3)(t+3)$$
$$t^2 + 3t - 2t + 6 = t^2 - 9 \quad \text{Simplify.}$$
$$t + 6 = -9$$
$$t = -15$$

Check The number -15 does not make any denominator have a value of 0.

The solution set is $\{-15\}$.

Example 4 Solve the equation:

$$\frac{z+2}{z-2} = \frac{1}{z} + \frac{8}{z(z-2)}$$

Solution The least common denominator of the fractions appearing in the equation is $z(z-2)$. Therefore, we multiply both sides of the equation by $z(z-2)$ and then simplify.

$$z(z-2)\left(\frac{z+2}{z-2}\right) = z(z-2)\left(\frac{1}{z} + \frac{8}{z(z-2)}\right)$$
$$z(z-2)\left(\frac{z+2}{z-2}\right) = z(z-2)\left(\frac{1}{z}\right) + z(z-2)\left(\frac{8}{z(z-2)}\right)$$
$$z\cancel{(z-2)}\left(\frac{z+2}{\cancel{z-2}}\right) = \cancel{z}(z-2)\left(\frac{1}{\cancel{z}}\right) + \cancel{z}\cancel{(z-2)}\left(\frac{8}{\cancel{z}\cancel{(z-2)}}\right)$$

$$z(z + 2) = (z - 2) + 8$$

$$z^2 + 2z = z + 6$$

$$z^2 + 2z - z - 6 = 0 \qquad \text{Subtract } z + 6 \text{ from both sides.}$$

$$z^2 + z - 6 = 0$$

To solve this last equation for z, factor the left side.

$$(z + 3)(z - 2) = 0$$

Either

$$z + 3 = 0 \qquad \text{or} \qquad z - 2 = 0$$

Therefore,

$$z = -3 \qquad \text{or} \qquad z = 2$$

Next, we check these two answers in the original equation. -3 does not make any denominator have a value of 0; it is a solution. However, 2 makes the denominator, $z - 2$, zero. So 2 is not a solution. The solution set is $\{-3\}$.

TEAM PROJECT

(3 or 4 Students)

SOLVING EQUATIONS WITH RATIONAL EXPRESSIONS

Course of Action: Each member selects one problem from Exercise Set 6.5, and then solves this problem. Next, each member assigns the selected problem to the person on the right, who solves it and returns the solution to the person on the left. All team members then compare their neighbors' solutions to their own. If the solutions differ, the two teammates work the problem together until both agree on the solution.

Now, each team member selects another problem from Exercise 6.5, solves it, and assigns the selected problem to the person on the left. Each person then solves the assigned problem and returns the solution to the person on the right. Team members compare their neighbors' solutions to their own. If the solutions differ, they work the problem together until both agree on the solution.

Group Report:

1. How successful was the team in solving equations?
2. Write a paragraph explaining how to solve equations with rational expressions. Include any special tips to help others solve such equations.

WARMING UP

Answer true or false.

1. The solution set of $\dfrac{2}{x} = 0$ is $\{0\}$.

2. If $\dfrac{6}{x} = \dfrac{1}{2}$, the solution set is $\{12\}$.

3. We can clear fractions in the equation $\dfrac{1}{x} + \dfrac{1}{2x} = 2$ by multiplying both sides by $2x$.

4. To solve $\dfrac{3}{x-2} + \dfrac{2}{x(x-2)} = 1$, start by multiplying the left side by $x(x-2)$.

5. The solution set of $\dfrac{1}{x+1} + \dfrac{1}{x-2} = 1$ is $\{-1, 2\}$.

EXERCISE SET 6.5

For Exercises 1–34, solve the equation.

1. $\dfrac{3x}{x-1} + \dfrac{4}{x-1} = 2$

2. $\dfrac{2}{y+1} + \dfrac{5y}{y+1} = 4$

3. $\dfrac{1}{x+2} + \dfrac{x}{x+2} = -1$

4. $\dfrac{2x}{x-3} + \dfrac{3}{x-3} = -2$

5. $\dfrac{3z}{2z+1} - \dfrac{4}{2z+1} = 2$

6. $\dfrac{5}{3t-2} - \dfrac{3t}{3t-2} = -2$

7. $\dfrac{2}{x} + \dfrac{1}{2} = 1$

8. $\dfrac{1}{3} + \dfrac{4}{x} = 1$

9. $\dfrac{2}{3} + \dfrac{1}{y} = 2$

10. $\dfrac{2}{m} + \dfrac{3}{2} = -1$

11. $\dfrac{3}{2r} - \dfrac{1}{2} = 1$

12. $\dfrac{1}{3x} - \dfrac{1}{3} = 1$

13. $\dfrac{2}{x-2} + \dfrac{1}{x} = \dfrac{3}{x(x-2)}$

14. $\dfrac{3}{x} + \dfrac{2}{x(x+1)} = -\dfrac{1}{x+1}$

15. $\dfrac{4}{x^2+2x} - \dfrac{3}{x} = -\dfrac{2}{x+2}$

16. $\dfrac{5}{2x^2+x} - \dfrac{1}{2x+1} = -\dfrac{1}{x}$

17. $\dfrac{x}{x-2} - 1 = \dfrac{4}{x-2}$

18. $2 - \dfrac{x}{x+3} = \dfrac{3}{x+3}$

19. $\dfrac{x+10}{x+6} + \dfrac{24}{x^2+6x} = \dfrac{5}{x}$

20. $\dfrac{x+8}{x+2} + \dfrac{12}{x^2+2x} = \dfrac{2}{x}$

21. $\dfrac{x+2}{x+1} - 2 = \dfrac{1}{x+1}$

22. $\dfrac{x+2}{x} + \dfrac{x-2}{x} = 3$

23. $\dfrac{14}{x+3} - \dfrac{1}{x} = 2$

24. $\dfrac{y+2}{y-12} - \dfrac{1}{y} = \dfrac{14}{y^2-12y}$

25. $\dfrac{6}{x+1} - \dfrac{2}{x} = 1$

26. $\dfrac{1}{x} - 1 = \dfrac{6}{x-2}$

27. $\dfrac{6}{x^2-4} + \dfrac{3}{x+2} = -1$

28. $\dfrac{4}{x^2-1} - 1 = \dfrac{1}{x+1}$

29. $\dfrac{1}{2-x} - 3 = \dfrac{8}{x^2-4}$

30. $\dfrac{1}{y^2-1} + \dfrac{1}{1-y} = \dfrac{2}{3}$

31. $\dfrac{4}{x^2+x-2} = \dfrac{3}{x-1} + 1$

32. $1 + \dfrac{3}{x^2+4x+3} = \dfrac{2}{x+1}$

33. $\dfrac{6-3t}{t-6} - 2t = \dfrac{12}{6-t}$

34. $\dfrac{25}{y-10} + 3y = \dfrac{25-5y}{10-y}$

SAY IT IN WORDS

35. Explain your method for solving equations containing algebraic fractions.

REVIEW EXERCISES

The following exercises review parts of Sections 2.4 and 5.6. Doing these exercises will help prepare you for the next section.

Section 2.4

For Exercises 36–39, solve each linear equation.

36. $2x + 7 = -1$

37. $1 - 4r = 5$

38. $\dfrac{t}{-2} = -\dfrac{3}{4}$

39. $\dfrac{3y-4}{3} = -1$

Section 5.6

For Exercises 40–42, solve each quadratic equation.

40. $s^2 + 6s - 7 = 0$

41. $2x^2 + 7x + 5 = 0$

42. $6t^2 - 5t + 1 = 0$

ENRICHMENT EXERCISES

For Exercises 1–4, solve the equation, where A and B are nonzero constants.

1. $\dfrac{x}{A} + \dfrac{2}{B} = \dfrac{1}{2}$

2. $\dfrac{2x}{A} + \dfrac{1}{B} = 2$

3. $\dfrac{x}{AB} - \dfrac{1}{B} = \dfrac{2}{A}$

4. $\dfrac{2x+1}{A} + \dfrac{1}{B} = \dfrac{1}{3}$

Answers to Enrichment Exercises begin on page A.1.

6.6 Ratio and Proportion; Similar Triangles

OBJECTIVES

- *To write a ratio as a fraction in lowest terms*
- *To solve proportions*
- *To use proportions to solve word problems*
- *To use proportions to solve problems involving similar triangles*

LEARNING ADVANTAGE

Is the answer reasonable? *A question on a test may relate to a practical situation in which the final answer will agree reasonably well with the real world. Suppose your answer is not reasonable. If you have time to review your calculations, consider the following possible error: If the answer is unreasonably large or unreasonably small, check the position of any decimal point in your calculations.*

Ratio and Proportion

We can use a fraction to compare a company's $4 million earnings last year to its $7 million earnings this year:

$$\begin{array}{r} \text{earnings last year} \rightarrow \\ \text{earnings this year} \rightarrow \end{array} \frac{4}{7}$$

We say that the earnings for the two years are in the *ratio* of 4 to 7.

DEFINITION

A **ratio** is the comparison of two numbers a and b, $b \neq 0$, using division. The ratio of two numbers can be written as

$$a \text{ to } b, \qquad a:b, \qquad \text{or} \qquad \frac{a}{b}$$

In algebra, the fraction $\frac{a}{b}$ is the most common way to write a ratio.

An equation, such as $\frac{2}{3} = \frac{4}{6}$, where two ratios are equal, is called a *proportion.*

DEFINITION

A **proportion** is a statement that two ratios are equal.

$$a:b = c:d \qquad \text{or} \qquad \frac{a}{b} = \frac{c}{d}$$

The proportion $a:b = c:d$ is read "a is to b as c is to d."

A proportion consists of *four* numbers. If three of the numbers are known, the proportion can be solved to find the fourth. To solve a proportion such as $\frac{2}{3} = \frac{6}{x}$ for x, we make use of the following property:

PROPERTY

The Cross-Multiplication Property of Proportions

If $\frac{a}{b} = \frac{c}{d}$, then

$$ad = bc$$

This property can be justified by multiplying both sides of the proportion $\frac{a}{b} = \frac{c}{d}$ by the common denominator bd. Here are the details:

$$\frac{a}{b} = \frac{c}{d}$$

$$\left(\frac{a}{b}\right)(bd) = \left(\frac{c}{d}\right)(bd) \quad \text{Multiply both sides by the common denominator, } bd.$$

$$ad = bc$$

Example 1 Solve each proportion.

(a) $\frac{2}{3} = \frac{6}{x}$ **(b)** $2:x = 4:1$

Solution **(a)** Using the cross-multiplication property for proportions, if $\frac{2}{3} = \frac{6}{x}$, then

$$2x = 3 \cdot 6$$

$$x = \frac{3 \cdot 6}{2} \quad \text{Divide both sides by 2.}$$

$$x = 9$$

Check to see that the proportion $\frac{2}{3} = \frac{6}{9}$ is true.

(b) We write the proportion as

$$\frac{2}{x} = \frac{4}{1}$$

$$2 \cdot 1 = 4x \quad \text{Cross-multiplication property.}$$

$$x = \frac{1}{2}$$

We can use proportions to solve certain word problems. Here are some examples.

Example 2 A commercial claims that three out of five dentists in a survey use ExtraBrite toothpaste. If there were 15,000 dentists in the survey, how many use this toothpaste?

Solution Let

$x =$ the number of dentists in the survey that use ExtraBrite toothpaste

Next, we write a proportion by setting the ratios equal. Be sure that the corresponding numbers appear in the numerator and denominator.

$$\frac{x}{15{,}000} = \frac{3}{5}$$

To solve for x, do not cross multiply, but multiply both sides by 15,000.

$$x = \left(\frac{3}{5}\right)(15{,}000)$$
$$= 9000$$

Therefore, 9000 dentists in the survey use ExtraBrite toothpaste.

Example 3 Naomi uses 3 skeins of cotton yarn to make 4 towels. How many skeins will be needed to make 10 towels?

Solution Let $x =$ the number of skeins. From the information $3:4 = x:10$ or $\frac{3}{4} = \frac{x}{10}$.

Solving for x,

$$3 \cdot 10 = 4 \cdot x$$

$$\frac{3 \cdot \overset{5}{\cancel{10}}}{\underset{2}{\cancel{4}}} = x$$

$$\frac{15}{2} = x$$

Therefore, $\frac{15}{2}$ or $7\frac{1}{2}$ skeins are needed.

Similar Triangles

Another application of proportions comes from a topic in geometry. Two triangles are called **similar triangles** if they have the same shape but not necessarily the same size. Examples of pairs of similar triangles are shown in the following figures.

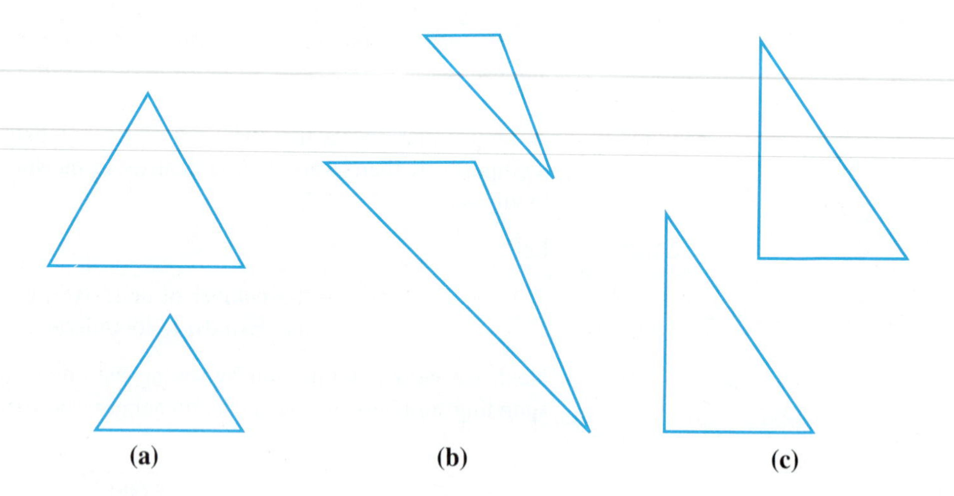

The two triangles in the third pair **(c),** not only have the same shape but also the same size. Two triangles that are both the same size and the same shape are called **congruent triangles.**

Similar triangles have the following property:

PROPERTY

Similar Triangles

If two triangles are similar, then corresponding angles are equal and corresponding sides are proportional.

Example 4 Find the lengths of the missing sides of the smaller triangle, given that the two triangles are similar. Assume the lengths are given in inches.

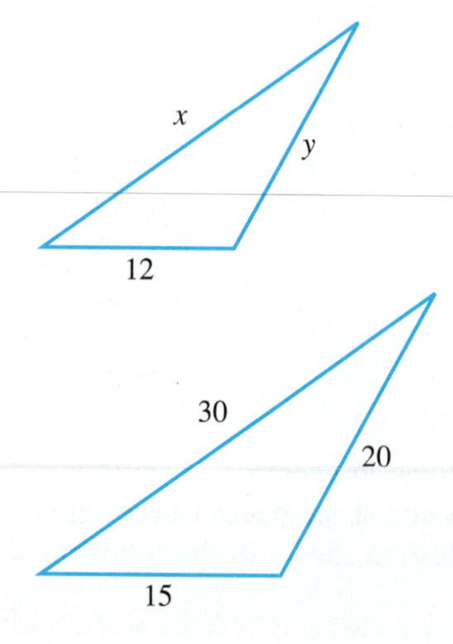

Solution Since the two triangles are similar, corresponding sides are proportional. To find x, we set up the proportion

$$\frac{x}{30} = \frac{12}{15}$$

Instead of using the cross-multiplication property of proportions to solve for x, it is quicker to simply multiply both sides by 30,

$$x = \left(\frac{12}{15}\right)(30)$$

$$= 24 \text{ inches}$$

To find the length y, set up the proportion:

$$\frac{y}{20} = \frac{12}{15}$$

Multiplying both sides by 20, we have

$$y = \left(\frac{12}{15}\right)(20)$$
$$= 16 \text{ inches}$$

WARMING UP

Answer true or false.

1. If $\frac{x}{3} = \frac{2}{27}$, then x is $\frac{2}{9}$.
2. If $2:x = 4:3$, then x is 6.
3. If 10 is to x as 5 is to 3, then x is 6.
4. A proportion is an equation equating two ratios.
5. If $\frac{a}{b} = \frac{c}{d}$, then $ad = bc$.
6. If $\frac{4}{x} = \frac{x}{25}$, then x is 10.

EXERCISE SET 6.6

For Exercises 1–16, solve each proportion.

1. $\frac{3}{x} = \frac{2}{5}$
2. $\frac{1}{x} = \frac{3}{2}$
3. $\frac{2}{7} = \frac{4}{x}$
4. $\frac{5}{2} = \frac{15}{x}$
5. $2:x = 3:4$
6. $1:x = 2:3$
7. 2 is to 4 as 8 is to x.
8. 3 is to x as 12 is to 5.
9. $\frac{24}{x} = \frac{x}{6}$
10. $\frac{1}{y+1} = \frac{y}{2}$
11. $\frac{x}{4} = \frac{1}{x}$
12. $\frac{18}{x} = \frac{x}{2}$
13. $4:x = x:4$
14. $x:32 = 2:x$
15. x is to 4 as 9 is to x.
16. x is to 7 as 7 is to x.

For Exercises 17–22, solve each word problem.

17. If 2 inches represents 15 miles on a map, how many miles does 5 inches represent?
18. If 3 cubic feet of shredded pine costs $4, how much does 18 cubic feet cost?
19. At State College, 4 out of 5 professors belong to a union. How many professors are union members if the total number of professors at the college is 450?
20. Don uses 9 gallons of gasoline to drive 200 miles. How far can he drive on a full tank of 15 gallons?

21. The sales tax on an item that costs \$100 to \$6.50. What is the sales tax on an item that costs \$350?

22. One serving of a reduced-fat cheese cracker contains 60 calories. One evening while watching television, Ralph ate one package of these cheese crackers. How many calories did Ralph consume if there are 11 servings per package?

For Exercises 23–26, find the lengths of the missing sides in each pair of similar triangles. (Assume the lengths are given in centimeters.)

23.

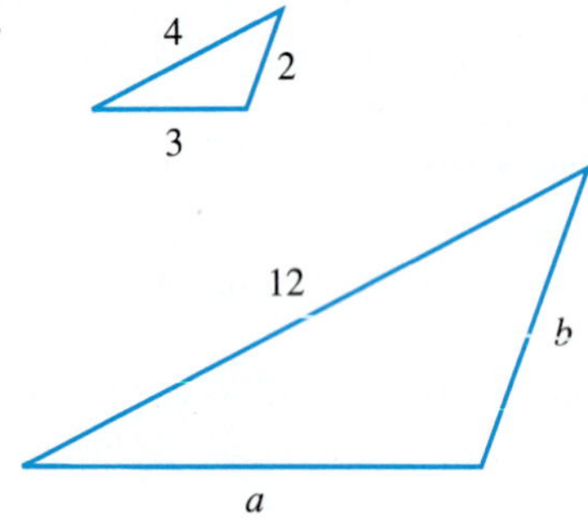

24.

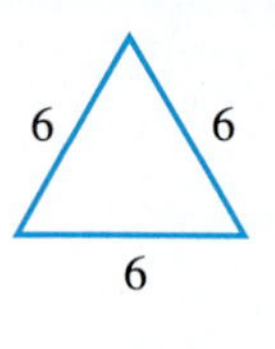

x
8
y

25.

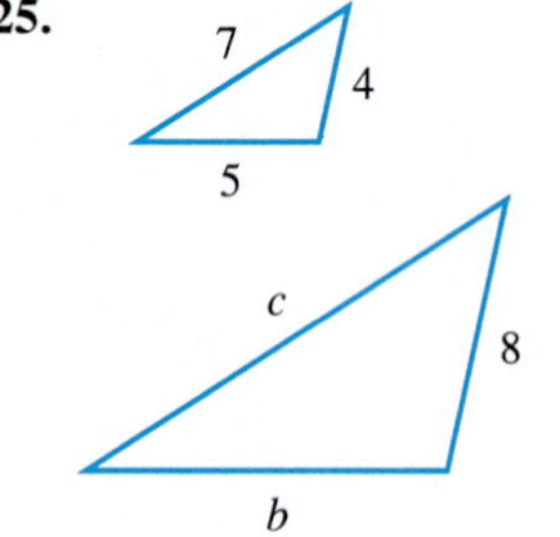

26.

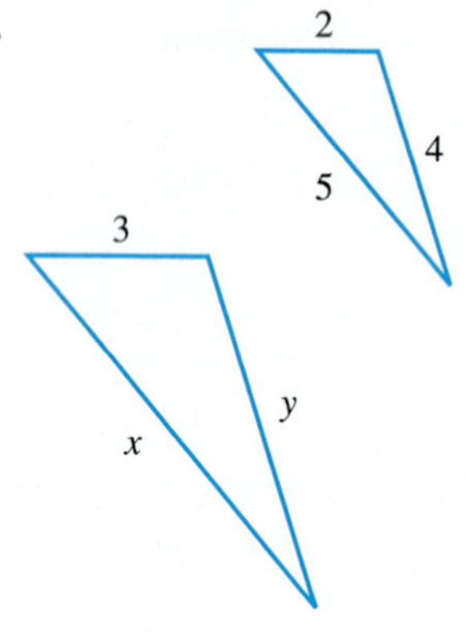

For Exercises 27 and 28, find x. (Assume the lengths are given in inches.)

27.

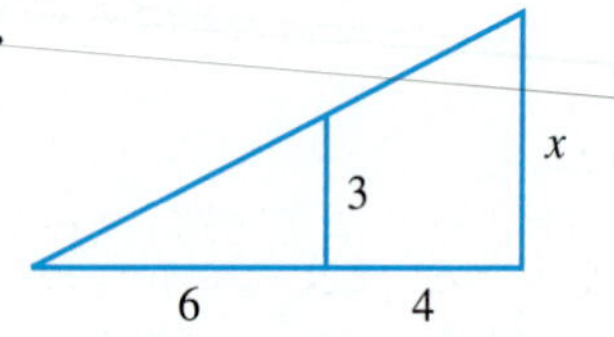

28.

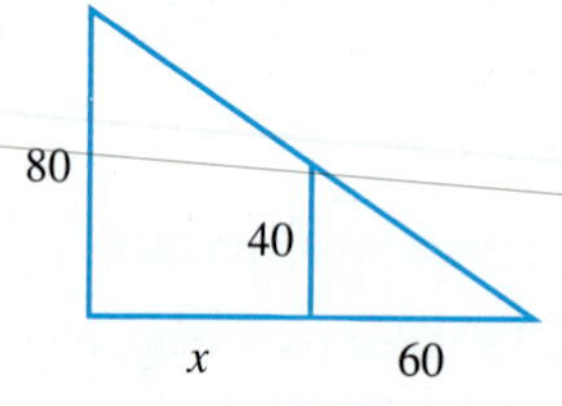

SAY IT IN WORDS

29. What is a proportion?

30. List the ways a proportion can be written.

REVIEW EXERCISES

The following exercises review parts of Section 2.1. Doing these exercises will help prepare you for the next section.

For Exercises 31–34, write each word phrase as an algebraic expression using x as the unknown number.

31. The sum of a number and $\frac{3}{4}$.

32. The difference of $\frac{2}{3}$ and a number.

33. Three-fifths of the sum of a number and $\frac{10}{3}$.

34. The reciprocal of the sum of a number and $\frac{4}{5}$.

ENRICHMENT EXERCISES

For Exercises 1–3, solve the proportion.

1. $\frac{2x - 13}{x + 4} = \frac{x - 4}{2x}$

2. $\frac{1 - x}{x} = \frac{5x + 1}{x + 1}$

3. $\frac{x}{x + 2} = \frac{2 - x}{x + 7}$

4. A board 15 feet long is to be cut into two pieces whose lengths have the ratio of 2 to 3. Find the length of each piece.

Answers to Enrichment Exercises begin on page A.1.

6.7 Applications of Rational Expressions

OBJECTIVE

▶ *To use rational expressions in applications*

LEARNING ADVANTAGE

Is your answer reasonable? *Suppose on a test, you realize that your answer is unreasonable, but you do not have enough time to review your calculations. Write the following statement for the teacher.*

> *I know my answer is unreasonable [here you explain why it is unreasonable]; however, I do not have time to locate my error.*

In this section, we will look at some other applications of rational expressions.

Our first example is solving for a variable in a formula.

Example 1 The formula $\frac{1}{R} = \frac{1}{R_1} + \frac{1}{R_2}$ concerns the resistance in an electrical circuit. Solve for R_1.

Solution

$$\frac{1}{R} = \frac{1}{R_1} + \frac{1}{R_2}$$

$$RR_1R_2\left(\frac{1}{R}\right) = RR_1R_2\left(\frac{1}{R_1} + \frac{1}{R_2}\right) \quad \text{Multiply both sides by the LCD.}$$

$$RR_1R_2\left(\frac{1}{R}\right) = RR_1R_2\left(\frac{1}{R_1}\right) + RR_1R_2\left(\frac{1}{R_2}\right)$$

$$R_1R_2 = RR_2 + RR_1$$

$$R_1R_2 - RR_1 = RR_2$$

$$(R_2 - R)R_1 = RR_2 \quad \text{Factor out } R_1.$$

$$R_1 = \frac{RR_2}{R_2 - R} \quad \text{Divide both sides by } R_2 - R.$$

Example 2 Solve for y: $\frac{y+1}{x-2} = \frac{1}{2}$.

Solution We want to isolate y to the left side. Our first step is to multiply both sides by $x - 2$.

$$\frac{y+1}{x-2} = \frac{1}{2}$$

$$(x-2)\frac{y+1}{x-2} = (x-2)\frac{1}{2}$$

$$y + 1 = \frac{1}{2}x - 1 \quad \text{Simplify.}$$

$$y = \frac{1}{2}x - 2 \quad \text{Subtract 1 from both sides.}$$

We have solved number problems before; however, this time the resulting equation will involve rational expressions.

Example 3 Find two numbers whose sum is 25 and the sum of their reciprocals is $\frac{1}{6}$.

Solution Let x be one of the unknown numbers. Since the sum of the two numbers is 25, the second number is $25 - x$. Now, the sum of their reciprocals is $\frac{1}{6}$. Therefore,

$$\frac{1}{x} + \frac{1}{25 - x} = \frac{1}{6}$$

To solve this equation, multiply both sides by the least common denominator, $6x(25 - x)$, of the three fractions appearing.

$$6x(25 - x)\left(\frac{1}{x} + \frac{1}{25 - x}\right) = 6x(25 - x)\left(\frac{1}{6}\right)$$

$$6(25 - x) + 6x = x(25 - x)$$

$$150 - 6x + 6x = 25x - x^2$$

$$x^2 - 25x + 150 = 0 \quad \text{Put in standard form.}$$

$$(x - 10)(x - 15) = 0 \quad \text{Factor.}$$

Therefore, either $x = 10$ or $x = 15$. These two values of x actually give the same pair of numbers. When $x = 10$, then $25 - x = 15$ and when $x = 15$, then $25 - x = 10$. We next check our pair of numbers 10 and 15 to see if they satisfy the requirements. Their sum, $10 + 15$, is 25. The sum of their reciprocals,

$$\frac{1}{10} + \frac{1}{15} = \frac{3}{30} + \frac{2}{30} = \frac{5}{30} = \frac{1}{6}$$

Therefore, 10 and 15 are the correct numbers. ◀

There is a group of applications called **work problems.** Any job requires time for completion. In these problems, we will assume that a person doing a job works at a constant rate.

Suppose Lisa can mow a lawn in 3 hours. Then in 1 hour, she can mow $\frac{1}{3}$ of the lawn. That is, her rate of work is $\frac{1}{3}$ of the lawn mowed per hour as shown in the diagram below.

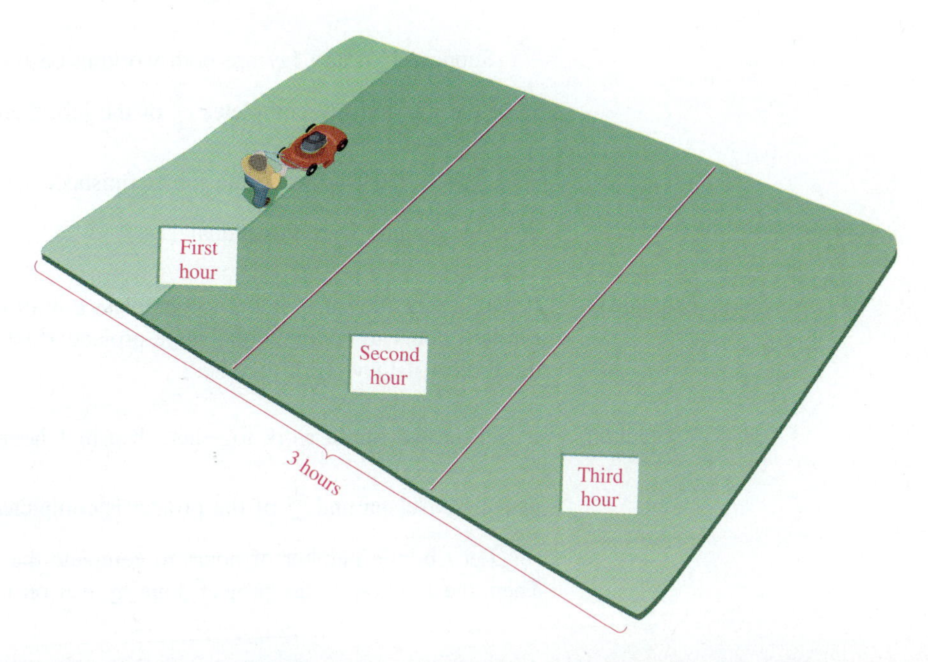

For example, after $\frac{1}{2}$ hour, she will have mowed $\left(\frac{1}{3}\right)\left(\frac{1}{2}\right)$ or $\frac{1}{6}$ of the lawn. That is, the fraction of the lawn mowed in $\frac{1}{2}$ hour is the product of the rate per hour and the time spent mowing. This is illustrated in the formula

The fraction of work completed = (rate per hour) × (number of hours worked)

Example 4 A pump can fill a tank in 18 hours. How full is the tank after

(a) 3 hours? **(b)** 8 hours? **(c)** t hours?

Solution We restate the formula for this problem.

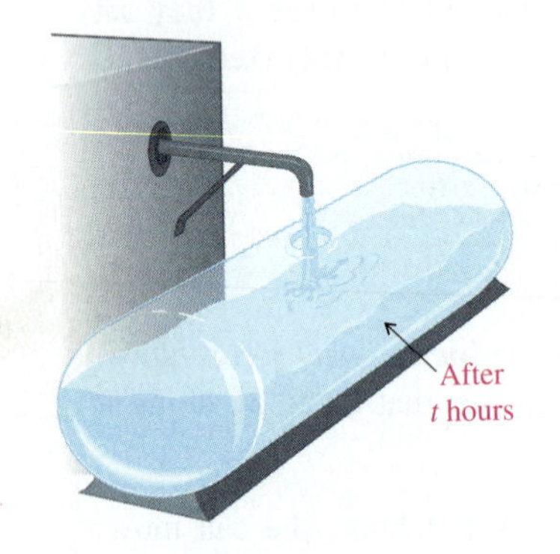

The fraction of the tank filled = (rate of filling)(the number of hours of filling)

Since the pump can fill the tank in 18 hours, then in 1 hour, $\frac{1}{18}$ of the tank is filled. That is, the rate per hour is $\frac{1}{18}$.

(a) The fraction of the tank filled in 3 hours is the product of $\frac{1}{18}$ and 3. Therefore, the tank is $\frac{1}{18}(3)$ or $\frac{1}{6}$ full after 3 hours.

(b) After 8 hours, the tank is $\frac{1}{18}(8)$ or $\frac{4}{9}$ full.

(c) After t hours, the tank is $\frac{1}{18}(t)$ or $\frac{t}{18}$ full. ◀

Suppose Lyn and Jay are both working on the same job. If Lyn completes $\frac{1}{3}$ of the job and Jay completes $\frac{2}{3}$ of the job, then the job is finished, since $\frac{1}{3} + \frac{2}{3} = 1$. That is, when the job is finished, the sum of the fractional parts done by Jay and Lyn must equal 1.

Example 5 It would take Nelson 3 hours to complete a drafting project. It would take Maria 2 hours to complete the same project. How long will it take to complete the project if they work together?

Solution If Nelson and Maria work together, then in 1 hour $\frac{1}{3}$ of the project is completed by Nelson and $\frac{1}{2}$ of the project is completed by Maria.

Let t be the number of hours to complete the project working together. Then, the fraction of the project done by Nelson is

$$\frac{1}{3}(t) = \frac{t}{3}$$

and the fraction of the project done by Maria is

$$\frac{1}{2}(t) = \frac{t}{2}$$

Now, the sum of the fractions of the project done by Nelson and Maria is the total project. That is,

$$\begin{matrix}\text{Fraction of the project} \\ \text{done by Nelson}\end{matrix} + \begin{matrix}\text{fraction of the project} \\ \text{done by Maria}\end{matrix} = \text{the completed project}$$

This gives the equation

$$\frac{t}{3} + \frac{t}{2} = 1$$

To solve this equation, multiply both sides by the least common denominator 6 and then simplify.

$$6\left(\frac{t}{3} + \frac{t}{2}\right) = 6(1)$$

$$6\left(\frac{t}{3}\right) + 6\left(\frac{t}{2}\right) = 6$$

$$\overset{2}{\cancel{6}}\left(\frac{t}{\cancel{3}}\right) + \overset{3}{\cancel{6}}\left(\frac{t}{\cancel{2}}\right) = 6 \qquad \text{Simplify each term on the left side.}$$

$$2t + 3t = 6$$

$$5t = 6$$

$$t = \frac{6}{5} \qquad \text{Divide both sides by 5.}$$

Therefore, it takes $\frac{6}{5}$ or $1\frac{1}{5}$ hours for Nelson and Maria to complete the project working together. ◀

The formula, $d = rt$, expresses the distance traveled d as the product of the rate r and the length of time t. This formula can be solved for t by dividing both sides by r.

$$\frac{d}{r} = \frac{rt}{r}$$

$$\frac{d}{r} = t$$

That is, the distance traveled divided by the rate is the amount of time for the trip. In the next example we show how this formula is used when the time is the same for two different trips.

Example 6 It took Fred the same time to drive 450 miles as it did Jeff to drive 400 miles. Jeff's speed (rate) was 7 miles per hour slower than Fred's speed. How fast did each person drive?

Solution Let x be Fred's rate. Then $x - 7$ is Jeff's rate. We make a table using the given information to fill in the first two columns labeled distance and rate. The third column (time) is found by using the formula $t = \frac{d}{r}$.

	Distance	Rate	Time
Fred	450	x	$\frac{450}{x}$
Jeff	400	$x - 7$	$\frac{400}{x - 7}$

Since

$$\text{Fred's time} = \text{Jeff's time}$$
$$\frac{450}{x} = \frac{400}{x - 7}$$

This equation contains rational expressions. To solve for x, first multiply both sides by the least common denominator, $x(x - 7)$, then simplify the result.

$$x(x - 7)\left(\frac{450}{x}\right) = x(x - 7)\left(\frac{400}{x - 7}\right)$$
$$450(x - 7) = 400x$$
$$450x - 450(7) = 400x$$
$$450x - 400x = 450(7)$$
$$50x = 450(7)$$
$$x = \frac{450 \cdot 7}{50}$$
$$x = \frac{\overset{9}{\cancel{450}} \cdot 7}{\cancel{50}}$$
$$x = 63$$

Therefore, Fred's rate is 63 miles per hour and Jeff's rate is $63 - 7$ or 56 miles per hour.

WARMING UP

Answer true or false.

1. If $\frac{1}{R} = \frac{1}{R_1} + \frac{1}{R_2}$, then $R_2 = \frac{RR_1}{R - R_1}$.

2. The sum of the reciprocals of 3 and 4 is $\frac{7}{12}$.

3. The reciprocal of $\frac{1}{3}$ is 3.

4. The reciprocal of $x + 1$ is $\frac{1}{x + 1}$.

5. If $\frac{y - 2}{x} = -1$, then $y = -x - 2$.

6. If $\frac{y + 1}{x - 3} = -\frac{1}{3}$, then $y = -\frac{1}{3}x$.

EXERCISE SET 6.7

For Exercises 1–8, solve the formula for the indicated variable.

1. $\frac{1}{R} = \frac{1}{R_1} + \frac{1}{R_2}$ for R.

2. $d = rt$ for t.

3. $V = \frac{1}{3}\pi r^2 h$ for h.

4. $V = \pi r^2 h$ for h.

5. $V = \frac{1}{3}Bh$ for B.

6. $A = \frac{1}{2}bh$ for b.

7. $A = \frac{1}{2}(b_1 + b_2)h$ for b_1.

8. $P = 2w + 2\ell$ for ℓ.

For Exercises 9–18, solve for y.

9. $\frac{y + 2}{x - 1} = 3$

10. $\frac{y - 1}{x + 1} = 1$

11. $\frac{y - 3}{x + 2} = -2$

12. $\frac{y + 5}{x - 3} = -4$

13. $\frac{y - 2}{x + 3} = m$

14. $\frac{y + 4}{x - 1} = m$

15. $\frac{y - 2}{x - 5} = \frac{2}{5}$

16. $\frac{y + 2}{x + 3} = \frac{1}{3}$

17. $\frac{y - h}{x - k} = -3$

18. $\frac{y - h}{x - k} = -1$

19. Find two numbers whose sum is 15 and the sum of their reciprocals is $\frac{5}{18}$.

20. Find two numbers whose sum is 14 and the sum of their reciprocals is $\frac{7}{24}$.

21. The sum of two numbers is $-\frac{1}{4}$ and the sum of their reciprocals is $\frac{2}{3}$. Find the two numbers.

22. The sum of two numbers is $-\frac{9}{20}$ and the sum of their reciprocals is $\frac{1}{2}$. Find the two numbers.

23. What number must be added to the numerator and denominator of $\frac{1}{4}$ to make a fraction equivalent to $\frac{2}{3}$?

24. The denominator of a fraction is 16 less than the numerator. Find this fraction, if it reduces to -3.

25. Kelly can finish her homework assignment in $4\frac{1}{2}$ hours. How much of the assignment does she complete after

(a) $\frac{1}{2}$ hour? **(b)** 2 hours? **(c)** $1\frac{1}{2}$ hours?

26. Harlan can write a science fiction short story in $3\frac{1}{2}$ hours. How much is completed after

(a) 1 hour? **(b)** $1\frac{1}{2}$ hours? **(c)** $2\frac{1}{4}$ hours?

27. Sam can rake a lawn in 3 hours and Walter can rake the same lawn in 2.4 hours. How long would it take the two of them working together to rake the lawn?

28. It would take George 5 hours to paint the master bedroom and it would take his wife Phyllis 7 hours. How long would it take if they both worked together?

29. It takes Ted 4.8 hours to run the daily blood tests in the hospital's laboratory. Another technician, Pam, can do it in 2.4 hours. How long will it take to run the blood tests if they work together?

30. One pipe can fill a swimming pool in 6 hours. A second pipe can do it in $4\frac{1}{2}$ hours. How long will it take to fill the pool using both sources of water?

31. Working together, Allison and John can landscape their property in 6 days. It would take Allison three times as long as it would take John to do it alone. Working alone, how long would it take each of them?

32. Working together, two fraternities at Faber College can build a football homecoming float in 8 days. It would take Lambda Lambda twice as long as it would take Delta House to do it alone. Working alone, how long would it take each club to build the float?

33. Mr. Campbell can prepare a company's federal income tax return in 15 hours and his associate can prepare it in 20 hours. If Mr. Campbell has been working on this tax return for 6 hours before he is joined by his associate, how long did it take the two of them working together to finish the tax return?

34. A farmer's new wheat combine can harvest his crop in 36 hours, and the old and new combines working together can complete the harvest in 26 hours. How long would it take the old combine, working alone, to harvest the crop?

35. At a vegetable canning factory, green peas flow through a pipe into a holding tank at a rate that would fill the tank in $\frac{1}{2}$ hour. Another pipe can empty a full tank in $\frac{3}{4}$ hour.

Starting with an empty tank, with both pipes in operation, how long will it take to fill the tank?

Hint: Let t be the time it takes to fill the empty tank. Then,

$$\begin{matrix}\text{Amount filled} \\ \text{in } t \text{ hours}\end{matrix} - \begin{matrix}\text{amount emptied} \\ \text{in } t \text{ hours}\end{matrix} = \begin{matrix}\text{the holding} \\ \text{tank is full}\end{matrix}$$

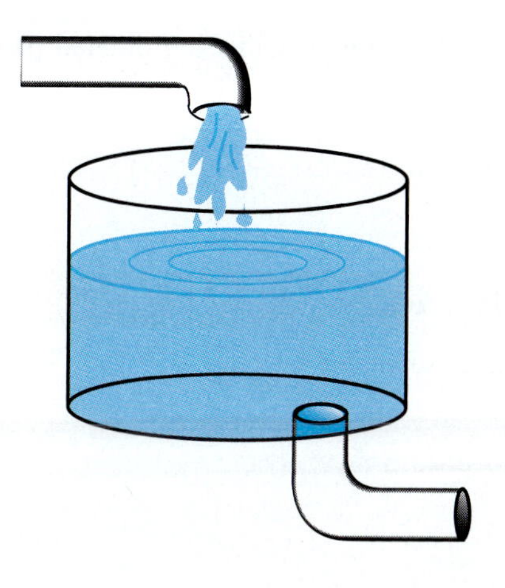

36. An oil storage tank can be filled by the intake pipe in $2\frac{1}{2}$ hours. An outtake pipe can empty the tank in 3 hours. Starting with an empty tank, with both pipes in operation, how long will it take to fill the tank?

37. It took Fran the same time to drive 21 miles as it took Carry to drive 15 miles. Fran's speed was 10 miles per hour faster than Carry's speed. How fast did Fran drive?

38. Harry hiked 12 miles in the same time it took Mick to hike 14 miles. Harry's rate was 1 mile per hour less than Mick's. How fast did each person walk?

39. The Delaware River has a current of 4 miles per hour. A canoeist takes as long to go 3 miles downriver as to go $1\frac{3}{4}$ miles upriver. What is the rate of the canoeist in still water?

	Distance	Rate	Time
Downriver	3	$r + 4$	
Upriver	$1\frac{3}{4}$	$r - 4$	

40. The Rock River in northern Illinois has a 3-kilometer-per-hour current. Jennifer can paddle her kayak 2 kilometers downriver in the same time that she can paddle $\frac{1}{2}$ kilometer upriver. What is Jennifer's rate in still water?

	Distance	Rate	Time
Downriver	2	$r + 3$	
Upriver	$\frac{1}{2}$	$r - 3$	

SAY IT IN WORDS

41. Can the reciprocal of a number be an integer? Explain.

REVIEW EXERCISES

The following Exercises review parts of Section 3.1. Doing these exercises will help prepare you for the next section.

For Exercises 42–44, determine whether the ordered pair is a solution of the given equation.

42. $2x + 3y = 6$; $(0, 2)$

43. $4x - 5y = 20$; $(5, 0)$

44. $3x - 4y = -12$; $(4, 0)$

ENRICHMENT EXERCISES

For Exercises 1 and 2, solve for y.

1. $\dfrac{y - \dfrac{1}{2}}{x + \dfrac{1}{4}} = \dfrac{2}{3}$

2. $\dfrac{y + \dfrac{3}{5}}{x - \dfrac{1}{3}} = -\dfrac{6}{5}$

3. Solve $\dfrac{2xy - x^2y'}{y^2} = 3x^2$ for y'.

Answers to Enrichment Exercises begin on page A.1.

CHAPTER 6 Summary and review

Examples

$$\frac{3(x-2)}{x^2-4} = \frac{3\cancel{(x-2)}}{(x+2)\cancel{(x-2)}} = \frac{3}{x+2}$$

Rational expressions (6.1)

A **rational expression** is an expression that can be put in the form $\dfrac{A}{B}$, where A and B are polynomials. A rational expression is in lowest terms if the numerator and denominator have no common factors.

Multiplying rational expressions (6.2)

If $\dfrac{A}{B}$ and $\dfrac{C}{D}$ are rational expressions, then

$$\frac{A}{B} \cdot \frac{C}{D} = \frac{A \cdot C}{B \cdot D}$$

$$\frac{x^2+x-6}{x^2-6x+9} \cdot \frac{x^2-3x}{x+3}$$

$$= \frac{(x+3)(x-2)}{(x-3)^2} \cdot \frac{x(x-3)}{x+3}$$

$$= \frac{\cancel{(x+3)}(x-2)x\cancel{(x-3)}}{\underbrace{\cancel{(x-3)^2}}_{x-3}\cancel{(x+3)}}$$

$$= \frac{x(x-2)}{x-3}$$

To multiply rational expressions:

Step 1 Factor the numerators and denominators.

Step 2 Divide out any common factors.

Step 3 Write the product of the factors that remain in the numerators over the product of the factors that remain in the denominators.

Dividing rational expressions (6.2)

To divide rational expressions:

$$\frac{A}{B} \div \frac{C}{D} = \frac{A}{B} \cdot \frac{D}{C}, \qquad \text{where } \frac{C}{D} \neq 0$$

That is, multiply the first fraction by the reciprocal of the second fraction.

Examples

$$\frac{2x-2}{x} \div \frac{x^2-1}{x+2}$$
$$= \frac{2x-2}{x} \cdot \frac{x+2}{x^2-1}$$
$$= \frac{2\cancel{(x-1)}}{x} \cdot \frac{x+2}{(x+1)\cancel{(x-1)}}$$
$$= \frac{2(x+2)}{x(x+1)}$$

Addition and subtraction of rational expressions (6.3)

To find the least common denominator:

Step 1 Factor each denominator completely.

Step 2 Form the product of each different factor taken the highest number of times that it occurs in any denominator.

The LCD is this product.

To add or subtract rational expressions with **like denominators,** simply add or subtract their numerators, keeping the same denominator.

For **unlike denominators:**

Step 1 Find the least common denominator.

Step 2 Rewrite each fraction using the least common denominator.

Step 3 Add or subtract the numerators, keeping the same denominator.

Step 4 Simplify the numerator, then write the fraction in lowest terms.

Find the LCD of $\frac{1}{x^3-4x}$ and $\frac{2}{x^3-2x^2}$.

$$x^3 - 4x = x(x^2-4)$$
$$= x(x+2)(x-2)$$
$$x^3 - 2x^2 = x^2(x-2)$$

Therefore, LCD = $x^2(x+2)(x-2)$.

$$\frac{4x}{2x^2+x-3} - \frac{3}{2x+3}$$
$$= \frac{4x}{(x-1)(2x+3)} - \frac{3}{2x+3}$$
$$= \frac{4x}{(x-1)(2x+3)} - \frac{3(x-1)}{(x-1)(2x+3)}$$
$$= \frac{4x-3(x-1)}{(x-1)(2x+3)}$$
$$= \frac{4x-3x+3}{(x-1)(2x+3)}$$
$$= \frac{x+3}{(x-1)(2x+3)}$$

Complex fractions (6.4)

A **complex fraction** is a fraction that has a fraction in the numerator, the denominator, or both.

There are two methods to simplify a complex fraction:

Method 1 Add and subtract the fractions in the numerator, add and subtract the fractions in the denominator, then divide.

$$\frac{\frac{1}{x} - \frac{3}{x}}{\frac{4}{x^2} + \frac{1}{x^2}} = \frac{-\frac{2}{x}}{\frac{5}{x^2}}$$

Examples

$$= \left(-\frac{2}{x}\right)\left(\frac{x^2}{5}\right)$$

$$= -\frac{2x}{5}$$

$$\frac{1 + \frac{3}{x}}{x - \frac{9}{x}} = \frac{\left(1 + \frac{3}{x}\right)(x)}{\left(x - \frac{9}{x}\right)(x)}$$

$$= \frac{x + 3}{x^2 - 9}$$

$$= \frac{\cancel{x + 3}}{(\cancel{x + 3})(x - 3)}$$

$$= \frac{1}{x - 3}$$

Method 2 Find the least common denominator of all the fractions appearing in the numerator and denominator, then multiply both the numerator and denominator by it to clear fractions.

Equations with rational expressions (6.5)

Solve:

$$\frac{3x}{x^2 - 1} - \frac{1}{x + 1} = 0$$

$$(x^2 - 1)\left(\frac{3x}{x^2 - 1} - \frac{1}{x + 1}\right) = (x^2 - 1) \cdot 0$$

$$3x - (x - 1) = 0$$

$$2x + 1 = 0$$

$$x = -\frac{1}{2}$$

To **solve** an equation with rational expressions, multiply both sides by the least common denominator of all fractions appearing. Solve the resulting equation using previous methods. Be sure to check all solutions in the original equation.

Ratio and proportion; Similar triangles (6.6)

A **ratio** is the comparison of two numbers a and b, $b \neq 0$, using division. The ratio of two numbers can be written as

$$a \text{ to } b, \qquad a{:}b, \qquad \text{or} \qquad \frac{a}{b}$$

An equation in which two ratios are equal is called a **proportion.** We can solve a proportion, such as $\frac{2}{5} = \frac{4}{x}$, by using the cross-multiplication property of proportions: If

$$\frac{a}{b} = \frac{c}{d}$$

then

$$ad = bc$$

Solve for x:

$$\frac{2}{5} = \frac{4}{x}$$

By the cross-multiplication property,

$$2x = 5 \cdot 4$$

$$x = \frac{5 \cdot 4}{2} \quad \text{or} \quad 10$$

Two triangles are **similar** if they have the same shape but not necessarily the same size.

Property of Similar Triangles

If two triangles are similar, then corresponding angles are equal and corresponding sides are proportional.

Examples

Solve for y:

$$\frac{y-2}{x+7} = -\frac{1}{7}$$

$$(x+7) \cdot \frac{y-2}{x+7} = (x+7)\left(-\frac{1}{7}\right)$$

$$y - 2 = -\frac{1}{7}x - 1$$

$$y = -\frac{1}{7}x + 1$$

Applications of rational expressions (6.7)

To solve a formula or equation for an indicated variable means to isolate that variable to the left side.

Check Section 6.7 for applications such as number problems, work problems, and distance problems.

CHAPTER 6 REVIEW EXERCISE SET

Section 6.1

1. Evaluate $\frac{2}{x+3}$ when

(a) x is 1 **(b)** x is -1 **(c)** x is 0

2. Evaluate $\frac{2y}{y^2+1}$ when

(a) y is 2 **(b)** y is 0 **(c)** y is -1

For Exercises 3–6, find the values of x, if any, for which the rational expression is undefined.

3. $\frac{2x}{x+1}$ **4.** $\frac{3x-1}{x^2-4}$ **5.** $\frac{x+2}{x^2+9}$ **6.** $\frac{x+5}{x^2-x}$

For Exercises 7–12, reduce to lowest terms.

7. $\frac{2x^5}{x^7}$ **8.** $\frac{y^2+4y-5}{y^2-25}$ **9.** $\frac{4x^2+8x+3}{4x^2-9}$

10. $\frac{3n-12}{7n-28}$ **11.** $\frac{2x^2-9x+4}{16-x^2}$ **12.** $\frac{12-w}{w-12}$

Section 6.2

For Exercises 13–20, multiply or divide and write the answer in lowest terms.

13. $\frac{8}{x^4} \cdot \frac{x^3}{4}$ **14.** $\frac{3x-1}{2x-1} \cdot \frac{2x-1}{3x^2+5x-2}$ **15.** $\frac{2x^2+xy-y^2}{2xy^2} \cdot \frac{3x^4y}{x^2-y^2}$

16. $\frac{6x^3}{7y^5} \div \frac{2x^2}{21y^4}$ **17.** $\frac{2x^2+5x-3}{x-2} \div (x+3)$ **18.** $\frac{2-y}{4y+2} \div \frac{y^2-4}{12y+6}$

19. $\dfrac{t^2 + t}{t^2 - 3t + 2} \div \dfrac{t^2}{t - 1}$

20. $\dfrac{2x^2 - 7x + 6}{9x^2 - 6x + 1} \div \dfrac{2x - 3}{3x^2 + 2x - 1}$

Section 6.3

For Exercises 21–29, find the least common denominator.

21. $\dfrac{2}{21}$ and $\dfrac{5}{28}$

22. $\dfrac{1}{6}$ and $\dfrac{2}{7a}$

23. $\dfrac{3}{x^4}$ and $\dfrac{2}{5x^2}$

24. $\dfrac{b + 1}{6b^4}$ and $\dfrac{2}{15b^3}$

25. $\dfrac{7}{4y^3}$ and $\dfrac{22}{3y^4 + 6y^3}$

26. $\dfrac{u + 1}{u^2 - 4}$ and $\dfrac{3u}{u + 2}$

27. $\dfrac{3n}{2n^2 + n - 1}$ and $\dfrac{2n - 5}{n^2 + 3n + 2}$

28. $\dfrac{3x}{2x - 7}$ and $\dfrac{1 - x}{7 - 2x}$

29. $\dfrac{7}{3p^2}$ and $5p + 2$

For Exercises 30–38, add or subtract. Write your answer in lowest terms.

30. $\dfrac{3}{x} + \dfrac{11}{x}$

31. $\dfrac{26}{a} - \dfrac{14}{a}$

32. $\dfrac{12r}{r + 1} - \dfrac{5r}{r + 1}$

33. $\dfrac{4y}{(y + 3)(2y + 1)} + \dfrac{12}{(y + 3)(2y + 1)}$

34. $\dfrac{5}{4x^2} + \dfrac{4}{5x^2}$

35. $\dfrac{r}{r + 3} - \dfrac{r^2}{r^2 - 9}$

36. $\dfrac{z + 3}{z^2 - 16} - \dfrac{z - 2}{z^2 + 3z - 4}$

37. $x + 1 + \dfrac{x - 3}{x - 2}$

38. $\dfrac{2}{x - 2} - \dfrac{11x + 2}{3x^3 - 6x^2} - \dfrac{5}{3x}$

Section 6.4

For Exercises 39–44, simplify each complex fraction.

39. $\dfrac{\frac{2}{7}}{\frac{4}{21}}$

40. $\dfrac{\frac{12}{a}}{\frac{15}{a^3}}$

41. $\dfrac{\frac{1}{5} - \frac{7}{10}}{\frac{3}{4} + \frac{1}{3}}$

42. $\dfrac{1 - \frac{2}{3t}}{\frac{1}{t} - \frac{1}{2}}$

43. $\dfrac{\frac{7}{y} + \frac{3}{4}}{\frac{9}{2y^2} - \frac{1}{y}}$

44. $\dfrac{\frac{3}{u + 2} + \frac{1}{u - 3}}{\frac{2}{u - 3} - \frac{1}{u + 2}}$

Section 6.5

For Exercises 45–48, solve the equation.

45. $\frac{3}{x} - \frac{2}{3} = -1$

46. $\frac{x-1}{x+1} + \frac{5x+1}{x^2+x} = \frac{3}{x}$

47. $\frac{3r}{r+5} - \frac{5}{r-5} = 3$

48. $\frac{x+2}{x+1} = \frac{5}{x} - \frac{1}{x(x+1)}$

Section 6.6

For Exercises 49–52, solve each proportion.

49. $\frac{2}{x} = \frac{3}{2}$

50. $3:x = 1:4$

51. $x:16 = 1:x$

52. 3 is to x as x is to 12

53. On a roadmap, if 4 inches represents 18 miles, how many miles does 10 inches represent?

For Exercises 54 and 55, find the lengths of the missing sides in each pair of similar triangles. Assume the lengths are given in feet.

54.

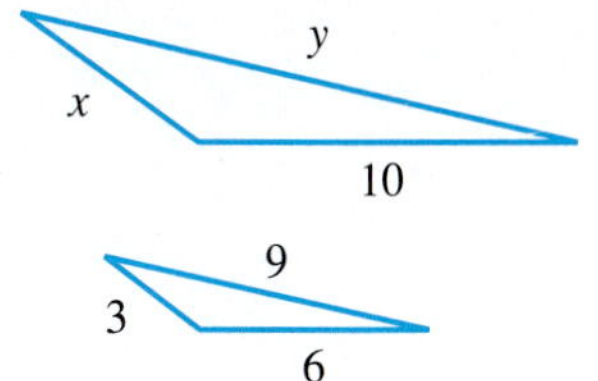

55.

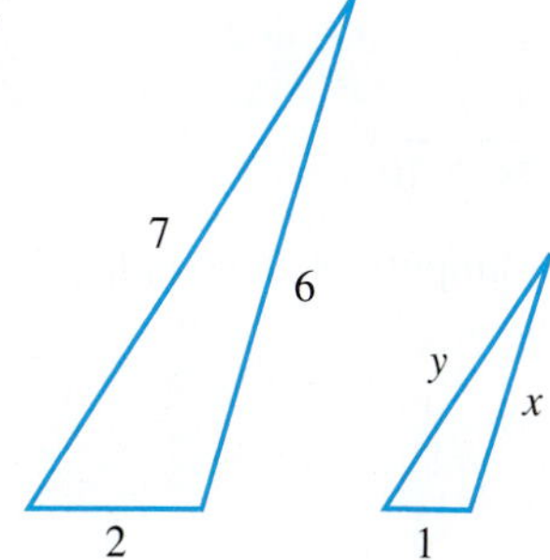

Section 6.7

56. Solve for t in the formula $P = \frac{a}{1 + rt}$.

57. Find two numbers whose sum is 15 and the sum of their reciprocals is $\frac{5}{12}$.

58. It takes one crew 3 weeks to clean the windows of a skyscraper. It takes another crew 4 weeks to clean the windows of the same building. Working together, how long would it take the two crews to clean the windows of the skyscraper?

59. Sue drove a four-wheel all-terrain vehicle 5 miles in the same time that it took Don to drive a three-wheel ATV 8 miles. If Sue drove 4 miles per hour slower than Don, how fast was each person traveling?

CHAPTER 6 TEST

For Problems 1–3, reduce the rational expression to lowest terms.

1. $\dfrac{6a^2b^3c}{14a^4bc}$

2. $\dfrac{3x-9}{x^2-9}$

3. $\dfrac{2z-4}{8-4z}$

For Problems 4–6, multiply or divide as indicated. Write the answer in lowest terms.

4. $\dfrac{5u^3}{6v} \div \dfrac{10u^5}{3v^2}$

5. $\dfrac{4a-a^3}{16a^3b} \cdot \dfrac{8a^2b^2}{a+2}$

6. $\dfrac{20x^2-17x+3}{25x^2-9} \div \dfrac{4x^2-9x+2}{5x^2-7x-6}$

For Problems 7 and 8, find the least common denominator.

7. $\dfrac{2c}{3c^2-4c+1}$ and $\dfrac{5}{c^2-1}$

8. $\dfrac{4}{2x-1}$ and $2x+1$

For Problems 9–11, add or subtract as indicated. Write the answer in lowest terms.

9. $\dfrac{12}{x^3} - \dfrac{7}{x^3}$

10. $\dfrac{5n}{25n^2-4} + \dfrac{2}{25n^2-4}$

11. $\dfrac{y+1}{y^2+y-6} - \dfrac{y-7}{y^2-7y+10}$

For Problems 12 and 13, simplify each complex fraction.

12. $\dfrac{\dfrac{3a^2}{b}}{\dfrac{12a^3}{b^3}}$

13. $\dfrac{\dfrac{3}{x+1} - \dfrac{1}{x^2-1}}{\dfrac{4}{x-1}}$

For Problems 14 and 15, solve the equation.

14. $\dfrac{5}{a^2-25} = \dfrac{3}{a+5} + \dfrac{2}{a-5}$

15. $\dfrac{4}{2x^2+5x-3} - \dfrac{x}{4x^2-1} = \dfrac{x}{2x^2+5x-3}$

For Problems 16 and 17, solve the proportion.

16. $\dfrac{5}{x} = \dfrac{15}{4}$

17. $x:16 = 4:x$

18. If it costs \$0.62 to make 400 nails, how much does it cost to make 1,400 nails?

19. Find the lengths of the missing sides in the pair of similar triangles. (Assume the lengths are given in centimeters.)

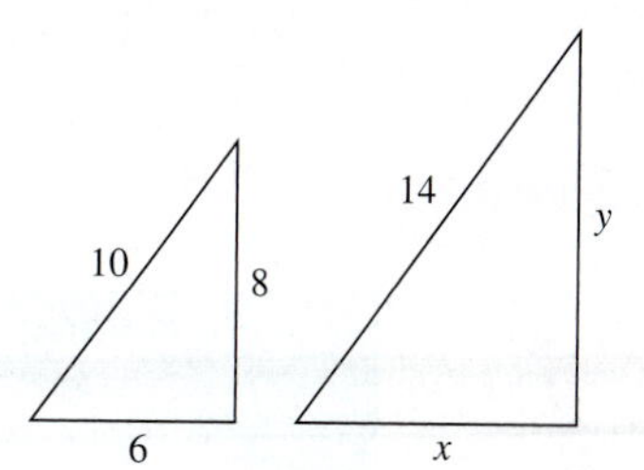

20. Solve for y:

$$\frac{y-4}{x-1} = -2$$

21. The sum of two numbers is 4, and the sum of their reciprocals is $\frac{16}{15}$. Find the two numbers.

22. Working together, Trevor and Terence can plow the field in 2 hours. It would take Trevor twice as long as it would take Terence to do it alone. Working alone, how long would it take each of them?

23. The current of a river is 3 miles per hour. It takes a boat as long to go 13 miles downriver as to go 10 miles upriver. What is the rate of the boat in still water?

Systems of Equations

CHAPTER 7

CONNECTIONS

Lyn works a summer job mowing lawns x hours a week for \$10 per hour and trimming hedges y hours a week for \$20 per hour. One week she worked 35 hours and made \$450. How many hours that week did she mow lawns and how many hours did she trim hedges? See Exercise 39 in Section 7.3.

Overview

In Chapter 3, we studied linear equations in two variables. In this chapter, we look at systems of linear equations. We will learn three methods for solving a system of equations. In the last section, we will study systems of linear inequalities.

7.1 Solutions by Graphing

OBJECTIVES

- *To find solutions of systems by graphing*
- *To determine when a system has no solution*
- *To determine when a system has infinitely many solutions*

LEARNING ADVANTAGE

Mathematics at Work in Society: Medical Careers. *Whenever people think of medical careers they usually think of doctors, nurses, and dentists. These important careers are the most visible careers in health care. But what are some other medical careers that are just as important? Here are two examples.*

A physician's associate is trained to assist a doctor in the general practice of medicine. A physician's associate may examine and diagnose patients under the doctor's supervision. Training: 2 years of college plus 2 years of specialized study in an associate's program.

A pharmacist dispenses drugs and medicines prescribed by physicians and dentists. Training: 4 years of college plus 1 year of specialized training.

These careers require good technical training. You are achieving a good start by taking this course to develop your mathematical skills.

Consider the linear equation

$$3x + 2y = 6$$

The graph of this equation is a straight line. Every point on this line is a solution of the equation. Now, consider a second equation

$$2x - y = 4$$

The graph of this equation is another straight line. The two equations together is called a **system of linear equations.** A **solution** of the system is a point that satisfies both equations. For example, let us show that the point (2, 0) is a solution of the system

$$3x + 2y = 6$$
$$2x - y = 4$$

Replacing x by 2 and y by 0 in the first equation, $3(2) + 2(0) = 6$. Replacing x by 2 and y by 0 in the second equation, $2(2) - 0 = 4$. This verifies that (2, 0) is a solution of the system.

Example 1 Determine if $(-1, 3)$ is a solution of the system of linear equations

$$-x + 4y = 13$$
$$5x - 2y = -11$$

Solution To see if $(-1, 3)$ is a solution of the system, replace x by -1 and y by 3 in each equation.

$-x + 4y = 13$	$5x - 2y = -11$
$-(-1) + 4(3) \stackrel{?}{=} 13$	$5(-1) - 2(3) \stackrel{?}{=} -11$
$1 + 12 \stackrel{?}{=} 13$	$-5 - 6 \stackrel{?}{=} -11$
$13 = 13$	$-11 = -11$

The ordered pair $(-1, 3)$ is a solution of both equations and, therefore, is a solution of the system. ◀

There are several methods for solving a system of linear equations. In the next example, we show how to solve a system by the **graphing method.**

Example 2 Solve the system of equations by the graphing method.

$$-x + y = 1$$
$$-2x + y = -1$$

Solution We graph each line in the same coordinate plane. We find the coordinates of three points from each equation.

$-x + y = 1$

x	y
0	1
−1	0
1	2

$-2x + y = -1$

x	y
0	−1
1	1
−1	−3

We plot these points and draw the two lines as shown in the figure below.

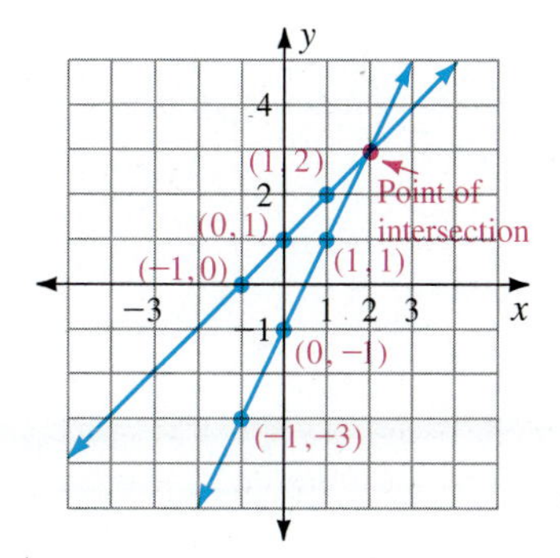

Notice that the two lines intersect at a point. The coordinates of this point of intersection are a solution of both equations and, therefore, a solution of the system of equations. We estimate this point to be (2, 3). To determine if this is the exact point of intersection, we check to see if (2, 3) is a solution of each equation:

Check (2, 3):

$$-x + y = 1 \qquad\qquad -2x + y = -1$$
$$-2 + 3 \stackrel{?}{=} 1 \qquad\qquad -2(2) + 3 \stackrel{?}{=} -1$$
$$1 = 1 \qquad\qquad -1 = -1$$

Therefore, (2, 3) is the solution of the system. ◀

The graphing method has its limitations. For example, it would be difficult to use if the coordinates of the point of intersection were not integers. The graphing method requires an extremely accurate graph and, therefore, you must use graph paper. In this book, when you are asked to solve a system by the graphing method, all points of intersection will be integers.

In Sections 7.2 and 7.3, we will discuss other methods that will give the solution in a precise way. The graphing method, however, does give us a useful visual picture concerning the nature of solutions.

Example 3 Use the graphing method to determine the solution set of the system

$$3x - 2y = 1$$
$$-3x + 2y = 6$$

Solution We graph each line using two convenient points.

$3x - 2y = 1$

x	y
1	1
3	4

$-3x + 2y = 6$

x	y
0	3
2	6

The two lines are shown in the figure below. Notice that the two lines are parallel—they do not intersect. No ordered pair of numbers could satisfy *both* equations and therefore this system has no solution. We conclude that the solution set is empty.

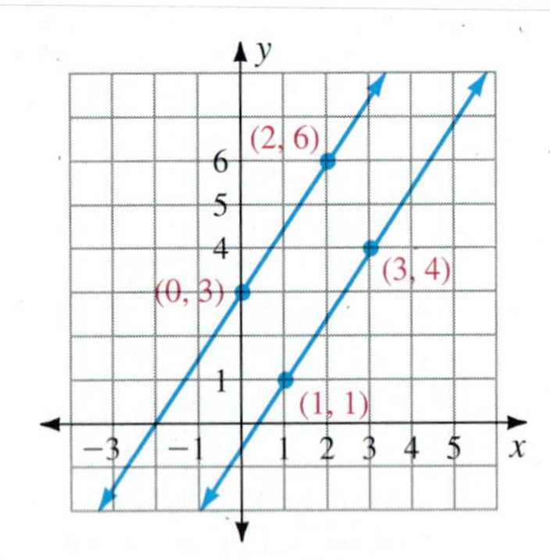

In Example 3, we were shown a system with no solutions. This system is called an **inconsistent system.**

The system in Example 4 is an illustration of the other extreme.

Example 4 Determine the solution set of the system

$$2x - 3y = 6$$
$$-6x + 9y = -18$$

Solution We find the intercepts from each equation.

x	y
0	−2
3	0

x	y
0	−2
3	0

Notice that the intercepts are the same. Therefore, each equation has the same line as its graph. That is, the graphs coincide. This means that any point on this line satisfies both equations and, therefore, is a solution of the system. Our conclusion is that the system has infinitely many solutions. (See the figure below.)

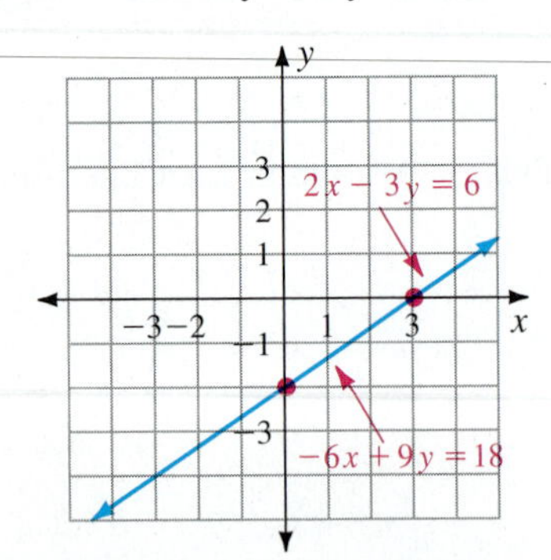

The system in Example 4, where the graphs coincide, is said to be **dependent.** There are infinitely many solutions of the system where any solution of one equation will be a solution of the other equation.

In Examples 2, 3 and 4 we have illustrated the three possibilities for the solution set of a system of equations. A system may have one solution (Example 2), no solution (Example 3), or infinitely many solutions (Example 4).

These three possibilities are shown graphically in the figures below.

Independent

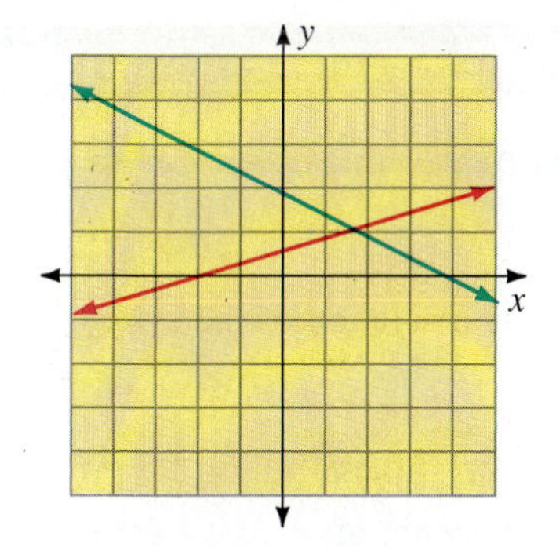

The Two Lines Intersesct (One Solution)

(a)

Inconsistent

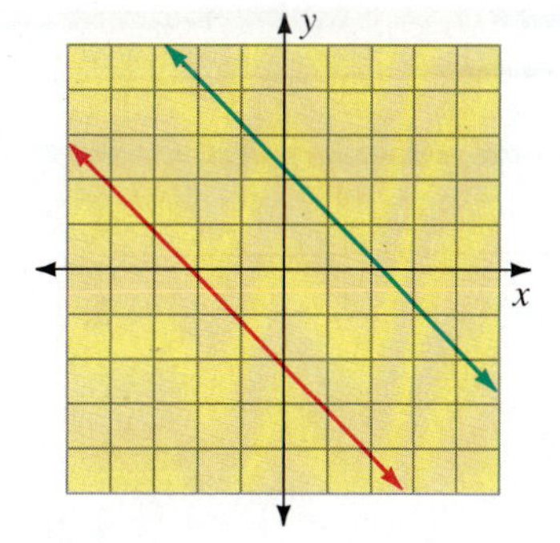

The Two Lines Are Parallel (No Solution)

(b)

Dependent

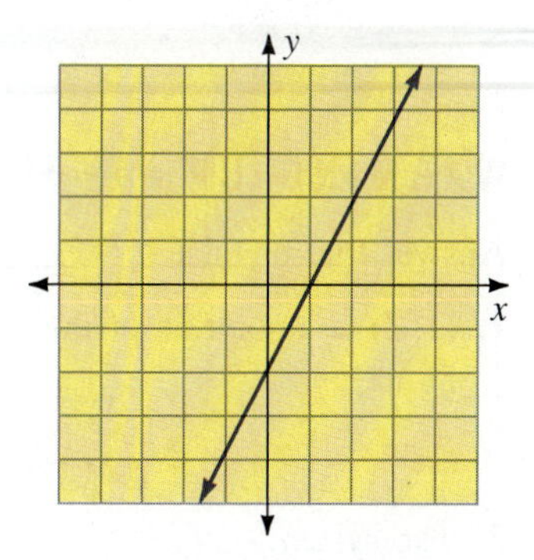

The Graphs Coincide (Infinitely many solutions)

(c)

GRAPHING CALCULATOR CROSSOVER

EXAMPLE Use a graphing calculator to estimate the solution to the system

$$y = 2x + 1$$
$$y = 4 - x$$

Solution Using the RANGE as $X_{min} = -10$, $X_{max} = 10$, $X_{scl} = 1$, $Y_{min} = -10$, $Y_{max} = 10$, $Y_{scl} = 1$, we GRAPH both equations. The viewing screen should look like this.

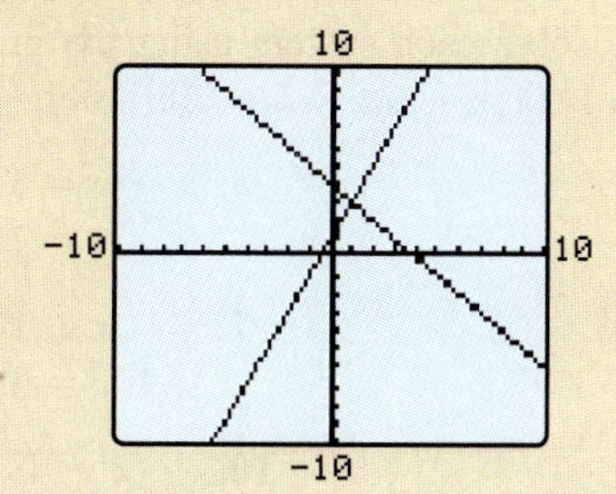

GRAPHING CALCULATOR CROSSOVER *(continued)*

From the graph, it appears that the solution to the system is the point (1, 3). We check to see if this is correct. Replacing x by 1 and y by 3 in each equation,

$$y = 2x + 1 \qquad\qquad y = 4 - x$$

$$3 = 2(1) + 1 \quad \text{true} \qquad\qquad 3 = 4 - 1 \quad \text{true}$$

The solution is (1, 3).

WARMING UP

Answer true or false.

1. (1, 2) is a solution of the system

$$x + 3y = 7$$
$$2x - y = 0$$

2. (3, −4) is a solution of the system

$$2x + 3y = -6$$
$$3x + 2y = -1$$

3. The system

$$2x + 4y = -10$$
$$-x - 2y = 5$$

has infinitely many solutions.

4. The solution set of

$$-3x + 4y = -12$$
$$3x - 4y = -12$$

is $\varnothing$.

5. An independent system has exactly one solution.

6. A dependent system has no solution.

EXERCISE SET 7.1

For Exercises 1 and 2, decide whether or not the ordered pairs are solutions of the given system of equations.

1. $6x - 5y = 1$
$2x - y = 1$
(a) (1, 1)
(b) (0, 2)
(c) (5, −1)

2. $-3x + 2y = -8$
$4x + 7y = 1$
(a) (2, −1)
(b) (−3, 0)
(c) (1, 2)

For Exercises 3–16, solve each system using the graphing method. If the system is inconsistent, write ''no solution.'' If the system is dependent, write ''infinitely many solutions.'' Use graph paper.

3. $y - 2x = 2$
$y + 2x = 6$

4. $x + y = 5$
$x - y = 7$

5. $2x + y = 4$
$x - y = -4$

6. $-x + y = -2$
$3x - 2y = 6$

7. $y = x + 1$
$x + y = 3$

8. $x - y = 7$
$2x + y = 2$

9. $2x + 2y = 3$
$3x + 3y = 4.5$

10. $x - 4y = 8$
$-2x + 8y = -16$

11. $x + y = -2$
$-4x - 4y = 8$

12. $3x - y = 3$
$-6x + 2y = -6$

13. $y = 2x + 1$
$y = 2x + 3$

14. $y = -x + 5$
$y = -x - 1$

15. $y = -3x + 5$
$y = x - 3$

16. $y = -x + 2$
$y = 2x - 1$

17. Determine if $(1.59, -7.27)$ is a solution of

$$2.8x - 3.1y = 26.989$$
$$7x + 4.1y = -18.677$$

18. Determine if $(-2.04, 6.15)$ is a solution of

$$9.1x + 0.8y = 13.644$$
$$-10.8x - 6.92y = 20.526$$

For Exercises 19–22, use a graphing calculator to estimate the solution of the system. Then, check your answer.

19. $y = 5 - x$
$y = 2x - 1$

20. $y = x + 2$
$y = 4 - x$

21. $y = x$
$y = 6 - 2x$

22. $y = 3 - x$
$y = x - 1$

SAY IT IN WORDS

23. Explain the graphing method for solving a system of equations.

24. Explain independent, dependent, and inconsistent systems.

REVIEW EXERCISES

The following exercises review parts of Section 2.3. Doing these exercises will help prepare you for the next section.

For Exercises 25–30, solve each equation.

25. $2x - 5 = 9$

26. $4t = -2 + 8$

27. $6y - 3 = -6$

28. $3x - 7 = 1 + 10$

29. $2x + 3(-1) = 2$

30. $6x - 3(-3) = 15$

ENRICHMENT EXERCISES

Find the relationship between c_1 and c_2 so that the system

$$2x - 3y = c_1$$
$$-12x + 18y = c_2$$

has

1. No solution.

2. Infinitely many solutions.

Answers to Enrichment Exercises begin on page A.1.

7.2 The Elimination Method

OBJECTIVE

▶ *To solve a system of equations by the elimination method*

LEARNING ADVANTAGE

Mathematics at Work in Society: Optometry. *An optometrist examines eyes for vision problems and disease, prescribes lenses to correct vision problems, and when disease is discovered, refers patients to the appropriate physician. Training: 4 years of college plus 2 to 3 years of specialized study at an accredited college.*

Optometry requires good technical training. If you are interested in this career, continue to improve your mathematics background.

In the last section, we solved systems of equations using the graphing method. This method is not adequate when systems have solutions that are fractions or decimals. In this section and the next, we will learn two algebraic methods that will take less time and will also give accurate solutions. The first algebraic method is called the **elimination method.** This method uses the addition property of equality.

Since we can add the same number to both sides of an equation, we can add two equations together.

Example 1 Use the addition property of equality to solve the system:

$$x - y = 18$$
$$x + y = -2$$

Solution Using the addition property, we add these two equations.

$$\begin{array}{rl} x - y = 18 & \\ \underline{x + y = -2} & \text{Add.} \\ 2x \quad\; = 16 & \end{array}$$

Solving this resulting equation,

$$\begin{array}{rl} 2x = 16 & \\ x = 8 & \text{Divide both sides by 2.} \end{array}$$

Next, we replace x by 8 in either of the original equations, then solve for y. We select the first equation.

$$\begin{array}{rl} x - y = 18 & \\ 8 - y = 18 & \text{Replace } x \text{ by 8.} \\ -y = 10 & \text{Subtract 8 from both sides.} \\ y = -10 & \text{Multiply both sides by } -1. \end{array}$$

The solution is $x = 8$ and $y = -10$. On your own, check this solution in the original system. ◀

Notice that in Example 1, the two equations of the system were added together to obtain a single equation with only one variable. That is, we *eliminated* one of the variables by adding equations.

The elimination method requires that when the equations are added together, one of the variables is eliminated. Consider the system

$$-3x + 2y = -7$$
$$6x + 5y = -4$$

If we add these two equations,

$$\begin{array}{r} -3x + 2y = -7 \\ \underline{6x + 5y = -4} \\ 3x + 7y = -11 \end{array} \quad \text{Add.}$$

The result, $3x + 7y = -11$, still contains both variables. However, if we multiply the first equation by 2 and *then* add, the resulting equation will have only one variable. The details are shown in the next example.

Example 2 Solve the system:

$$-3x + 2y = -7$$
$$6x + 5y = -4$$

Solution Adding the two equations does not eliminate either variable. However, we can eliminate x if we multiply the first equation by 2, then add the result to the second equation.

$$\begin{array}{rcr} -3x + 2y = -7 & \xrightarrow{\text{multiply both sides by 2}} & -6x + 4y = -14 \\ 6x + 5y = -4 & \xrightarrow{\text{leave as is}} & \underline{6x + 5y = -4} \\ & & 9y = -18 \end{array}$$

Solving $9y = -18$ for y, we have $y = -2$. Next, replace y by -2 in either of the original equations. We select the first equation.

$$-3x + 2y = -7$$
$$-3x + 2(-2) = -7$$
$$-3x - 4 = -7$$
$$-3x = -3 \quad \text{Add 4 to both sides.}$$
$$x = 1 \quad \text{Divide both sides by } -3.$$

The solution is $x = 1$ and $y = -2$. Check this.

NOTE *When multiplying an equation by a constant, be sure to multiply* both *sides of the equation. In Example 2, we multiplied* $-3x + 2y = -7$ *by 2. That is,*

$$2(-3x + 2y) = 2(-7)$$

Sometimes it is necessary to multiply each equation by appropriate numbers before adding. We illustrate this situation in the next example.

Example 3

$$3x + 4y = -5$$
$$2x + 3y = -4$$

Solution We choose to eliminate x. Multiply the first equation by 2 and the second by -3, then add.

$$\begin{aligned} 3x + 4y &= -5 \xrightarrow{\text{multiply by } 2} & 6x + 8y &= -10 \\ 2x + 3y &= -4 \xrightarrow{\text{multiply by } -3} & -6x - 9y &= 12 \\ & & -y &= 2 \\ & & y &= -2 \end{aligned}$$

Next, replace y by -2 in either equation of the original system. We choose the first equation.

$$\begin{aligned} 3x + 4y &= -5 \\ 3x + 4(-2) &= -5 \\ 3x &= 3 \\ x &= 1 \end{aligned}$$

The solution is $(1, -2)$. Check.

NOTE *After finding the value of one of the variables, we use one of the original equations to solve for the other variable. Since it makes no difference which equation we pick, choose the equation that is easier to solve.*

Before we proceed with more examples, here is a summary of the elimination method. Use this summary as you work through the exercise set.

STRATEGY

The Elimination Method

Step 1 If necessary, transform each equation to the standard form $Ax + By = C$. Also, clear fractions by multiplying by the least common denominator.

Step 2 Decide which variable is easier to eliminate.

Step 3 If necessary, multiply one or both equations by appropriate numbers so that the coefficients of the variable to be eliminated are opposites.

Step 4 Add the two equations to obtain an equation with only one variable appearing.

Step 5 Solve this equation for the variable.

Step 6 Take the answer from Step 5 and use either equation from the original system to find the value of the other variable.

Step 7 Check your solution in the original system.

Recall from the graphing method, a system has no solution when the two lines are parallel. The elimination method will also work on this kind of system.

Example 4 Solve the system:

$$2x - 3y = -1$$
$$-4x + 6y = -2$$

Solution We choose to eliminate x by multiplying the first equation by 2, then adding.

$$2x - 3y = -1 \xrightarrow{\text{multiply by 2}} 4x - 6y = -2$$
$$-4x + 6y = 1 \xrightarrow{\text{leave as is}} \underline{-4x + 6y = 1}$$
$$0 = -1$$

The assumption that the system has a solution leads to the contradiction that $0 = -1$. Therefore, the system has no solution—the system is inconsistent.

NOTE *In the previous example, the elimination method was used to conclude that the system had no solution. Keep in mind that geometrically, the two lines are parallel.*

The other "unusual" case is when a system has infinitely many solutions. Graphically, the two equations have the same line for their respective graphs.

Example 5 Solve the system:

$$x + 4y = -4$$
$$-2x - 8y = 8$$

Solution We choose to eliminate the variable x. Multiply the first equation by 2, then add.

$$x + 4y = -4 \xrightarrow{\text{multiply by 2}} 2x + 8y = -8$$
$$-2x - 8y = 8 \xrightarrow{\text{leave as is}} \underline{-2x - 8y = 8}$$
$$0 = 0$$

The statement $0 = 0$ means that the two equations describe the same line. Therefore, the system has infinitely many solutions—it is dependent.

As indicated in these examples, by using the elimination method we can automatically tell if the system has one solution, no solution, or infinitely many solutions.

STRATEGY

The Number of Solutions of a System—Using the Elimination Method

1. If the elimination method ends with a statement of the form $x = a$ and $y = b$, then the system is *independent* and has exactly **one solution,** namely, (a, b). (See Examples 1 through 3.)
2. If the elimination method ends with a false statement of the form such as $0 = 3$, then the system is *inconsistent* and has **no solution.** (See Example 4.)
3. If the elimination method ends with a true statement of the form such as $0 = 0$, then the system is *dependent* and has **infinitely many solutions.** (See Example 5.)

TEAM PROJECT

(3 or 4 Students)

SOLVING SYSTEMS OF EQUATIONS USING THE ELIMINATION METHOD

Course of Action: Each team member selects one problem from the first 15 problems in Exercise Set 7.2 and solves it. Give the selected problem number to the group member on your left. After that person solves your problem, both of you compare your individual solutions. If the solutions do not agree, ask the other team members to help decide which solution is correct.

Next, each team member selects one problem from the next 15 problems in Exercise Set 7.2 and solves it. Give the selected problem number to the group member on your right. After that person solves your problem, both of you compare your individual solutions. If the solutions do not agree, ask the other team members to help decide which solution is correct.

Team Report:

1. How successful was the group in solving these systems of equations by the elimination method?
2. Write a paragraph giving tips and advice on how to solve such systems.

WARMING UP

Answer true or false.

1. To eliminate x from the system

$$\begin{aligned} x - 3y &= -1 \\ -3x + 2y &= -4, \end{aligned}$$

multiply the first equation by 3 and add the result to the second equation.

2. To eliminate y from the system

$$-2x + y = 7$$
$$3x + y = -2,$$

add the two equations.

3. $(-3, -4)$ is a solution of the system

$$-3x + 2y = 1$$
$$x + y = -7$$

4. The system

$$2x - y = 7$$
$$-4x + 2y = -14$$

is dependent.

5. The system

$$-x + 3y = -1$$
$$3x - 9y = 2$$

is dependent.

6. We can eliminate either variable when using the elimination method.

7. If a system is independent, it has infinitely many solutions.

8. If a system is inconsistent, then its solution set is empty.

EXERCISE SET 7.2

For Exercises 1–30, solve the system by the elimination method.

1. $\begin{aligned} 2x + 3y &= 8 \\ -2x + 4y &= 6 \end{aligned}$

2. $\begin{aligned} x + 2y &= 1 \\ -x + 3y &= 4 \end{aligned}$

3. $\begin{aligned} x + y &= 7 \\ -x + 4y &= 3 \end{aligned}$

4. $\begin{aligned} -3x - 2y &= 2 \\ 3x + y &= -4 \end{aligned}$

5. $\begin{aligned} 2x - y &= -3 \\ -3x + y &= 4 \end{aligned}$

6. $\begin{aligned} 3x + 2y &= 1 \\ 5x - 2y &= 7 \end{aligned}$

7. $\begin{aligned} 4x + 3y &= 2 \\ -5x - 3y &= -4 \end{aligned}$

8. $\begin{aligned} x + 4y &= 0 \\ -3x - 4y &= 8 \end{aligned}$

9. $\begin{aligned} x - 3y &= -2 \\ -2x + y &= -1 \end{aligned}$

10. $\begin{aligned} x + 2y &= 0 \\ -3x - 5y &= 1 \end{aligned}$

11. $\begin{aligned} -2x + y &= -7 \\ 3x - 2y &= 11 \end{aligned}$

12. $\begin{aligned} 2x + 3y &= 6 \\ -3x + y &= 2 \end{aligned}$

13. $\begin{aligned} 5x + 3y &= -1 \\ x + y &= 1 \end{aligned}$

14. $\begin{aligned} 4x + 5y &= -2 \\ 2x + 3y &= 0 \end{aligned}$

15. $\begin{aligned} 3x + y &= 2 \\ x - 2y &= 10 \end{aligned}$

16. $\begin{aligned} -2x - 3y &= 7 \\ 3x - y &= 6 \end{aligned}$

17. $\begin{aligned} -x + 2y &= 10 \\ 3x + 4y &= 0 \end{aligned}$

18. $\begin{aligned} -2x - 3y &= 2 \\ 3x + y &= 4 \end{aligned}$

19. $3x + 4y = -1$
$2x - 5y = 7$

20. $4x - 2y = 2$
$3x - 5y = 12$

21. $2x + 3y = -9$
$5x + 4y = -5$

22. $-3x + 4y = 3$
$2x - 3y = -1$

23. $2x + 5y = -8$
$5x - 6y = -20$

24. $3x + 2y = 6$
$5x - 7y = -21$

25. $8x - 2y = 14$
$-4x + y = 6$

26. $3x - 2y = -4$
$6x - 4y = 7$

27. $4x - 5y = -3$
$-16x + 20y = 12$

28. $6x - 15y = 9$
$-2x + 5y = -3$

29. $3x + 2y = -4$
$5x + 4y = -2$

30. $4x + 3y = 5$
$-3x - 2y = -2$

For Exercises 31 and 32, solve each system by the elimination method.

31. $2.1x - 3.9y = 7.6$
$x + 4.5y = 8$

32. $0.7x + 0.9y = 2.45$
$-0.6x + 1.01y = 7.9$

SAY IT IN WORDS

33. Explain the elimination method.

34. Make up a system of equations where the solution is $(1, -1)$.

REVIEW EXERCISES

The following exercises review parts of Section 2.3. Doing these exercises will help prepare you for the next section.

For Exercises 35–39, solve each equation.

35. $3x + (2x + 1) = 3$

36. $2(1 - 3y) + 4 = -1$

37. $1 - 4x - (2 - 7x) = -2$

38. $3\left(\frac{1}{3} - x\right) + 4x = -5$

39. $8z + 2(4 - 3z) = -10$

ENRICHMENT EXERCISES

1. Solve the system by letting $x = \frac{1}{a}$ and $y = \frac{1}{b}$.

$$4\left(\frac{1}{a}\right) + 3\left(\frac{1}{b}\right) = -1$$
$$2\left(\frac{1}{a}\right) - 3\left(\frac{1}{b}\right) = -5$$

2. Solve the system by letting $x = s - 2$, $y = t + 1$.

$$3(s - 2) + 2(t + 1) = 0$$
$$4(s - 2) + 3(t + 1) = 1$$

3. Solve the system by letting $x = -2u + 1$ and $y = 3v - 2$.

$$3(-2u + 1) + (3v - 2) = 1$$
$$-5(-2u + 1) - 2(3v - 2) = 1$$

Answers to Enrichment Exercises begin on page A.1.

7.3 The Substitution Method

OBJECTIVE

▶ *To solve a system using the substitution method*

LEARNING ADVANTAGE

Mathematics at Work in Society: Consumerism. *"Consumer" is the name given to people who use goods and services to satisfy personal needs. You are a consumer and will continue to be a consumer for the rest of your life. Mathematics will aid you. For example,*

1. *Math prevents being overcharged or shortchanged.*
2. *By calculating the unit price, math determines the most economical buy.*
3. *Math helps you to read graphs and to interpret data in newspapers and magazines.*
4. *Math helps with do-it-yourself building and repair projects.*

Consider the system

$$x + y = 7$$
$$y = x + 1$$

Notice that the second equation is solved for y. If y is $x + 1$ in the second equation, it must be $x + 1$ in the first equation. Therefore, we replace y by $x + 1$ in the first equation.

$$x + (x + 1) = 7$$

We can now solve this equation for x.

$$2x + 1 = 7$$
$$2x = 6$$
$$x = 3$$

Since the x-value of the solution of the system is 3, the y-value can be found from the second equation.

$$y = x + 1$$
$$= 3 + 1$$
$$= 4$$

Therefore, the solution is $(3, 4)$.

We call this method the **substitution method.**

Example 1 Use the substitution method to solve the system.

$$3x - 5y = -12$$
$$2x - 4y = -10$$

Solution Neither equation is already solved for a variable, so we select the equation and the variable to be solved. Our selection depends upon the coefficients involved. There are four possible choices. We choose to solve for x in the second equation as this choice will not introduce fractions into the problem.

$$2x - 4y = -10$$
$$x - 2y = -5 \qquad \text{Divide both sides by 2.}$$
$$x = 2y - 5$$

Next, we substitute $2y - 5$ for x in the first equation and solve for y.

$$3x - 5y = -12$$
$$3(2y - 5) - 5y = -12$$
$$6y - 15 - 5y = -12$$
$$y = 3$$

To find x, use the equation $x = 2y - 5$.

$$x = 2(3) - 5$$
$$x = 1$$

Our solution is $(1, 3)$. Check this answer in the original system.

We summarize the substitution method.

STRATEGY

The Substitution Method

Step 1 Solve one of the equations for one of the variables, if neither equation is already solved for one of its variables.

Step 2 Substitute for that variable in the other equation. We now have one equation in one unknown.

Step 3 Solve the resulting equation.

Step 4 Find the value of the other variable by substituting the number from Step 3 into the equation of Step 1.

Step 5 Check your answer.

As we have seen before, some systems have no solution. Geometrically, the lines are parallel and the system is inconsistent. If we attempt to solve such systems using the substitution method, we will arrive at a contradiction.

Example 2 Use the substitution method to solve

$$x - 3y = 6$$
$$-2x + 6y = 1$$

Solution Assuming that there is a solution, we use the substitution method to find it. We solve the first equation for x.

$$x - 3y = 6$$
$$x = 3y + 6$$

Replace x by $3y + 6$ in the second equation.

$$-2x + 6y = 1$$
$$-2(3y + 6) + 6y = 1$$
$$-6y - 12 + 6y = 1$$
$$-12 = 1 \quad \text{Contradiction.}$$

We had assumed that the system had a solution. This gave us the contradiction $-12 = 1$. Therefore, the system has no solution.

Recall that a system may have infinitely many solutions. In the next example, we show how the substitution method can be used on systems that are dependent.

Example 3 Use the substitution method to solve

$$-5x + 3y = 6$$
$$y = \frac{5}{3}x + 2$$

Solution Replace y by $\frac{5}{3}x + 2$ in the first equation, then solve for x.

$$-5x + 3y = 6$$
$$-5x + 3\left(\frac{5}{3}x + 2\right) = 6$$
$$-5x + 5x + 6 = 6$$
$$6 = 6$$

Notice that the last statement, $6 = 6$, is true regardless what number is used for x. Therefore, the system is dependent and has infinitely many solutions.

RULE

Inconsistent and Dependent Systems

Using either the elimination or substitution method:

(1) Inconsistent systems give a contradiction. For example, $0 = 1$.
(2) Dependent systems give an identity. For example, $0 = 0$.

We next look at word problems that can be solved using the substitution method.

Example 4 A rectangular sign has a length y that is 6 feet longer than the width x. If the perimeter is 36 feet, find the dimensions of the sign.

Solution Since the length is 6 feet longer than the width, $y = x + 6$. Using the formula for perimeter,

$$2(\text{width}) + 2(\text{length}) = 36$$
$$2x + 2y = 36$$

To find the dimensions of the sign, we solve the system

$$y = x + 6$$
$$2x + 2y = 36$$

We use the substitution method and replace y by $x + 6$ in the second equation:

$$2x + 2(x + 6) = 36$$
$$2x + 2x + 12 = 36 \quad \text{Distributive property.}$$
$$4x + 12 = 36 \quad \text{Simplify.}$$
$$4x = 24 \quad \text{Subtract 12 from both sides.}$$
$$x = 6 \quad \text{Divide both sides by 4.}$$

Now, to find the length y, replace x by 6 in the first equation.

$$y = x + 6$$
$$y = 6 + 6 = 12$$

The dimensions of the sign are 6 feet by 12 feet.

TEAM PROJECT

(3 or 4 Students)

SOLVING SYSTEMS OF EQUATIONS USING THE SUBSTITUTION METHOD

Course of Action: Each team member selects one problem from the first 10 problems in Exercise Set 7.3 and solves it. Give the selected problem number to the group member on your left. After that person solves your problem, both of you compare your individual solutions. If the solutions do not agree, ask the other team members to help decide which solution is correct.

Next, each team member selects one problem from the next 10 problems in Exercise Set 7.3 and solves it. Give the selected problem number to the group member on your right. After that person solves your problem, both of you compare your individual solutions. If the solutions do not agree, ask the other team members to help decide which solution is correct.

Team Report:

1. How successful was the group in solving these systems of equations by the substitution method?
2. Write a paragraph giving tips and advice on how to solve such systems.

WARMING UP

Answer true or false.

1. To solve the system

$$y = 2x - 1$$
$$x + 2y = 7$$

replace y in the second equation by $2x - 1$.

2. The solution of the system

$$y = x + 2$$
$$x + y = 6$$

is $(2, 4)$.

3. The system

$$y = x + 1$$
$$-2x + 2y = 2$$

is inconsistent.

4. The system

$$2x - 3y = 2$$
$$3x + 2y = -1$$

can be solved by the substitution method.

5. The system

$$y = -2x + 3$$
$$2x + y = 4$$

is dependent.

6. If $y = 2$ and $2x + y = 8$, then the x-value is 3.

EXERCISE SET 7.3

For Exercises 1–20, solve the system using the substitution method.

1. $x + y = 6$
$y = x$

2. $x + y = -8$
$y = 3x$

3. $y = x + 1$
$x + y = 3$

4. $y = x - 3$
$x + y = 5$

5. $x + 2y = 3$
$y = x$

6. $x + 3y = -1$
$y = x - 3$

7. $y = 2x + 5$
$3x + y = 0$

8. $2x - y = 2$
$y = 3x - 5$

9. $x = y + 2$
$2x - 3y = 5$

10. $-3x + 2y = -4$
$x = 3y - 1$

11. $2x + 3y = -7$
$y = -2x - 9$

12. $x = -2y - 7$
$3x - 2y = 3$

13. $x + y = 1$
$3x - 2y = 18$

14. $x + y = -3$
$15x + 8y = -80$

15. $5x - y = -20$
$x - y = -8$

16. $x - 6y = 16$
$x - y = 6$

17. $x - 2y = -7$
$3x - 6y = 10$

18. $x + 3y = -2$
$-2x - 6y = 4$

19. $3x - y = 2$
$-9x + 3y = -6$

20. $-x + 4y = 3$
$4x - 16y = -13$

For Exercises 21–32, solve the system by either the elimination method or the substitution method.

21. $2x - 3y = 0$
$-2x + 4y = -2$

22. $3x + 2y = 11$
$x - 2y = 1$

23. $-4x + y = 8$
$4x + 3y = 8$

24. $x + 2y = -9$
$x + y = -7$

25. $3x - y = -2$
$-2x + y = 6$

26. $4x + 2y = -14$
$4x - 2y = 6$

27. $3x - 2y = 3$
$4x - 3y = 3$

28. $2x - y = -10$
$3x + 4y = -4$

29. $3x - y = 5$
$6x - 2y = 10$

30. $x - 2y = -1$
$-4x + 8y = 3$

31. $4x + y = 7$
$5x - 2y = -14$

32. $x - 2y = -8$
$-2x + 3y = 16$

For Exercises 33–40, solve the word problem using systems of equations.

33. The rectangular side of a plastic tent used to protect plants from light frosts has a length that is 2 feet more than the width. If the perimeter is 20 feet, find the dimensions of the sides.

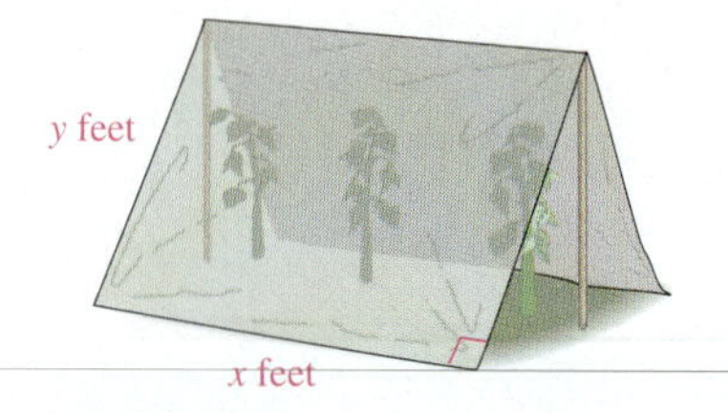

34. A rectangular appointment book is being designed. It must have a width that is 3 inches less than the length. If the perimeter is to be 26 inches, find the dimensions of the book.

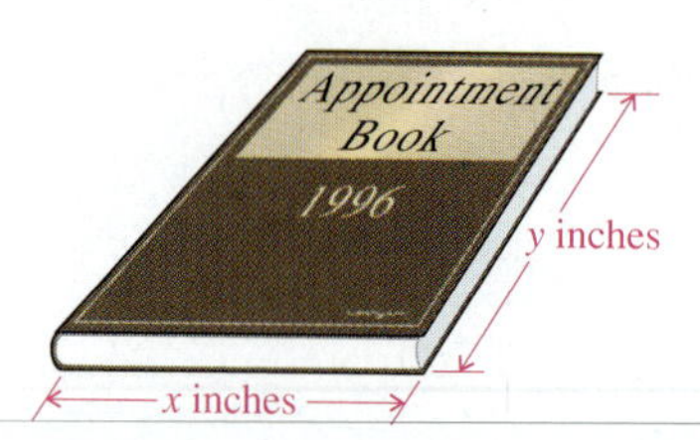

35. Two numbers, x and y, have a sum of 8. If one number is 10 more than the other number, find the two numbers.

36. Two numbers, x and y, have a sum of 4. If one number is 12 more than the other number, find the two numbers.

37. A soccer field has a perimeter of 400 yards. If the length is 40 yards longer than the width, find the dimensions of the field.

38. A rectangular playground is 20 meters longer than its width. If it is enclosed by 120 meters of fencing, what are the dimensions of the playground?

39. Lyn works a summer job mowing lawns x hours a week for \$10 per hour and trimming hedges y hours a week for \$20 per hour. One week she worked 35 hours and made \$450. How many hours that week did she mow lawns and how many hours did she trim hedges?

40. Alex, a tennis professional, wants to spend x hours a week teaching clinics and y hours a week giving private lessons, for a total of 40 hours. He makes \$100 per hour for the clinics and \$50 per hour for private lessons. If he wants to make \$3,500 a week, how many hours each week should he spend in clinics and how many hours in private lessons?

For Exercises 41 and 42, solve each system. Round off your answers to two decimal places.

41. $0.21x - 3.5y = 7$
$x + 0.21y = 0.56$

42. $0.34m - 3.92n = 5.44$
$n = 0.98m - 6.19$

SAY IT IN WORDS

43. Given the system to solve:

$$2x + 3y = -1$$
$$x + 2y = -1,$$

which method—elimination or substitution—do you prefer to use? Why?

44. Explain the substitution method.

REVIEW EXERCISES

The following exercises review parts of Section 2.6. Doing these exercises will help prepare you for the next section.

For Exercises 45–48, solve the inequality.

45. $3x < 12$

46. $4x \geq -8$

47. $-2x > 6$

48. $-3 \leq 2x + 1 \leq 1$

ENRICHMENT EXERCISES

Use the substitution method to solve the system.

1. $y = x + C_1$
   $x + y = C_2$

2. $y = mx$
   $x + y = 3$, where $m \neq -1$

3. $y = mx$
   $x + y = 3$, where $m = -1$

Answers to Enrichment Exercises begin on page A.1.

7.4 Graphing Linear Inequalities

OBJECTIVES

- *To graph linear inequalities*
- *To use linear inequalities in applications*

LEARNING ADVANTAGE

Mathematics at Work in Society: Life Scientist. *Life science is the study of living organisms and the relationship of plants and animals to their surroundings. Those who work in this area are called life scientists. They can be biologists, botanists, agronomists, zoologists, or ecologists.*

Much of their work involves measurement and estimation. To perform these tasks, they need mathematics.

So far in this chapter, we have studied linear equations such as $-x + 2y = 6$. In this section, we investigate **linear inequalities** such as

$$-x + 2y \geq 6$$

This is read "negative x plus $2y$ is greater than or equal to 6." The inequality means two things: either (a) $-x + 2y = 6$ or (b) $-x + 2y > 6$.

The **graph** of a linear inequality is the set of points (x, y) in the plane that satisfy the inequality.

Now, the graph of $-x + 2y \geq 6$ includes the graph of the *equation* $-x + 2y = 6$. Furthermore, the graph of this equation is a straight line. Using the x- and y-intercepts, we graph the line as shown in the figure below.

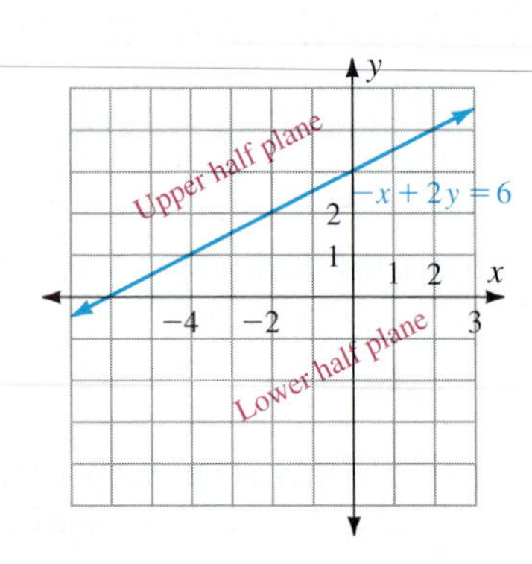

The line separates the plane into two regions—one above the line and the other below the line. These regions are called **half-planes.**

Exactly one of these half-planes belongs to the graph of the inequality. To determine the correct half-plane, we first solve $-x + 2y \geq 6$ for y.

$$2y \geq x + 6 \qquad \text{Add } x \text{ to both sides.}$$

$$y \geq \frac{1}{2}x + 3 \qquad \text{Divide by 2.}$$

Therefore, either $y = \frac{1}{2}x + 3$ or $y > \frac{1}{2}x + 3$. We already know that if the coordinates of a point (x, y) are related by the equation $y = \frac{1}{2}x + 3$, then the point lies on the line. Now, if $y > \frac{1}{2}x + 3$, then the y-coordinate is larger than $\frac{1}{2}x + 3$. This means that the point (x, y) lies in the half-plane above the line as shown in the figure below. For example, $(2, 4)$ lies on the line, but $(2, 5)$ and $(2, 6)$ lie above the line.

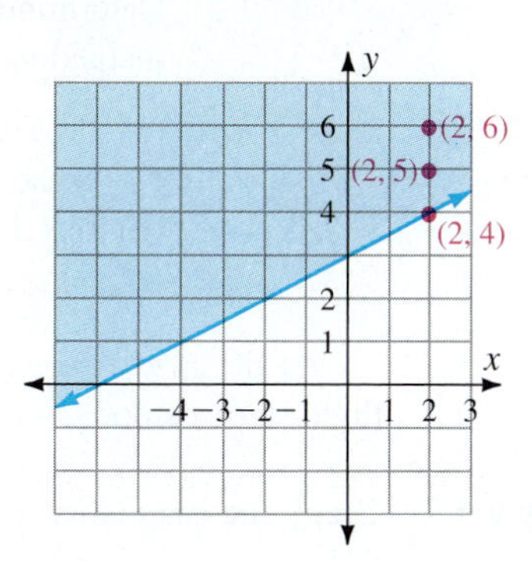

To indicate the graph of $-x + 2y \geq 6$, we shade the region above the line. Remember that the graph includes not only the upper half-plane but also the boundary line.

There is a faster way to determine which half-plane belongs to the graph of a linear inequality. Once the line has been drawn, select a convenient test point *not on the line,* and see whether or not the coordinates of the point satisfy the original inequality. For example, using $(0, 0)$ as our test point for the inequality $-x + 2y \geq 6$, we replace x by 0 and y by 0 in the original inequality.

$$-x + 2y \geq 6$$

$$-0 + 2(0) \geq 6$$

$$0 \geq 6 \qquad \text{False.}$$

Since 0 is not greater than or equal to 6, $(0, 0)$ is not a solution to the inequality $-x + 2y \geq 6$. Since $(0, 0)$ lies in the lower half-plane, no point from the lower half-plane satisfies the inequality. Our conclusion is that the graph includes the *upper* half-plane. This answer is the same as found before. (See the figure above.)

NOTE *It is* not *automatic that the inequality symbol* $\geq$ *means that the graph will include the upper half-plane. A test point must be used to determine the correct half-plane.*

STRATEGY

To Graph a Linear Inequality

Step 1 Replace the inequality symbol ($>$, $<$, $\geq$, or $\leq$) by an equal sign and draw the resulting straight line. If the inequality symbol is $\geq$ or $\leq$, use a solid line. If the symbol is $>$ or $<$, use dashes for the line. This line separates the plane into two regions called half-planes.

Step 2 Choose a convenient test point that is *not on the line.* Use (0, 0), when possible.

Step 3 Determine whether or not the test point is a solution of the inequality.

(a) If it is a solution, shade the half-plane in which the test point lies.

(b) If it is not a solution, shade the other half-plane.

In the next two examples, we illustrate graphing inequalities by using the three-step method.

Example 1 Graph the inequality $x + 3y \leq 9$.

Solution Step 1 Replace the "less than or equal to" sign by an equal sign to obtain the equation $x + 3y = 9$. We use the intercepts to graph the line. When $x = 0$, $y = 3$ and when $y = 0$, $x = 9$. We plot the two points (0, 3) and (9, 0) and draw the solid line as shown in Part (a) of the figure below.

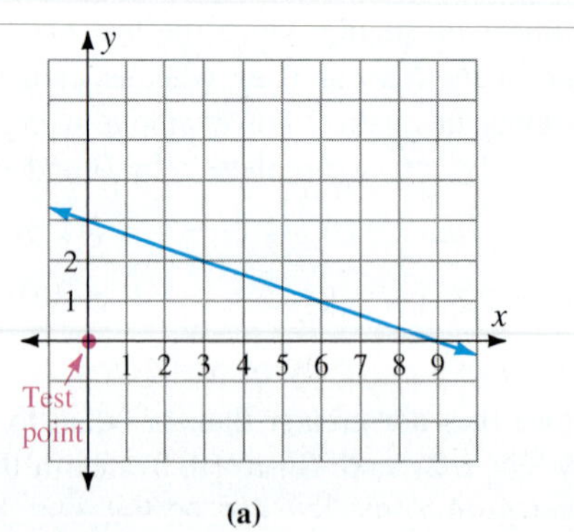

(a)

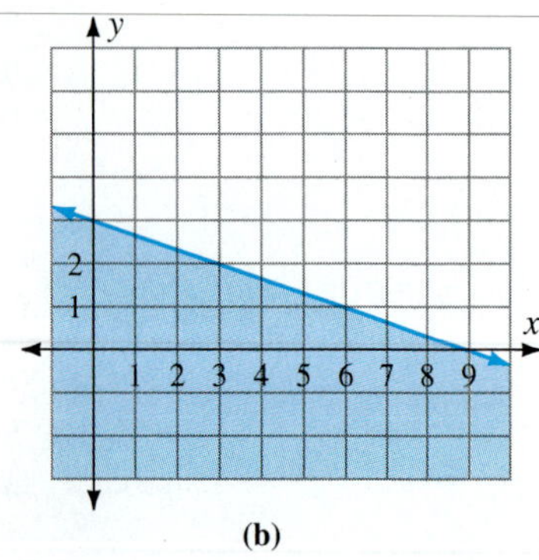

(b)

Step 2 We select (0, 0) for our test point, as it is not on the line.

Step 3 Replace x by 0 and y by 0 in the inequality.

$$x + 3y \leq 9$$
$$0 + 3(0) \overset{?}{\leq} 9$$

Since 0 is less than 9, the test point (0, 0) is a solution of the inequality. Since (0, 0) lies in the lower half-plane, we shade that part of the plane. The graph is shown in Part (b) of the figure on the preceding page.

Example 2 Graph the inequality $x > y$.

Solution First, replace the "greater than" sign by an equal sign to obtain the equation $x = y$. Letting $x = 0$, then $y = 0$, so the origin (0, 0) is on the line. Notice that we cannot let $y = 0$ to find a second point, since $y = 0$ means that $x = 0$. Therefore, use another value for x, say $x = 5$. So, (5, 5) is another point on the line. The line is drawn in Part (a) of the figure below using dashes. Next, choose a test point not on the line. In this case, (0, 0) is on the line, so we must choose another point. We use (1, 0). Replacing x by 1 and y by 0 in the original inequality,

$$x > y$$
$$1 > 0 \quad \text{This is a true statement.}$$

Therefore, (1, 0) is a solution and it lies below the line. Therefore, we shade the lower half-plane. The graph is shown in Part (b) of the figure below.

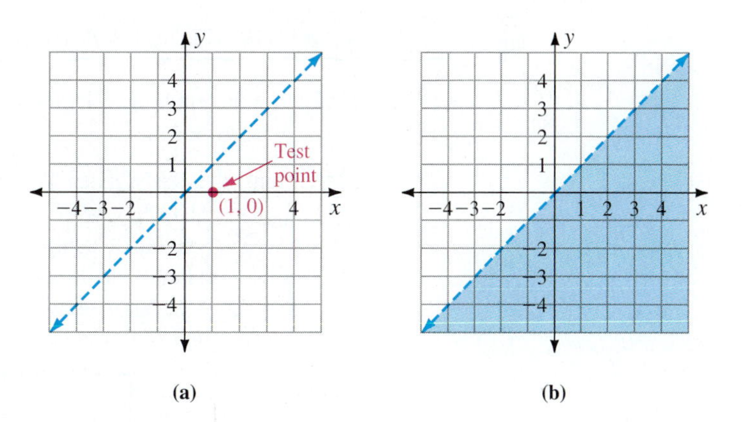

(a) (b)

Example 3 Graph the inequality.

(a) $x \leq 4$ **(b)** $y > -3$

Solution **(a)** The line $x = 4$ is a vertical line through (4, 0). The inequality describes the region to the left and on this line. See Figure (a).

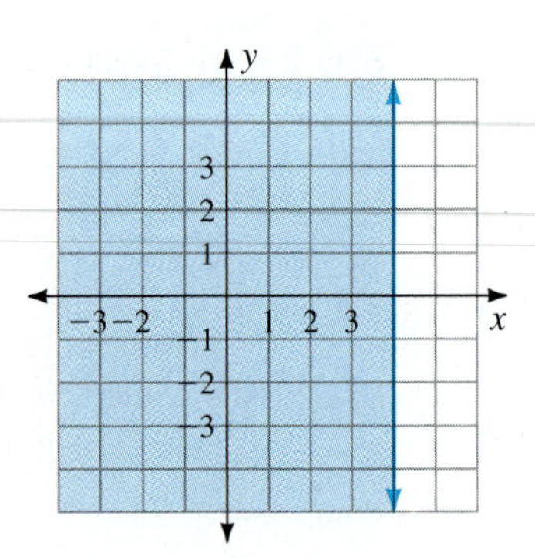

(b) The equation $y = -3$ is a horizontal line 3 units below the x-axis. Draw a dashed line. (Why?). Next, shade the region above this line. See Figure (b).

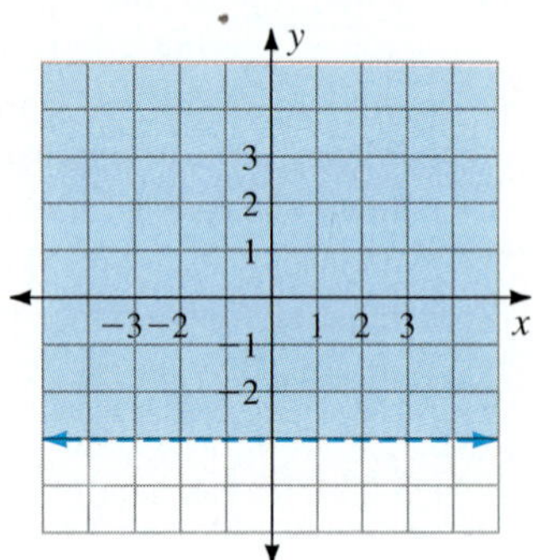

We finish this section with a graphing calculator example.

GRAPHING CALCULATOR CROSSOVER

EXAMPLE Use a graphing calculator to graph the inequality $y \leq 2x + 6$.

Solution Using the RANGE as $X_{min} = -10$, $X_{max} = 10$, $X_{scl} = 1$, $Y_{min} = -10$, $Y_{max} = 10$, $Y_{scl} = 1$, we GRAPH the equation $y = 2x + 6$. Next, using the shading instructions in your calculator manual, shade the area above $Y = -10$ and below $Y = 2X + 6$. Your viewing screen should look like this:

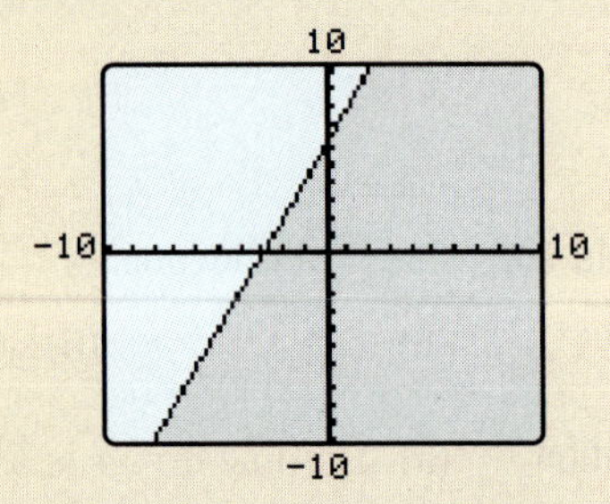

WARMING UP

Answer true or false.

1. The point (1, 2) satisfies the inequality

$$2x - 3y \geq 1$$

2. The point $(-1, 3)$ satisfies the inequality $x + y \leq 2$.

3. The graph of $x < -2$ is the region to the left of the vertical line $x = -2$.

4. The point $(1, -3)$ is on the graph of $3x + y < 0$.

5. To graph $x < y + 1$, use a dashed line.

6. The origin is on the graph of $2x < y + 1$.

EXERCISE SET 7.4

For Exercises 1–18, graph by using the three-step strategy for graphing linear inequalities.

1. $x + 2y \leq 8$

2. $2x + y \geq 4$

3. $x - 3y > 9$

4. $4x - y < 8$

5. $3x - y \leq -3$

6. $5x + 2y \geq -10$

7. $x < 2y$

8. $y > 3x$

9. $y \leq 0$

10. $x \leq 0$

11. $x + y < 0$

12. $x + 2y > 0$

13. $4x - 3y \geq 12$

14. $3x - 5y \leq 15$

15. $x \geq -2$

16. $x < 4$

17. $y \leq 3$

18. $y \geq 1$

For Exercises 19–26, write an inequality from the information, then graph it.

19. The sum of the x-coordinate and y-coordinate exceeds 2.

20. The difference of the x-coordinate and y-coordinate does not exceed 50.

21. The difference of twice the x value and three times the y value is no more than 18.

22. Five times the x value minus three times the y value is greater than -15.

23. The y-coordinate reduced by four times the x-coordinate is negative.

24. Ten times the x-coordinate minus 25 times the y-coordinate is positive.

25. Twice the x value is at most three times the y value.

26. Five times the y value is at least four times the x value.

For Exercises 27–30, use a graphing calculator to graph the inequality.

27. $y \leq x + 4$

28. $y \leq x$

29. $y \geq 3 - x$

30. $y \geq -x - 1$

SAY IT IN WORDS

31. Explain your method for graphing a linear inequality.

32. When do you use a solid line or dashed line when graphing a linear inequality?

REVIEW EXERCISES

The following exercises review parts of Section 7.1. Doing these exercises will help prepare you for the next section.

For Exercises 33–36, draw the graph of the system.

33. $y = x + 1$
$y = 2x$

34. $y = x$
$y = -x + 4$

35. $x + 2y = 6$
$-x + y = 0$

36. $x - 2y = 2$
$x - y = -1$

ENRICHMENT EXERCISES

1. In a coordinate plane, find the intersection of the graphs of $x \geq 0$ and $y \leq 3$.

2. In a coordinate plane, find the intersection of the graphs of $x \leq 5$ and $y \geq 2$.

3. Graph the inequalities $0 \leq x \leq 1$ and $0 \leq y \leq 4$.

Answers to Enrichment Exercises begin on page A.1.

7.5 Systems of Linear Inequalities

OBJECTIVE

▶ *To solve systems of linear inequalities by the graphing method*

LEARNING ADVANTAGE

Mathematics at Work in Society: Transportation Careers. *The job of a traffic control engineer involves monitoring traffic conditions; determining speed limits, the placement of stop signs and traffic signals; as well as timing traffic signals to permit the best flow of cars and trucks. Mathematics is used to study how traffic signals should be timed with respect to each other to permit a car traveling at a constant speed to pass through each (green) signal at the same time of the cycle. Setting the signals like this permits a better flow of traffic and it encourages drivers to obey the posted speed limit.*

A solution to a system of equations is a point that satisfies both equations. Similarly, a solution to a system of *inequalities* is a point that is a solution to both inequalities.

Example 1 Determine if the point is a solution to the system.

(a) $\begin{aligned} x + y &\geq 0 \\ x + 2y &\geq 4 \end{aligned}$, (2, 1) **(b)** $\begin{aligned} x + y &< 6 \\ 2x - 3y &> 12 \end{aligned}$, (0, 0)

Solution **(a)** The point (2, 1) is a solution to the system if it satisfies both inequalities. We let $x = 2$ and $y = 1$ in each inequality.

$$\begin{aligned} x + y &\geq 0 \\ 2 + 1 &\overset{?}{\geq} 0 \\ 3 &\geq 0 \quad \text{true} \end{aligned} \qquad \begin{aligned} x + 2y &\geq 4 \\ 2 + 2(1) &\overset{?}{\geq} 4 \\ 4 &\geq 4 \quad \text{true} \end{aligned}$$

Since both inequalities are satisfied, (2, 1) is a solution to the system.

(b) To see if (0, 0) is a solution to the system, replace x and y by 0 in each inequality.

$$\begin{aligned} x + y &< 6 \\ 0 + 0 &\overset{?}{<} 6 \\ 0 &< 6 \quad \text{true} \end{aligned} \qquad \begin{aligned} 2x - 3y &> 12 \\ 2(0) - 3(0) &\overset{?}{>} 12 \\ 0 &> 12 \quad \text{false} \end{aligned}$$

Since only one inequality is satisfied, (0, 0) is not a solution to the system.

Example 2 Graph the solution set of the system

$$\begin{aligned} x + 2y &\geq 4 \\ -x + y &\leq 1 \end{aligned}$$

Solution We graph both inequalities in the same coordinate plane as shown in the figure below. The solution set is the region common to both half-planes. Notice that this region includes parts of the two boundary lines.

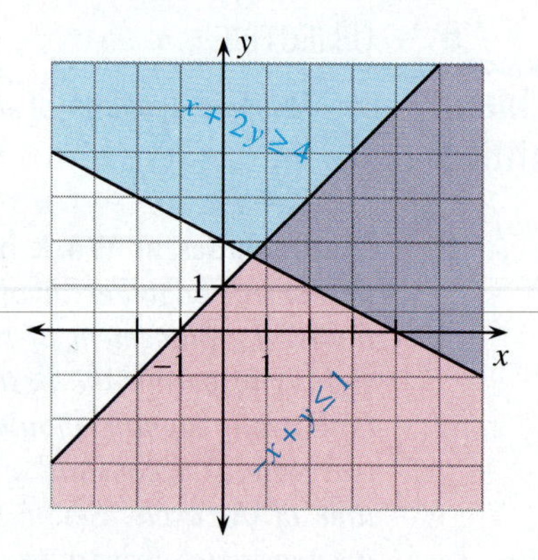

Example 3 Graph the solution set of the system

$$\begin{aligned} -x + y &< 0 \\ 2x + y &> 4 \end{aligned}$$

Solution Graphing each inequality in the same coordinate system, the region common to both half-planes is the solution of the system as shown in the figure that follows. Notice that the boundary lines are dashed since they are not part of the solution set.

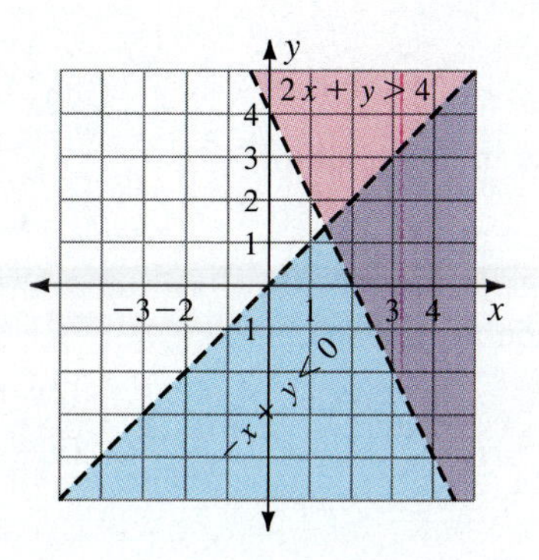

Example 4 Graph the solution set of

$$x \leq 4$$
$$y \geq -3$$

Solution First, graph the vertical line $x = 4$ and the horizontal line $y = -3$. The points that satisfy both inequalities are in the region to the left of the line $x = 4$ and above the line $y = -3$. Notice also that the boundary points are in the graph.

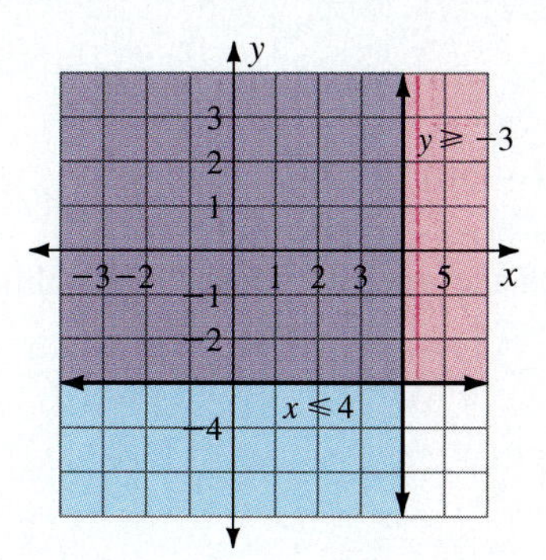

WARMING UP

Answer true or false.

1. The graph of a system of inequalities is the set of all points in the plane that are solutions to each inequality of the system.

2. The point (1, 2) is a solution to the system

$$x + y \leq 3$$
$$x + 2y \geq 1$$

3. The point $(-1, 1)$ is a solution to the system

$$\begin{aligned} 2x + 3y &\geq 0 \\ x + y &\geq -1 \end{aligned}$$

4. The origin is in the graph of the system

$$\begin{aligned} x + y &> 0 \\ 2x - y &< 1 \end{aligned}$$

5. The point $(-2, -3)$ is in the graph of

$$\begin{aligned} x &> -2 \\ y &< -3 \end{aligned}$$

6. The origin is a solution to the system

$$\begin{aligned} 2x + y &\leq 1 \\ 3x - 2y &\geq -1 \end{aligned}$$

EXERCISE SET 7.5

For Exercises 1–8, determine if the point is a solution to the system. Do not graph.

1. $\begin{aligned} x + y &\geq 1 \\ 2x + 3y &\leq 6 \end{aligned}$, $(1, 1)$

2. $\begin{aligned} 3x + y &\leq 9 \\ x + 2y &\geq 3 \end{aligned}$, $(2, 2)$

3. $\begin{aligned} 2x + y &< 10 \\ x + y &> 5 \end{aligned}$, $(5, 1)$

4. $\begin{aligned} 5x + 3y &> 2 \\ 4x + y &< 6 \end{aligned}$, $(0, 1)$

5. $\begin{aligned} x &\geq 3 \\ y &\leq 2 \end{aligned}$, $(4, -1)$

6. $\begin{aligned} x &< 5 \\ y &> -2 \end{aligned}$, $(3, -2)$

7. $\begin{aligned} x - 2y &\geq -1 \\ 3x + 5y &\leq 4 \end{aligned}$, $(-1, 0)$

8. $\begin{aligned} 4x - 3y &\leq 5 \\ -x + y &\geq -1 \end{aligned}$, $(0, 0)$

For Exercises 9–24, graph the system of linear inequalities.

9. $\begin{aligned} x + y &\leq 3 \\ -x + y &\geq 1 \end{aligned}$

10. $\begin{aligned} 2x + y &\leq 4 \\ 3x - y &\leq 1 \end{aligned}$

11. $\begin{aligned} -2x + 3y &< 6 \\ x + 2y &> 4 \end{aligned}$

12. $\begin{aligned} x + 3y &< 3 \\ -x + y &< 5 \end{aligned}$

13. $y \leq 4$
$x + y \geq 0$

14. $-x + y \leq 1$
$y \geq -2$

15. $2x + y < 3$
$y > -3$

16. $y < 2$
$x + 2y < 4$

17. $x > -2$
$2x - y < -1$

18. $x < 3$
$-x + 2y > 3$

19. $3x + y \leq -5$
$x \geq -3$

20. $x - 3y \geq -5$
$x \leq 4$

21. $x \geq -3$
$y \leq 2$

22. $x \leq 4$
$y \leq 3$

23. $x \leq 0$
$y \geq 0$

24. $x > -2$
$y > -2$

SAY IT IN WORDS

25. Explain your method for graphing a system of inequalities.

26. Give an example of a system of inequalities that has an empty solution set.

REVIEW EXERCISES

The following exercises review parts of Section 4.1. Doing these exercises will help prepare you for the next section.

For Exercises 27–34, simplify.

27. 2^3 **28.** 4^2 **29.** 12^1 **30.** 5^0

31. $x^3 \cdot x^4$ **32.** $(y^2)^3$ **33.** $(3a)^3$ **34.** $(-z)^5$

ENRICHMENT EXERCISES

Graph the following systems of inequalities.

1. $-2 \leq x \leq 2$
$-2 \leq y \leq 2$

2. $x + y \leq 4$
$x \geq 0$
$y \geq 0$

3. $2x + 3y \leq 6$
$x \geq 0$
$y \geq 0$

Answers to Enrichment Exercises begin on page A.1.

CHAPTER 7

Summary and review

Examples

Systems of linear equations (7.1)

An **ordered pair** (x, y) is a solution of a system of linear equations, if it is a solution of *all* equations in the system.

Since the graph of a linear equation is a straight line, we can graph the two equations of a system in the same coordinate plane. The **nature of the solution set** can be determined by these two lines.

1. If the two lines intersect in a point, the solution set of the system contains that one ordered pair.
2. If the two lines are parallel, the solution set is empty and the system is inconsistent.
3. If the two lines are coincident (the same), there are infinitely many solutions and the system is dependent.

There are **three methods** available to solve a system of linear equations: the graphing method, the elimination method, and the substitution method.

The graphing method (7.1)

For this method, graph the two equations and estimate the coordinates of the point of intersection. Be sure to use graph paper.

The system

$$x + 3y = 9$$
$$2x - 3y = 0$$

is solved by graphing.

The solution is $x = 3$ and $y = 2$.

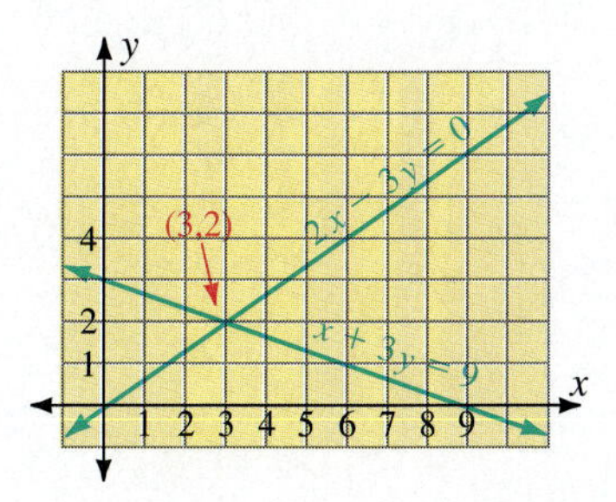

The elimination method (7.2)

Solve $2x - 3y = -4$
$3x - y = 1$

by the elimination method.

We choose the variable y to be eliminated.

Step 1 If necessary, transform each equation to the standard form $Ax + By = C$. Also, clear fractions by multiplying by the least common denominator.

Step 2 Decide which variable is easier to eliminate.

Step 3 If necessary, multiply one or both equations by appropriate numbers so that the coefficients of the variable to be eliminated are opposites.

Examples

$2x - 3y = -4$

$3x - \quad y = 1$

——leave as is——→

——multiply by -3——→

$$\begin{aligned} 2x - 3y &= -4 \\ -9x + 3y &= -3 \\ \hline -7x \quad &= -7 \\ x \quad &= 1 \end{aligned}$$

Replacing x by 1 in either of the original two equations, we have $y = 2$. The solution is (1, 2).

Step 4 Add the two equations to obtain an equation with only one variable appearing.

Step 5 Solve this equation for the variable.

Step 6 Take the answer from Step 4 and use either equation from the original system to find the value of the other variable.

Step 7 Check your solution in the original system.

The substitution method (7.3)

Solve $x - 2y = -1$

$y = 3 - 2x$

by the substitution method.

Replace y by $3 - 2x$ in the first equation and solve for x.

$$\begin{aligned} x - 2(3 - 2x) &= -1 \\ x - 6 + 4x &= -1 \\ 5x &= 5 \\ x &= 1 \end{aligned}$$

Next, replace x by 1 in the second equation of the system

$$\begin{aligned} y &= 3 - 2x \\ &= 3 - 2(1) \\ &= 1 \end{aligned}$$

The solution is (1, 1).

Step 1 Solve one of the equations for one of the variables, if neither equation is already solved for one of its variables.

Step 2 Substitute for that variable in the other equation. We now have one equation in one unknown.

Step 3 Solve the resulting equation.

Step 4 Find the value of the other variable by substituting the number from Step 3 into the equation of Step 1.

Step 5 Check your answer.

Graphing linear inequalities (7.4)

To graph $2x - 3y < 6$, first draw the line $2x - 3y = 6$ using dashes. Since (0, 0) is a solution of the inequality, we shade the half-plane as shown in the figure on the next page.

To graph a linear inequality:

1. Replace the inequality symbol by an equal sign and draw the resulting straight line. If the inequality is either $\leq$ or $\geq$, use a solid line; otherwise, use dashes for the line. This line divides the plane into half-planes.
2. Choose a convenient test point that is not on the line. Use (0, 0), when possible.
3. Determine whether or not the test point is a solution of the inequality. If it is a solution, shade the half-plane in which the test point lies. If it is not a solution, shade the other half-plane.

Examples

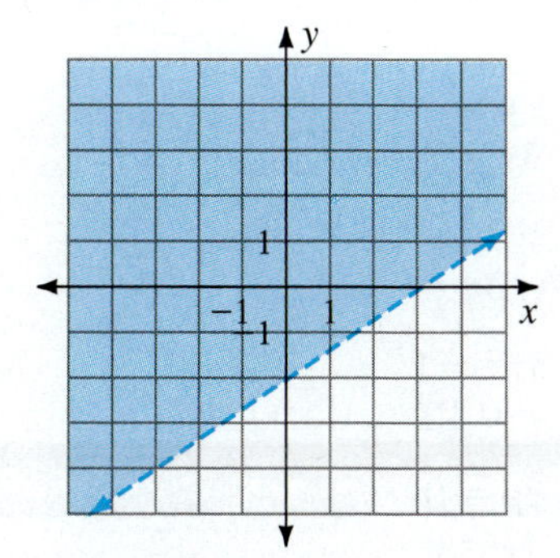

Systems of linear inequalities (7.5)

To **graph the solution set** of a system of linear inequalities, graph each inequality of the system in the same coordinate plane and shade the intersection.

Graph of the solution set of the system

$$x + 2y < 4$$
$$-x + 5y < 5$$

We graph each inequality using dashed lines, then shade the intersection as shown in the figure.

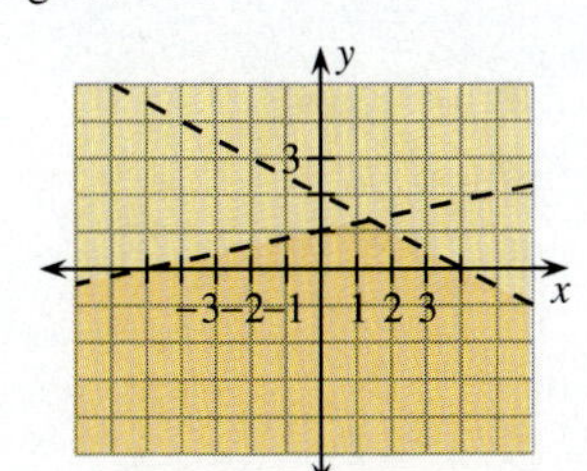

CHAPTER 7 REVIEW EXERCISE SET

Section 7.1

1. Determine if $(1, -2)$ is a solution of the system of linear equations

$$3x + y = 1$$
$$-2x + 3y = -8$$

2. Determine if $(-3, -4)$ is a solution of the system of linear equations

$$x - 2y = 5$$
$$3x + y = -12$$

For Exercises 3–6, use the graphing method to find the solution to the system of linear equations.

3. $x - y = -2$
$2x + y = 8$

4. $4y = x + 2$
$x - 4y = 4$

5. $-x + 2y = 3$
$2x - 4y = -6$

6. $2x + y = 4$
$-x + y = 1$

Sections 7.2 and 7.3

For Exercises 7–18, use the elimination method or substitution method to solve the system of linear equations.

7. $3x + y = 3$
$x - y = 17$

8. $3x - 2y = 5$
$4x + 2y = 2$

9. $4x + y = -4$
$2x + 3y = 8$

10. $2x + 3y = 4$
$3x + 4y = 5$

11. $x + 2y = 2$
$2x + 4y = 3$

12. $-6x + 9y = 18$
$2x - 3y = -6$

13. $-3x + 4y = -6$
$-2x + 3y = -4$

14. $2x + y = -5$
$3x + 4y = 0$

15. $x - y = -1$
$2x + y = 10$

16. $-x + 2y = 2$
$2x + 3y = -4$

17. $5x + 3y = -3$
$2x - y = 1$

18. $x + 2y = 0$
$3x - 4y = 0$

19. The rectangular top of a cold frame (see the figure) has a length y that is 4 feet longer than the width x. If the perimeter is 32 feet, find the dimensions of the top.

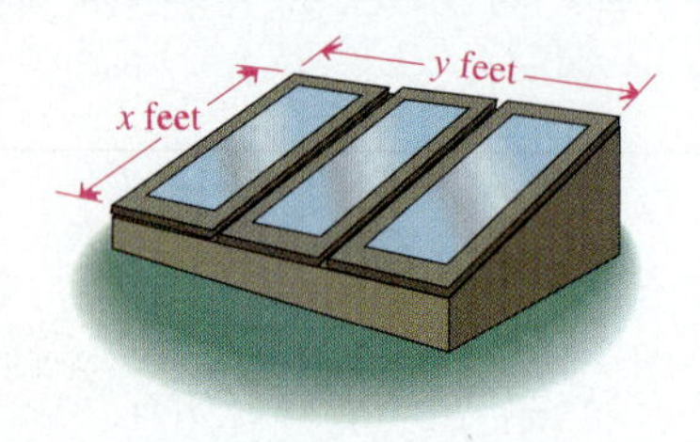

Section 7.4

For Exercises 20–23, graph the inequality.

20. $x + y \leq 6$

21. $2x - 4y > 8$

22. $x \geq 0$

23. $x - 2y < 4$

Section 7.5

For Exercises 24–27, graph the solution set of the system.

24. $x + y \leq 4$
$x - 2y \leq -2$

25. $x \geq -2$
$y \leq 5$

26. $x + 2y > 6$
$-x + y < 3$

27. $x + y > -3$
$x < 2$

CHAPTER 7 TEST

1. Determine if (6, 1) is a solution of the system of linear equations

$$3x - 5y = 13$$
$$x + 3y = 9$$

For Problems 2 and 3, use the graphing method to solve the system of linear equations. If the system is inconsistent, write ''no solution.'' If the system is dependent, write ''infinitely many solutions.''

2. $x + y = -4$
$3x + 3y = 6$

3. $y = 2x + 1$
$-x + y = 3$

For Problems 4–10, use the elimination or substitution method to solve the system of linear equations.

4. $x + 5y = 12$
$2x - 5y = 9$

5. $8x - 2y = 12$
$-4x + y = -6$

6. $2x + 3y = 4$
$3x + 4y = 5$

7. $x + y = -4$
$2x + y = -3$

8. $3x + 2y = 22$
$x - 3y = 0$

9. $2x + y = 5$
$x - 4 = 0$

10. $3x + 2y = 2$
$4x + y = 1$

11. The width of a rectangular piece of cardboard is 6 centimeters less than twice the length. If the perimeter is 18 centimeters, find the length and width of the cardboard.

For Problems 12–14, graph the inequality.

12. $2x + y < 4$

13. $x \leq 2$

14. $3x + 2y \geq -6$

For Problems 15 and 16, graph the solution set of the system.

15. $3x - 2y \leq 6$
$2x + y \leq 4$

16. $x \geq -2$
$x + y \leq 4$

CHAPTER 8

Roots and Radicals

CONNECTIONS

The period T of a pendulum is related to the length ℓ by the formula

$$T = 2\pi\sqrt{\frac{\ell}{32}}$$

Find the period of a pendulum 8 feet in length. See Example 9 in Section 8.1.

Overview

So far, we have raised numbers to integer powers and studied properties such as $a^m a^n = a^{m+n}$, $(a^m)^n = a^{m \cdot n}$, and $a^{-n} = \frac{1}{a^n}$. In this chapter, we define numbers raised to rational powers. We shall connect this concept with taking the nth root of a number. An expression involving taking a root is called a radical. We will then learn the algebra of radical expressions and solve radical equations. The final topic is the set of complex numbers that will be used to solve the general quadratic equation in the next chapter.

8.1 Roots

OBJECTIVES

- ▶ *To find the square roots of a number*
- ▶ *To find the nth roots of a number*
- ▶ *To use the notation for nth roots*

Given a number such as 7, we know how to find its square, 7^2,

$$7^2 = 7 \cdot 7 = 49$$

We say that the square of 7 is 49. We can also say that a *square root* of 49 is 7, since squaring 7 yields 49. Notice also that $(-7)^2 = (-7)(-7) = 49$. Therefore, -7 is also a square root of 49. The square roots of 49 are 7 and -7. This leads us to the following definition.

DEFINITION

A number b is a **square root** of a if $b^2 = a$.

Any positive number has two square roots, one positive and the other negative.

Example 1 Find the two square roots of each number.

(a) 25 **(b)** 100 **(c)** $\frac{4}{9}$ **(d)** 0.16

Solution **(a)** Since $5^2 = 5 \cdot 5 = 25$ and $(-5)^2 = (-5)(-5) = 25$, the square roots of 25 are 5 and -5.

(b) Since $10^2 = 100$ and $(-10)^2 = 100$, the square roots of 100 are 10 and -10.

(c) Since $\left(\frac{2}{3}\right)^2 = \frac{2}{3} \cdot \frac{2}{3} = \frac{4}{9}$ and $\left(-\frac{2}{3}\right)^2 = \left(-\frac{2}{3}\right)\left(-\frac{2}{3}\right) = \frac{4}{9}$, the two square roots of $\frac{4}{9}$ are $\frac{2}{3}$ and $-\frac{2}{3}$.

(d) Since $(0.4)(0.4) = 0.16$ and $(-0.4)(-0.4) = 0.16$, the two square roots of 0.16 are 0.4 and -0.4.

The *positive square root* of a number is symbolized by using the symbol $\sqrt{}$, which is called a **radical sign.** For example,

$$\sqrt{49} = 7, \qquad \sqrt{36} = 6, \qquad \sqrt{121} = 11$$

To express the *negative square root,* we use $-\sqrt{}$. For example, $-\sqrt{49} = -7$.

The expression within the radical sign is called the **radicand.** For example, the radicand is 4 in the expression $\sqrt{4}$. A **radical** is the total expression, made up of the radical sign together with the radicand. Using this symbolism, we have the following statements.

1. If a is a positive real number,

$\sqrt{a}$ is the positive square root of a, and

$$\sqrt{a}\sqrt{a} = a$$

$-\sqrt{a}$ is the negative square root of a, and

$$(-\sqrt{a})(-\sqrt{a}) = a$$

Furthermore,

$$\sqrt{a^2} = a, \text{ for } a \geq 0$$

2. If a is zero, $\sqrt{0} = 0$.

Consider trying to find a square root of a negative number. For example, does -4 have square roots? If -4 had a square root, say b, then $b^2 = -4$. However, recall that the square of any real number is nonnegative. That is, b^2 is nonnegative and, therefore, could not be equal to -4. We conclude that

A negative number has no real square roots.

In particular, we have

If $a > 0$, then $\sqrt{-a}$ is not a real number.

Example 2 Evaluate each of the following.

(a) $\sqrt{64}$ (b) $-\sqrt{\frac{25}{4}}$ (c) $\sqrt{(0.79)^2}$

(d) $\sqrt{(-1)^2}$ (e) $\sqrt{6}\sqrt{6}$ (f) $\sqrt{-25}$

Solution (a) $\sqrt{64}$ means the positive square root of 64. Therefore, $\sqrt{64} = 8$.

(b) $-\sqrt{\frac{25}{4}} = -\frac{5}{2}$, since $\left(-\frac{5}{2}\right)^2 = \frac{25}{4}$.

(c) $\sqrt{(0.79)^2} = 0.79$ $\sqrt{a^2} = a$ for $a \geq 0$.

(d) $\sqrt{(-1)^2} = \sqrt{1} = 1$

(e) $\sqrt{6}\sqrt{6} = 6$ $\sqrt{a}\sqrt{a} = a$.

(f) Since -25 is negative, it has no real square roots. Therefore, $\sqrt{-25}$ is not a real number.

The numbers 144 and $\frac{4}{9}$ have rational square roots, since the square roots of 144 are 12 and -12 and the square roots of $\frac{4}{9}$ are $\frac{2}{3}$ and $-\frac{2}{3}$. The numbers 144 and $\frac{4}{9}$ are called *perfect squares.* In general, a number with rational square roots is called a **perfect square.** Here is a list of positive integers that are perfect squares.

$$4, 9, 16, 25, 36, 49, 64, 81, 100, 121, 144, 169, \ldots$$

If a positive number, a, is not a perfect square then its square roots, $\sqrt{a}$ and $-\sqrt{a}$, are irrational numbers. Recall from Chapter 1 that the real numbers are split into two groups—rational and irrational numbers. A rational number can be expressed as a fraction in which the numerator and denominator are integers, whereas an irrational number cannot be expressed as a quotient of integers. Furthermore, the irrational numbers have decimal expansions that have neither an ending nor a repeating block of digits.

Higher roots such as cube roots, fourth roots, and so on, are defined in a manner similar to square roots.

DEFINITION

Let n be a positive integer. A number b is an ***n*th root** of a if $b^n = a$.

For example, 2 is a cube root of 8, since $2^3 = 8$. We use the notation $\sqrt[n]{}$ to represent nth roots. For example, 2 is the cube root of 8, which is written $2 = \sqrt[3]{8}$.

The positive *square root* of a, $\sqrt{a}$, has the property that $(\sqrt{a})^2 = a$.

The *cube root* of a, $\sqrt[3]{a}$, has the property that $(\sqrt[3]{a})^3 = a$.

Similarly, the *fourth root* of a, $\sqrt[4]{a}$, has the property that $(\sqrt[4]{a})^4 = a$.

In general,

$$(\sqrt[n]{a})^n = a$$

Notice that we can take the cube root of a negative number. For example, $\sqrt[3]{-8} = -2$. This is true for any *odd* integer n. However, the even roots, square, fourth, sixth, . . . , are not defined for negative numbers. Even roots of negative numbers are not real numbers.

The following are examples of common roots.

Square Roots		***Cube Roots***		***Fourth Roots***	
$\sqrt{0} = 0$	$\sqrt{1} = 1$	$\sqrt[3]{0} = 0$	$\sqrt[3]{1} = 1$	$\sqrt[4]{0} = 0$	$\sqrt[4]{1} = 1$
$\sqrt{4} = 2$	$\sqrt{9} = 3$	$\sqrt[3]{-1} = -1$	$\sqrt[3]{8} = 2$	$\sqrt[4]{16} = 2$	$\sqrt[4]{81} = 3$
$\sqrt{16} = 4$	$\sqrt{25} = 5$	$\sqrt[3]{-8} = -2$	$\sqrt[3]{27} = 3$	$\sqrt[4]{256} = 4$	

In the symbol for the nth root, $\sqrt[n]{}$, n is called the **index.** Note that when $n = 2$, the index is usually not written and we write $\sqrt{}$ for $\sqrt[2]{}$.

How many roots does a number have? It depends on the index as well as the number itself.

The Number of Real nth Roots of a Real Number a		
	a is Positive	*a is Negative*
n even	Two real nth roots -3 and 3 are both square roots of 9.	No real nth roots -9 has no real square roots.
n odd	One real nth root 2 is the only cube root of 8.	One real nth root -2 is the only cube root of -8.

For example, 11 has two square roots, two fourth roots, two sixth roots, and so on. One of the roots is positive and the other is negative. In fact, the two roots are opposites of each other. The positive root is called the **principal *n*th root.** The number -5 has one cube root, one fifth root, one seventh root, and so on. The number -5 has no square root, no fourth root, no sixth root, and so on.

Example 3 **(a)** $\sqrt[3]{8} = 2$, because $2^3 = 8$.

(b) $\sqrt[3]{-8} = -2$, because $(-2)^3 = -8$.

(c) $\sqrt{-16}$ is not a real number, since there is no real number whose square is -16.

(d) $-\sqrt{16} = -4$, since $-\sqrt{16}$ is the negative square root of 16.

(e) $\sqrt[4]{-45}$ is not a real number, since there is no real number we can raise to the fourth power and obtain -45.

(f) $\sqrt[5]{-32} = -2$, because $(-2)^5 = -32$. ◀

Since we cannot take an even root of a negative number, we will assume that all variables appearing under a radical sign represent nonnegative numbers.

Example 4 Find $\sqrt{81x^2}$.

Solution We want to find a quantity whose square is $81x^2$. Since $9^2 = 81$ and the square of x is x^2, then the square of $9x$ is the desired quantity. That is,

$$\sqrt{81x^2} = 9x$$

since $(9x)^2 = 81x^2$. ◀

Example 5 Find $\sqrt{25x^4y^2}$.

Solution We want a quantity whose square is $25x^4y^2$. That quantity is $5x^2y$. That is,

$$\sqrt{25x^4y^2} = 5x^2y$$

since $(5x^2y)^2 = 25x^4y^2$. ◀

Example 6 Find $\sqrt[3]{-27a^6b^{15}}$.

Solution We want an expression we cube to obtain $-27a^6b^{15}$. That expression is $-3a^2b^5$. That is,

$$\sqrt[3]{-27a^6b^{15}} = -3a^2b^5$$

since $(-3a^2b^5)^3 = -27a^6b^{15}$. ◀

Example 7 Find $\sqrt[4]{16s^4t^8r^{16}}$.

Solution We want an expression we raise to the fourth power to obtain $16s^4t^8r^{16}$. That expression is $2st^2r^4$. That is,

$$\sqrt[4]{16s^4t^8r^{16}} = 2st^2r^4$$

since $(2st^2r^4)^4 = 16s^4t^8r^{16}$. ◀

In Chapter 5, we introduced the Pythagorean Theorem. We restate it here for convenience.

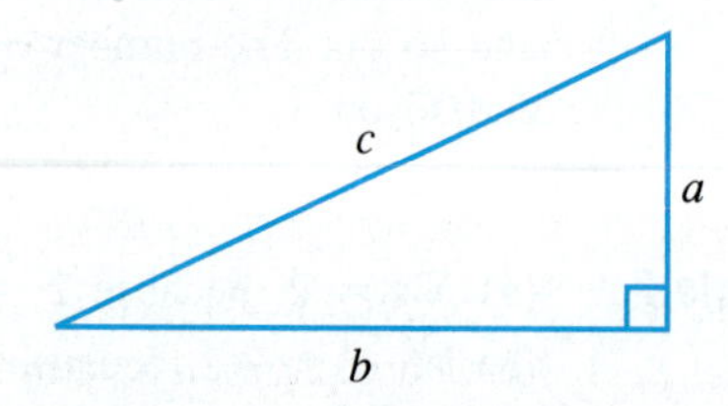

Consider a right triangle with lengths a, b, and c, where a and b are the length of the two legs and c is the length of the hypotenuse. Then

$$c^2 = a^2 + b^2$$

Example 8 For each right triangle, find the missing length.

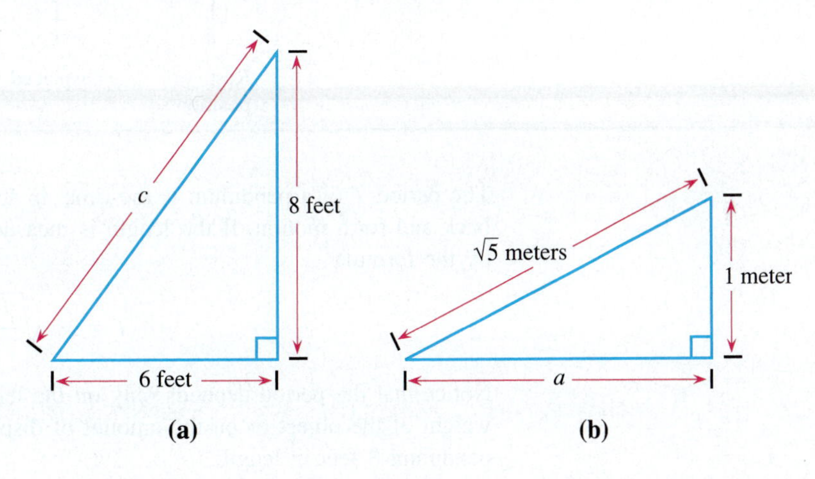

Solution **(a)** In this triangle, we are given lengths $a = 6$ feet and $b = 8$ feet. From this information, we find length c by using the Pythagorean Theorem.

$$\begin{aligned} c^2 &= a^2 + b^2 \\ &= 6^2 + 8^2 \\ &= 36 + 64 \\ &= 100 \end{aligned}$$

Since c is the length of a side of a triangle, c is positive. Therefore, c is the positive square root of 100; that is, $c = \sqrt{100}$ or 10 feet.

(b) For this triangle, we are given lengths $b = 1$ meter and $c = \sqrt{5}$ meters. To find length a, use the Pythagorean Theorem.

$$\begin{aligned} c^2 &= a^2 + b^2 \\ (\sqrt{5})^2 &= a^2 + 1^2 \\ 5 &= a^2 + 1 \\ 5 - 1 &= a^2 \\ 4 &= a^2 \\ 2 &= a \qquad a \text{ is the positive square root of 4.} \end{aligned}$$

Therefore, a is 2 meters.

Example 9 A pendulum of length ℓ is shown in the rest position. (See the figure on the following page). The pendulum is displaced, then released.

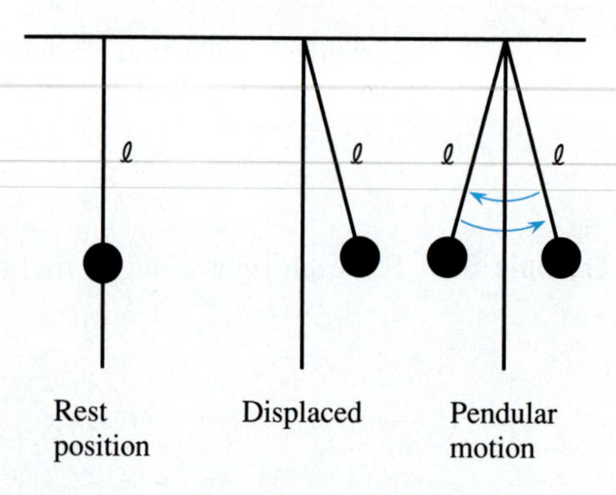

The period T of a pendulum is the time, in seconds, required for one complete back and forth motion. If the length is measured in feet, the period T is given by the formula

$$T = 2\pi\sqrt{\frac{\ell}{32}}$$

Notice that the period depends only on the length ℓ and does not depend on the weight of the object or on the amount of displacement. Find the period of a pendulum 8 feet in length.

Solution Replace ℓ by 8 and simplify.

$$T = 2\pi\sqrt{\frac{\ell}{32}}$$

$$T = 2\pi\sqrt{\frac{8}{32}}$$

$$= 2\pi\sqrt{\frac{1}{4}}$$

$$= 2\pi\left(\frac{1}{2}\right)$$

$$= \pi \text{ seconds}$$

Since π is approximately 3.14, the period is approximately 3.14 seconds. ◀

WARMING UP

Answer true or false.

1. The two square roots of 4 are 2 and -2.

2. The number 0 has only one square root.

3. $\sqrt{-9} = -3$

4. $\sqrt[3]{8} = 2$

5. $\sqrt[3]{-27} = -3$

6. $\sqrt{4+9} = \sqrt{4} + \sqrt{9}$

7. 81 is a perfect square

8. If x is a positive number, the negative square root of $9x^2$ is $-3x$.

EXERCISE SET 8.1

For Exercises 1–8, find the two square roots of each number.

1. 49 **2.** 64 **3.** 121 **4.** 81

5. $\frac{25}{9}$ **6.** $\frac{81}{16}$ **7.** 0.81 **8.** 0.25

For Exercises 9–18, evaluate each expression.

9. $\sqrt{49}$ **10.** $\sqrt{16}$ **11.** $-\sqrt{\frac{4}{25}}$ **12.** $-\sqrt{\frac{100}{81}}$

13. $\sqrt{(0.18)^2}$ **14.** $\sqrt{\left(\frac{2}{3}\right)^2}$ **15.** $\sqrt{(-3)^2}$ **16.** $\sqrt{(-5)^2}$

17. $\sqrt{-25}$ **18.** $\sqrt{-4}$

For Exercises 19–36, find the value of each expression, if it exists.

19. $\sqrt[3]{-8}$ **20.** $\sqrt{121}$ **21.** $\sqrt[4]{16}$ **22.** $\sqrt[3]{64}$

23. $\sqrt[7]{-1}$ **24.** $\sqrt[5]{1}$ **25.** $\sqrt{-16}$ **26.** $\sqrt[4]{-81}$

27. $\sqrt[7]{128}$ **28.** $\sqrt[3]{216}$ **29.** $-\sqrt[3]{8}$ **30.** $-\sqrt[4]{81}$

31. $\sqrt[3]{-27}$ **32.** $\sqrt[3]{-1}$ **33.** $\sqrt[6]{-1}$ **34.** $\sqrt{-81}$

35. $-\sqrt[4]{16}$ **36.** $-\sqrt[3]{-8}$

For Exercises 37–48, find the root. Assume that all variables are positive.

37. $\sqrt{16s^4t^2}$ **38.** $\sqrt{81w^6u^4}$ **39.** $\sqrt[3]{27x^6y^9}$

40. $\sqrt[3]{a^3c^{12}}$ **41.** $\sqrt[4]{16a^8b^4}$ **42.** $\sqrt[4]{81p^{16}q^{20}}$

43. $\sqrt[3]{-8a^9c^3z^{15}}$ **44.** $\sqrt[3]{-r^6s^3t^{12}}$ **45.** $\sqrt[3]{-r^6t^{15}}$

46. $\sqrt[4]{81x^{16}y^{12}}$ **47.** $\sqrt{64m^2n^4}$ **48.** $\sqrt{121p^8q^4r^2}$

For Exercises 49–52, find the missing length of each right triangle.

49.

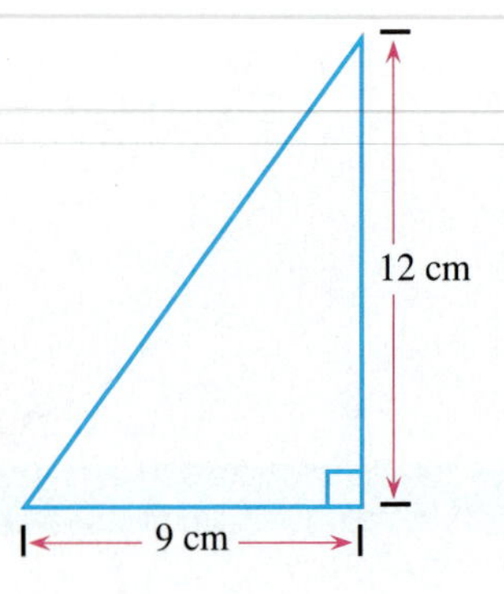

50.

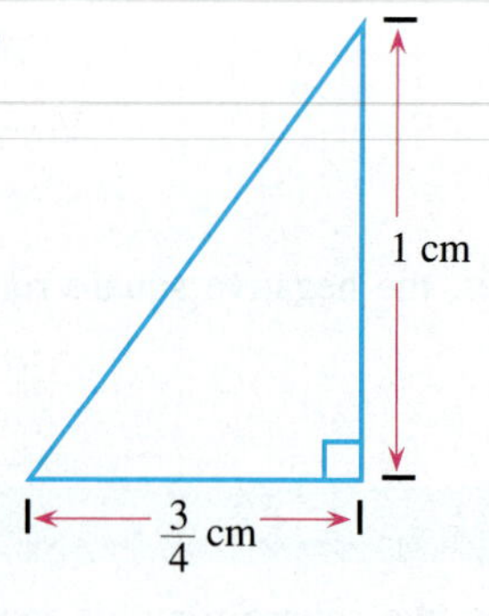

51.

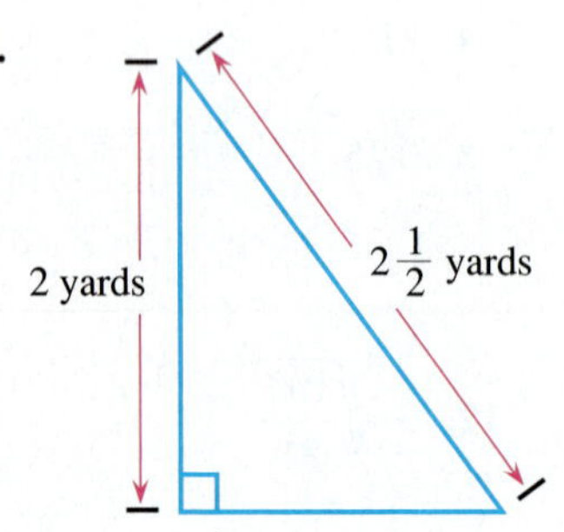

52.

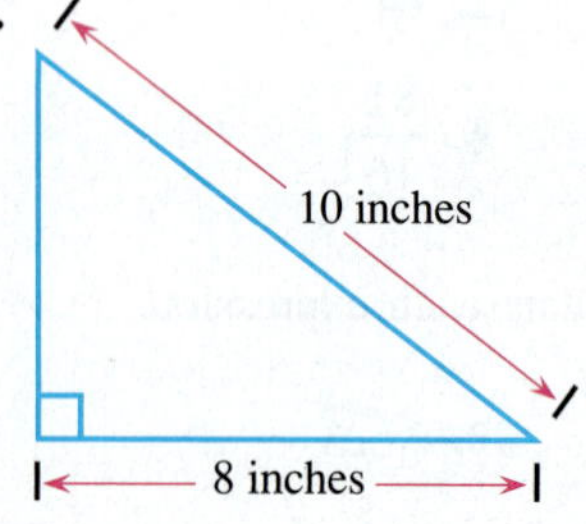

53. Find the period of a pendulum $4\frac{1}{2}$ feet long. (See Example 9.) Find an approximate decimal answer by replacing π by 3.14.

54. The formula for the period T of a pendulum when the length ℓ is measured in meters is given by

$$T = 2\pi\sqrt{\frac{\ell}{9.8}}$$

Find the period of a pendulum 2.45 meters in length. Find an approximate decimal answer by replacing π by 3.14.

55. Suppose an object is dropped from a tall building. After falling d feet, the object acquires a velocity v, in feet per second, that is given by $v = \sqrt{64d}$. What is the velocity of an object that has fallen 100 feet?

56. The radius r of a right circular cylinder with height h and volume V is given by $r = \sqrt{\frac{V}{\pi h}}$. A can in the shape of a right circular cylinder is to be constructed to have a volume of 6.28 cubic inches and a height of 8 inches. What should be the radius of this can? Find an approximate decimal answer by replacing π by 3.14.

A calculator with a $\sqrt{x}$ button can be used to approximate numbers like $\sqrt{45}$. First enter 45, then press the $\sqrt{x}$ button. The (8 digit) display at the top is 6.7082039. If we round this decimal off to three decimal places, then

$\sqrt{45}$ is approximately equal to 6.708, and we write

$$\sqrt{45} \approx 6.708$$

For Exercises 57–61, use the $\sqrt{x}$ button on your calculator to find approximate values on the radicals. Round off to three decimal places. Then, plot each radical on the same number line.

57. $\sqrt{33}$ **58.** $\sqrt{76.3}$ **59.** $\sqrt{12.91}$

60. $\sqrt{\dfrac{40}{7}}$ **61.** $-\sqrt{23.82}$

SAY IT IN WORDS

62. What are perfect squares?

63. Why doesn't $\sqrt{-4}$ exist as a real number?

REVIEW EXERCISES

The following exercises review parts of Section 4.1. Doing these exercises will help prepare you for the next section.

Simplify and write your answer without negative exponents.

64. $\dfrac{2^2}{2^3}$ **65.** x^4x^6 **66.** $\dfrac{x^4}{x^2}$ **67.** $\left(\dfrac{c}{d}\right)^3$

68. $(st)^5$ **69.** $(v^2u^3)^4$ **70.** $\left(\dfrac{s^2}{t^4v^2}\right)^4$ **71.** x^{-3}

72. $(a^{-1}b^2)^{-2}$ **73.** $\left(\dfrac{a}{b^2}\right)^{-1}$ **74.** $(x^{-3})^{-2}$ **75.** $(xz^{-2})^{-1}$

ENRICHMENT EXERCISES

Simplify the following expressions.

1. $\sqrt{x^{2n}}$ **2.** $\sqrt[3]{a^{3r}}$ **3.** $\sqrt[4]{z^{4m}}$

4. $\sqrt[m]{x^{mn}}$ **5.** $\sqrt[3]{-\sqrt[4]{1}}$ **6.** $\sqrt{\sqrt[3]{64}}$

Answers to Enrichment Exercises begin on page A.1.

8.2 Simplifying Radical Expressions

OBJECTIVES

- *To use the product rule for radicals to simplify radicals*
- *To use the quotient rule for radicals to simplify radicals*
- *To rationalize the denominator of a radical expression*

In this section, we consider properties that will help us simplify radical expressions. We introduce the first property with the following illustration: Notice that

$$\sqrt{4 \cdot 25} = \sqrt{100} = 10$$

and that

$$\sqrt{4} \cdot \sqrt{25} = 2 \cdot 5 = 10$$

Since both $\sqrt{4 \cdot 25}$ and $\sqrt{4} \cdot \sqrt{25}$ are equal to 10,

$$\sqrt{4 \cdot 25} = \sqrt{4} \cdot \sqrt{25}$$

Thus, we have illustrated the *product rule for radicals.*

RULE

The Product Rule for Radicals

If a and b are nonnegative numbers, then

$$\sqrt{a \cdot b} = \sqrt{a} \cdot \sqrt{b}$$

That is, the square root of the product is the product of the square roots.

Example 1 Multiply.

(a) $\sqrt{5}\sqrt{3} = \sqrt{5 \cdot 3} = \sqrt{15}$ (b) $\sqrt{2}\sqrt{a} = \sqrt{2a}$ ◀

NOTE *The pattern of the product rule applies for the product of a and b but not for the sum of a and b.*

$$\sqrt{a + b} \neq \sqrt{a} + \sqrt{b}$$

For example,

$$\sqrt{9 + 16} = \sqrt{25} = 5$$

whereas,

$$\sqrt{9} + \sqrt{16} = 3 + 4 = 7$$

Therefore,

$$\sqrt{9 + 16} \neq \sqrt{9} + \sqrt{16}$$

There is also a *quotient rule for radicals* whose proof is similar to that of the product rule.

RULE

Quotient Rule for Radicals

If a and b are nonnegative real numbers, with $b \neq 0$, then

$$\sqrt{\frac{a}{b}} = \frac{\sqrt{a}}{\sqrt{b}}$$

That is, the square root of the quotient is the quotient of the square roots.

Example 2 Rewrite each expression by using the quotient rule for radicals.

(a) $\sqrt{\frac{9}{49}} = \frac{\sqrt{9}}{\sqrt{49}} = \frac{3}{7}$ **(b)** $\sqrt{\frac{11}{100}} = \frac{\sqrt{11}}{\sqrt{100}} = \frac{\sqrt{11}}{10}$ ◀

The product and quotient rules for radicals also work for higher roots.

RULE

Use of Product and Quotient Rules for Higher Roots

If a and b are real numbers for which the nth roots are defined,

$$\sqrt[n]{a \cdot b} = \sqrt[n]{a} \cdot \sqrt[n]{b}$$

$$\sqrt[n]{\frac{a}{b}} = \frac{\sqrt[n]{a}}{\sqrt[n]{b}}, \qquad \text{where } b \neq 0$$

Example 3 Rewrite each expression.

(a) $\sqrt[3]{3}\sqrt[3]{9} = \sqrt[3]{27} = 3$ **(b)** $\sqrt[4]{\frac{ab}{16}} = \frac{\sqrt[4]{ab}}{\sqrt[4]{16}} = \frac{\sqrt[4]{ab}}{2}$ ◀

Using the product and quotient rules for radicals, we can change the form and simplify radical expressions according to the following criteria.

DEFINITION

Simplifying Radical Expressions

A radical expression is *simplified* if:

1. The radicand (the quantity under the radical sign) contains no power greater than or equal to the index. This means that no perfect squares are factors under a square root sign; no perfect cubes are factors under a cube root sign; and so on.

Example: $\sqrt{x^5}$ does not satisfy this condition.

DEFINITION *(continued)*

2. The radicand has no fractions.

   Example: $\sqrt[3]{\frac{2}{7}}$ does not satisfy this condition.

3. There are no radicals in denominators.

   Example: $\frac{4}{\sqrt{3}}$ does not satisfy this condition.

In the next two examples, we illustrate how to simplify radical expressions.

Example 4 Simplify $\sqrt{75}$.

Solution The largest perfect square that is a factor of 75 is 25. We write 75 as $25 \cdot 3$, then apply the product rule.

$$\sqrt{75} = \sqrt{25 \cdot 3} = \sqrt{25} \cdot \sqrt{3} = 5\sqrt{3}$$

Example 5 Simplify $\sqrt{32x^4y^7}$.

Solution The largest perfect square factor of the radicand is $16x^4y^6$. Therefore,

$$\sqrt{32x^4y^7} = \sqrt{16x^4y^6 \cdot 2y} = \sqrt{16x^4y^6}\sqrt{2y} = 4x^2y^3\sqrt{2y}$$

Example 6 Simplify $\sqrt[3]{16r^7s^5}$.

Solution We want to factor out the largest perfect cube from the radicand. Namely, write $16r^7s^5$ as $8r^6s^3 \cdot 2rs^2$ and then use the product rule for radicals.

$$\begin{aligned} \sqrt[3]{16r^7s^5} &= \sqrt[3]{8r^6s^3 \cdot 2rs^2} \\ &= \sqrt[3]{8r^6s^3}\sqrt[3]{2rs^2} \\ &= 2r^2s\sqrt[3]{2rs^2} \end{aligned}$$

Example 7 Simplify $\sqrt{\frac{3x}{8y}}$.

Solution This radical does not satisfy Condition 2 for simplifying radicals, as the radicand is a fraction. To simplify, multiply numerator and denominator by a factor that makes the denominator a perfect square. If we multiply $8y$ by $2y$, then $8y \cdot 2y = 16y^2$, which is a perfect square. Therefore, we multiply numerator and denominator by $2y$.

$$\begin{aligned} \sqrt{\frac{3x}{8y}} &= \sqrt{\frac{3x \cdot 2y}{8y \cdot 2y}} \\ &= \sqrt{\frac{6xy}{16y^2}} \end{aligned}$$

$$= \frac{\sqrt{6xy}}{\sqrt{16y^2}}$$

$$= \frac{\sqrt{6xy}}{4y}$$

We could have multiplied numerator and denominator by $8y$ instead of $2y$. We would still attain the correct answer; however, it would have resulted in more steps. Here are the details:

$$\sqrt{\frac{3x}{8y}} = \sqrt{\frac{3x \cdot 8y}{8y \cdot 8y}}$$

$$= \sqrt{\frac{24xy}{64y^2}}$$

$$= \frac{\sqrt{4 \cdot 6xy}}{\sqrt{64y^2}}$$

$$= \frac{2\sqrt{6xy}}{8y}$$

$$= \frac{\sqrt{6xy}}{4y}$$

Example 8 Simplify $\frac{3w}{\sqrt{6w}}$.

Solution This radical does not satisfy Condition 3 for simplifying radicals. Multiply numerator and denominator by the smallest expression that will make the denominator the square root of a perfect square.

$$\frac{3w}{\sqrt{6w}} = \frac{3w\sqrt{6w}}{\sqrt{6w}\sqrt{6w}}$$

$$= \frac{3w\sqrt{6w}}{6w}$$

$$= \frac{\sqrt{6w}}{2}$$

In Example 8, the process of eliminating the radical in the denominator is called *rationalizing the denominator.*

DEFINITION

To **rationalize the denominator** of an expression is to rewrite the expression so that there are no radicals in the denominator.

In general, the process is to multiply both the numerator and denominator by an expression so that the product in the denominator is a perfect square, cube, and so on. For example, if the denominator is $\sqrt{8x}$, multiply by $\sqrt{2x}$:

$$\sqrt{8x}\sqrt{2x} = \sqrt{16x^2} = 4x$$

Example 9 Rationalize the denominator.

(a) $\dfrac{3}{\sqrt{5}} = \dfrac{3}{\sqrt{5}} \cdot \dfrac{\sqrt{5}}{\sqrt{5}} = \dfrac{3\sqrt{5}}{(\sqrt{5})^2} = \dfrac{3\sqrt{5}}{5}$

(b) $\dfrac{4}{\sqrt{18z}} = \dfrac{4}{\sqrt{18z}} \cdot \dfrac{\sqrt{2z}}{\sqrt{2z}} = \dfrac{4\sqrt{2z}}{\sqrt{36z^2}}$

$= \dfrac{4\sqrt{2z}}{6z} = \dfrac{2\sqrt{2z}}{3z}$

Many times a radical expression can be simplified by one of several ways. Therefore, it is important to think of a method before moving the pencil.

TEAM PROJECT

(3 or 4 Students)

SIMPLIFYING RADICAL EXPRESSIONS

Course of Action: As a team, simplify the following radical expressions:

$$\sqrt{72} \qquad \sqrt{147} \qquad \sqrt{405} \qquad \sqrt{363}$$

Also, simplify the following radical expressions:

$$\sqrt{x^6y^{10}} \qquad \sqrt{a^7b^{16}} \qquad \sqrt{288r^9s^{15}t^{14}}$$

As a team, make up three radical expressions and simplify each. In Exercise Set 8.2, each member selects five problems and solves them. As a team, help any group members having difficulty. In Exercise Set 8.2, each team member selects two problems for the person to the right to simplify. As a team, help any member having problems.

Group Report: List the problems encountered when simplifying radical expressions.

WARMING UP

Answer true or false.

1. $\sqrt{3}\sqrt{3} = 3$
2. $\sqrt{2}\sqrt{7} = \sqrt{14}$
3. $\sqrt{\dfrac{3}{16}} = \dfrac{\sqrt{3}}{8}$
4. $\sqrt{8} = 2\sqrt{3}$
5. $\sqrt{3^4} = 9$
6. $\sqrt{48} = 4\sqrt{3}$
7. $\sqrt[3]{2^6} = 4$
8. $\sqrt[3]{\dfrac{5}{8}} = \dfrac{\sqrt[3]{5}}{8}$

EXERCISE SET 8.2

For Exercises 1–10, use the product rule for radicals to multiply. Assume that the variables represent nonnegative real numbers.

1. $\sqrt{5}\sqrt{2}$ **2.** $\sqrt{3}\sqrt{5}$ **3.** $\sqrt{2}\sqrt{2}$ **4.** $\sqrt{27}\sqrt{3}$

5. $\sqrt[3]{16}\sqrt[3]{4}$ **6.** $\sqrt[4]{9}\sqrt[4]{9}$ **7.** $\sqrt{x}\sqrt{3y}$ **8.** $\sqrt{2r}\sqrt{t}$

9. $\sqrt[4]{3p}\sqrt[4]{2q}$ **10.** $\sqrt[3]{5s}\sqrt[3]{7t}$

For Exercises 11–18, rewrite each expression by using the quotient rule for radicals. Assume all variables are positive.

11. $\sqrt{\dfrac{5}{4}}$ **12.** $\sqrt{\dfrac{5}{49}}$ **13.** $\sqrt{\dfrac{2}{x^4}}$ **14.** $\sqrt{\dfrac{7}{b^6}}$

15. $\sqrt[3]{\dfrac{11}{64}}$ **16.** $\sqrt[4]{\dfrac{23}{16}}$ **17.** $\sqrt[3]{\dfrac{x}{8}}$ **18.** $\sqrt[3]{\dfrac{a}{27b^3}}$

For Exercises 19–44, simplify. Assume the variables represent positive real numbers.

19. $\sqrt{32}$ **20.** $\sqrt{245}$ **21.** $\sqrt{12}$

22. $\sqrt{18}$ **23.** $\sqrt[3]{270}$ **24.** $\sqrt[4]{32}$

25. $\sqrt[3]{48c}$ **26.** $\sqrt[3]{80r}$ **27.** $-2\sqrt{8az}$

28. $-5\sqrt{45pq}$ **29.** $\sqrt{4^6}$ **30.** $\sqrt{5^4}$

31. $\sqrt{x^6}$ **32.** $\sqrt{a^4}$ **33.** $\sqrt[3]{16s^{10}}$

34. $\sqrt[3]{81t^{14}}$ **35.** $\sqrt{u^4v^2}$ **36.** $\sqrt{x^6z^{12}}$

37. $\sqrt[4]{32s^{18}}$ **38.** $\sqrt[4]{48c^{10}}$ **39.** $\sqrt[3]{x^4}$

40. $\sqrt[3]{z^7}$ **41.** $\sqrt{b^3}$ **42.** $\sqrt{t^9}$

43. $\sqrt{49u^{11}}$ **44.** $\sqrt{81v^5}$

For Exercises 45–50, simplify by rationalizing the denominator. Assume that the variables represent positive real numbers.

45. $\dfrac{2}{\sqrt{3}}$ **46.** $\dfrac{7}{\sqrt{5}}$ **47.** $\dfrac{x}{\sqrt{5x}}$

48. $\dfrac{a^2}{\sqrt{7a}}$ **49.** $\dfrac{c^3}{\sqrt{3c}}$ **50.** $-\dfrac{3}{\sqrt{z}}$

51. Find the value of each expression.
(a) $\sqrt{6^2}$ (b) $\sqrt{7^2}$ (c) $\sqrt{(-6)^2}$
(d) $\sqrt{(-7)^2}$

52. Find the value of each expression.
(a) $\sqrt[4]{3^4}$ (b) $\sqrt[4]{2^4}$ (c) $\sqrt[4]{(-3)^4}$
(d) $\sqrt[4]{(-2)^4}$

SAY IT IN WORDS

53. How do you simplify a radical expression?

54. Explain the product rule for radicals.

REVIEW EXERCISES

The following exercises review parts of Sections 4.4 and 6.3. Doing these exercises will help prepare you for the next section.

Section 4.4

Combine the like terms.

55. $4x^2 + 7x^2 - 3x^2$

56. $5a^3 - 4a^3 + 9a^3 - a^3$

57. $3s^2 - 4s + 1 - (5s^2 - 2s - 4)$

Section 6.3

Combine.

58. $\frac{3}{x} - \frac{5}{x}$

59. $-\frac{6z}{x^2r} + \frac{5z}{x^2r} - \frac{12z}{x^2r}$

60. $\frac{4}{2x+3} - \frac{7}{2x+3}$

61. $\frac{3}{9-4x^2} - \frac{2x}{9-4x^2}$

62. $\frac{2}{y-3} + \frac{3}{y+2} - \frac{5y}{y^2-y-6}$

ENRICHMENT EXERCISES

For Exercises 1–3, for what values of x will the radical be a real number?

1. $\sqrt{x^2 - 2x - 3}$

2. $\sqrt{2x^2 + 5x - 18}$

3. $\sqrt{x^4 + 2x^2 + 1}$

For Exercises 4–6, let $D = \sqrt{b^2 - 4ac}$, where a, b, and c are the coefficients from the trinomial $ax^2 + bx + c$. Evaluate D for the following trinomials.

4. $x^2 - 3x - 5$

5. $x^2 + 4x + 1$

6. $2x^2 - x - 1$

Answers to Enrichment Exercises begin on page A.1.

8.3 The Addition and Subtraction of Radical Expressions

OBJECTIVES

- ▶ *To simplify radical expressions by combining like radicals*
- ▶ *To combine the sum or difference of radical expressions into a single fraction*

Recall that the terms $4x^2$ and $-12x^2$ are like terms since their variable parts are the same. Similarly, $4\sqrt[3]{5}$ and $-12\sqrt[3]{5}$ are *like radicals* because they have the same index and radicand.

DEFINITION

Two radicals are said to be **like or similar radicals** if they have the same index and the same radicand.

For example, $2\sqrt{x}$, $-5\sqrt{x}$, and $\frac{3}{2}\sqrt{x}$ are like radicals, whereas $11\sqrt{7}$ and $6\sqrt{28}$ are not like radicals in their present form.

To combine like radicals we use the distributive property. For example,

$$\begin{aligned} 5\sqrt{2} + 6\sqrt{2} &= (5 + 6)\sqrt{2} \\ &= 11\sqrt{2} \end{aligned}$$

Example 1 Simplify by combining like radicals.

(a) $3\sqrt{5} - 7\sqrt{5} - 8\sqrt{5}$ **(b)** $16\sqrt[3]{11} + 2\sqrt{7} - \sqrt[3]{11}$

(c) $4\sqrt{3} - 6\sqrt{27}$ **(d)** $2\sqrt{50} + 14\sqrt{8} - 20\sqrt{18}$

Solution **(a)** All three radicals are similar. Using the distributive property,

$$\begin{aligned} 3\sqrt{5} - 7\sqrt{5} - 8\sqrt{5} &= (3 - 7 - 8)\sqrt{5} \\ &= -12\sqrt{5} \end{aligned}$$

(b) The first and third radicals are similar. Therefore, we combine only those two.

$$\begin{aligned} 16\sqrt[3]{11} + 2\sqrt{7} - \sqrt[3]{11} &= 16\sqrt[3]{11} - \sqrt[3]{11} + 2\sqrt{7} && \text{Commutative property of addition.} \\ &= (16 - 1)\sqrt[3]{11} + 2\sqrt{7} && \text{Distributive property.} \\ &= 15\sqrt[3]{11} + 2\sqrt{7} \end{aligned}$$

Notice that we cannot combine any further, since $15\sqrt[3]{11}$ and $2\sqrt{7}$ are not like radicals.

(c) At first glance, it appears that we cannot combine the two terms. However, the square root in the second term is not simplified. Therefore, we first simplify, then combine the resulting similar radicals.

$$\begin{aligned} 4\sqrt{3} - 6\sqrt{27} &= 4\sqrt{3} - 6\sqrt{9 \cdot 3} \\ &= 4\sqrt{3} - 6\sqrt{9}\sqrt{3} && \text{9 is a perfect square factor of 27.} \\ &= 4\sqrt{3} - 6 \cdot 3\sqrt{3} \\ &= 4\sqrt{3} - 18\sqrt{3} \\ &= -14\sqrt{3} && \text{Combine like radicals.} \end{aligned}$$

(d) First simplify the square roots, then combine any like radicals.

$$2\sqrt{50} + 14\sqrt{8} - 20\sqrt{18} = 2\sqrt{25 \cdot 2} + 14\sqrt{4 \cdot 2} - 20\sqrt{9 \cdot 2}$$

Where possible, factor using numbers that are perfect squares.

$= 2\sqrt{25}\sqrt{2} + 14\sqrt{4}\sqrt{2} - 20\sqrt{9}\sqrt{2}$ $\sqrt{a \cdot b} = \sqrt{a} \cdot \sqrt{b}$

$= 2 \cdot 5\sqrt{2} + 14 \cdot 2\sqrt{2} - 20 \cdot 3\sqrt{2}$

$= 10\sqrt{2} + 28\sqrt{2} - 60\sqrt{2}$

$= (10 + 28 - 60)\sqrt{2}$ Distributive property.

$= -22\sqrt{2}$

NOTE *In Example 1, we used the distributive property to combine like radicals. We can combine like radicals immediately as we would like terms, as in $3x + 5x = 8x$. For example, $3\sqrt{2} + 5\sqrt{2} = 8\sqrt{2}$.*

COMMON ERROR

When adding two similar radicals, be sure not to add the radicands. For example,

$$2\sqrt{5} + 6\sqrt{5} \neq 8\sqrt{10}$$

The correct answer is

$$2\sqrt{5} + 6\sqrt{5} = 8\sqrt{5}$$

In the next example, we show how to combine like radicals when the radicands involve variables.

Example 2 Simplify.

(a) $4\sqrt{x} - 3\sqrt{x} + 12\sqrt{x}$ **(b)** $5a^3\sqrt[3]{a} - 8a^3\sqrt[3]{a}$

(c) $\sqrt{z^5} + 3z^2\sqrt{z}$ **(d)** $r^2\sqrt{r^3} - 3r\sqrt{4r^5} + 2\sqrt{r^7}$

Solution **(a)** $4\sqrt{x} - 3\sqrt{x} + 12\sqrt{x} = (4 - 3 + 12)\sqrt{x} = 13\sqrt{x}$

(b) $5a^3\sqrt[3]{a} - 8a^3\sqrt[3]{a} = (5a^3 - 8a^3)\sqrt[3]{a} = -3a^3\sqrt[3]{a}$

(c) First simplify $\sqrt{z^5}$, then combine.

$\sqrt{z^5} + 3z^2\sqrt{z} = \sqrt{z^4 \cdot z} + 3z^2\sqrt{z}$

$= \sqrt{z^4}\sqrt{z} + 3z^2\sqrt{z}$ $\sqrt{a \cdot b} = \sqrt{a} \cdot \sqrt{b}$.

$= z^2\sqrt{z} + 3z^2\sqrt{z}$

$= (z^2 + 3z^2)\sqrt{z}$ Distributive property.

$= 4z^2\sqrt{z}$

(d) We start by simplifying each of the three radicals, then combine any similar radicals.

$$\begin{aligned} r^2\sqrt{r^3} - 3r\sqrt{4r^5} + 2\sqrt{r^7} &= r^2\sqrt{r^2 \cdot r} - 3r\sqrt{4r^4 \cdot r} + 2\sqrt{r^6 \cdot r} \\ &= r^2\sqrt{r^2}\sqrt{r} - 3r\sqrt{4r^4}\sqrt{r} + 2\sqrt{r^6}\sqrt{r} \\ &= r^2r\sqrt{r} - 3r \cdot 2r^2\sqrt{r} + 2r^3\sqrt{r} \\ &= r^3\sqrt{r} - 6r^3\sqrt{r} + 2r^3\sqrt{r} \\ &= -3r^3\sqrt{r} \end{aligned}$$

In Example 2, we indicate how to simplify radicals first before attempting to add or subtract them. We summarize this method as the following strategy.

STRATEGY

Adding or Subtracting Radicals

1. Put each radical in simplified form.
2. Apply the distributive property to combine similar radicals.

In radical expressions involving fractions, we may be able to combine terms after first simplifying by rationalizing the denominator.

Example 3 Combine.

(a) $\sqrt{3} + \dfrac{1}{\sqrt{3}}$ **(b)** $2\sqrt{\dfrac{5}{4}} + 3\sqrt{20}$

(c) $\dfrac{2x}{\sqrt{x^3}} + \dfrac{3x^2}{\sqrt{x^5}}$, $x > 0$

Solution **(a)** We start by rationalizing the denominator of $\dfrac{1}{\sqrt{3}}$.

$$\begin{aligned} \sqrt{3} + \frac{1}{\sqrt{3}} &= \sqrt{3} + \frac{1 \cdot \sqrt{3}}{\sqrt{3} \cdot \sqrt{3}} \\ &= \sqrt{3} + \frac{\sqrt{3}}{3} && \sqrt{3} \cdot \sqrt{3} = 3. \\ &= 1 \cdot \sqrt{3} + \frac{1}{3}\sqrt{3} \\ &= \left(1 + \frac{1}{3}\right)\sqrt{3} && \text{Distributive property.} \\ &= \frac{4}{3}\sqrt{3} \quad \text{or} \quad \frac{4\sqrt{3}}{3} \end{aligned}$$

(b) We first simplify the two radicals.

$$2\sqrt{\frac{5}{4}} + 3\sqrt{20} = \frac{2\sqrt{5}}{\sqrt{4}} + 3\sqrt{4 \cdot 5} \qquad \sqrt{\frac{a}{b}} = \frac{\sqrt{a}}{\sqrt{b}}.$$

$$= \frac{2\sqrt{5}}{2} + 3\sqrt{4}\sqrt{5} \qquad \sqrt{a \cdot b} = \sqrt{a} \cdot \sqrt{b}.$$

$$= \sqrt{5} + 3 \cdot 2\sqrt{5}$$

$$= \sqrt{5} + 6\sqrt{5}$$

$$= 7\sqrt{5} \qquad \text{Combine the two terms.}$$

(c) We start by simplifying each of the two radicals by factoring out the largest power that is a perfect square.

$$\frac{2x}{\sqrt{x^3}} + \frac{3x^2}{\sqrt{x^5}} = \frac{2x}{\sqrt{x^2 \cdot x}} + \frac{3x^2}{\sqrt{x^4 \cdot x}}$$

$$= \frac{2x}{x\sqrt{x}} + \frac{3x^2}{x^2\sqrt{x}} \qquad \sqrt{x^2} = x, \text{ since } x > 0.$$

$$= \frac{2}{\sqrt{x}} + \frac{3}{\sqrt{x}}$$

$$= \frac{2+3}{\sqrt{x}} \qquad \text{Add the two fractions.}$$

$$= \frac{5}{\sqrt{x}}$$

$$= \frac{5 \cdot \sqrt{x}}{\sqrt{x} \cdot \sqrt{x}} \qquad \text{Rationalize the denominator.}$$

$$= \frac{5\sqrt{x}}{x}$$

In algebra, it is sometimes necessary to combine two radical expressions as a single fraction. We illustrate this technique in the next example.

Example 4 Combine as a single fraction and simplify. Assume the variables represent positive real numbers.

$$\frac{3}{\sqrt{5}} - \frac{2}{\sqrt{5}}$$

Solution

$$\frac{3}{\sqrt{5}} - \frac{2}{\sqrt{5}} = \frac{3-2}{\sqrt{5}} \qquad \text{Subtract the two fractions.}$$

$$= \frac{1}{\sqrt{5}}$$

$$= \frac{1\sqrt{5}}{\sqrt{5}\sqrt{5}} \quad \text{Rationalize the denominator.}$$

$$= \frac{\sqrt{5}}{5}$$

WARMING UP

Answer true or false.

1. $2\sqrt{3} + \sqrt{3} = 3\sqrt{3}$
2. $3\sqrt{7} - 4\sqrt{7} = \sqrt{-7}$
3. $3\sqrt{3} + \sqrt{27} = 6\sqrt{3}$
4. $2\sqrt{x} + 3\sqrt{x} = 5\sqrt{x}$
5. $\frac{1}{\sqrt{2}} - \frac{3}{\sqrt{2}} = -\sqrt{2}$
6. $\frac{2}{\sqrt{5}} + \frac{1}{\sqrt{5}} = \frac{3\sqrt{5}}{\sqrt{5}}$

EXERCISE SET 8.3

For Exercises 1–18, simplify the expressions by combining like radicals. Assume that the variables represent positive real numbers.

1. $2\sqrt{3} - 4\sqrt{3} - \sqrt{3}$
2. $6\sqrt{2} - 12\sqrt{2} + 19\sqrt{2}$
3. $5\sqrt{8} - 7\sqrt{8}$
4. $9\sqrt{12} + \sqrt{12}$
5. $5\sqrt{18} - 2\sqrt{2}$
6. $8\sqrt{6} - 2\sqrt{24}$
7. $3\sqrt{14} - \sqrt{56} + \sqrt{7}$
8. $\sqrt{12} - 7\sqrt{3} + 2\sqrt{6}$
9. $4\sqrt[3]{24} - 20\sqrt[3]{3}$
10. $2\sqrt[3]{32} + \sqrt[3]{108}$
11. $5\sqrt[4]{32} + \sqrt[4]{2} - \sqrt[4]{162}$
12. $\sqrt[5]{64} - 2\sqrt[5]{486} - 3\sqrt[5]{2}$
13. $4\sqrt{c} + 3\sqrt{c}$
14. $7\sqrt{x} + 5\sqrt{x}$
15. $5\sqrt[3]{ax} - 8\sqrt[3]{ax}$
16. $6\sqrt[4]{z^3} - 11\sqrt[4]{z^3}$
17. $3a\sqrt{a^3} - 4a^2\sqrt{a} - 2a^2$
18. $-3\sqrt{b^5} + b\sqrt{b^3} + b^2$

For Exercises 19–26, combine as a single fraction and simplify. Assume that the variables represent positive real numbers.

19. $\frac{1}{\sqrt{3}} - \frac{5}{\sqrt{3}}$
20. $\frac{2}{\sqrt{5}} + \frac{7}{\sqrt{5}}$
21. $\frac{3\sqrt{2}}{\sqrt{7}} + \frac{4\sqrt{8}}{\sqrt{7}}$
22. $\frac{5\sqrt{3}}{\sqrt{11}} - \frac{2\sqrt{27}}{\sqrt{11}}$
23. $\frac{\sqrt{8}}{3} - \frac{\sqrt{18}}{6}$
24. $\frac{\sqrt{12}}{4} + \frac{\sqrt{27}}{6}$
25. $\frac{x\sqrt{x^3}}{12} - \frac{2\sqrt{x^5}}{9}$
26. $\frac{c\sqrt{c^5}}{20} + \frac{\sqrt{c^7}}{15}$

For Exercises 27–30, simplify.

27. $\frac{-4 + \sqrt{16 - 4(2)}}{2}$
28. $\frac{2 - \sqrt{4 - 4(-1)}}{2}$

29. $\dfrac{7 - \sqrt{(-7)^2 - 4(2)(6)}}{2 \cdot 2}$

30. $\dfrac{10 + \sqrt{(-10)^2 - 4(3)(3)}}{2 \cdot 3}$

SAY IT IN WORDS

For Exercises 31 and 32, explain why each equation is incorrect.

31. $13\sqrt{7} - 4\sqrt{2} = 9\sqrt{5}$

32. $3\sqrt{2} + 5\sqrt{3} = 8\sqrt{5}$

REVIEW EXERCISES

The following exercises review parts of Section 4.6. Doing these exercises will help prepare you for the next section.

Multiply.

33. $2ab(3az - 4bz)$

34. $x^3y^2(3xy - 2x^2y - 5)$

35. $(2x + 3)(x - 1)$

36. $(5b + d)(7b - 3d)$

37. $(2a - 2b)^2$

ENRICHMENT EXERCISES

For Exercises 1 and 2, rewrite each expression without radicals in the denominator, then simplify. [*Hint:* Make use of the fact that $(a + b)(a - b) = a^2 - b^2$.]

1. $\dfrac{1}{\sqrt{7} + 1}$

2. $\dfrac{2}{\sqrt{3} - \sqrt{2}}$

For Exercises 3–8, evaluate $\dfrac{-b - \sqrt{b^2 - 4ac}}{2a}$ for each trinomial $ax^2 + bx + c$.

3. $x^2 - 3x + 2$

4. $x^2 - x - 6$

5. $x^2 - 8$

6. $4x^2 - 4x - 17$

7. $x^2 - 4x - 8$

8. $4x^2 - 12x + 9$

Answers to Enrichment Exercises begin on page A.1.

8.4 Products and Quotients of Radical Expressions

OBJECTIVES

- *To multiply expressions that contain radicals*
- *To rationalize denominators containing sums and differences of radicals*
- *To factor expressions involving radicals*

In this section, we multiply and divide expressions that contain radicals. The techniques we use to multiply radical expressions will be similar to those for multiplying polynomials. We start with finding the product of expressions having one term.

Example 1 Multiply.

(a) $(4\sqrt{3})(5\sqrt{2})$ **(b)** $(2\sqrt{t})(7\sqrt{t})$, where $t \geq 0$.

Solution **(a)** We use the commutative and associative properties of multiplication to rearrange the order and grouping of the factors in this product.

$$(4\sqrt{3})(5\sqrt{2}) = (4 \cdot 5)(\sqrt{3}\sqrt{2})$$ Associative and commutative properties of multiplication.

$$= 20(\sqrt{3 \cdot 2})$$ $\sqrt{a} \cdot \sqrt{b} = \sqrt{a \cdot b}$.

$$= 20\sqrt{6}$$

(b) $(2\sqrt{t})(7\sqrt{t}) = (2 \cdot 7)(\sqrt{t})(\sqrt{t})$ Associative and commutative properties of multiplication.

$$= 14t$$ $(\sqrt{t})(\sqrt{t}) = t$. ◀

Recall from Chapter 4, we used the distributive property to multiply algebraic expressions such as a monomial times a binomial,

$$a(b + c) = ab + ac \quad \text{and} \quad (b + c)a = ba + ca$$

We may use the distributive property to multiply expressions that contain radicals. Keep in mind that radicands that contain variables are assumed to be nonnegative.

Example 2 Multiply and then simplify. Assume that all variables represent nonnegative real numbers.

(a) $3(\sqrt{5} + \sqrt{2})$ **(b)** $\sqrt{5}(\sqrt{15} - \sqrt{2})$
(c) $(\sqrt{6} - \sqrt{3})\sqrt{3}$ **(d)** $\sqrt{a}(\sqrt{b} + 5\sqrt{a})$

Solution **(a)** We distribute the 3 over the two terms in the parentheses.

$$3(\sqrt{5} + \sqrt{2}) = 3\sqrt{5} + 3\sqrt{2}$$

(b) After applying the distributive property, we simplify the resulting two products.

$$\begin{aligned}\sqrt{5}(\sqrt{15} - \sqrt{2}) &= \sqrt{5}\sqrt{15} - \sqrt{5}\sqrt{2}\\ &= \sqrt{5}\sqrt{5 \cdot 3} - \sqrt{5 \cdot 2}\\ &= \sqrt{5}\sqrt{5}\sqrt{3} - \sqrt{10}\\ &= 5\sqrt{3} - \sqrt{10}\end{aligned}$$ $\sqrt{a}\sqrt{a} = a$.

(c)
$$\begin{aligned}(\sqrt{6} - \sqrt{3})\sqrt{3} &= \sqrt{6}\sqrt{3} - \sqrt{3}\sqrt{3}\\ &= \sqrt{3 \cdot 2}\sqrt{3} - 3\\ &= \sqrt{3}\sqrt{2}\sqrt{3} - 3\\ &= 3\sqrt{2} - 3\end{aligned}$$

(d) $\sqrt{a}(\sqrt{b} + 5\sqrt{a}) = \sqrt{a}\sqrt{b} + \sqrt{a}(5\sqrt{a})$

$= \sqrt{ab} + 5\sqrt{a}\sqrt{a}$ — $\sqrt{a}\sqrt{b} = \sqrt{ab}.$

$= \sqrt{ab} + 5a$ — $\sqrt{a}\sqrt{a} = a.$

Example 3 Multiply and then simplify. Assume the variables represent nonnegative real numbers.

(a) $(5\sqrt{2} - \sqrt{3})(\sqrt{2} + 4\sqrt{3})$ **(b)** $(3\sqrt{r} + \sqrt{t})^2$

(c) $(\sqrt{5} + \sqrt{3})(\sqrt{5} - \sqrt{3})$ **(d)** $(\sqrt{x} + 4)(\sqrt{x} - 4)$

Solution **(a)** We expand the product using the distributive property.

$$\begin{aligned}(5\sqrt{2} - \sqrt{3})(\sqrt{2} + 4\sqrt{3}) &= 5\sqrt{2}(\sqrt{2} + 4\sqrt{3}) - \sqrt{3}(\sqrt{2} + 4\sqrt{3})\\ &= 5\sqrt{2}\sqrt{2} + 5\sqrt{2}(4\sqrt{3}) - \sqrt{3}\sqrt{2} - \sqrt{3}(4\sqrt{3})\\ &= 5 \cdot 2 + 20\sqrt{2}\sqrt{3} - \sqrt{3}\sqrt{2} - 4\sqrt{3}\sqrt{3}\\ &= 10 + 20\sqrt{6} - \sqrt{6} - 4 \cdot 3\\ &= 10 + 19\sqrt{6} - 12\\ &= -2 + 19\sqrt{6}\end{aligned}$$

(b) Recall the formula for squaring a binomial, $(a + b)^2 = a^2 + 2ab + b^2$. Setting $a = 3\sqrt{r}$ and $b = \sqrt{t}$,

$$\begin{aligned}(3\sqrt{r} + \sqrt{t})^2 &= (3\sqrt{r})^2 + 2 \cdot 3\sqrt{r}\sqrt{t} + (\sqrt{t})^2\\ &= 3^2 \cdot (\sqrt{r})^2 + 6\sqrt{r}\sqrt{t} + (\sqrt{t})^2\\ &= 9r + 6\sqrt{rt} + t\end{aligned}$$

$(\sqrt{a})^2 = a.$

(c) For this one, we make use of the special product formula $(a + b)(a - b) = a^2 - b^2$, where a is $\sqrt{5}$ and b is $\sqrt{3}$.

$$\begin{aligned}(\sqrt{5} + \sqrt{3})(\sqrt{5} - \sqrt{3}) &= (\sqrt{5})^2 - (\sqrt{3})^2\\ &= 5 - 3\\ &= 2\end{aligned}$$

(d) Again, we use the formula $(a + b)(a - b) = a^2 - b^2$.

$$\begin{aligned}(\sqrt{x} + 4)(\sqrt{x} - 4) &= (\sqrt{x})^2 - 4^2\\ &= x - 16\end{aligned}$$

In Parts (c) and (d) of Example 3, we suggest a way to rationalize a denominator that is the sum or difference of terms involving radicals. The first step is to form the *conjugate* of the denominator.

DEFINITION

The expressions $a + b$ and $a - b$ are called **conjugates** of each other.

The following expressions are some examples of binomial expressions and their conjugates.

Expressions	***Conjugates***
$\sqrt{2} + 5$	$\sqrt{2} - 5$
$\sqrt{3} + \sqrt{10}$	$\sqrt{3} - \sqrt{10}$
$\sqrt{x} + 1$	$\sqrt{x} - 1$
$\sqrt{2s} + 4\sqrt{t}$	$\sqrt{2s} - 4\sqrt{t}$
$5 - \sqrt{12}$	$5 + \sqrt{12}$
$\sqrt{x} - 4$	$\sqrt{x} + 4$
$\sqrt{y} - \sqrt{2z}$	$\sqrt{y} + \sqrt{2z}$
$3u - \sqrt{6v}$	$3u + \sqrt{6v}$

Now, consider a quotient such as $\dfrac{5}{\sqrt{3} + 1}$. In order to rationalize the denominator, we make use of the fact that

$$\frac{a}{b} = \frac{a \cdot c}{b \cdot c}$$

That is, we may multiply the numerator and denominator by the same quantity c and the quotient remains unchanged in value. In particular, we *let c be the conjugate of the denominator.* Therefore, setting $c = \sqrt{3} - 1$,

$$\begin{aligned} \frac{5}{\sqrt{3} + 1} &= \frac{5(\sqrt{3} - 1)}{(\sqrt{3} + 1)(\sqrt{3} - 1)} \\ &= \frac{5(\sqrt{3} - 1)}{(\sqrt{3})^2 - 1^2} \\ &= \frac{5(\sqrt{3} - 1)}{3 - 1} \\ &= \frac{5(\sqrt{3} - 1)}{2} \end{aligned}$$

Example 4 Rationalize the denominator and simplify.

(a) $\dfrac{3}{\sqrt{5} + 4}$ **(b)** $\dfrac{\sqrt{5}}{3\sqrt{3} - 2\sqrt{5}}$

Solution **(a)** The conjugate of $\sqrt{5} + 4$ is $\sqrt{5} - 4$. Therefore, we multiply numerator and denominator by $\sqrt{5} - 4$ and simplify.

$$\frac{3}{\sqrt{5}+4} = \frac{3(\sqrt{5}-4)}{(\sqrt{5}+4)(\sqrt{5}-4)}$$
$$= \frac{3(\sqrt{5}-4)}{(\sqrt{5})^2-4^2}$$
$$= \frac{3(\sqrt{5}-4)}{5-16}$$
$$= \frac{3(\sqrt{5}-4)}{-11}$$
$$= -\frac{3(\sqrt{5}-4)}{11}$$

(b) The conjugate of $3\sqrt{3}-2\sqrt{5}$ is $3\sqrt{3}+2\sqrt{5}$. Therefore,

$$\frac{\sqrt{5}}{3\sqrt{3}-2\sqrt{5}} = \frac{\sqrt{5}(3\sqrt{3}+2\sqrt{5})}{(3\sqrt{3}-2\sqrt{5})(3\sqrt{3}+2\sqrt{5})}$$
$$= \frac{3\sqrt{5}\sqrt{3}+2\sqrt{5}\sqrt{5}}{(3\sqrt{3})^2-(2\sqrt{5})^2}$$
$$= \frac{3\sqrt{15}+2\cdot 5}{9\cdot 3-4\cdot 5} \quad (ab)^2 = a^2b^2.$$
$$= \frac{3\sqrt{15}+10}{27-20}$$
$$= \frac{3\sqrt{15}+10}{7}$$

WARMING UP

Answer true or false.

1. $(2\sqrt{2})(3\sqrt{3}) = 6\sqrt{5}$
2. $\sqrt{2}\sqrt{5} = \sqrt{10}$
3. $\sqrt{3}(1-\sqrt{3}) = \sqrt{3}-3$
4. $\sqrt{5}(\sqrt{5}+2) = 7$
5. $(2\sqrt{3}-\sqrt{2})^2 = 14-4\sqrt{6}$
6. $(2\sqrt{5}+1)(2\sqrt{5}-1) = 19$

EXERCISE SET 8.4

For Exercises 1–26, multiply and then simplify. Assume that the variables represent nonnegative real numbers.

1. $(6\sqrt{3})(2\sqrt{5})$
2. $(4\sqrt{2})(7\sqrt{3})$
3. $(2\sqrt{x})(3\sqrt{x})$
4. $(-7\sqrt{y})(3\sqrt{y})$
5. $\sqrt{2}(\sqrt{5}+\sqrt{11})$
6. $\sqrt{7}(\sqrt{2}+\sqrt{3})$

7. $(\sqrt{3} - 2\sqrt{5})\sqrt{13}$ **8.** $(2\sqrt{2} - 4\sqrt{7})\sqrt{3}$ **9.** $\sqrt{2}(\sqrt{2} - 3)$

10. $\sqrt{5}(1 - 6\sqrt{5})$ **11.** $(\sqrt{15} - 1)\sqrt{3}$ **12.** $(\sqrt{12} + 3)\sqrt{3}$

13. $\sqrt{12}(\sqrt{3} + \sqrt{6})$ **14.** $\sqrt{18}(\sqrt{9} - \sqrt{2})$ **15.** $(3\sqrt{5} - 2)(2\sqrt{5} + 1)$

16. $(4\sqrt{2} + 3)(\sqrt{2} + 1)$ **17.** $(\sqrt{r} + 1)(2\sqrt{r} + 3)$ **18.** $(\sqrt{a} - 2)(3\sqrt{a} + 1)$

19. $(\sqrt{3} - \sqrt{2})^2$ **20.** $(\sqrt{5} + \sqrt{6})^2$ **21.** $(\sqrt{x} + 2\sqrt{z})^2$

22. $(3\sqrt{a} - 4\sqrt{b})^2$ **23.** $(\sqrt{3} + \sqrt{5})(\sqrt{3} - \sqrt{5})$ **24.** $(\sqrt{7} + \sqrt{2})(\sqrt{7} - \sqrt{2})$

25. $(\sqrt{x} - 1)(\sqrt{x} + 1)$ **26.** $(\sqrt{r} - 5)(\sqrt{r} + 5)$

For Exercises 27–34, rationalize the denominator and simplify. Assume that all variables are positive.

27. $\dfrac{2}{\sqrt{3} - 4}$ **28.** $\dfrac{1}{\sqrt{7} - 2}$ **29.** $\dfrac{\sqrt{6} - 2}{1 - \sqrt{6}}$

30. $\dfrac{5 - \sqrt{3}}{\sqrt{3} - 4}$ **31.** $\dfrac{3\sqrt{2} - 1}{\sqrt{2} + 3}$ **32.** $\dfrac{4\sqrt{3} + 3}{\sqrt{3} + 4}$

33. $\dfrac{3\sqrt{5}}{\sqrt{5} + \sqrt{2}}$ **34.** $\dfrac{\sqrt{3}}{\sqrt{3} + \sqrt{6}}$

For Exercises 35–38, each equation is incorrect. Change the right side so that the equation becomes true.

35. $(\sqrt{x} + 3)^2 = x + 9$ **36.** $(\sqrt{a} + \sqrt{b})^2 = a + b$

37. $3(2\sqrt{5}) = 6\sqrt{15}$ **38.** $\sqrt{2}(3\sqrt{2}) = 6\sqrt{2}$

SAY IT IN WORDS

39. What does "conjugate" mean?

40. Explain how to rationalize a denominator.

REVIEW PROBLEMS

The following exercises review parts of Section 5.6. Doing these problems will help prepare you for the next section.

Solve the following equations.

41. $x(x - 3) = 0$ **42.** $(2t - 1)(t + 4) = 0$ **43.** $3y^2 - 4y + 1 = 0$

44. $2x^2 - 7x + 6 = 0$ **45.** $a^2 - 2a + 1 = 0$ **46.** $9b^2 + 6b + 1 = 0$

ENRICHMENT EXERCISES

Multiply and then simplify.

1. $\sqrt[3]{x}(\sqrt[3]{x^2} - \sqrt[3]{x^5})$

2. $(\sqrt[3]{4} + 1)(\sqrt[3]{4} - 1)$

3. $(2\sqrt{7} - 3\sqrt{6})(2\sqrt{7} + 3\sqrt{6})$

4. $(\sqrt[3]{9x} + 1)(\sqrt[3]{9x} - 1)$

Answers to Enrichment Exercises begin on page A.1.

8.5 Solving Radical Equations

OBJECTIVE

▶ *To solve equations containing radicals*

An equation that contains one or more radicals with variables in the radicands is called a **radical equation.** Examples of radical equations are

$$\sqrt{x} = 7, \qquad \sqrt{y^2 + 1} = y, \qquad \text{and} \qquad \sqrt[3]{r} - 1 = 2\sqrt[3]{r + 9} + 3$$

To solve a radical equation requires rewriting the equation to one that does not contain radicals. For radical equations containing square roots, this is done using the *squaring property.*

PROPERTY

The Squaring Property

Suppose A and B are two expressions. If

$$A = B$$

then

$$A^2 = B^2$$

All solutions of the original equation will be solutions of the squared equation. However, when squaring, we may introduce *extraneous solutions.* Extraneous solutions satisfy the squared equation but do not satisfy the original equation.

For example, consider the equation $x = 3$. The number 3, of course, is the only solution to this equation. Applying the squaring property to this equation, $x^2 = 9$. However, $x^2 = 9$ has *two* solutions, 3 and -3. But -3 is not a solution of the original equation and is, therefore, extraneous. Therefore, it is very important to check the numbers in the original equation for extraneous roots.

Example 1 Solve the equation $\sqrt{x} = 6$.

Solution We use the squaring property,

$$\sqrt{x} = 6$$

implies that

$$(\sqrt{x})^2 = 6^2$$

Since $(\sqrt{x})^2$ can be replaced by x [recall $(\sqrt{a})^2 = a$], we have $x = 6^2$ or 36. As we have done in all equation solving, we check our solution in the original equation.

Check 36:

$$\sqrt{x} = 6$$
$$\sqrt{36} \stackrel{?}{=} 6$$
$$6 = 6$$

Therefore, $\{36\}$ is the solution set.

Example 2 Solve the equation $\sqrt{a - 1} = 5$.

Solution We use the squaring property,

$$\sqrt{a - 1} = 5$$
$$(\sqrt{a - 1})^2 = 5^2$$
$$a - 1 = 25$$
$$a = 26 \qquad \text{Add 1 to both sides.}$$

We check our solution in the original equation.

Check 26:

$$\sqrt{a - 1} = 5$$
$$\sqrt{26 - 1} \stackrel{?}{=} 5$$
$$\sqrt{25} \stackrel{?}{=} 5$$
$$5 = 5$$

Therefore, the solution set is $\{26\}$.

Example 3 Solve the equation $\sqrt{3x + 2} = 4$.

Solution Using the squaring property,

$$\sqrt{3x + 2} = 4$$
$$(\sqrt{3x + 2})^2 = 4^2$$
$$3x + 2 = 16$$
$$3x = 14 \qquad \text{Subtract 2 from both sides.}$$
$$x = \frac{14}{3} \qquad \text{Divide both sides by 3.}$$

Check $\frac{14}{3}$:

$$\sqrt{3x + 2} = 4$$
$$\sqrt{3\left(\frac{14}{3}\right) + 2} \stackrel{?}{=} 4$$
$$\sqrt{14 + 2} \stackrel{?}{=} 4$$
$$\sqrt{16} \stackrel{?}{=} 4$$
$$4 = 4$$

Therefore, $\left\{\frac{14}{3}\right\}$ is the solution set.

Example 4 Solve the equation $\sqrt{2x+5}+10=4$.

Solution First, subtract 10 from both sides before using the squaring property.

$$\sqrt{2x+5}+10=4$$
$$\sqrt{2x+5}+10-10=4-10$$
$$\sqrt{2x+5}=-6$$
$$(\sqrt{2x+5})^2=(-6)^2$$
$$2x+5=36$$
$$2x+5-5=36-5$$
$$2x=31$$
$$x=\frac{31}{2}$$

Next, we check $\frac{31}{2}$ in the original equation.

Check $\frac{31}{2}$:

$$\sqrt{2x+5}+10=4$$
$$\sqrt{2\left(\frac{31}{2}\right)+5}+10\stackrel{?}{=}4$$
$$\sqrt{31+5}+10\stackrel{?}{=}4$$
$$\sqrt{36}+10\stackrel{?}{=}4$$
$$6+10\stackrel{?}{=}4$$
$$16\neq 4$$

Since $16\neq 4$, $\frac{31}{2}$ is not a solution of the original equation. We conclude that $\sqrt{2x+5}+10=4$ has no real solution.

Example 5 Solve the equation $\sqrt{r}=-8$.

Solution Before using our pencil, let us think what this equation means. We are to find a number whose *nonnegative* square root is -8. Since $\sqrt{r}$ is nonnegative no matter what r is, certainly $\sqrt{r}$ could never be equal to -8. Therefore, $\sqrt{r}=-8$ has no solution.

Example 6 Solve the equation $\sqrt{10y} = 2\sqrt{6}$.

Solution

$$\sqrt{10y} = 2\sqrt{6}$$

$$(\sqrt{10y})^2 = (2\sqrt{6})^2$$

$$10y = 4 \cdot 6 \qquad (ab)^2 = a^2 \cdot b^2.$$

$$y = \frac{4 \cdot 6}{10}$$

$$= \frac{12}{5}$$

Check $\frac{12}{5}$:

$$\sqrt{10y} = 2\sqrt{6}$$

$$\sqrt{10\left(\frac{12}{5}\right)} \stackrel{?}{=} 2\sqrt{6}$$

$$\sqrt{2 \cdot 12} \stackrel{?}{=} 2\sqrt{6}$$

$$\sqrt{2 \cdot 2 \cdot 6} \stackrel{?}{=} 2\sqrt{6}$$

$$\sqrt{2^2 \cdot 6} \stackrel{?}{=} 2\sqrt{6}$$

$$2\sqrt{6} = 2\sqrt{6}$$

Therefore, $\left\{\frac{12}{5}\right\}$ is the solution set. ◀

Example 7 Solve the equation $\sqrt{c^2 + 2c + 6} = c$.

Solution

$$\sqrt{c^2 + 2c + 6} = c$$

$$(\sqrt{c^2 + 2c + 6})^2 = c^2$$

$$c^2 + 2c + 6 = c^2$$

$$2c + 6 = 0 \qquad \text{Subtract } c^2 \text{ from both sides.}$$

$$c = -3$$

Check -3:

$$\sqrt{c^2 + 2c + 6} = c$$

$$\sqrt{(-3)^2 + 2(-3) + 6} \stackrel{?}{=} -3$$

$$\sqrt{9 - 6 + 6} \stackrel{?}{=} -3$$

$$\sqrt{9} \stackrel{?}{=} -3$$

$$3 \neq -3$$

Therefore, -3 is not a solution and our original equation has no real solution. ◀

In the next example, squaring both sides of the radical equation results in a quadratic equation. The solutions of the quadratic equation must be checked in the original equation.

Example 8 Solve the equation $3t - 1 = \sqrt{4t}$.

Solution

$$(3t - 1)^2 = (\sqrt{4t})^2$$

$$9t^2 - 6t + 1 = 4t \qquad (a - b)^2 = a^2 - 2ab + b^2.$$

$$9t^2 - 10t + 1 = 0 \qquad \text{Subtract } 4t \text{ from both sides.}$$

$$(9t - 1)(t - 1) = 0 \qquad \text{Factor.}$$

Therefore, either

$$9t - 1 = 0 \quad \text{or} \quad t - 1 = 0$$

Recall: $A \cdot B = 0$ implies either $A = 0$ or $B = 0$.

Thus, either $t = \frac{1}{9}$ or $t = 1$.

Check $\frac{1}{9}$: $3t - 1 = \sqrt{4t}$

$$3\left(\frac{1}{9}\right) - 1 \stackrel{?}{=} \sqrt{4\left(\frac{1}{9}\right)}$$

$$\frac{1}{3} - 1 \stackrel{?}{=} \sqrt{\frac{4}{9}}$$

$$-\frac{2}{3} \neq \frac{2}{3}$$

Check 1: $3t - 1 = \sqrt{4t}$

$$3(1) - 1 \stackrel{?}{=} \sqrt{4(1)}$$

$$2 = 2$$

Therefore, $\frac{1}{9}$ is not a solution and 1 is a solution. ◀

Example 9 Solve the equation $r - 2 = \sqrt{r - 2}$.

Solution We square both sides and solve the resulting equation.

$$(r - 2)^2 = (\sqrt{r - 2})^2$$

$$r^2 - 4r + 4 = r - 2$$

$$r^2 - 5r + 6 = 0$$

$$(r - 2)(r - 3) = 0 \qquad \text{Factor.}$$

Therefore, either $r = 2$ or $r = 3$.

Check 2: $r - 2 = \sqrt{r - 2}$

$$2 - 2 \stackrel{?}{=} \sqrt{2 - 2}$$

$$0 \stackrel{?}{=} \sqrt{0}$$

$$0 = 0$$

Check 3: $r - 2 = \sqrt{r - 2}$

$$3 - 2 \stackrel{?}{=} \sqrt{3 - 2}$$

$$1 \stackrel{?}{=} \sqrt{1}$$

$$1 = 1$$

Therefore, both 2 and 3 are solutions of the original equation. ◀

In the next example, we have an equation where all the radicals cannot be eliminated by applying the squaring property. To obtain a radical-free equation, we must isolate the remaining radical and square again.

Example 10 Solve the equation $\sqrt{1 - 3z} = \sqrt{3z} + 1$.

Solution We begin by squaring both sides. When we do this, we eliminate the radical on the left side, but still have a radical on the right side.

$$\sqrt{1 - 3z} = \sqrt{3z} + 1$$
$$(\sqrt{1 - 3z})^2 = (\sqrt{3z} + 1)^2$$
$$1 - 3z = 3z + 2\sqrt{3z} + 1$$

Next, we isolate the remaining radical and square again.

$$-6z = 2\sqrt{3z}$$
$$-3z = \sqrt{3z} \quad \text{Divide both sides by 2.}$$
$$(-3z)^2 = (\sqrt{3z})^2$$
$$9z^2 = 3z$$
$$9z^2 - 3z = 0$$
$$3z(3z - 1) = 0$$

Therefore, either $z = 0$ or $z = \frac{1}{3}$.

Next, we check our two answers in the original equation.

Check 0: $\sqrt{1 - 3z} = \sqrt{3z} + 1$

$$\sqrt{1 - 3(0)} \stackrel{?}{=} \sqrt{3(0)} + 1$$
$$1 = 1$$

Check $\frac{1}{3}$: $\sqrt{1 - 3z} = \sqrt{3z} + 1$

$$\sqrt{1 - 3\left(\frac{1}{3}\right)} \stackrel{?}{=} \sqrt{3\left(\frac{1}{3}\right)} + 1$$
$$\sqrt{1 - 1} \stackrel{?}{=} \sqrt{1} + 1$$
$$0 \neq 2$$

Therefore, 0 is the only solution.

Certain equations involving radicals where the index is greater than 2 can be solved by raising both sides to a power equal to the index. We need to check for extraneous solutions only when the power is even. However, it is always a good idea to check solutions in the original equation.

A strategy for solving a radical equation follows. Keep this strategy in mind as you solve the equations in the exercise set.

STRATEGY

Solving a Radical Equation

Step 1 If necessary, isolate a radical on one side of the equation.

Step 2 Raise both sides of the equation to the power equal to the index. If the resulting equation is free of radicals, go to Step

STRATEGY (*continued*)

3. If the resulting equation is not free of radicals, go to Step 1.

Step 3 Solve the resulting equation.

Step 4 Check all solutions from Step 3 in the original equation.

WARMING UP

Answer true or false.

1. If $\sqrt{x} = 9$, then $x = 3$.

2. The equation $\sqrt{x} = -1$ has no solution.

3. The solution set of $\sqrt{-x} = 4$ is $\varnothing$.

4. -1 is the solution of $\sqrt{x + 2} = 1$.

5. The solution of $\sqrt{x} = 2\sqrt{3}$ is 12.

6. $x^2 = 16$ and $\sqrt{x} = 2$ have the same solution set.

EXERCISE SET 8.5

For Exercises 1–32, solve using the strategy for solving a radical equation. If an equation has no solution, write "no solution."

1. $\sqrt{x} = 4$

2. $\sqrt{y} = 5$

3. $\sqrt{a} = -1$

4. $\sqrt{z} = -\dfrac{2}{3}$

5. $\sqrt{z - 1} = 3$

6. $\sqrt{c + 2} = 1$

7. $\sqrt{s - 1} = 1$

8. $\sqrt{x + 1} = 2$

9. $4 + 3\sqrt{s} = 13$

10. $5 + 2\sqrt{y} = 9$

11. $7 = \sqrt{p + 12}$

12. $5 = \sqrt{r - 2}$

13. $\sqrt{3u + 1} = \dfrac{1}{2}$

14. $\sqrt{4b + 2} = \dfrac{2}{3}$

15. $\sqrt{10c} = 2\sqrt{3}$

16. $\sqrt{6r} = 3\sqrt{5}$

17. $\sqrt{-2x} = 4x + 1$

18. $\sqrt{3t} = 2 - 3t$

19. $\sqrt{x + 3} = x + 1$

20. $2a + 1 = \sqrt{6a + 13}$

21. $2\sqrt{3y} + 1 = 7$

22. $3\sqrt{2s} + 1 = 5$

23. $x - 1 = \sqrt{x^2 + 3}$

24. $z + 4 = \sqrt{z^2 + 10}$

25. $2 - t = \sqrt{t^2 + 2}$

26. $4 - v = \sqrt{v^2 + 6}$

27. $3x + 2 = \sqrt{21x + 2}$

28. $a + 2 = \sqrt{8a + 1}$

29. $\sqrt{4w + 1} - 1 = 2w$

30. $3 = 4v + \sqrt{3 - 2v}$

31. $\sqrt{3 - b} + b = 3$

32. $7z = \sqrt{7z + 2} - 2$

SAY IT IN WORDS

33. What is the squaring property?

34. Explain your strategy for solving a radical equation.

REVIEW EXERCISES

The following exercises review parts of Sections 4.1 and 4.2. Doing these exercises will help prepare you for the next section.

For Exercises 35–42, simplify. Write your answer without negative exponents.

35. $\frac{x^3}{x^2}$

36. $r^5 \cdot r$

37. $(xy)^4$

38. $(r^2t)^2$

39. a^{-5}

40. $(b^{-3})^{-2}$

41. $(c^3d^{-1})^{-1}$

42. $\left(\frac{p^2}{q^3}\right)^4$

ENRICHMENT EXERCISES

Solve the given equation.

1. $\sqrt{9 - 3x} = \sqrt{3x} - 3$

2. $\sqrt{16 - 5x} = 4 - \sqrt{x}$

3. $\sqrt[3]{x - 5} = 2$

4. $\sqrt[4]{x - 2} = 1$

Answers to Enrichment Exercises begin on page A.1.

8.6 Rational Exponents and Radicals

OBJECTIVES

- ▶ *To convert from radical notation to exponent notation and vice versa*
- ▶ *To evaluate numbers of the form $a^{1/n}$*
- ▶ *To evaluate numbers of the form $a^{m/n}$*

There is another notation to use for nth roots. This involves the use of exponents. In Chapter 1, we defined a^n where n is an integer. Let us now consider a power such as $8^{1/3}$. If the property for exponents $(a^m)^n = a^{mn}$ still holds, then

$$(8^{1/3})^3 = 8^{(1/3)3} = 8^1 = 8$$

This tells us that the number $8^{1/3}$ cubed is 8 and thus

$$8^{1/3} = \sqrt[3]{8}.$$

Therefore, we define exponents of the form $\frac{1}{n}$, where n is a positive integer in terms of nth roots.

DEFINITION

If a is a real number and n is a positive integer, then,

$$a^{1/n} = \sqrt[n]{a}, \qquad \text{where } a \geq 0 \text{ when } n \text{ is even}$$

If a is positive and n is even, recall that there are two nth roots of a. The symbol $a^{1/n}$ represents the positive or principal nth root of a.

Using this definition, we can evaluate numbers raised to powers of the form $\frac{1}{n}$.

Example 1 Evaluate each expression.

(a) $64^{1/3} = \sqrt[3]{64} = 4$

(b) $100^{1/2} = \sqrt{100} = 10$

(c) $(-16)^{1/4} = \sqrt[4]{-16}$ is not a real number.

(d) $(-32)^{1/5} = \sqrt[5]{-32} = -2$

(e) $\left(\frac{125}{27}\right)^{1/3} = \sqrt[3]{\frac{125}{27}} = \frac{5}{3}$

We now define the more general power $a^{m/n}$, where $\frac{m}{n}$ is any rational number.

DEFINITION

If m and n are positive integers with $\frac{m}{n}$ written in lowest terms, then

$$a^{m/n} = (a^{1/n})^m$$

provided that $a^{1/n}$ exists. If $a^{1/n}$ does not exist, then $a^{m/n}$ does not exist.

We define exponents of negative rational numbers by the equation

$$a^{-m/n} = \frac{1}{a^{m/n}}, \qquad a \neq 0$$

provided that $a^{m/n}$ exists.

Example 2

(a) $16^{3/2} = (16^{1/2})^3 = (\sqrt{16})^3 = 4^3 = 64$

(b) $27^{2/3} = (27^{1/3})^2 = (\sqrt[3]{27})^2 = 3^2 = 9$

(c) $-9^{5/2} = -(9^{5/2}) = -(9^{1/2})^5 = -(\sqrt{9})^5 = -3^5 = -243$

(d) $(-64)^{4/3} = [(-64)^{1/3}]^4 = [\sqrt[3]{-64}]^4 = [-4]^4 = 256$

(e) $(-25)^{7/2}$ is not a real number since $(-25)^{1/2}$ is not a real number.

(f) $81^{-3/4} = \frac{1}{81^{3/4}} = \frac{1}{(81^{1/4})^3} = \frac{1}{(\sqrt[4]{81})^3} = \frac{1}{3^3} = \frac{1}{27}$

(g) $(-8)^{-2/3} = \frac{1}{(-8)^{2/3}} = \frac{1}{[(-8)^{1/3}]^2} = \frac{1}{(\sqrt[3]{-8})^2} = \frac{1}{(-2)^2} = \frac{1}{4}$ ◀

As seen in Example 2, given a rational exponent m/n, the denominator indicates the root to be taken and the numerator indicates the power. For example,

power

root

$$32^{3/5} = (\sqrt[5]{32})^3 = 2^3 = 8$$

In Chapter 4, we developed properties of exponents where the exponents were integers. Using the definition of rational exponents, it can be shown that all the properties still hold for rational exponents. We restate the properties of exponents without proving them.

STRATEGY

Exponents

Suppose a and b are real numbers, r and s rational numbers, where a and b are positive when even roots are taken. Then

1. $a^r a^s = a^{r+s}$

2. $(a^r)^s = a^{rs}$

3. $(ab)^r = a^r b^r$

4. $\left(\frac{a}{b}\right)^r = \frac{a^r}{b^r}, b \neq 0$

5. $\frac{a^r}{a^s} = a^{r-s}, a \neq 0$

We have defined $a^{m/n}$ to mean $(a^{1/n})^m$. However, using the properties of exponents, we have an alternate way of evaluating $a^{m/n}$.

$$a^{m/n} = a^{m(1/n)}$$
$$= (a^m)^{1/n} \quad \text{Property 2.}$$

For example, we can evaluate $8^{2/3}$ in two different ways.

First way: $8^{2/3} = (8^{1/3})^2 = 2^2 = 4$
Second way: $8^{2/3} = (8^2)^{1/3} = 64^{1/3} = 4$

In the next three examples, we illustrate how the properties of exponents can be used to simplify expressions.

Example 3 Use the properties of exponents to simplify. Assume all variables represent positive real numbers. Write the answer with only positive exponents.

(a) $3^{1/4} \cdot 3^{1/2} = 3^{1/4+1/2}$
$= 3^{1/4+2/4}$
$= 3^{3/4}$

(b) $\dfrac{7^{3/5}}{7^{9/5}} = 7^{3/5-9/5}$
$= 7^{-6/5}$
$= \dfrac{1}{7^{6/5}}$

(c) $\dfrac{(a^{2/3}b^{1/2})^6}{a^2b} = \dfrac{a^{(2/3)6}b^{(1/2)6}}{a^2b}$
$= \dfrac{a^4b^3}{a^2b}$
$= a^{4-2}b^{3-1}$
$= a^2b^2$

(d) $\left(\dfrac{a^{-5/2}}{b^{3/4}}\right)^8 = \dfrac{(a^{-5/2})^8}{(b^{3/4})^8}$
$= \dfrac{a^{-20}}{b^6}$
$= \dfrac{1}{a^{20}b^6}$ ◀

Example 4 Replace all radicals with rational exponents and simplify. Assume all variables represent positive real numbers.

(a) $\sqrt[4]{x^3} = (x^3)^{1/4}$
$= x^{3/4}$

(b) $\dfrac{\sqrt[3]{a^5}}{\sqrt{a^3}} = \dfrac{(a^5)^{1/3}}{(a^3)^{1/2}}$
$= \dfrac{a^{5/3}}{a^{3/2}}$
$= a^{5/3-3/2}$
$= a^{10/6-9/6}$
$= a^{1/6}$

(c) $\sqrt[3]{\sqrt{a}} = \sqrt[3]{a^{1/2}}$
$= (a^{1/2})^{1/3}$
$= a^{1/6}$ ◀

WARMING UP

Answer true or false.

1. $9^{1/2} = \pm 3$

2. $49^{1/2} = 7$

3. $1^{-1/2}$ is not a real number.

4. $9^{-1/2} = \frac{1}{3}$

5. $4^{3/2} = 8$

6. $1^{-3/2} = 1$

EXERCISE SET 8.6

For Exercises 1–16, evaluate each expression.

1. $100^{1/2}$ **2.** $64^{1/2}$ **3.** $(-1)^{1/5}$ **4.** $(-8)^{1/3}$

5. $(-12)^{1/2}$ **6.** $(-24)^{1/4}$ **7.** $16^{3/2}$ **8.** $9^{5/2}$

9. $(-32)^{2/5}$ **10.** $(-8)^{4/3}$ **11.** $\left(\frac{1}{27}\right)^{2/3}$ **12.** $(-25)^{3/2}$

13. $-25^{3/2}$ **14.** $\left(\frac{1}{32}\right)^{-1/5}$ **15.** $\left(\frac{1}{16}\right)^{-1/4}$ **16.** $\left(-\frac{1}{32}\right)^{-2/5}$

For Exercises 17–44, use properties of exponents to simplify. Assume the variables represent positive real numbers. Write the answer using only positive exponents.

17. $4^{2/3} \cdot 4^{1/6}$ **18.** $5^{3/8} \cdot 5^{1/4}$ **19.** $t^{3/2} \cdot t^{-5/2}$ **20.** $z^{-3/4} \cdot z^{-7/4}$

21. $a^{1/12} \cdot a^{-1/12}$ **22.** $x^{-3/8} \cdot x^{3/8}$ **23.** $\frac{1}{a^{-1/2}}$ **24.** $\frac{2}{z^{-2/3}}$

25. $\frac{s^{4/5}}{s^{2/5}}$ **26.** $\frac{v^{5/4}}{v^{3/4}}$ **27.** $x^{1/3} \cdot x$ **28.** $b^{3/2} \cdot b^{5/2}$

29. $a^{4/5} \cdot a^{6/5}$ **30.** $d^{4/5} \cdot d^{1/5}$ **31.** $w^{3/4} \cdot w^{1/4}$ **32.** $\frac{s^{3/5}}{s^{1/10}}$

33. $(a^{3/2})^{4/9}$ **34.** $(z^{5/2})^{4/15}$ **35.** $\frac{z^2}{z^{9/4}}$ **36.** $\frac{c^{3/2}}{c^{9/4}}$

37. $(x^{3/7})^{-21/2}$ **38.** $(8x^6y^2)^{1/3}$ **39.** $(4t^8r^4)^{1/2}$ **40.** $(4^{3/10}d^{-2/5})^{-5}$

41. $(3s^{1/2}t^{5/4})^2$ **42.** $(-2x^{2/3}y^2)^3$ **43.** $\left(\frac{a^{-2/7}}{b^{1/7}}\right)^{14}$ **44.** $\left(\frac{m^{1/3}}{n^{-2/3}}\right)^6$

For Exercises 45–48, replace all radicals with rational exponents and simplify. Assume all variables represent positive real numbers.

45. $\sqrt[3]{x^5}$ **46.** $\sqrt[4]{a^3}$ **47.** $\sqrt[3]{\sqrt{z^5}}$ **48.** $\sqrt[4]{\sqrt[3]{x^5}}$

A calculator can be used to find powers where the exponent is a rational number. For example, to find $1.4^{1/3}$, follow the sequence:

$$1.4 \boxed{y^x}\ 3\ \boxed{1/x}\ \boxed{=}$$

or alternately:

$$1.4 \boxed{y^x}\ \boxed{(}\ 1\ \boxed{\div}\ 3\ \boxed{)}\ \boxed{=}$$

Either method gives $1.4^{1/3} \approx 1.1187$.

Here is another example. To find $52^{2/3}$, follow the sequence:

$$52 \boxed{y^x}\ \boxed{(}\ 2\ \boxed{\div}\ 3\ \boxed{)}\ \boxed{=}$$

The answer, accurate to two decimal places: $52^{2/3} \approx 13.93$.

For Exercises 49–52, find each.

49. $842^{1/3}$ **50.** $5.92^{1/4}$ **51.** $190^{4/5}$ **52.** $8262^{-3/8}$

SAY IT IN WORDS

53. Explain how to evaluate $a^{1/n}$, where a is a real number and n is a positive integer and $a \geq 0$ when n is even.

54. Is $(-4)^{-1/2}$ a real number? Why?

REVIEW EXERCISES

The following exercises review parts of Section 8.4. Doing these problems will help prepare you for the next section.

Multiply and then simplify.

55. $\sqrt{3}(\sqrt{2} - \sqrt{5})$ **56.** $(\sqrt{x} - 2)(3\sqrt{x} - 1)$ **57.** $(\sqrt{a} - 3)^2$ **58.** $(\sqrt{z} - \sqrt{2})(\sqrt{z} + \sqrt{2})$

ENRICHMENT EXERCISES

Simplify each expression using the properties of exponents.

1. $\left(\frac{x^{-2/5}}{y^{3/10}}\right)^{15}$ **2.** $\left(\frac{a^{3/2}}{b^{-4/3}}\right)^{12}$ **3.** $\frac{\sqrt[4]{x^5}}{\sqrt[3]{x^4}}$ **4.** $\frac{\sqrt[3]{x^4}}{\sqrt[5]{x^3}}$

Answers to Enrichment Exercises begin on page A.1.

8.7 Complex Numbers

OBJECTIVES

- *To write the square roots of negative numbers in terms of i*
- *To learn the algebra of complex numbers*

The equation $x^2 = -9$ has no solution in the set of real numbers, since any real number squared is nonnegative. Therefore, we need a set of numbers, called *complex numbers,* so that every quadratic equation will have solutions. We start with the following definition.

DEFINITION

The number $\boldsymbol{i}$ is defined as a square root of -1.

$$i = \sqrt{-1}$$

Since i is a square root of -1,

$$i^2 = -1$$

Notice that i is not a real number, since its square is negative. This contradicts the real number property that the product of a number with itself is nonnegative. Any nonzero real number multiple of i is called an **imaginary number.** Examples of imaginary numbers are the following:

$$-i, \quad 12i, \quad -\frac{3}{4}i, \quad i\sqrt{2}, \quad 5i\sqrt{15}$$

We can rewrite the square root of a negative number as an imaginary number. For example, if a is a positive real number, the number $\sqrt{-a}$ must have the property that its square is $-a$,

$$(\sqrt{-a})^2 = -a$$

Now, consider the imaginary number $i\sqrt{a}$. Keeping in mind that $i^2 = -1$,

$$\begin{aligned} (i\sqrt{a})^2 &= (i\sqrt{a})(i\sqrt{a}) \\ &= i^2(\sqrt{a})^2 && \text{Rearrange factors.} \\ &= (-1)a && i^2 = -1. \\ &= -a \end{aligned}$$

Therefore, both $\sqrt{-a}$ and $i\sqrt{a}$ are square roots of $-a$, and

$$\sqrt{-a} = i\sqrt{a}$$

Example 1 Write each expression as an imaginary number.

(a) $\sqrt{-5} = i\sqrt{5}$

(b) $-\sqrt{-121} = -i\sqrt{121}$

$$= -i \cdot 11$$

$$= -11i$$

(c) $-\sqrt{-15} = -i\sqrt{15}$

(d) $\sqrt{-24} = i\sqrt{24}$

$$\begin{aligned} &= i\sqrt{4 \cdot 6} && \text{Factor 24 into } 4 \cdot 6. \\ &= i\sqrt{4}\sqrt{6} && \sqrt{a \cdot b} = \sqrt{a}\sqrt{b}. \\ &= i \cdot 2\sqrt{6} \\ &= 2i\sqrt{6} \end{aligned}$$

◄

NOTE *It is common practice to write i before the radical in imaginary numbers, such as $\sqrt{2}i$, so that we do not mistake $\sqrt{2}i$ for $\sqrt{2i}$. Therefore, we write $\sqrt{2}i$ as $i\sqrt{2}$.*

We now define the set of complex numbers.

DEFINITION

A **complex number** is any number that can be put in the form

$$a + bi$$

where a and b are real numbers.

The quantity $a + bi$ is called the **standard form** for a complex number.

Example 2 Express each complex number in the standard form $a + bi$.

(a) $3 - \sqrt{-121} = 3 - i\sqrt{121}$
$= 3 - 11i$

(b) $\sqrt{-20} = i\sqrt{20}$
$= (2\sqrt{5})i$
$= 0 + (2\sqrt{5})i$

(c) $6 = 6 + 0i$

(d) $0 = 0 + 0i$

Since any real number a is of the form $a + 0i$, the set of real numbers is a subset of the set of complex numbers. Recall in Chapter 1, we studied the set of real numbers along with its subsets of integers and rational and irrational numbers. Now, with the complex numbers, we have the following relationships existing in the family tree of numbers.

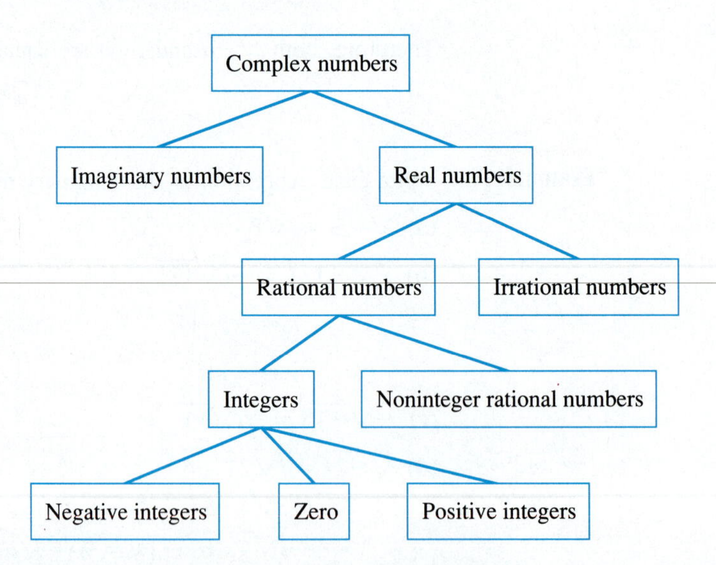

An expression containing a sum or difference of two complex numbers can be simplified by combining like terms.

Example 3 Add or subtract as indicated.

(a) $(4 - 2i) + (2 + 5i) = (4 + 2) + (-2 + 5)i$
$= 6 + 3i$

(b) $(9 - 4i) - (5 - 2i) = (9 - 5) + (-4 + 2)i$
$= 4 - 2i$

(c) $(-10 + \sqrt{-3}) - (-7 + \sqrt{-12}) = (-10 + i\sqrt{3}) - (-7 + 2i\sqrt{3})$ $\quad \sqrt{-a} = i\sqrt{a}.$
$= -3 - i\sqrt{3}$ ◀

Multiplying complex numbers is done in the same way as multiplying polynomials. The answer may be simplified by using the fact that $i^2 = -1$.

Example 4 Multiply: $3i(7 - 2i)$.

Solution We start by using the distributive property.

$$
\begin{aligned}
3i(7 - 2i) &= 3i(7) + 3i(-2i) \\
&= 21i - 6i^2 \\
&= 21i - 6(-1) && i^2 = -1. \\
&= 21i + 6 \\
&= 6 + 21i && \text{Write in standard form.}
\end{aligned}
$$
◀

Example 5 Multiply: $(3 - 2i)(5 + 6i)$.

Solution We multiply each term in the second complex number by each term in the first and then simplify the result.

$$
\begin{aligned}
(3 - 2i)(5 + 6i) &= 3 \cdot 5 + 3 \cdot 6i + (-2i)5 + (-2i)(6i) \\
&= 15 + 18i - 10i - 12i^2 \\
&= 15 + 18i - 10i - 12(-1) && i^2 = -1. \\
&= (15 + 12) + (18i - 10i) \\
&= 27 + 8i
\end{aligned}
$$

Therefore, the product of $3 - 2i$ and $5 + 6i$ is the complex number $27 + 8i$. ◀

In the next two examples, we show how to use formulas, such as $(a + b)^2 = a^2 + 2ab + b^2$, to multiply complex numbers.

Example 6 Find $(3 - 5i)^2$.

Solution Using the formula $(a - b)^2 = a^2 - 2ab + b^2$,

$$
\begin{aligned}
(3 - 5i)^2 &= 3^2 - 30i + (5i)^2 \\
&= 9 - 30i - 25 \\
&= -16 - 30i
\end{aligned}
$$
◀

Example 7 Multiply: $(4 + 3i)(4 - 3i)$.

Solution Recall the formula for binomials: $(a + b)(a - b) = a^2 - b^2$. Letting $a = 4$ and $b = 3i$,

$$\begin{aligned}(4 + 3i)(4 - 3i) &= 4^2 - (3i)^2 \\ &= 16 + 9 \\ &= 25\end{aligned}$$

Notice in Example 7 that the product of the two complex numbers $4 + 3i$ and $4 - 3i$ is the real number 25. The two numbers $4 + 3i$ and $4 - 3i$ are called *complex conjugates.*

DEFINITION

The complex numbers $a + bi$ and $a - bi$ are called **(complex) conjugates.**

Examples of complex conjugates are the following pairs of numbers.

Number	Complex Conjugate
$1 - 4i$	$1 + 4i$
$8 + 3i$	$8 - 3i$
$2 - \frac{2}{5}i$	$2 + \frac{2}{5}i$
14	14
$-i$	i

A valuable property of complex conjugates is the fact that their product is a real number.

THEOREM The complex conjugates $a + bi$ and $a - bi$ have the property that

$$(a + bi)(a - bi) = a^2 + b^2$$

Proof It is easy to verify this formula.

$$(a + bi)(a - bi) = a^2 - (bi)^2 = a^2 + b^2$$

For example,

$$(1 + 8i)(1 - 8i) = 1^2 + 8^2 = 65$$

We use this theorem to convert division problems involving complex numbers into the standard form.

Example 8 Write $\frac{4 - 3i}{1 + 5i}$ in standard form.

Solution We want to find a complex number in the standard form $a + bi$ that is equal to the quotient $\frac{4 - 3i}{1 + 5i}$. To find this number, we multiply numerator and denominator by the complex conjugate of the denominator and then simplify.

$$\begin{aligned}\frac{4 - 3i}{1 + 5i} &= \frac{(4 - 3i)(1 - 5i)}{(1 + 5i)(1 - 5i)}\\ &= \frac{4 - 23i - 15}{1^2 + 5^2}\\ &= \frac{-11 - 23i}{26}\\ &= -\frac{11}{26} - \frac{23}{26}i\end{aligned}$$

Therefore, the division of $4 - 3i$ by $1 + 5i$ results in the complex number $-\frac{11}{26} - \frac{23}{26}i$.

WARMING UP

Answer true or false.

1. $\sqrt{-4} = \pm 2i$
2. $-\sqrt{-9} = -3i$
3. $1 + \sqrt{-1} = 1 + i$
4. $(1 + i)(1 - i) = 2$
5. $-\sqrt{9} = 3i$
6. $\frac{i}{1 - i} = -\frac{1}{2} + \frac{1}{2}i$

EXERCISE SET 8.7

For Exercises 1–8, write each as an imaginary number.

1. $\sqrt{-9}$
2. $\sqrt{-64}$
3. $-\sqrt{-17}$
4. $-\sqrt{-21}$
5. $\sqrt{-28}$
6. $\sqrt{-32}$
7. $-\sqrt{-50}$
8. $-\sqrt{-27}$

For Exercises 9–14, express the complex number in the form $a + bi$.

9. $4 - \sqrt{-100}$
10. $-5 + \sqrt{-9}$
11. $2 + \sqrt{-8}$
12. $-1 + \sqrt{-24}$
13. $1 - \sqrt{20}$
14. $8 + \sqrt{12}$

For Exercises 15–34, perform the indicated operations and write the answer in the form $a + bi$.

15. $(4 - 2i) + (6 + 3i)$

16. $(7 + 3i) + (-1 - 5i)$

17. $(5 - 2i) - (15 - 2i)$

18. $(7 - 4i) - (8 - 11i)$

19. $(4 - 2i) - 6i$

20. $(5 - 3i) - 7i$

21. $(1 + 2i)(2 - 5i)$

22. $3i(7 - 3i)$

23. $(3 - 2i)^2$

24. $(1 + 6i)^2$

25. $(1 + 4i)(1 - 4i)$

26. $\dfrac{2 - 3i}{1 - 2i}$

27. $\dfrac{2}{1 + i}$

28. $\dfrac{1}{1 - i}$

29. $\dfrac{1 - 3i}{i}$

30. $\dfrac{7 - 8i}{i}$

31. $\dfrac{2 - i}{3i}$

32. $\dfrac{3 - 12i}{4i}$

33. $\dfrac{i}{1 - 2i}$

34. $\dfrac{3i}{1 + 2i}$

SAY IT IN WORDS

35. What is a complex number?

36. Explain your method to convert a fraction of complex numbers to standard form.

REVIEW EXERCISES

The following exercises review parts of Section 5.6. Doing these exercises will help prepare you for the next section.

Solve each equation.

37. $(5t - 1)(3t + 7) = 0$

38. $(4z + 6)(8z - 3) = 0$

39. $x^2 - 49 = 0$

40. $z^2 = 121$

ENRICHMENT EXERCISES

Multiply.

1. $(1 - i)^3$

2. $(1 + i)^3$

Answers to Enrichment Exercises begin on page A.1.

CHAPTER 8

Summary and review

Examples

Square roots (8.1)

A **square root** of a given number is any number whose square is the given number. Any positive real number a has two square roots. The **positive square root** is given by $\sqrt{a}$, and the **negative square root** is given by $-\sqrt{a}$.

Both square roots of a, when squared, give a:

$$(\sqrt{a})^2 = a \quad \text{and} \quad (-\sqrt{a})^2 = a, \text{ for } a > 0$$

The square roots of 64 are 8 and -8.

$\sqrt{64} = 8$ and $-\sqrt{64} = -8$

The number zero has one square root, $\sqrt{0} = 0$.

A negative number has no real square roots.

Higher roots (8.1)

Let n be a positive integer. A number b is an ***n*th root** of a if $b^n = a$.

The symbol $\sqrt[n]{a}$ means the nth root of a. The number n is the **index** and the number a is the **radicand.**

We have $(\sqrt[n]{a})^n = a$ where $a \geq 0$ when n is even.

Two real fourth roots of 81:

$\sqrt[4]{81} = 3$ and $-\sqrt[4]{81} = -3$

since $3^4 = 81$ and $(-3)^4 = 81$.

Also,

$\sqrt[3]{-8} = -2$

since $(-2)^3 = -8$.

Simplifying radical expressions (8.2)

The following rules are used to simplify real radical expressions.

The **product rule** for radicals: $\sqrt{a \cdot b} = \sqrt{a} \cdot \sqrt{b}$

The **quotient rule** for radicals: $\sqrt{\dfrac{a}{b}} = \dfrac{\sqrt{a}}{\sqrt{b}}$

The product and quotient rules work also for higher roots.

A radical expression is **simplified** if

1. The radicand contains no power greater than or equal to the index.
2. The radicand has no fractions.
3. There are no radicals in denominators.

$$\sqrt{\frac{9}{7}} = \frac{\sqrt{9}}{\sqrt{7}} = \frac{3}{\sqrt{7}} = \frac{3\sqrt{7}}{\sqrt{7}\sqrt{7}} = \frac{3\sqrt{7}}{7}$$

Rationalizing the denominator (8.2 and 8.4)

If an expression contains a radical in the denominator, we **rationalize the denominator.** There are two cases.

Case 1. There is one term in the denominator.

Examples

$$\frac{2}{\sqrt{3}} = \frac{2\sqrt{3}}{\sqrt{3}\sqrt{3}} = \frac{2\sqrt{3}}{3}$$

Case 2. There is the sum or difference of terms in the denominator.

$$\frac{4}{5-\sqrt{2}} = \frac{4(5+\sqrt{2})}{(5-\sqrt{2})(5+\sqrt{2})} = \frac{4(5+\sqrt{2})}{5^2-(\sqrt{2})^2} = \frac{4(5+\sqrt{2})}{23}$$

Addition and subtraction of radical expressions (8.3)

We add and subtract radical expressions by combining like radicals.

$$4\sqrt[3]{5} - 7\sqrt[3]{5} = -3\sqrt[3]{5}$$

$$5\sqrt{12} + 2\sqrt{3} = 5 \cdot 2\sqrt{3} + 2\sqrt{3} = 10\sqrt{3} + 2\sqrt{3} = 12\sqrt{3}$$

Products and quotients of radical expressions (8.4)

We can multiply radical expressions by using the distributive property and then simplifying.

$$\sqrt{5}(4 - 2\sqrt{10}) = 4\sqrt{5} - 2\sqrt{5}\sqrt{10} = 4\sqrt{5} - 10\sqrt{2}$$

Solving radical equations (8.5)

To solve radical equations, we use the **squaring property:** We may square both sides of an equation provided we check the resulting solutions in the original equation.

$$\sqrt{3x-2} = 5$$

$$(\sqrt{3x-2})^2 = 5^2$$

$$3x - 2 = 25$$

$$x = 9$$

Check:

$$\sqrt{3x-2} = 5$$

$$\sqrt{3(9)-2} \stackrel{?}{=} 5$$

$$\sqrt{27-2} \stackrel{?}{=} 5$$

$$\sqrt{25} \stackrel{?}{=} 5$$

$$5 = 5$$

Therefore, $\{9\}$ is the solution set of $\sqrt{3x-2} = 5$.

Rational exponents (8.6)

Rational exponents are another way to indicate roots. If a is a real number and n is a positive integer, then

$$a^{1/n} = \sqrt[n]{a}, \quad \text{where } a \geq 0 \text{ when } n \text{ is even}$$

If m and n are positive integers, then

$$a^{m/n} = (a^{1/n})^m, \text{ where } a \geq 0 \text{ when } n \text{ is even}$$

We define

$$a^{-m/n} = \frac{1}{a^{m/n}}$$

All the properties of exponents still hold when the exponents are rational numbers.

$$a^r a^s = a^{r+s}, \qquad (a^r)^s = a^{rs}, \qquad (ab)^r = a^r b^r,$$

$$\left(\frac{a}{b}\right)^r = \frac{a^r}{b^r}, \qquad \frac{a^r}{a^s} = a^{r-s}$$

Using these properties, we can simplify expressions involving rational exponents.

Examples

$100^{1/2} = 10$

$(-8)^{1/3} = -2$

$16^{3/2} = (16^{1/2})^3 = 4^3 = 64$

$(-27)^{2/3} = [(-27)^{1/3}]^2 = [-3]^2 = 9$

$8^{-4/3} = \frac{1}{8^{4/3}} = \frac{1}{2^4} = \frac{1}{16}$

$x^{2/3}(3x^{1/6}) = 3x^{2/3+1/6} = 3x^{5/6}$

Complex numbers (8.7)

A **complex number** is any number that can be put in the form $a + bi$, where a and b are real numbers and $i = \sqrt{-1}$.

We **add and subtract** complex numbers by adding and subtracting like terms.

We **multiply** complex numbers using the same techniques for multiplying polynomials.

To write a division problem in the **standard form $a + bi$,** rationalize the denominator using the complex conjugate.

Examples

$12 - \sqrt{-20} = 12 - i\sqrt{20} = 12 - 2i\sqrt{5}$

$(4 - 2i) + (5 - i) = 9 - 3i$

$(5 + 7i) - (4 - 3i) = 1 + 10i$

$(2 - 5i)(3 + 4i) = 6 + 8i - 15i - 20i^2 = 26 - 7i$

$$\frac{1 - 4i}{2 + 3i} = \frac{(1 - 4i)(2 - 3i)}{(2 + 3i)(2 - 3i)} = \frac{-10 - 11i}{13} = -\frac{10}{13} - \frac{11}{13}i$$

CHAPTER 8 REVIEW EXERCISE SET

Assume all variables are positive.

Section 8.1

For Exercises 1–6, simplify each expression.

1. $\sqrt{64}$

2. $-\sqrt{\frac{9}{16}}$

3. $\sqrt[3]{-27}$

4. $-\sqrt[4]{-16}$

5. $\sqrt{9x^8y^4}$

6. $\sqrt[3]{-8a^6b^9c^{12}}$

Section 8.2

For Exercises 7–10, simplify.

7. $\sqrt{24}$

8. $\sqrt[3]{-16}$

9. $\sqrt[4]{z^8t^{12}}$

10. $-\sqrt{3s^7r^3}$

Sections 8.2 and 8.4

For Exercises 11–20, rationalize the denominator.

11. $\frac{2}{\sqrt{6}}$

12. $\frac{3}{\sqrt{x}}$

13. $\frac{4s}{\sqrt{2s}}$

14. $\frac{3k}{\sqrt[3]{k^2}}$

15. $\frac{4v}{3\sqrt[4]{v}}$

16. $\frac{3}{2-\sqrt{5}}$

17. $\frac{4}{1+\sqrt{3}}$

18. $\frac{\sqrt{x}-1}{\sqrt{x}+1}$

19. $\frac{\sqrt{s}+2\sqrt{t}}{2\sqrt{st}}$

20. $\frac{c-3}{\sqrt{c}-\sqrt{3}}$

Section 8.3

For Exercises 21–26, simplify by combining like radicals.

21. $4\sqrt{3}+5\sqrt{3}-2\sqrt{3}$

22. $3\sqrt[4]{32}+5\sqrt[4]{2}$

23. $2\sqrt{a^5}-5a\sqrt{a^3}$

24. $5\sqrt[3]{c^2}-19\sqrt[3]{c^2}$

25. $\sqrt[3]{u^4v^5}-4uv\sqrt[3]{uv^2}$

26. $2x\sqrt{x^3}-x^2\sqrt{x}+3\sqrt{x^5}$

Section 8.4

For Exercises 27–31, multiply and then simplify.

27. $\sqrt{3}(\sqrt{3}-2\sqrt{6})$

28. $\sqrt{s}(2\sqrt{s}-3\sqrt{s^3})$

29. $\sqrt[3]{x}(\sqrt[3]{x^5}-2\sqrt[3]{x^7})$

30. $(\sqrt{a}-2\sqrt{b})^2$

31. $(\sqrt[3]{9y^2}+2)(\sqrt[3]{9y^2}-2)$

Section 8.5

For Exercises 32–40, solve each equation.

32. $\sqrt{x} = 8$

33. $\sqrt{t} = 2\sqrt{3}$

34. $\sqrt{z + 3} = 2$

35. $\sqrt{4z^2 - 2z + 9} = 2z$

36. $\sqrt{2x - 3} = 3\sqrt{x}$

37. $3y - 1 = \sqrt{3y + 1}$

38. $\sqrt{2x} + 5 = \sqrt{3x + 25}$

39. $\sqrt{z + 1} = \sqrt{2z} - 1$

40. $\sqrt[3]{3s + 12} = -2$

Section 8.6

For Exercises 41–46, simplify.

41. $3^{2/3} \cdot 3^{5/6}$

42. $z^{-4/5} \cdot z^{3/10}$

43. $(x^{3/5})^{-15/9}$

44. $(s^{2/3}t^{1/6})^{12}$

45. $\dfrac{w^{2/5}}{w^{1/15}}$

46. $\dfrac{(a^{-2/7})^{14}}{a^3}$

Section 8.7

For Exercises 47–51, express each in the form $a + bi$.

47. $\sqrt{-100}$

48. $2 - \sqrt{-9}$

49. $4 + \sqrt{-40}$

50. $1 - \sqrt{20}$

51. $3 + \sqrt{-\dfrac{4}{9}}$

For Exercises 52–59, perform the indicated operations and write the answer in the form $a + bi$.

52. $(4 - 2i) + (6 - 4i)$

53. $(6 + 9i) - (-2 - 4i)$

54. $2i(1 - i)$

55. $(3 - 2i)(1 - 5i)$

56. $(3 - 5i) \div (1 - i)$

57. $\dfrac{4 + 9i}{2i}$

58. $\dfrac{6 + 2i}{1 + 3i}$

59. $(6 - 2i)^2$

CHAPTER 8 TEST

Assume all variables are positive.

For Problems 1–3, simplify each expression.

1. $\sqrt{144}$

2. $\sqrt[3]{-\dfrac{8}{27}}$

3. $\sqrt{25x^4y^2z^{10}}$

Simplify.

4. $-\sqrt{80}$

For Problems 5–7, rationalize the denominator.

5. $\dfrac{2}{\sqrt{7}}$ **6.** $\dfrac{3}{\sqrt{5}-2}$ **7.** $\dfrac{\sqrt{r}+3}{2\sqrt{r}+5}$

For Problems 8 and 9, simplify by combining like radicals.

8. $3\sqrt{48}-5\sqrt{3}+\sqrt{20}$ **9.** $2a\sqrt{a^3}-5\sqrt{a^5}+a^2$

For Problems 10 and 11, combine as a single fraction and simplify.

10. $\dfrac{3}{\sqrt{5}}-3$ **11.** $\dfrac{4\sqrt{5}}{\sqrt{8}}-\dfrac{5\sqrt{2}}{\sqrt{20}}$

For Problems 12–14, multiply and then simplify.

12. $2\sqrt{a}(\sqrt{a^3}-5a\sqrt{a})$ **13.** $(2\sqrt{x}-3\sqrt{y})^2$

14. $(3\sqrt{2}-5\sqrt{3})(4\sqrt{2}+3\sqrt{3})$

For Problems 15–17, solve each equation. Be sure to check your answers.

15. $\sqrt{w-3}=7$ **16.** $\sqrt{z^2-8}=4-z$ **17.** $\sqrt{5x-4}=3\sqrt{x}$

For Problems 18 and 19, simplify.

18. $x^{3/7}\cdot x^{4/7}$ **19.** $\left(\dfrac{u^{-3/4}}{u^{1/2}}\right)^4$

20. Express $3+\sqrt{-24}$ in the form $a+bi$.

For Problems 21–23, perform the indicated operations and write the answer in the form $a+bi$.

21. $(6-2i)-(3-4i)$ **22.** $(6-5i)(6+5i)$ **23.** $\dfrac{5i}{4-3i}$

Quadratic Equations

CHAPTER 9

CONNECTIONS

A zoo plans to build a rectangular rhinoceros mudhole. The length is to be 6 yards longer than the width. If the area must be 210 square yards, what are the dimensions of the mudhole? See Exercise 40 in Section 9.3.

Overview

One of the major topics of this book is studying techniques for solving equations. We have solved linear equations, factorable quadratic equations, rational equations, and radical equations. We return to the problem of solving the general quadratic equation $ax^2 + bx + c = 0$ and will develop methods for solving this equation when the trinomial is not factorable.

9.1 The Square Root Method to Solve Quadratic Equations

OBJECTIVES

- ▶ *To solve quadratic equations of the form $x^2 = a$*
- ▶ *To solve quadratic equations of the form $(ax + b)^2 = c$*

In Chapter 4, we solved quadratic equations such as

$$3x^2 + 7x + 2 = 0$$

by factoring. Many quadratic equations, however, cannot be factored but may still have real solutions. In this chapter, we will develop techniques of solving a general quadratic equation.

A quadratic equation like $x^2 = 49$ can be solved two ways:

Method 1. Solve it by the *factoring technique.*

$$x^2 - 49 = 0$$

$$(x - 7)(x + 7) = 0$$

Either $x = 7$ or $x = -7$.

Method 2. By using *square root properties* from the last chapter, $x^2 = 49$ means that either $x = \sqrt{49}$ or $x = -\sqrt{49}$. That is, either $x = 7$ or $x = -7$.

Notice that Method 2 is the faster way to solve $x^2 = 49$. The general solution of an equation of the form $x^2 = a$ is summarized as follows.

STRATEGY

The Square Root Method of Solving $x^2 = a$

Let a be a nonnegative constant. If

$$x^2 = a$$

then either

$$x = \sqrt{a} \quad \text{or} \quad x = -\sqrt{a}$$

The phrase "either $x = \sqrt{a}$ or $x = -\sqrt{a}$" can be written as $x = \pm\sqrt{a}$.

When we use this method, it is common to say that we are "taking square roots" of the equation $x^2 = a$. This method is not only faster than the factoring method but will also work even when the factoring method does not apply.

Example 1 Solve $z^2 = 32$.

Solution We use the square root method.

$$\begin{aligned} z^2 &= 32 && z^2 = a. \\ z &= \pm\sqrt{32} && z = \pm\sqrt{a}. \\ &= \pm\sqrt{16 \cdot 2} && \text{Factor 32 as } 16 \cdot 2. \\ &= \pm\sqrt{16} \cdot \sqrt{2} && \sqrt{a \cdot b} = \sqrt{a} \cdot \sqrt{b}. \\ &= \pm 4\sqrt{2} \end{aligned}$$

The solution set is $\{4\sqrt{2}, -4\sqrt{2}\}$. ◀

The square root method can be used on equations such as $(t + 7)^2 = 4$, if we think of $t + 7$ as a single quantity.

Example 2 Solve $(t + 7)^2 = 4$.

Solution Thinking of $t + 7$ as a single quantity, by the square root method,

$$\begin{aligned} t + 7 &= \pm\sqrt{4} \\ &= \pm 2 \end{aligned}$$

If $t + 7 = 2$, then $t = 2 - 7$ or -5.

If $t + 7 = -2$, then $t = -2 - 7$ or -9.

We check our two answers in the original equation.

Check -5: $(t + 7)^2 = 4$

$$\begin{aligned} (-5 + 7)^2 &\stackrel{?}{=} 4 \\ 2^2 &\stackrel{?}{=} 4 \\ 4 &= 4 \end{aligned}$$

Check -9: $(t + 7)^2 = 4$

$$\begin{aligned} (-9 + 7)^2 &\stackrel{?}{=} 4 \\ (-2)^2 &\stackrel{?}{=} 4 \\ 4 &= 4 \end{aligned}$$

The two solutions are -5 and -9. ◀

Example 3 Solve $(z - 1)^2 = 75$.

Solution Using the square root method,

$$z - 1 = \pm\sqrt{75}$$
$$= \pm\sqrt{25 \cdot 3}$$
$$= \pm 5\sqrt{3}$$

Therefore, $z = 1 \pm 5\sqrt{3}$.

Check $1 + 5\sqrt{3}$:

$$(z - 1)^2 = 75$$
$$(1 + 5\sqrt{3} - 1)^2 \stackrel{?}{=} 75$$
$$(5\sqrt{3})^2 \stackrel{?}{=} 75$$
$$5^2(\sqrt{3})^2 \stackrel{?}{=} 75$$
$$25 \cdot 3 \stackrel{?}{=} 75$$
$$75 = 75$$

Check $1 - 5\sqrt{3}$:

$$(z - 1)^2 = 75$$
$$(1 - 5\sqrt{3} - 1)^2 \stackrel{?}{=} 75$$
$$(-5\sqrt{3})^2 \stackrel{?}{=} 75$$
$$(-5)^2(\sqrt{3})^2 \stackrel{?}{=} 75$$
$$25 \cdot 3 \stackrel{?}{=} 75$$
$$75 = 75$$

The two solutions are $1 \pm 5\sqrt{3}$.

In the next example, we will rationalize the denominator of our answer for the final simplified form.

Example 4 Solve $3y^2 = 2$.

Solution Before using the square root method, we first divide both sides by 3.

$$3y^2 = 2$$
$$y^2 = \frac{2}{3}$$
$$y = \pm\sqrt{\frac{2}{3}} \quad \text{Apply the square root method.}$$
$$= \pm\frac{\sqrt{2}}{\sqrt{3}}$$
$$= \pm\frac{\sqrt{2}\sqrt{3}}{\sqrt{3}\sqrt{3}} \quad \text{Rationalize the denominator.}$$
$$= \pm\frac{\sqrt{6}}{3}$$

It is up to you to check these two answers.

The solutions are $\frac{\sqrt{6}}{3}$ and $-\frac{\sqrt{6}}{3}$.

We now consider some applications from geometry.

DEFINITION

An **isosceles** triangle is one that has two sides of equal length.

Example 5 Consider the isosceles right triangle as shown in the figure below. Denote the common length of the two equal sides by s and the length of the hypotenuse by c.

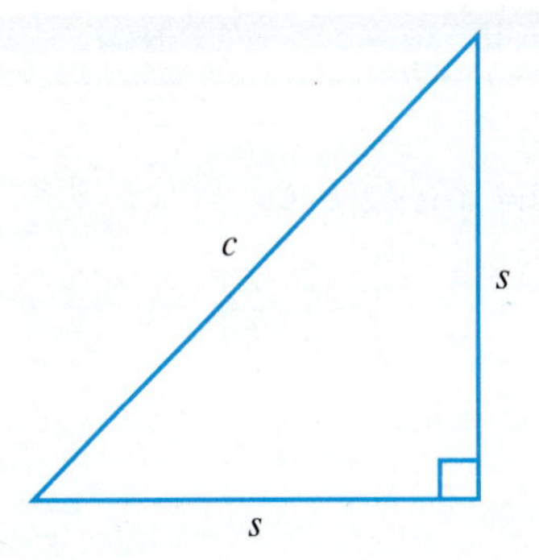

(a) Express s in terms of c.

(b) If the hypotenuse is 8 inches long, find s.

Solution **(a)** By the Pythagorean Theorem, $a^2 + b^2 = c^2$. For this particular right triangle, a and b are both equal to s. Therefore,

$$s^2 + s^2 = c^2$$
$$2s^2 = c^2$$
$$s^2 = \frac{c^2}{2}$$

We apply the square root method on this equation. We only consider the positive square root, since s represents a length and therefore must be positive.

$$s = \sqrt{\frac{c^2}{2}}$$

$$= \frac{\sqrt{c^2}}{\sqrt{2}}$$

$$= \frac{c}{\sqrt{2}} \qquad \sqrt{c^2} = c, \text{ since } c > 0.$$

$$= \frac{c\sqrt{2}}{\sqrt{2}\sqrt{2}} \qquad \text{Rationalize the denominator.}$$

$$= \frac{c\sqrt{2}}{2} \text{ or } \frac{\sqrt{2}}{2}c$$

(b) We replace c by 8 in the formula developed in Part (a).

$$s = \left(\frac{\sqrt{2}}{2}\right)c$$
$$= \frac{\sqrt{2}}{2}(8)$$
$$= 4\sqrt{2} \text{ inches}$$

WARMING UP

Answer true or false.

1. The solution set of $x^2 = 4$ is $\{2\}$.

2. -3 is a solution of $x^2 = 9$.

3. The solution set of $x^2 = \frac{4}{9}$ is $\left\{-\frac{2}{3}, \frac{2}{3}\right\}$.

4. $-\frac{1}{4}$ satisfies the equation $x^2 = \frac{1}{4}$.

5. $2\sqrt{2}$ is a solution of $x^2 = 8$.

6. If a number squared is 3, that number is $\sqrt{3}$.

EXERCISE SET 9.1

For Exercises 1–26, solve each equation. Remember to check your answers.

1. $z^2 = 9$

2. $b^2 = 100$

3. $x^2 = 121$

4. $u^2 = 36$

5. $w^2 = \frac{4}{9}$

6. $r^2 = \frac{16}{81}$

7. $d^2 = 18$

8. $x^2 = 28$

9. $t^2 = 125$

10. $u^2 = 50$

11. $(s - 6)^2 = 16$

12. $(t + 15)^2 = 9$

13. $y^2 = \frac{32}{9}$

14. $a^2 = \frac{45}{16}$

15. $(q + 2)^2 = 28$

16. $(u - 4)^2 = 72$

17. $5y^2 = 30$

18. $8w^2 = 56$

19. $10k^2 = 200$

20. $11z^2 = 198$

21. $(2z - 3)^2 = 81$

22. $(3y + 1)^2 = 100$

23. $\frac{(2c - 8)^2}{4} = 8$

24. $\frac{(6x + 12)^2}{3} = 12$

25. $(4 - s)^2 = 56$

26. $(9 - q)^2 = 48$

27. A positive number squared is 45. Find the number.

28. A negative number squared is 88. Find the number.

29. Twice a number is reduced by 3. The result, squared, is equal to 18. Find the number. (There are two answers.)

30. Five is added to three times a number. The result, squared, is equal to 20. Find the number. (There are two answers.)

For Exercises 31–34, from the given right triangle, express s in terms of c.

31.

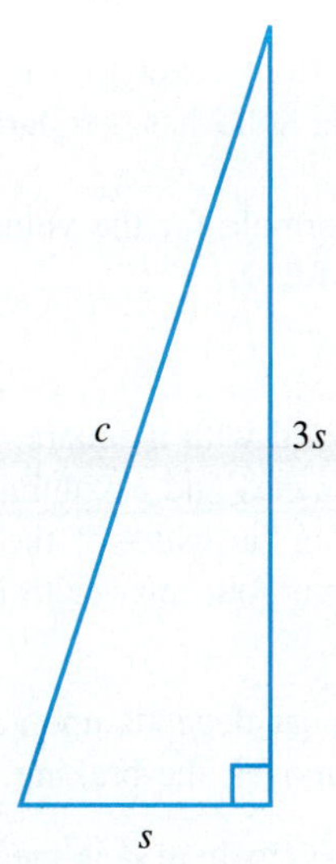

32.

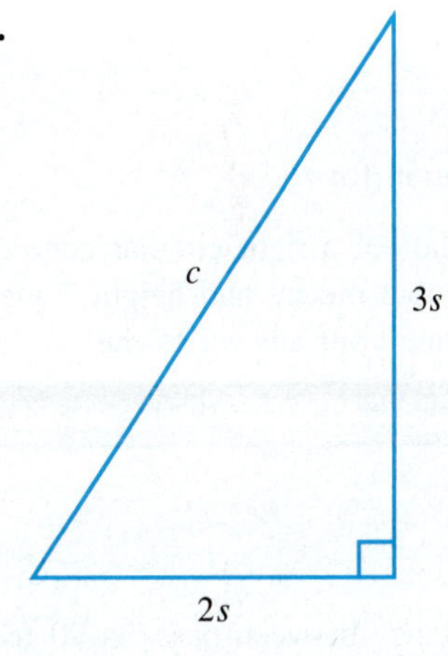

33.

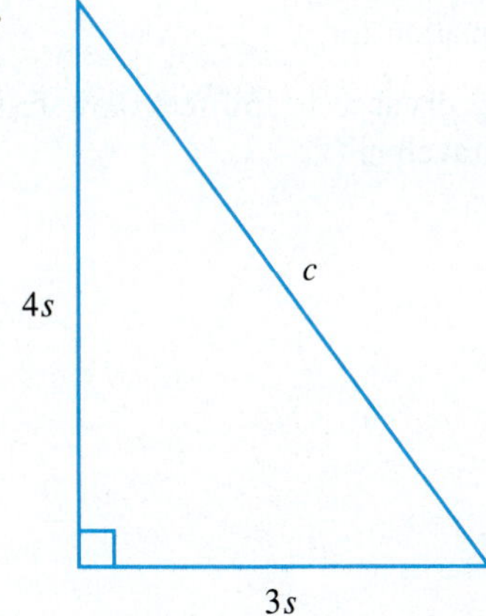

34.

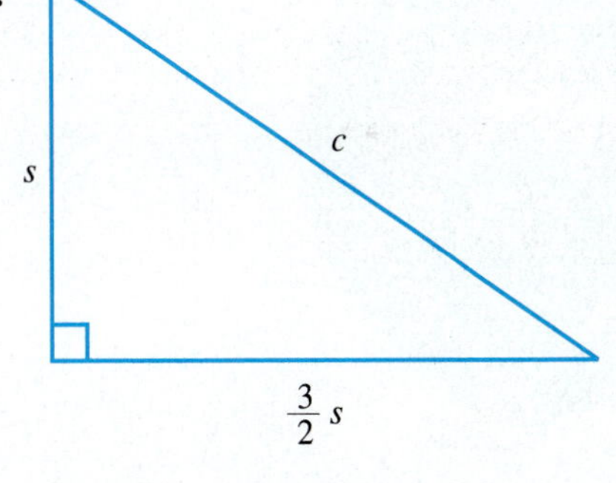

35. The area A of a circle of radius r is $A = \pi r^2$.

(a) Solve this equation for r.

(b) Find r, if the area is 20π square inches.

36. The volume V of a right circular cylinder of radius r and height h is $V = \pi r^2 h$.

(a) Solve this equation for r.

(b) What is the radius of a right circular cylinder of volume 450π cubic feet and height 10 feet?

37. The volume V of a right circular cone of radius r and height h is $V = \frac{1}{3}\pi r^2 h$. See the inside front cover for the figure.

(a) Solve this equation for r.

(b) What is the radius of a right circular cone of volume 252π cubic meters and height 7 meters? Approximate your answer to the thousandth.

38. The volume V of a rectangular solid is given by $V = lwh$, where l is the length, w the width, and h the height. See the inside front cover for the figure. Suppose a rectangular solid has a square base; that is, $l = w$.

(a) Replace l by w in the formula for the volume, then solve this formula for w.

(b) Suppose a rectangular solid with a square base has a height of 5 inches and a volume of 120 cubic inches. What is the width of the square base? Approximate your answer to the nearest thousandth.

39. In baseball, the distance between bases is 90 feet. If the catcher throws the ball from home plate to second base, how far must the ball travel?

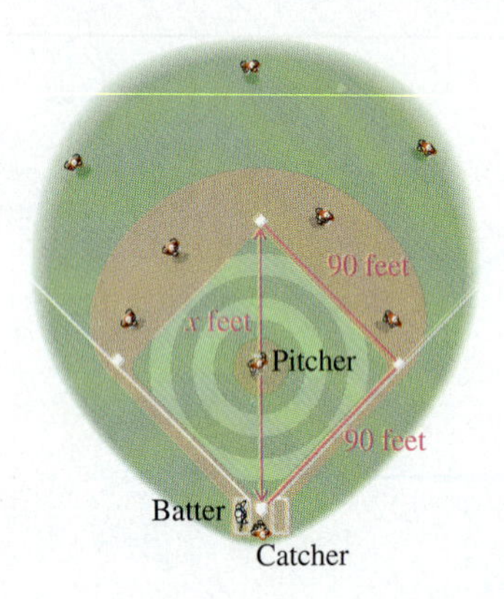

40. The braking distance d of a car depends upon its speed v. For one particular model, the braking distance is given by $d = \frac{3}{50}v^2$, where d is measured in feet and v in mph.

(a) Solve this equation for v.

(b) If the braking distance is 150 feet, how fast was the car traveling?

For Exercises 41 and 42, solve the equation using the square root method.

41. $(2.4x - 8.34)^2 = 30.8$

42. $(5.92v + 7.82)^2 = 12.6$

If P dollars is deposited in a savings account at an annual interest rate of r (expressed as a decimal) and compounded twice a year, then the amount A in the account at the end of one year is given by the formula

$$A = P\left(1 + \frac{r}{2}\right)^2$$

43. Solve this formula for r.

44. If $P = \$500$ and $A = \$530.45$, find r.

SAY IT IN WORDS

45. Explain the square root method of solving $x^2 = a$.

46. Give examples of isosceles triangles in the real world.

REVIEW EXERCISES

The following exercises review parts of Section 4.7. Doing these exercises will help prepare you for the next section.

Multiply.

47. $(x + 3)^2$

48. $(x + 7)^2$

49. $(x - \sqrt{2})^2$

50. $(y + \sqrt{5})^2$

ENRICHMENT EXERCISES

For Exercises 1–5, solve using the square root property.

1. $(x + 1)^2 = (2x - 3)^2$

2. $(4a - 5)^2 = (2a + 3)^2$

3. $(3w - 5)^2 = (5 - 3w)^2$

4. $(3t + 7)^2 = (8 + 3t)^2$

5. A television manufacturer wants to design a TV set in the 16:9 format with a 50-inch diagonal as shown in the figure. What must be the length and width of the screen?

Answers to Enrichment Exercises begin on page A.1.

9.2 Completing the Square to Solve Quadratic Equations

OBJECTIVE

▶ *To complete the square to solve a quadratic equation*

In the last section, we used the square root method to solve a quadratic equation where the left side was a perfect square. For example, in the equation $(x + 3)^2 = 16$, the left side, $(x + 3)^2$, is a perfect square—it is the square of the binomial $x + 3$. In this section, we analyze perfect squares of the form $(x + b)^2$ and $(x - b)^2$.

From our knowledge of special products (Section 4.7), we have

$$(x + b)^2 = x^2 + 2bx + b^2$$

and

$$(x - b)^2 = x^2 - 2bx + b^2$$

The two trinomials from the right sides of the above equations,

$$x^2 + 2bx + b^2$$

and

$$x^2 - 2bx + b^2$$

are *perfect square trinomials.* Here are some examples of perfect square trinomials.

$$x^2 + 4x + 4, \qquad r^2 - 8r + 16, \qquad \text{and} \qquad z^2 + \frac{8}{3}z + \frac{16}{9}$$

Can you see the relationship between the coefficient of the middle term, $2b$ or $-2b$, and the constant term b^2?

For a perfect square trinomial, the **constant term** is

$$\left(\frac{\textbf{coefficient of the middle term}}{\textbf{2}}\right)^{2}$$

Now, suppose we are given a binomial like $x^2 + 14x$. What constant should we add to this expression so that we obtain a perfect square trinomial? That is,

$$x^2 + 14x + ?$$

is a perfect square. The middle term of the proposed perfect square will be $14x$. Therefore, divide the coefficient of the middle term, 14, by 2, then square. The correct constant is $\left(\frac{14}{2}\right)^2 = (7)^2 = 49$. Thus, 49 must be added to $x^2 + 14x$ to make the perfect square trinomial $x^2 + 14x + 49$.

Example 1 What constant must be added to $x^2 - 6x$ to make a perfect square trinomial?

Solution The coefficient of the x term is -6. Divide -6 by 2, then square. The correct constant is

$$\left(-\frac{6}{2}\right)^2 = (-3)^2 = 9$$

We can use this technique to solve quadratic equations. ◀

Example 2 Solve $r^2 + 8r = -7$.

Solution We first find the correct constant to make the binomial on the left side a perfect square trinomial. The coefficient of r is 8 and $\left(\frac{8}{2}\right)^2 = (4)^2 = 16$. Therefore, we add 16 to *both* sides.

$$r^2 + 8r + 16 = -7 + 16$$

$$(r + 4)^2 = 9 \qquad r^2 + 8r + 16 \text{ is a perfect square.}$$

The equation is now in the form in which the square root method of the last section applies.

$$r + 4 = \pm 3 \qquad \text{The square root method.}$$

$$r = -4 \pm 3 \qquad \text{Subtract 4 from both sides.}$$

Therefore, either

$$r = -4 + 3 \quad \text{or} \quad r = -4 - 3$$
$$= -1 \qquad\qquad = -7$$

Check -1:

$$r^2 + 8r = -7$$
$$(-1)^2 + 8(-1) \stackrel{?}{=} -7$$
$$1 - 8 \stackrel{?}{=} -7$$
$$-7 = -7$$

Check -7:

$$r^2 + 8r = -7$$
$$(-7)^2 + 8(-7) \stackrel{?}{=} -7$$
$$49 - 56 \stackrel{?}{=} -7$$
$$-7 = -7$$

The solutions are -1 and -7. ◀

The method used in Example 2 is called **completing the square.** Completing the square will work on any quadratic equation.

Example 3 Solve $y^2 - 2y = 2$ by completing the square.

Solution To complete the square, the correct constant to add is $\left(\frac{-2}{2}\right)^2 = (-1)^2 = 1$.

Therefore, we add 1 to both sides.

$$y^2 - 2y + 1 = 2 + 1$$
$$(y - 1)^2 = 3$$
$$y - 1 = \pm\sqrt{3} \qquad \text{Take square roots.}$$
$$y = 1 \pm \sqrt{3} \qquad \text{Add 1 to both sides.}$$

The two answers are $y = 1 + \sqrt{3}$ and $y = 1 - \sqrt{3}$.

We will check the first answer, $1 + \sqrt{3}$.

Check $1 + \sqrt{3}$:

$$y^2 - 2y = 2$$
$$(1 + \sqrt{3})^2 - 2(1 + \sqrt{3}) \stackrel{?}{=} 2$$
$$1 + 2\sqrt{3} + (\sqrt{3})^2 - 2 - 2\sqrt{3} \stackrel{?}{=} 2$$
$$1 + 2\sqrt{3} + 3 - 2 - 2\sqrt{3} \stackrel{?}{=} 2$$
$$2 = 2$$

On your own, check the second answer in the original equation. ◀

In order for the completing the square method to work, the coefficient of the squared term must be 1. If the coefficient is not 1, we first divide both sides of the equation by this number before completing the square.

Example 4 Solve $4z^2 + 8z = 21$ by completing the square.

Solution To make the coefficient of z^2 equal to 1, divide both sides by 4.

$$4z^2 + 8z = 21$$

$$\frac{4z^2 + 8z}{4} = \frac{21}{4} \qquad \text{Divide both sides by 4.}$$

$$z^2 + 2z = \frac{21}{4} \qquad \text{Simplify.}$$

$$z^2 + 2z + 1 = \frac{21}{4} + 1 \qquad \text{Complete the square.}$$

$$(z + 1)^2 = \frac{21}{4} + \frac{4}{4}$$

$$(z + 1)^2 = \frac{25}{4}$$

$$z + 1 = \pm\sqrt{\frac{25}{4}} \qquad \text{Take square roots.}$$

$$z + 1 = \pm\frac{5}{2}$$

$$z = -1 \pm \frac{5}{2}$$

Therefore, either

$$z = -1 + \frac{5}{2} \quad \text{or} \quad z = -1 - \frac{5}{2}$$

$$= \frac{3}{2} \qquad\qquad = -\frac{7}{2}$$

The solution set is $\left\{\frac{3}{2}, -\frac{7}{2}\right\}$.

Example 5 Solve $4v^2 - 24v = -40$ by completing the square.

Solution

$$4v^2 - 24v = -40$$

$$v^2 - 6v = -10 \qquad \text{Divide both sides by 4.}$$

$$v^2 - 6v + 9 = -10 + 9 \qquad \text{Complete the square.}$$

$$(v - 3)^2 = -1$$

Notice that this equation cannot be solved in the real number system, since the square root of -1 is not a real number. Therefore, the original equation has no solution that is a real number.

Sometimes the equation must first be rewritten so that the variable terms are on one side and the constant is on the other side.

We summarize the method of completing the square to solve quadratic equations.

STRATEGY

Solving Quadratic Equations by Completing the Square

Step 1 If necessary, rewrite the equation so that the variable terms are on one side and the constant is on the other side.

Step 2 If the coefficient a of the squared term is not one, divide both sides by a.

Step 3 Complete the square on the side having the variables. Be sure to add the number that completes the square to *both* sides of the equation.

Step 4 Use the square root property to solve the equation in Step 3.

Step 5 Check your answers in the original equation.

WARMING UP

Answer true or false.

1. The solution set of $x^2 + 2x = 0$ is $\{-2, 0\}$.

2. The solution set of $x^2 + x = 0$ is $\{-1\}$.

3. $x^2 + 4x - 4$ is a perfect square trinomial.

4. If 1 is added to $x^2 - 2x$, the result is a perfect square trinomial.

5. The solution set of $x^2 - 2x = -2$ is $\varnothing$.

6. The number 3 is a solution of $x^2 - 3x = 0$.

EXERCISE SET 9.2

1. What constant must be added to $x^2 + 8x$ to make a perfect square trinomial?

2. What constant must be added to $y^2 - 22y$ to make a perfect square trinomial?

3. What constant must be added to $r^2 + 3r$ to make a perfect square trinomial?

4. What constant must be added to $z^2 - \frac{3}{2}z$ to make a perfect square trinomial?

For Exercises 5–24, solve the equation by completing the square.

5. $s^2 + 8s = 33$

6. $y^2 + 10y = -21$

7. $t^2 - 12t = -11$

8. $r^2 - 2r = 8$

9. $q^2 + 16q = -9$

10. $p^2 - 4p = 11$

11. $s^2 - 14s = 72$

12. $a^2 + 3a = -2$

13. $z^2 + 5z = -\frac{11}{2}$

14. $y^2 - 7y = -11$

15. $d^2 + d = -\frac{3}{16}$

16. $9b^2 - 3b = 12$

17. $5r^2 + 3r + \frac{2}{5} = 0$

18. $4k^2 - 7k + 3 = 0$

19. $z^2 - 100 = 6z + 12$

20. $x^2 - 64 = -12x - 36$

21. $2x^2 - 6x = -x + 7$

22. $32 = 9x^2 + 18x + 35$

23. $(x + 2)^2 = 2(x + 5)$

24. $y(y - 4) = 5$

SAY IT IN WORDS

25. What is your method to solve a quadratic equation by completing the square?

26. Explain the steps you would take to solve $2x^2 + 4x = 3$ by completing the square.

REVIEW EXERCISES

The following exercises review parts of Section 8.4. Doing these exercises will help prepare you for the next section.

Simplify.

27. $\frac{4 + \sqrt{8}}{10}$

28. $\frac{9 - \sqrt{27}}{6}$

29. $\frac{12 - \sqrt{48}}{6}$

30. $\frac{14 + \sqrt{98}}{21}$

31. $\frac{-4 - \sqrt{16 - 4}}{2}$

32. $\frac{-10 + \sqrt{100 - 4(4)(2)}}{4}$

ENRICHMENT EXERCISES

Solve by completing the square.

1. $x^4 - 8x^2 = -7$

2. $r^4 + 6r^2 = 7$

3. $x^4 + 2x^2 = -1$

Answers to Enrichment Exercises begin on page A.1.

9.3 The Quadratic Formula

OBJECTIVE

▶ *To solve a quadratic equation using the quadratic formula*

In the last section, we solved quadratic equations by completing the square. If we apply this method on the general equation, we will obtain a formula that will automatically generate any solutions. The **standard form** for a quadratic equation is

$$ax^2 + bx + c = 0, \quad a > 0$$

In order to solve a quadratic equation using the quadratic formula, it is important to write the quadratic equation in standard form and then to identify the values of a, b, and c.

Example 1 If necessary, write the quadratic equation in standard form. Identify the values of a, b, and c.

(a) $3x^2 - 5x + 2 = 0$ **(b)** $t^2 + t - 21 = 0$

(c) $3 + 7r = 2r^2$

Solution

(a) The equation is already in standard form with $a = 3$, $b = -5$, and $c = 2$.

(b) Since $t^2 + t - 21 = 0$ is the same as $1 \cdot t^2 + 1 \cdot t - 21 = 0$, $a = 1$, $b = 1$, and $c = -21$.

(c) This equation is not in standard form, since one side is not equal to zero. Therefore, we subtract $2r^2$ from both sides of the equation.

$$3 + 7r - 2r^2 = 2r^2 - 2r^2$$
$$3 + 7r - 2r^2 = 0$$
$$-2r^2 + 7r + 3 = 0 \qquad \text{Rearrange terms.}$$

Since the coefficient of the squared term is negative, our final step is to multiply both sides by -1.

$$(-1)(-2r^2 + 7r + 3) = (-1)0$$
$$2r^2 - 7r - 3 = 0 \qquad \text{Distribute } -1.$$

The equation is now in standard form where $a = 2$, $b = -7$, and $c = -3$. ◀

The quadratic formula is derived by completing the square on the general quadratic equation. We start with the standard form.

$$ax^2 + bx + c = 0, \quad a > 0$$

If a is not one, divide both sides by a.

$$x^2 + \frac{b}{a}x + \frac{c}{a} = 0$$

Next, subtract $\frac{c}{a}$ from both sides of the equation.

$$x^2 + \frac{b}{a}x = -\frac{c}{a}$$

Now, to complete the square, divide $\frac{b}{a}$ by 2, then square.

$$\frac{\frac{b}{a}}{2} = \frac{b}{2a} \quad \text{and} \quad \left(\frac{b}{2a}\right)^2 = \frac{b^2}{4a^2}$$

Therefore, we add $\frac{b^2}{4a^2}$ to both sides of the equation.

$$x^2 + \frac{b}{a}x + \frac{b^2}{4a^2} = -\frac{c}{a} + \frac{b^2}{4a^2}$$

The left side is now a perfect square trinomial.

$$\left(x + \frac{b}{2a}\right)^2 = -\frac{c}{a} + \frac{b^2}{4a^2}$$

Next, we combine the two terms of the right side as one fraction.

$$\left(x + \frac{b}{2a}\right)^2 = -\frac{4ac}{4a^2} + \frac{b^2}{4a^2} \qquad \text{The common denominator is } 4a^2.$$

$$\left(x + \frac{b}{2a}\right)^2 = \frac{-4ac + b^2}{4a^2}$$

$$\left(x + \frac{b}{2a}\right)^2 = \frac{b^2 - 4ac}{4a^2} \qquad -4ac + b^2 = b^2 - 4ac.$$

Taking square roots,

$$x + \frac{b}{2a} = \pm\sqrt{\frac{b^2 - 4ac}{4a^2}}$$

$$x + \frac{b}{2a} = \pm\frac{\sqrt{b^2 - 4ac}}{\sqrt{4a^2}}$$

$$x + \frac{b}{2a} = \pm\frac{\sqrt{b^2 - 4ac}}{2a}$$

$$x = \frac{-b}{2a} \pm \frac{\sqrt{b^2 - 4ac}}{2a} \qquad \text{Subtract } \frac{b}{2a} \text{ from both sides.}$$

$$x = \frac{-b \pm \sqrt{b^2 - 4ac}}{2a}$$

This result is called the *quadratic formula.*

RULE

Quadratic Formula

If $ax^2 + bx + c = 0$, where $a > 0$, then

$$x = \frac{-b \pm \sqrt{b^2 - 4ac}}{2a}$$

In the next three examples, we show how to find solutions of quadratic equations by using the quadratic formula.

Example 2 Solve $x^2 + 5x - 24 = 0$ by using the quadratic formula.

Solution This equation is in standard form with $a = 1$, $b = 5$, and $c = -24$. Using the quadratic formula,

$$\begin{aligned} x &= \frac{-b \pm \sqrt{b^2 - 4ac}}{2a} \\ &= \frac{-5 \pm \sqrt{5^2 - 4(1)(-24)}}{2(1)} \\ &= \frac{-5 \pm \sqrt{25 + 96}}{2} \\ &= \frac{-5 \pm \sqrt{121}}{2} \\ &= \frac{-5 \pm 11}{2} \end{aligned}$$

Therefore, either

$$\begin{aligned} x &= \frac{-5 + 11}{2} & \text{or} \quad x &= \frac{-5 - 11}{2} \\ &= \frac{6}{2} & &= \frac{-16}{2} \\ &= 3 & &= -8 \end{aligned}$$

As a final step, check these numbers in the original equation. ◀

Example 3 Solve $2t^2 = t + 3$ by the quadratic formula.

Solution First rewrite the equation in standard form.

$$2t^2 - t - 3 = 0$$

Now, $a = 2$, $b = -1$, and $c = -3$. Therefore,

$$\begin{aligned} t &= \frac{-b \pm \sqrt{b^2 - 4ac}}{2a} \\ &= \frac{-(-1) \pm \sqrt{(-1)^2 - 4(2)(-3)}}{2(2)} \\ &= \frac{1 \pm \sqrt{1 + 24}}{4} \\ &= \frac{1 \pm \sqrt{25}}{4} \\ &= \frac{1 \pm 5}{4} \end{aligned}$$

Therefore, either

$$t = \frac{1 + 5}{4} = \frac{6}{4} = \frac{3}{2} \qquad \text{or} \qquad t = \frac{1 - 5}{4} = \frac{-4}{4} = -1$$

Example 4 Solve by using the quadratic formula.

(a) $2x^2 + 2x - 3 = 0$ **(b)** $-x^2 + 2x - 2 = 0$

Solution **(a)** For this equation, $a = 2$, $b = 2$, and $c = -3$. Using the quadratic formula,

$$\begin{aligned} x &= \frac{-b \pm \sqrt{b^2 - 4ac}}{2a} \\ &= \frac{-2 \pm \sqrt{2^2 - 4(2)(-3)}}{2(2)} \\ &= \frac{-2 \pm \sqrt{28}}{4} && 2^2 - 4(2)(-3) = 28. \\ &= \frac{-2 \pm 2\sqrt{7}}{2} && \sqrt{28} = \sqrt{4 \cdot 7} = 2\sqrt{7}. \\ &= -1 \pm \sqrt{7} && \text{Divide both terms in the numerator by 2.} \end{aligned}$$

The two solutions are $-1 + \sqrt{7}$ and $-1 - \sqrt{7}$.

(b) Since the coefficient of $-x^2$ is negative, multiply both sides by -1 to obtain the standard form,

$$x^2 - 2x + 2 = 0$$

Therefore, $a = 1$, $b = -2$, and $c = 2$. Using the quadratic formula,

$$x = \frac{-(-2) \pm \sqrt{(-2)^2 - 4(1)(2)}}{2(1)}$$

$$= \frac{2 \pm \sqrt{-4}}{2}$$

Since $\sqrt{-4}$ is not a real number, the equation has no real solutions. ◀

Example 5 A car that was involved in an accident left skid marks measuring 120 feet. The braking distance d for this particular make of car can be approximated by

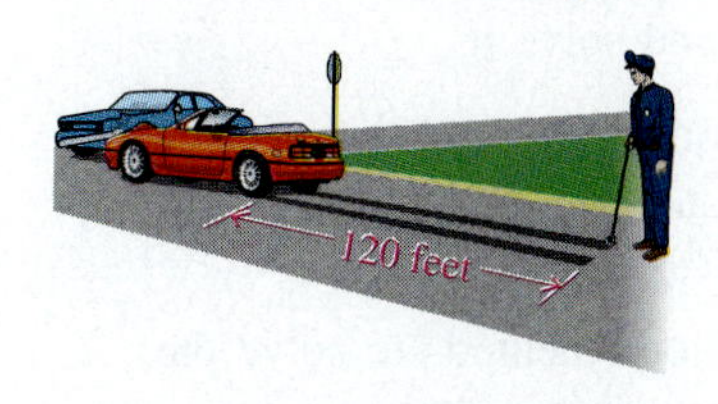

$$d = \frac{1}{10}v^2 + v$$

where v is the speed of the car. What was the speed of the car at the time of the accident?

Solution We replace d by 120 and solve for v.

$$d = \frac{1}{10}v^2 + v$$

$$120 = \frac{1}{10}v^2 + v \qquad \text{Replace } d \text{ by 120.}$$

$$10(120) = 10\left(\frac{1}{10}v^2 + v\right) \qquad \text{To clear the fraction, multiply by 10.}$$

$$1{,}200 = v^2 + 10v \qquad \text{Distributive property.}$$

$$v^2 + 10v - 1{,}200 = 0 \qquad \text{Write in standard form.}$$

For this quadratic equation, $a = 1$, $b = 10$, and $c = -1{,}200$. Therefore,

$$v = \frac{-10 \pm \sqrt{10^2 - 4(1)(-1{,}200)}}{2(1)}$$

$$= \frac{-10 \pm \sqrt{4{,}900}}{2}$$

$$= \frac{-10 \pm 70}{2}$$

We discard the negative solution. Why?

$$v = \frac{-10 + 70}{2} = \frac{60}{2} = 30$$

The car was traveling 30 mph at the time of the accident. ◀

TEAM PROJECT

(3 or 4 Students)

SOLVING QUADRATIC EQUATIONS

Course of Action: Each team member makes up a quadratic equation and solves it.

1. How many members used the quadratic formula?
2. How many members used the square root method?
3. How many members used completing the square?
4. How many members used the factoring method?
5. How many members had no solutions in the set of real numbers?

Individually, make up another quadratic equation and solve it.

6. Trade equations in the group. Compare solutions.
7. If solutions do not agree, have the team help decide which solution, if any, is correct.
8. Make a chart of how many members used various methods to solve the quadratic equations.
9. What method was used most frequently?
10. How many solutions were not in the set of real numbers?

Group Report: Make a list of suggestions for students solving quadratic equations.

WARMING UP

Answer true or false.

1. $2x^2 - 3x + 1 = 0$ is in standard form.
2. To write $-x^2 + 4x - 5 = 0$ in standard form, multiply the first term on the left by -1.
3. In the quadratic formula, $-b$ is always a negative number.
4. The solution set of $x^2 + x - 1 = 0$ is $\left\{-\frac{1}{2} \pm \frac{1}{2}\sqrt{5}\right\}$.
5. For the equation $2x^2 - 5 = 0$, $b = 0$.
6. The equation $3x^2 - 2x + 3 = 0$ has no real solutions.

EXERCISE SET 9.3

For Exercises 1–10, write the quadratic equation in standard form. Identify the values of a, b, and c. Do not solve the equation.

1. $2x^2 - 3x + 4 = 0$
2. $3x^2 + 2x - 1 = 0$
3. $-4x^2 + x - 2 = 0$

4. $-5x^2 - x + 1 = 0$

5. $3 - 2x = 2x^2$

6. $1 + 5x = 4x^2$

7. $2x - 1 = -7x^2$

8. $-x + 2 = -x^2$

9. $x(x + 1) = 4$

10. $2x(x - 1) = -5$

For Exercises 11–36, use the quadratic formula to solve.

11. $x^2 + 6x - 27 = 0$

12. $t^2 + 12t + 35 = 0$

13. $s^2 - 3s + 2 = 0$

14. $u^2 - 2u - 8 = 0$

15. $2v^2 + 7v - 4 = 0$

16. $3z^2 + 2z - 5 = 0$

17. $y^2 + y + 1 = 0$

18. $x^2 - 2x + 2 = 0$

19. $3m^2 + m - 1 = 0$

20. $4k^2 - k - 1 = 0$

21. $s^2 + 2s - 1 = 0$

22. $x^2 + 4x - 2 = 0$

23. $v^2 - 4v - 2 = 0$

24. $y^2 - 6y + 1 = 0$

25. $x^2 + 6x + 9 = 0$

26. $r^2 - 8r + 16 = 0$

27. $x^2 - 2x = -3$

28. $x^2 + 30 = -6$

29. $4z^2 - 7 = 4z$

30. $4x^2 + 12x = 5$

31. $0 = 20y^2 - 4y - 1$

32. $1 = 8y^2 - 2y$

33. $4z^2 + 7z = 0$

34. $s^2 - 12 = 0$

35. $x^2 - x - 1 = 0$

36. $5y^2 - 12y - 8 = 0$

37. What is wrong with the following solution?

$$x^2 - 3x - 5 = 0$$

$$x = \frac{-3 \pm \sqrt{9 + 20}}{2}$$

$$= \frac{-3 \pm \sqrt{29}}{2}$$

What is the correct solution?

38. What is wrong with the following solution?

$$x^2 + 2x - 4 = 0$$

$$x = -2 \pm \frac{\sqrt{4 + 16}}{2}$$

$$= -2 \pm \frac{\sqrt{20}}{2}$$

$$= -2 \pm \sqrt{5}$$

What is the correct solution?

39. A rectangular piece of sheet metal is 2 inches longer than it is wide. If the area is 50 square inches, what are the exact dimensions of the piece? Also, approximate the dimensions using a calculator or the table on the inside backcover.

40. A rectangular rhinoceros mudhole is to be constructed in a zoo. The length is to be 6 yards longer than its width. If the area must be 210 square yards, what are the exact dimensions of the mudhole? Approximate the dimensions using a calculator or the table on the inside back cover.

41. Ellen is tied to a 100-foot-long bungee cord as shown in the figure. When she jumps off the bridge, she will be in free-fall for the length of the cord. How many seconds will it take her to fall the 100 feet? (Hint: Use the formula $d = 16t^2$, where d is the distance fallen, in feet, after t seconds.)

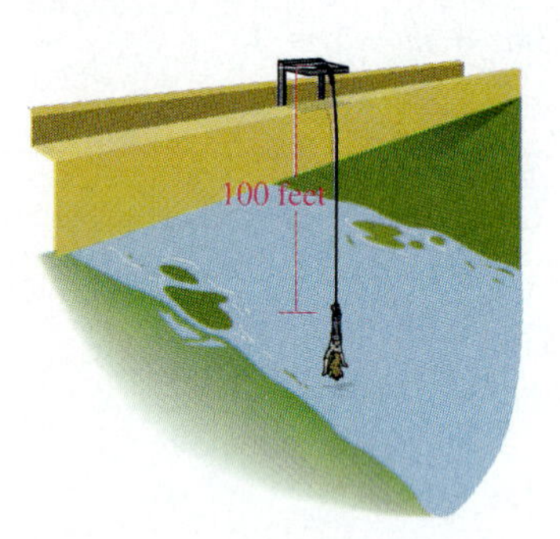

42. The number of diagonals N of a polygon of n sides is given by $N = \dfrac{n^2 - 3n}{2}$. Squares and rectangles are examples of four-sided polygons. The figures below show four- and five-sided polygons with the diagonals drawn in red. If a polygon has 20 diagonals, how many sides does it have? (*Hint:* Set $N = 20$ and solve for n.)

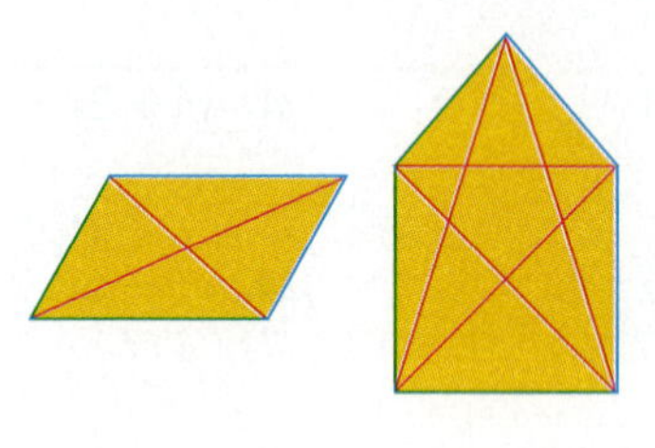

For Exercises 43 and 44, solve, then round off your answers to one decimal place.

43. $2.4x^2 + 6.3x - 4.5 = 0$

44. $7.1a^2 - 9.8a + 1.2 = 0$

SAY IT IN WORDS

45. Solve $x(2x + 3) = 2$ in three ways: (1) the factoring method, (2) completing the square, and (3) the quadratic formula. Which method did you prefer? Why?

46. Solve $3(x^2 - 1) = 8x$ in three ways: (1) the factoring method, (2) completing the square, and (3) the quadratic formula. Which method did you prefer? Why?

REVIEW EXERCISES

The following exercises review parts of Section 8.7. Doing these exercises will help prepare you for the next section.

Express each complex number in the form $a + bi$.

47. $\sqrt{-121}$

48. $1 + \sqrt{-4}$

49. $-3 - \sqrt{-6}$

50. $-12 + \sqrt{-5}$

51. $14 + \sqrt{-18}$

52. $-1 - \sqrt{-24}$

ENRICHMENT EXERCISES

Solve using the quadratic formula.

1. $x^4 - 10x^2 + 9 = 0$

2. $\frac{2}{y^2} + \frac{4}{y} - 3 = 0$

3. $x^2 + 2\sqrt{3}x - 9 = 0$

4. $r^2 - 4r\sqrt{2} - 2 = 0$

Answers to Enrichment Exercises begin on page A.1.

Which Method to Use?

We now have four methods available to solve a quadratic equation: factoring, the square root method, completing the square, and the quadratic formula. The completing the square method generally is not used to solve quadratic equations. However, it did enable us to verify the quadratic formula and it does have uses in other parts of mathematics.

The following is a strategy for solving quadratic equations.

STRATEGY

Solving the Quadratic Equation $ax^2 + bx + c = 0$

1. If $b = 0$, use the square root method.
2. If $b \neq 0$, think first of the factoring method. If it appears that the trinomial does not factor, or if the factors are not apparent, use the quadratic formula.

To sharpen your ability to solve quadratic equations by whatever method seems appropriate, the following exercise set has been created. Keep in mind that if everything else fails, the quadratic formula will always work.

MISCELLANEOUS QUADRATIC EQUATIONS

Use the strategy for solving quadratic equations to solve each given equation.

1. $x^2 - 4x - 1 = 0$

2. $x^2 + 5x - 3 = 0$

3. $x^2 - 6x = 0$

4. $y^2 + 4y = 0$

5. $x^2 - x + 2 = 0$

6. $x^2 - 5x + 6 = 0$

7. $6r^2 + 7r - 3 = 0$

8. $3y^2 + 5y + 2 = 0$

9. $4x^2 - 4x - 5 = 0$

10. $8s^2 + s - 1 = 0$

11. $z^2 - 16 = 0$

12. $v^2 + 9 = 0$

13. $4(c^2 - 5c) + 25 = 0$

14. $2x^2 + x - 4 = 0$

15. $\frac{1}{4}m^2 = m$

16. $2w = w^2$

17. $t(t + 1) = 20$

18. $m(2m + 3) = 5$

19. $x^2 + 11x + 3 = 0$

20. $3n^2 - 4n + 5 = 0$

21. $0 = 10y^2 + 11y + 3$

22. $0 = -12x^2 + 11x - 2$

23. $y\left(3y + \frac{17}{2}\right) = 7$

24. $3(5k^2 + 2) = 19k$

25. $3x^2 = 27x$

26. $2z^2 + 1 = 3z$

27. $d(d + 2) = 4$

28. $2(v^2 - 2) = -v$

29. $3c^2 + 8c - 1 = 0$

30. $9t^2 - 3t - 1 = 0$

31. $(2s + 1)(s - 5) = -2(s + 5)$

32. $(8u + 1)(u - 2) = 5u + 1$

33. $2s^2 - 4s = 0$

34. $6y^2 + 3y = 0$

35. $x^2 + 2x = 2$

36. $2x^2 + 4x = 5$

9.4 Quadratic Equations with Complex Solutions

OBJECTIVE

► *To solve quadratic equations using complex numbers*

The equation $x^2 = -9$ has no solution in the set of real numbers, since any real number squared is either zero or positive. If we enlarge our number system to the set of complex numbers, however, we can solve quadratic equations like $x^2 = -9$ that have no real solutions. We can still use the square root method from Section 9.1 as well as the quadratic formula in Section 9.3.

Example 1 Solve $(3t + 2)^2 = -8$.

Solution The equation is of the form to use the square root method.

$$(3t + 2)^2 = -8$$

$$3t + 2 = \pm\sqrt{-8} \qquad \text{Take square roots.}$$

$$3t + 2 = \pm\sqrt{8(-1)} \qquad -8 = 8(-1).$$

$$3t + 2 = \pm\sqrt{8}\sqrt{-1} \qquad \sqrt{ab} = \sqrt{a}\sqrt{b}.$$

$$3t + 2 = \pm(2\sqrt{2})i \qquad \text{Simplify } \sqrt{8} \text{ and } \sqrt{-1} = i.$$

$$3t = -2 \pm (2\sqrt{2})i \qquad \text{Subtract 2 from both sides.}$$

$$t = -\frac{2}{3} \pm \frac{2\sqrt{2}}{3}i \qquad \text{Divide by 3.}$$

The two numbers are $-\frac{2}{3} + \frac{2\sqrt{2}}{3}i$ and $-\frac{2}{3} - \frac{2\sqrt{2}}{3}i$. Check these numbers in the original equation. ◄

Example 2 Solve $r^2 + 6r + 11 = 0$.

Solution Since $r^2 + 6r + 11$ cannot be factored, we use the quadratic formula with $a = 1$, $b = 6$, and $c = 11$.

$$r = \frac{-6 \pm \sqrt{6^2 - 4(1)(11)}}{2(1)}$$

$$= \frac{-6 \pm \sqrt{36 - 44}}{2}$$

$$= \frac{-6 \pm \sqrt{-8}}{2}$$

$$= \frac{-6 \pm \sqrt{8}i}{2}$$

$$= \frac{-6 \pm 2\sqrt{2}i}{2}$$

$$= -\frac{6}{2} \pm \frac{2\sqrt{2}i}{2}$$

$$= -3 \pm \sqrt{2}i$$

The two complex solutions are $-3 + i\sqrt{2}$ and $-3 - i\sqrt{2}$. ◀

In the quadratic formula $x = \dfrac{-b \pm \sqrt{b^2 - 4ac}}{2a}$, the expression $b^2 - 4ac$ is called the **discriminant.** The discriminant gives information about the nature and number of the solutions of the quadratic equation $ax^2 + bx + c = 0$.

RULE

For the quadratic equation $ax^2 + bx + c = 0$, the discriminant gives the following information:

If $b^2 - 4ac > 0$, there are two real solutions.
If $b^2 - 4ac = 0$, there is one real solution.
If $b^2 - 4ac < 0$, there are two complex solutions.

Example 3 Use the discriminant to determine if the solutions are real or complex. Also, determine the number of solutions.

(a) $2x^2 - x - 6 = 0$ **(b)** $4x^2 - 4x + 1 = 0$
(c) $x^2 - 2x + 3 = 0$

Solution **(a)** Since $b^2 - 4ac = (-1)^2 - 4(2)(-6) = 1 + 48 = 49$ is positive, there are two real solutions. Verify that they are 2 and $-\dfrac{3}{2}$.

(b) Since $b^2 - 4ac = (-4)^2 - 4(4)(1) = 0$, there is one real solution. Verify that it is $\dfrac{1}{2}$.

(c) Since $b^2 - 4ac = (-2)^2 - 4(1)(3) = 4 - 12 = -8$ is negative, there are two complex solutions. Verify that they are $1 + i\sqrt{2}$ and $1 - i\sqrt{2}$. ◀

WARMING UP

Answer true or false.

1. The equation $3x^2 - 2x - 1 = 0$ has two complex solutions.
2. The equation $x^2 + 6x + 9 = 0$ has one real solution.
3. The solution set of $x^2 + 1 = 0$ is $\{i, -i\}$.
4. Any quadratic equation will have at least one real solution.
5. If a quadratic equation has one complex solution, then it has two complex solutions.

EXERCISE SET 9.4

For Exercises 1–20, use the strategy for solving the quadratic equation on page 482 to solve each equation. Write all solutions that are complex numbers in standard form.

1. $x^2 + 49 = 0$
2. $y^2 + 121 = 0$
3. $(r + 1)^2 = -4$
4. $(t - 2)^2 = -9$
5. $(3s - 5)^2 - 128 = 0$
6. $(7q + 3)^2 - 8 = 0$
7. $v^2 - 3v - 1 = 0$
8. $2u^2 - 4u + 1 = 0$
9. $2p^2 - 5p + 3 = 0$
10. $2z^2 - z - 6 = 0$
11. $x^2 + 3x + 4 = 0$
12. $y^2 - 5y + 6 = 0$
13. $x^2 - 4x + 5 = 0$
14. $w^2 + 3w + 6 = 0$
15. $z^2 - 4z + 1 = 0$
16. $t^2 - 2t + 3 = 0$
17. $n^2 - 21 = 2n$
18. $p^2 - 4p = -11$
19. $(u - 1)^2 = 7u$
20. $(q - 2)(q + 3) = 6$

For Exercises 21–30, use the discriminant to determine if the solutions are real or complex. Also, determine the number of solutions. Do not solve the equation.

21. $4x^2 + 4x - 3 = 0$
22. $2y^2 - 3y - 2 = 0$
23. $y^2 + 6y + 9 = 0$
24. $z^2 - 10z + 25 = 0$
25. $x^2 + 1 = 0$
26. $v^2 + 3 = 0$
27. $x^2 - 1 = 0$
28. $v^2 - 3 = 0$
29. $w^2 - 3w + 3 = 0$
30. $2x^2 - x + 1 = 0$

SAY IT IN WORDS

31. Explain how the discriminant determines the nature and number of solutions of a quadratic equation.

ENRICHMENT EXERCISES

Solve.

1. $(4x - 1)(x + 3) = 13x - 4$

2. $(3x - 1)(4x + 1) = 11x$

3. $16x(2x - 1) = -11$

4. $(x + 1)(3x + 7) = 13x$

Answers to Enrichment Exercises begin on page A.1.

CHAPTER 9 Summary and review

The square root method to solve quadratic equations (9.1)

We solve equations of the form $(ax + b)^2 = c$ by **taking square roots** of both sides.

Examples

$$(2x - 1)^2 = 16$$
$$2x - 1 = \pm 4$$
$$x = \frac{1 \pm 4}{2}$$
$$x = -\frac{3}{2} \quad \text{or} \quad x = \frac{5}{2}$$

Completing the square (9.2)

To solve a quadratic equation by **completing the square,** use the following strategy:

Step 1. If necessary, rewrite the equation so that the variable terms are on one side and the constant is on the other side.

Step 2. If the coefficient, a, of the squared term is not 1, divide both sides by a.

Step 3. Complete the square on the side having the variables. Be sure to add the number that completes the square to *both* sides of the equation.

Step 4. Use the square root property to solve the equation in Step 3.

Step 5. Check your answers in the original equation.

$$x^2 + 8x - 4 = 0$$
$$x^2 + 8x = 4$$
$$x^2 + 8x + 16 = 4 + 16$$
$$(x + 4)^2 = 20$$
$$x + 4 = \pm\sqrt{20}$$
$$x = -4 \pm 2\sqrt{5}$$

The quadratic formula (9.3)

The quadratic equation $ax^2 + bx + c = 0$, where $a \neq 0$, has **solutions** given by

$$x = \frac{-b \pm \sqrt{b^2 - 4ac}}{2a}$$

Examples

Given: $2x^2 - 4x - 3 = 0$.
Then,

$$x = \frac{4 \pm \sqrt{(-4)^2 - 4(2)(-3)}}{2(2)}$$
$$= \frac{4 \pm \sqrt{16 + 24}}{4}$$
$$= \frac{4 \pm \sqrt{40}}{4}$$
$$= \frac{2 \pm \sqrt{10}}{2}$$

Complex solutions of quadratic equations (9.4)

If $b^2 - 4ac$ is negative, the quadratic equation $ax^2 + bx + c = 0$ has complex solutions.

Given: $x^2 - 2x + 5 = 0$.
Then,

$$x = \frac{2 \pm \sqrt{(-2)^2 - 4(1)(5)}}{2(1)}$$
$$= \frac{2 \pm \sqrt{4 - 20}}{2}$$
$$= \frac{2 \pm \sqrt{-16}}{2}$$
$$= \frac{2 \pm 4i}{2}$$
$$= 1 \pm 2i$$

CHAPTER 9 REVIEW EXERCISE SET

Section 9.1

For Exercises 1–5, solve the equation.

1. $r^2 = 27$

2. $(t - 5)^2 = 9$

3. $(y - 3)^2 = 18$

4. $5c^2 = 7$

5. $\dfrac{(3x-2)^2}{4} = \dfrac{8}{9}$

6. In a right triangle one leg is twice as long as the other leg as shown in the figure below.

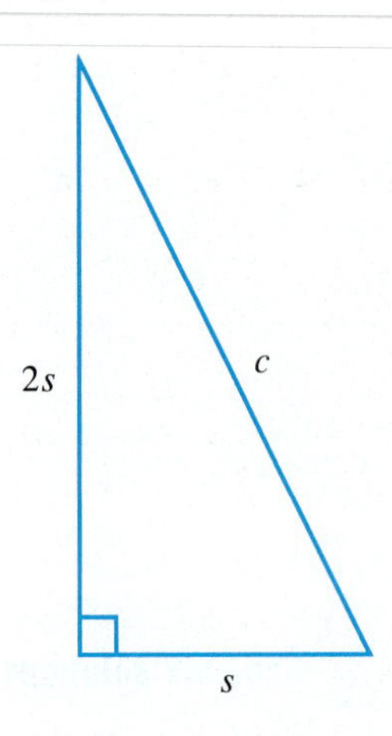

(a) Express s in terms of c.

(b) Find s if c is 15 centimeters.

Section 9.2

7. What constant must be added to $y^2 - 16y$ to make a perfect square trinomial?

For Exercises 8–12, solve by completing the square.

8. $z^2 - 10z = 11$
9. $x^2 + 12x = 13$
10. $4c^2 - 16c = -7$
11. $3t^2 - 8t = -3$
12. $k^2 + 10k = -25$

Section 9.3

For Exercises 13–16, if necessary, write the quadratic equation in the standard form $ax^2 + bx + c = 0$, with $a > 0$. Identify the values of a, b, and c.

13. $-2x^2 + 7x + 5 = 0$
14. $z^2 - z + 12 = 0$
15. $4y + y^2 = 7$
16. $(3x - 1)(7x - 4) = 5$

For Exercises 17–20, find all real solutions using the quadratic formula.

17. $x^2 - 3x - 28 = 0$
18. $3r^2 = 3 - 8r$
19. $3y^2 + 2y - 12 = 0$
20. $\dfrac{z^2}{3} + \dfrac{z}{6} + 2 = 0$

For Exercises 21–26, use the strategy for solving quadratic equations to solve each equation.

21. $2x^2 - x - 15 = 0$
22. $z^2 - z - 3 = 0$
23. $4c^2 = 100$

24. $x^2 + 2x = 4$

25. $(3y - 2)^2 = 18$

26. $4x(x + 3) = 11$

Section 9.4

27. The base of a triangle is 5 inches longer than the height. If the area is 50 square inches, find the length of the base and the height. Approximate your answers to one decimal place using the table on the inside back cover.

28. The braking distance d, in feet, for a particular car traveling v mph is approximated by

$$d = \frac{3}{10}v^2 + \frac{9}{10}v$$

The car required 516 feet to stop. Was the car traveling within the posted speed limit of 50 mph?

For Exercises 29–32, solve the quadratic equation.

29. $(2t - 3)^2 = -4$

30. $z^2 + 20 = 0$

31. $3v^2 - 2v + 4 = 0$

32. $4x(x + 1) = -49$

CHAPTER 9 TEST

For Problems 1 and 2, use the square root method to solve the equation.

1. $x^2 = \frac{40}{81}$

2. $\frac{(3x - 1)^2}{3} = 4$

For Problems 3 and 4, solve by completing the square.

3. $x^2 - 6x = -3$

4. $3y^2 + 12y = 36$

For Problems 5–7, find all real solutions using the quadratic formula.

5. $3x^2 - x + 5 = 0$

6. $y^2 + 10y + 6 = 0$

7. $2x^2 - 8x = -3$

For Problems 8–10, use the strategy for solving quadratic equations to solve each equation.

8. $4z(z - 4) + 15 = 0$

9. $\frac{x^2}{4} = 5$

10. $3(3x + 1) + x = x(3x + 1)$

For Problems 11–13, solve the quadratic equation. Write all solutions that are complex numbers in standard form.

11. $t^2 + 81 = 0$

12. $(2w - 5)^2 = -121$

13. $(6c - 1)(c + 2) = 2c$

14. One side of a rectangle is 12 centimeters longer than the other side. If the area is 30 square centimeters, find the dimensions of the rectangle. Approximate your answers to one decimal place using a calculator or the table on the inside back cover.

15. A ride at an amusement park features a free fall of 160 feet as shown in the figure. How many seconds do the riders experience free fall? Give the exact answer and an approximation to three decimal places. *Hint:* $d = 16t^2$

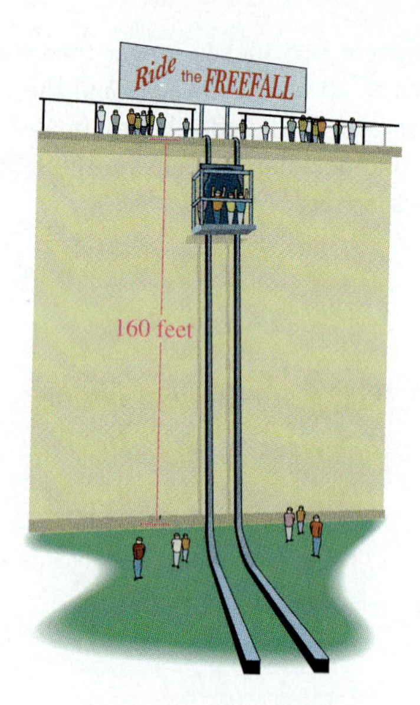

APPENDIX A

A Review of Decimals and Percents

OBJECTIVES

- ▶ *To convert a decimal to a fraction*
- ▶ *To convert a fraction to a decimal*
- ▶ *To round decimal numbers*
- ▶ *To combine decimals using the four basic operations of addition, subtraction, multiplication, and division*
- ▶ *To change a percent to a decimal*
- ▶ *To change a decimal to a percent*
- ▶ *To find percentages*

A **decimal** or **decimal fraction** is a number written using a decimal point. For example, the decimal 0.6 is equal to the fraction $\frac{6}{10}$. Likewise, the decimal 0.905 is equal to the fraction $\frac{905}{1000}$ and 0.0001 is another way to write $\frac{1}{10{,}000}$.

To convert a decimal to a fraction, we write the decimal as a fraction in lowest terms.

Example 1

(a) $0.6 = \frac{6}{10} = \frac{3}{5}$

(b) $0.45 = \frac{45}{100} = \frac{9}{20}$

(c) $2.3 = 2 + \frac{3}{10} = \frac{20}{10} + \frac{3}{10} = \frac{23}{10}$ ◀

To convert a fraction into a decimal, divide the denominator into the numerator.

Example 2 Convert $\frac{7}{8}$ into a decimal.

Solution

```
    .875
 8)7.000     Add a decimal point and three zeros after the 7.
   64
    60
    56
     40
     40
      0
```

Therefore, $\frac{7}{8} = 0.875$. ◀

Every real number can be written as a decimal. Irrational numbers have decimal expansions that do not end and do not repeat. For example,

$$\pi = 3.1415927\ldots$$

For calculation purposes, it is useful **to round decimals** to the nearest integer, tenth, hundredth, and so on.

> **RULE**
>
> **The Rounding Process**
>
> To round a decimal to the nearest tenth, consider the digit in the hundredths place (one digit to the right).
>
> **1.** If it is 5 or more, add 1 to the digit in the tenths place, dropping all digits after it.
>
> **2.** If it is less than 5, keep the digit in the tenths place, dropping all digits after it.

We use this technique to round decimals to the nearest hundredth (two decimal places), thousandth (three decimal places), or any number of decimal places. Look at the digit directly to the right of the desired position and follow Steps 1 and 2 in the rounding process.

Example 3

$\pi = 3.1$ to one decimal place

$\pi = 3.14$ to two decimal places

$\pi = 3.142$ to three decimal places

$\pi = 3.1416$ to four decimal places ◀

Adding or subtracting decimals can be done in column form as shown in the next two examples. Keep in mind that the decimal points must be in a straight line.

Example 4 Find the sum: 2.6 + 0.49 + 4.081.

Solution We place these numbers in a column; then, add in the usual manner. Zeros may be supplied as shown to help keep the digits in the correct column.

$$\begin{array}{r} 2.600 \\ 0.490 \\ \underline{4.081} \\ 7.171 \end{array}$$ ◀

Example 5 Subtract: 6.4 − 2.91.

Solution We write the two numbers in column form, making sure to line up the decimal points. We then subtract in the usual manner.

$$\begin{array}{r} 6.40 \\ -2.91 \\ \hline 3.49 \end{array}$$

In the next example, we review **multiplying two decimal numbers.**

Example 6 Multiply: 4.83×3.7.

Solution We first write in column form without regard to the positions of the decimal points. Then, multiply as if they were two whole numbers.

$$\begin{array}{rl} 4.83 & \text{2 decimal places.} \\ 3.7 & \text{1 decimal place.} \\ \hline 3381 & \\ 1449 & \\ \hline 17.871 & \text{3 decimal places.} \end{array}$$

The number of decimal places in the product is the sum of the number of decimal places in the two original decimals.

For **division of decimal numbers,** divide as if they were whole numbers. The placement of the decimal point in the quotient is explained in the next example.

Example 7 Divide 562.1 by 3.97. Round the quotient to one decimal place.

Solution

$$\begin{array}{r} 141.5 \\ 3.97\overline{)562.10.0} \\ 397 \\ \hline 1651 \\ 1588 \\ \hline 630 \\ 397 \\ \hline 2330 \\ 1985 \\ \hline 345 \end{array}$$

Since the remainder 345 is more than half the divisor 397, we round our answer by adding 1 in the tenths position.

The answer is 141.6

One important application of decimals involves the concept of percent. The word **percent** means ''per hundred,'' or hundredths. The symbol of percent is %. For example, 1% means 1 per hundred:

$$1\% = \frac{1}{100} = 0.01$$

The phrase ''per hundred'' means to divide by 100. For example, 24% means $\frac{24}{100}$ or 0.24. That is, 24% = 0.24.

Example 8 Change each percent to a decimal.

(a) 5% **(b)** 325% **(c)** 8.02%

Solution

(*a*) $5\% = \frac{5}{100} = 0.05$

(*b*) $325\% = \frac{325}{100} = 3.25$

(*c*) $8.02\% = \frac{8.02}{100} = 0.0802$ ◀

Example 9 Change each decimal to a percent.

(a) 0.03 **(b)** 2 **(c)** 0.925

Solution To convert a decimal to a percent, multiply by 100.

(a) $0.03 = (0.03)(100)\% = 3\%$
(b) $2 = 2(100)\% = 200\%$
(c) $0.925 = (0.925)(100)\% = 92.5\%$ ◀

To find a **percentage** or part of a number, we use multiplication. For example, $50\% = \frac{50}{100}$ or $\frac{1}{2}$, so to find 50% of a number, multiply by $\frac{1}{2}$. Thus, 50% of 24 is 12, 50% of 300 is 150, and so on.

Example 10 Find each percentage.

(a) 30% of 60 **(b)** 160% of 40

Solution The word ''of'' implies multiplication. Remember to first change the percent to a decimal before multiplying.

(a) $30\% \cdot 60 = (0.3)(60) = 18$
(b) $160\% \cdot 40 = (1.6)(40) = 64$ ◀

APPENDIX A EXERCISE SET

For Exercises 1–6, convert each decimal to a fraction in lowest terms.

1. 0.7 **2.** 0.55 **3.** 2.4

4. 6.8 **5.** 2.45 **6.** 9.05

For Exercises 7–12, convert each fraction into a decimal.

7. $\frac{2}{5}$

8. $\frac{3}{4}$

9. $\frac{7}{2}$

10. $\frac{11}{4}$

11. $\frac{3}{8}$

12. $\frac{9}{8}$

For Exercises 13–18, round each number to the indicated number of decimal places.

13. 4.934; two decimal places.

14. 51.839; one decimal place.

15. 10.20091; three decimal places.

16. 72.872065; four decimal places.

17. 4.8587; to the nearest tenth.

18. 65.0348; to the nearest hundredth.

For Exercises 19–42, perform the indicated operations.

19. $4.8 + 2.9$

20. $1.72 + 0.43$

21. $3.8 + 5.21 + 0.714$

22. $7.901 + 3.2 + 4.83$

23. $\begin{array}{r} 6.92 \\ -3.71 \\ \hline \end{array}$

24. $\begin{array}{r} 9.75 \\ -6.24 \\ \hline \end{array}$

25. $7.83 - 2.51$

26. $12.57 - 8.16$

27. $5.14 - 3.87$

28. $6.14 - 4.65$

29. $8.2 - 4.36$

30. $4.24 - 3.67$

31. 3.9×2.8

32. 0.4×8.12

33. $(0.62)(0.51)$

34. $(0.35)(4.8)$

35. $(0.03)(45.7)$

36. $(2.1)(39.8)$

37. $0.441 \div 0.3$

38. $0.392 \div 0.8$

39. $1.535 \div 0.5$

40. $2.646 \div 0.06$

41. $0.5428 \div 0.92$

42. $76.68 \div 1.8$

43. Divide 443.6 by 2.54. Round the quotient to the nearest tenth.

44. Divide 208.20 by 61.30. Round the quotient to the nearest hundredth.

For Exercises 45–52, change each percent to a decimal.

45. 36%

46. 10%

47. 57%

48. 69%

49. 100%

50. 200%

51. 3.8%

52. 7.92%

For Exercises 53–60, change each decimal to a percent.

53. 0.61

54. 0.83

55. 0.02

56. 0.10

57. 0.924

58. 0.461

59. 2.8

60. 1.25

For Exercises 61–70, find each percentage.

61. 5% of 240

62. 8% of 350

63. 12% of 1000

64. 10% of 3480

65. 200% of 16

66. 150% of 40

67. 1.6% of 100

68. 2.5% of 30

69. $\frac{1}{2}$% of 800

70. $\frac{3}{4}$% of 4000

71. What percent of 260 is 52?

72. What percent of 580 is 928?

73. 135 is what percent of 540?

74. 156 is what percent of 120?

75. A software graphics program sells for $49.75. What is the price to the nearest dollar?

76. A steel cable measures 1.367 cm in diameter. What is the length of the diameter to the nearest tenth of a centimeter?

77. The price of chocolate pecans, which previously cost $4.50 per pound, is increased by 8%. What is the new price?

78. A person who earns $1250 a week receives a 6% increase. What is the new weekly salary?

Answers To Selected Exercises

CHAPTER 1

WARMING UP

1. T **2.** T **3.** T **4.** F **5.** F **6.** F **7.** T **8.** T

EXERCISE SET 1.1

1. $\frac{2}{3}$ **3.** $\frac{1}{3}$ **5.** 2 **7.** $\frac{6}{5}$ **9.** 2 **11.** $\frac{2}{9}$ **13.** $\frac{15}{16}$ **15.** $\frac{9}{20}$ **17.** $\frac{4}{3}$ **19.** $\frac{2}{3}$ **21.** $\frac{3}{2}$ **23.** $\frac{1}{2}$

25. $\frac{4}{3}$ **27.** $\frac{2}{3}$ **29.** 6 **31.** $\frac{3}{4}$ **33.** $\frac{1}{2}$ **35.** $\frac{8}{3}$ **37.** 2 **39.** $\frac{1}{12}$ **41.** $\frac{1}{2}$ **43.** 3 **45.** $\frac{1}{12}$ **47.** $\frac{3}{5}$

49. $\frac{3}{2}$ **51.** 2 **53.** $\frac{1}{3}$ **55.** $\frac{2}{3}$ **57.** $\frac{3}{5}$ **59.** $\frac{5}{6}$ **61.** $\frac{9}{8}$ **63.** $\frac{1}{4}$ **65.** $\frac{2}{3}$ **67.** $\frac{1}{9}$ **69.** $\frac{11}{6}$ **71.** $\frac{17}{12}$

73. $\frac{1}{6}$ **75.** $\frac{11}{10}$ **77.** $\frac{5}{4}$ **79.** $\frac{3}{2}$ **81.** $\frac{5}{3}$ **83.** 4 **85.** $6\frac{3}{4}$ **87.** $4\frac{1}{2}$ **89.** $9\frac{1}{6}$ **91.** $7\frac{2}{3}$ feet

93. $\frac{5}{8}$ of the pizza **95.** 5 teaspoons **97.** $\frac{8}{12}+\frac{12}{18}=\frac{\overset{2}{\cancel{8}}}{\underset{3}{\cancel{12}}}+\frac{\overset{2}{\cancel{12}}}{\underset{3}{\cancel{18}}}=\frac{2}{3}+\frac{2}{3}=\frac{4}{3}$

SAY IT IN WORDS

99. ans. vary

ENRICHMENT EXERCISES

1. $\frac{11}{24}$ **2.** $\frac{3}{26}$ **3.** $\frac{17}{24}$ **4.** $\frac{1}{2}$ **5.** $\frac{3}{8}$ **6.** $\frac{1}{9}$ **7.** \$6.30

WARMING UP

1. T **2.** F **3.** T **4.** F **5.** F **6.** F **7.** T **8.** T **9.** F **10.** F

EXERCISE SET 1.2

1. the sum of 2 and 3 **3.** 6 is less than 7 **5.** the quotient of 9 and 3 **7.** the difference of 5 and 2 is 3 **9.** $7 - 4$ **11.** $8 \div 4$ **13.** $5 > 4$ **15.** $15 \div 5 = 3$ **17.** 64 **19.** 9 **21.** 2 **23.** 2500 **25.** 10 **27.** 3 **29.** 11 **31.** 17 **33.** 10 **35.** 34 **37.** 13 **39.** 6 **41.** $3(5 + 2)$ **43.** $(7 - 4)^2$ **45.** $\frac{1}{3}$ **47.** 3 **49.** 1 **51.** $\frac{3}{2}$ **53.** 16,807 **55.** 16,777,216 **57.** 4,782,969 **59.** 1,048,576

SAY IT IN WORDS

61. no; $(2 + 3)^2 = 25$, whereas $2^2 + 3^2 = 13$ **63.** no; $4 \div 2 = 2$, whereas $2 \div 4 = \frac{1}{2}$ **65.** ans. vary

ENRICHMENT EXERCISES

1. 2^7 **2.** 2^8 **3.** 2^9 **4.** no; $\left(\frac{1}{2}\right)^2 = \frac{1}{4}$ and $\frac{1}{4} < \frac{1}{2}$

WARMING UP

1. T **2.** F **3.** F **4.** T **5.** T **6.** T **7.** F **8.** F **9.** T **10.** F

EXERCISE SET 1.3

1. 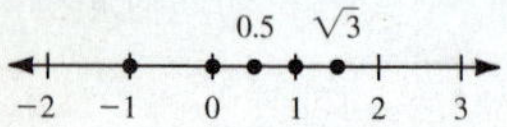

3. **(a)** 6 **(b)** 0, 6 **(c)** $-\frac{10}{5} = -2$ **(d)** $-\frac{10}{5}$, 0, 6 **(e)** $-\frac{1}{2}, \frac{1}{2}$, 7.8, $8\frac{1}{2}$ **(f)** $-\frac{10}{5}, -\frac{1}{2}$, 0, $\frac{1}{2}$, 6, 7.8, $8\frac{1}{2}$ **(g)** $\sqrt{3}$, π **(h)** all of them **5.** 45 **7.** $-10{,}000$ **9.** -234 **11.** 100 miles west **13.** **(a)** 12 **(b)** 42 **(c)** 0.1 **(d)** 0 **(e)** 15 **(f)** $\frac{2}{3}$ **(g)** $\sqrt{7}$ **15.** $<$ **17.** $>$ **19.** $>$ **21.** $>$ **23.** $<$ **25.** $\{0, 1, 2, 3\}$ 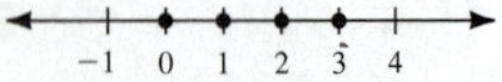

27. $\{-4, -3, -2, -1\}$ 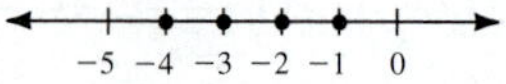

29. 5, 7, 9 **31.** $-2, 0, 2$ **33.** 0.4831561 **35.** $5^{2.047}$

SAY IT IN WORDS

37. ans. vary

ENRICHMENT EXERCISES

1. 3 **2.** x **3.** $-x$ **4.** if $x \geq 0$, $|-x| = x$ and $|x| = x$; if $x < 0$, $|-x| = -x$ and $|x| = -x$; In either case, $|-x| = |x|$ **5.** $a - b = -(b - a)$; so, $|a - b| = |-(b - a)| = |b - a|$ by the previous problem

WARMING UP

1. T **2.** T **3.** F **4.** T **5.** T **6.** F **7.** T **8.** F **9.** F **10.** F

EXERCISE SET 1.4

1. 5 **3.** 2 **5.** 6 **7.** -6 **9.** -10 **11.** -24 **13.** 1 **15.** 10 **17.** 6 **19.** 24 **21.** 0 **23.** $\frac{1}{2}$ **25.** -4 **27.** 6 **29.** -0.4 **31.** 0 **33.** -4 **35.** 12 mph **37.** 765 feet **39.** 5 **41.** -8 **43.** 14

45. 5 **47.** -1 **49.** 9 **51.** 0 **53.** 2 **55.** 3 **57.** -7 **59.** 6 **61.** -12 **63.** 4 **65.** 5 **67.** -6
69. -8 **71.** $-13°$ **73.** 15 days **75.** 1.36 parsecs **79.** -1.04 **81.** 0.0242 **83.** 0.238

SAY IT IN WORDS

85. ans. vary

ENRICHMENT EXERCISES

1. 1,009,999 **2.** 19 **3.** 0.0877 **4.** on the home team's own 41-yard line

WARMING UP

1. T **2.** T **3.** F **4.** T **5.** F **6.** T **7.** T **8.** F **9.** F **10.** T

EXERCISE SET 1.5

1. 35 **3.** -24 **5.** -24 **7.** -3 **9.** -1.2 **11.** 17 **13.** 6 **15.** $-\frac{3}{4}$ **17.** $\frac{7}{3}$ **19.** $\frac{3}{8}$ **21.** $\frac{1}{2}$

23. 2 **25.** $\frac{21}{2}$ **27.** 12 **29.** $\frac{1}{7}$ **31.** $-\frac{1}{7}$ **33.** $\frac{8}{7}$ **35.** $\frac{2}{11}$ **37.** $-\frac{3}{5}$ **39.** $\frac{3}{10}$ **41.** $\frac{3}{2}$ **43.** 0

45. $\frac{2}{5}$ **47.** not defined **49.** -6 **51.** -33 **53.** $\frac{1}{2}$ **55.** $-\frac{7}{9}$ **57.** 5 **59.** $-\frac{1}{12}$ **61.** -20 **63.** 18

65. -6 **67.** $-\frac{2}{3}$ **69.** -1 **71.** $=$ **73.** $\neq$ **75.** $=$ **77.** $=$ **79.** \$0.40 **81.** 880 mg.

83. 11,420,400 households **85.** 0.4 **87.** -250

SAY IT IN WORDS

89. ans. vary

ENRICHMENT EXERCISES

1. 90 **2.** $\frac{100 \cdot 101}{2} = 5050$

WARMING UP

1. T **2.** F **3.** F **4.** T **5.** F **6.** T **7.** F **8.** T **9.** T **10.** T

EXERCISE SET 1.6

1. 7, -8 **3.** 0, -15 **5.** 1, -3 **7.** -4, -4 **9.** 23 **11.** 2 **13.** $-\frac{1}{3}$ **15. (a)** 4 **(b)** 4 **(c)** -4 **(d)** -4

17. (a) twice x **(b)** 4 times the difference of 5 and m **(c)** the sum of a and 2, the result cubed **19. (a)** $x + 11$
(b) $a - 12$ **(c)** $5 + 2y$ **(d)** $3b - 5$ **21. (a)** $7x$ **(b)** $2n$ **(c)** $1.56x$ **23.** 2 is a solution **25.** 1 is not a solution

27. -1 is a solution **29.** $-\frac{1}{2}$ is a solution **31.** 3 is not a solution **33.** 1 is a solution **35.** -5 is not a solution

37. 2 is a solution **39.** $x + 10 = 15$ **41.** $6x = 11$ **43.** $\frac{12}{2x} = 34$ **45.** $10x^2 = 5$

SAY IT IN WORDS

47. ans. vary

ENRICHMENT EXERCISES

1. $2[n + (n + 2)] = (n + 4) + (n + 6) + 6$ **2.** $n = 6$ **3.** 6, 8, 10, 12

WARMING UP

1. T **2.** T **3.** F **4.** T **5.** T **6.** T **7.** F **8.** T **9.** F **10.** T

EXERCISE SET 1.7

1. $5 + x$ **3.** $x + 8$ **5.** $-1 + 3x$ **7.** $-3 + a$ **9.** $6x$ **11.** $-12y$ **13.** $-21c$ **15.** $15x$ **17.** $3z - 3$ **19.** x **21.** $2a$ **23.** x **25.** a **27.** $3z$ **29.** t **31.** $6 + 2x$ **33.** $8 + 4a$ **35.** $6 - 12y$ **37.** $-2 + 2y$ **39.** $4y - 4$ **41.** $-2 - x$ **43.** $-1 + y$ **45.** $x + 2$ **47.** $1 - 2b$ **49.** $4 + s$ **51.** $6x + 2$ **53.** $-9 - 3x$ **55.** $1 - x$ **57.** $14 - 4x$ **59.** $0.408 - 1.598x$ **61.** associative property of addition **63.** multiplicative inverse property **65.** additive inverse property **67.** multiplicative identity property **69.** commutative property of addition **71.** additive identity property **73.** distributive property **75.** $4 + 2x$ **77.** $x + y$ **79.** 0 **81.** not commutative **83.** commutative

SAY IT IN WORDS

85. ans. vary

ENRICHMENT EXERCISES

1. yes; yes **2.** no; for example, $4 - 3 = 1$, but $3 - 4 = -1$.

CHAPTER 1 REVIEW EXERCISE SET

Section 1.1

1. $\frac{1}{2}$ **2.** $\frac{1}{5}$ **3.** $\frac{6}{5}$ **4.** $\frac{3}{7}$ **5.** $\frac{1}{3}$ **6.** $\frac{8}{3}$ **7.** $\frac{5}{6}$ **8.** $\frac{3}{4}$ **9.** $\frac{3}{2}$ **10.** 1 **11.** $\frac{5}{7}$ **12.** $\frac{5}{4}$ **13.** $\frac{2}{3}$ **14.** $\frac{5}{16}$

Section 1.2

15. $x > 2$ **16.** $z \leq 9$ **17.** $a \neq 5$ **18.** 2 **19.** 51 **20.** 15 **21.** $\frac{25}{3}$ **22.** 3

Section 1.3

23. (a)–(f) [number line with points at $-2\frac{1}{2}$, 0.25, 1.5, π; ticks −4 −3 −2 −1 0 1 2 3 4] **24.** < **25.** < **26.** < **27.** > **28.** > **29.** 7 **30.** 5.65 **31.** 8 **32.** $\frac{1}{4}$ **33.** π

Section 1.4

34. 3 **35.** 0 **36.** -2 **37.** 1 **38.** -8

Section 1.5

39. 5.2 **40.** -7 **41.** 0 **42.** not defined **43.** 9

Section 1.6

44. 13 **45.** 6 **46.** $\frac{1}{8}$ **47.** -9 **48.** $y - 12$ **49.** $z + 9$ **50.** $6(2 - x)$ **51.** $x - 6 = 10$ **52.** $x \cdot 4 = 36$ **53.** $\frac{30}{x^2} = 5$ **54.** no solutions from the set **55.** $2x + 3 = 7; x = 2$ **56.** $3 - |x| = 1; x = 2$ **57.** $4x = x + 3;$ $x = 1$

Section 1.7

58. $18y$ **59.** $2 - 7b$ **60.** a **61.** $17 - y$ **62.** commutative property of multiplication **63.** multiplicative identity property **64.** commutative property of addition **65.** multiplicative inverse property **66.** distributive property

CHAPTER 1 TEST

1. $\frac{4}{5}$ **2.** $\frac{2}{35}$ **3.** $\frac{11}{12}$ **4.** $x + 3 \geq 5$ **5.** 8 **6.** 16 **7.** 1 **8.** -120 **9.** $>$ **10.** $<$ **11.** 4.1 **12.** -5 **13.** $\frac{3}{5}$ **14.** -21 **15.** 12 **16.** -21 **17.** $\frac{3}{8}$ **18.** $\frac{1}{7}$ **19.** not defined **20.** -8 **21.** 0 **22.** 13 **23.** -27 **24.** $2x - 3 = 11$ **25.** commutative property of addition **26.** $4x - 20$ **27.** $x - 10$ **28.** $4u - 2$

CHAPTER 2

WARMING UP

1. F **2.** T **3.** T **4.** T **5.** T **6.** T **7.** F **8.** T **9.** T **10.** F

EXERCISE SET 2.1

1. x; 2 **3.** x^2; -3 **5.** y; 1 **7.** x^2y; -1 **9.** a; $\frac{1}{3}$ **11.** x; $\frac{3}{5}$ **13.** $8x$ **15.** $2t$ **17.** c **19.** $8x$ **21.** $-8z$ **23.** $-2b$ **25.** $5x - 4$ **27.** $10 - 8x^2$ **29.** $-xy + 3a$ **31.** $4x - 4$ **33.** $z - 6$ **35.** $-\frac{y^2}{6} + \frac{3x}{2}$ **37.** $-5.59x + 2.93$ **39.** $-4x + 6$ **41.** $1 - 13y^2$ **43.** $x^2 + 2$ **45.** $x - 7y$ **47.** $-8y + 5x^2$ **49.** $3x + 8y - 3$ **51.** $-u - 8$ **53.** $23.17a - 47.002$ **55.** $3(x + 4)$; $3x + 12$ **57.** $6 - 3(a + 12)$; $-30 - 3a$ **59.** $6(2 + x) - x$; $5x + 12$ **61.** $3x - 2(x - 5)$; $x + 10$

SAY IT IN WORDS

63. ans. vary

REVIEW EXERCISES

65. -10 **67.** 2 **69.** 1

ENRICHMENT EXERCISES

1. $-1 - 2x$ **2.** $-3 + 3x + 28y$ **3.** $-19x + 18x^2$ **4.** ans. vary

WARMING UP

1. F **2.** T **3.** T **4.** F **5.** T **6.** F **7.** T **8.** T **9.** F **10.** T

EXERCISE SET 2.2

1. $\{-1\}$ **3.** $\{-3\}$ **5.** $\{8\}$ **7.** $\left\{-\frac{7}{4}\right\}$ **9.** $\{17\}$ **11.** $\{-1\}$ **13.** $\{2\}$ **15.** $\{3\}$ **17.** $\{-8\}$ **19.** $\{-6\}$ **21.** $\{-7\}$ **23.** $\left\{\frac{2}{5}\right\}$ **25.** $\{-4\}$ **27.** $\left\{\frac{3}{2}\right\}$ **29.** $\left\{\frac{2}{3}\right\}$ **31.** $\left\{-\frac{3}{2}\right\}$ **33.** $\left\{\frac{2}{3}\right\}$ **35.** $\{16\}$ **37.** $\{-7\}$ **39.** $\{-10\}$ **41.** $\{2\}$ **43.** $\left\{\frac{1}{2}\right\}$ **45.** $\{0.4\}$ **47.** $\{1.1\}$ **49.** $5x + 7 = 4x - 10$; $\{-17\}$ **51.** $12 - 6x = 11 - 5x$; $\{1\}$ **53.** $\frac{x}{3} = 5$; $\{15\}$

SAY IT IN WORDS

55. ans. vary

REVIEW EXERCISES

57. $-2x - 6$ **59.** $12 - 2a$ **61.** $1 - 12t$

ENRICHMENT EXERCISES

1. $5(4 + 2x) = 3(3x - 2)$; $x = -26$ **2.** $a = -1$ **3.** $a = -8$ **4.** Method 1: $3x + 5 = 14$; $3x + 5 - 5 = 14 - 5$; $3x = 9$; $x = 3$; Method 2: $3x + 5 = 14$; $x + \frac{5}{3} = \frac{14}{3}$; $x + \frac{5}{3} - \frac{5}{3} = \frac{14}{3} - \frac{5}{3}$; $x = \frac{9}{3}$ or 3

WARMING UP

1. T **2.** T **3.** F **4.** T **5.** F **6.** T **7.** T **8.** T

EXERCISE SET 2.3

1. $\{-3\}$ **3.** $\left\{\frac{1}{2}\right\}$ **5.** $\{1\}$ **7.** $\left\{-\frac{3}{2}\right\}$ **9.** $\{1\}$ **11.** $\{4\}$ **13.** $\left\{\frac{8}{3}\right\}$ **15.** $\{3\}$ **17.** $\{1\}$ **19.** $\{1\}$ **21.** $\{1\}$ **23.** $\{2\}$ **25.** $\left\{-\frac{1}{6}\right\}$ **27.** $\{-5\}$ **29.** $\left\{\frac{7}{2}\right\}$ **31.** $\{-5\}$ **33.** $\left\{-\frac{1}{3}\right\}$ **35.** $\{1\}$ **37.** $\varnothing$ **39.** the set of real numbers **41.** $\{1.4\}$ **43.** $\{-597\}$ **45.** $4x + 5 = 3$; $x = -\frac{1}{2}$ **47.** $2x + \frac{4}{3} = 6$; $x = \frac{7}{3}$ **49.** $2 + \frac{3}{4}x = -\frac{1}{4}$; $x = -3$

SAY IT IN WORDS

51. ans. vary

REVIEW EXERCISES

53. $x - 14$ **55.** $\frac{x}{10}$ **57.** $6x$

ENRICHMENT EXERCISES

1. $\left\{\frac{4}{3}\right\}$ **2.** $x = -\frac{b}{a}$ **3.** $x = \frac{d-b}{a-c}$ **4.** $x = d$ **5.** $x = \frac{3}{1-2a}$

WARMING UP

1. T **2.** F **3.** T **4.** F **5.** F **6.** T **7.** F **8.** T

EXERCISE SET 2.4

1. $x + 6$ **3.** $x - 7$ **5.** $4x$ **7.** $5(x + 1)$ **9.** $0.65x$ **11.** $\frac{1}{x+25}$ **13.** $4(x - 7)$ **15.** -2 **17.** -2 **19.** -16 **21.** 6 **23.** 10 **25.** 0 **27.** 6 **29.** 50 **31.** 500 **33.** 10 years old **35.** 16 **37.** $91

SAY IT IN WORDS

39. ans. vary

REVIEW EXERCISES

41. $3y$ **43.** $14t$

ENRICHMENT EXERCISES

1. 46 and 48 **2.** -39 and -37 **3.** 17, 19, and 21

WARMING UP

1. T **2.** F **3.** F **4.** T **5.** T **6.** F **7.** T **8.** T

EXERCISE SET 2.5

1. Let x = one of the numbers, then $6 - x$ = the other number **3.** Let x = one of the numbers, then $-7 - x$ = the other number **5.** Let x = the selling price, then $.55x$ = the cost **7.** Let x = the regular price, then $.8x$ = the sale price **9.** Let x = the length of one piece, then $14 - x$ = the length of the other piece **11.** Let x = the number of apples, then $50 - x$ = the number of oranges **13.** Let x = Mr. Abbot's income, then $110{,}000 - x$ = Mrs. Abbot's income. **15.** $137 **17.** $210 **19.** $200 **21.** $80 **23.** 38 and 40 **25.** -15, -13, and -11 **27.** 5 feet and 10 feet **29.** 10 feet and 6 feet **31.** 10 dimes and 16 quarters **33.** 7 five-dollar bills and 9 ten-dollar bills **35.** 10 gallons of the 20% solution and 15 gallons of the 30% solution **37.** 8 gallons of 1% milk and 4 gallons of 4% milk **39.** 60 liters **41.** $13,750 **43.** 8 fish

REVIEW EXERCISES

45. $>$ **47.** $>$ **49.** $<$

ENRICHMENT EXERCISES

1. 4 nickels, 16 dimes, and 7 quarters **2.** $-5, -3, -1, 1, 3$

WARMING UP

1. T **2.** F **3.** F **4.** T **5.** T **6.** F **7.** F **8.** T

EXERCISE SET 2.6

1. $x > 10$ **3.** $x \geq 9$ **5.** $x \leq 7$ **7.** −1 0 1 2 3 4 5 6 7 **9.** −1 0 1 2 3

11. −2 −1 0 1 2 **13.** −1 0 1 **15.** −1 0 1

17. $x \leq 10$ 0 1 2 3 4 5 6 7 8 9 10 **19.** $x > 0$ −1 0 1 2 **21.** $x < 20$ −10 0 10 20

23. $x < -1$ −1 0 1 **25.** $-2 < x < 3$ −2 −1 0 1 2 3

27. $0 \leq x < 3$ −1 0 1 2 3 4 **29.** $-2 < x < 0$ −2 −1 0

31. $x \geq -2$ −2 −1 0 1 **33.** $x < 6$ −1 0 1 2 3 4 5 6 **35.** $x \leq -3$ −4 −3 −2 −1 0 1

37. $x > 1$ −1 0 1 2 **39.** $x < -2$ −4 −3 −2 −1 0 1 **41.** $t \leq 3$ −1 0 1 2 3 4

43. $x > 2$ −1 0 1 2 **45.** $y \leq \frac{3}{2}$ 0 1 $\frac{3}{2}$ 2 **47.** $-3 < x < -2$ −4 −3 −2 −1 0

49. $-1 < x < -\frac{1}{2}$ −1 $-\frac{1}{2}$ 0 **51.** $-1 \leq x < 2$ −1 0 1 2

53. $x \leq -\frac{15}{4}$ **55.** $-\frac{5}{3} < w \leq 3$ **57.** $\frac{1}{2} < x < 2$ **59.** $\frac{1}{2} \leq y < 1$ **61.** $x > -0.1$ **63.** $n \geq 1$

65. $x + 2 < 1$; $x < -1$ −2 −1 0 1 2 **67.** $x - 3 \leq 1$; $x \leq 4$ −2 −1 0 1 2 3 4 5

69. at least 185 **71.** at least 200,000 boxes

73. No. x cannot be less than −4 and greater than −1 at the same time. **75.** No. x cannot be less than 2 and greater than 5 at the same time.

SAY IT IN WORDS

77. ans. vary

REVIEW EXERCISES

79. 14 **81.** 4 **83.** 12.56

ENRICHMENT EXERCISES

1. $-2 < x < 2$ **2.** $-3 < x < 3$ **3.** $1 < x < 3$ **4.** $-9 \leq x \leq 1$ **5.** $\varnothing$ **6.** The set of real numbers except 0

WARMING UP

1. T **2.** F **3.** T **4.** F **5.** T

EXERCISE SET 2.7

1. 1 hour **3.** $1\frac{1}{2}$ hours **5.** 3 hours **7.** 1 hour **9.** 5 hours **11.** \$120 **13.** \$700 **15.** \$5,000 at 4% and

\$3,000 at 5% **17.** \$12,000 at 5% and \$8,000 at 8% **19.** 5 feet by 9 feet **21.** 9 inches by 12 inches **23.** 6 feet **25.** $r = \dfrac{C}{2\pi}$ **27.** $I = d^2E$ **29.** $w = \dfrac{P - 2l}{2}$ **31.** $x = \dfrac{y - 1}{3}$ **33.** $x = \dfrac{3y + 1}{2}$ **35.** $x = \dfrac{2y + 1}{4}$ **37.** $x = \dfrac{y - b}{m}$ **39.** $y = \dfrac{2x + 1}{3}$ **41.** $y = 3x - 5$ **43.** $y = \dfrac{4}{3}x - 4$ **45.** $y = -2$ **47.** $x = 15$ **49.** 22.464 square feet **51.** \$11,200 at 11% and \$12,400 at 9.5%

ENRICHMENT EXERCISES

1. $y = -\dfrac{A}{B}x - \dfrac{C}{B}$ **2.** $y = -\dfrac{b}{a}x + b$ **3.** $T_3 = \dfrac{T_2V_3}{V_2}$

CHAPTER 2 REVIEW EXERCISE SET

Section 2.1

1. $5y$ **2.** $2st$ **3.** $-1\frac{1}{2}$ **4.** $30z + 21$ **5.** $21x^2yz - 10$ **6.** $-9ab^2 - a^2b$

Section 2.2

7. $\{1\}$ **8.** $\{-6\}$ **9.** $\{-14\}$ **10.** $\{1\}$ **11.** $\{-12\}$ **12.** $\{-14\}$ **13.** $\{-30\}$ **14.** $\{-7\}$ **15.** $\left\{\dfrac{7}{9}\right\}$ **16.** $6x + 3 = 5x$, $x = -3$

Section 2.3

17. $\{-2\}$ **18.** $\{-5\}$ **19.** $\{1\}$ **20.** $\left\{\dfrac{2}{5}\right\}$

Section 2.4

21. $x - 17$ **22.** $5 - 5x$ **23.** $3x + 9$ **24.** $\dfrac{3x}{5}$ **25.** $\dfrac{1}{x + 10}$ **26.** $\dfrac{1}{x} + 4$ **27.** $x - 0.12x$ **28.** 15

Section 2.5

29. let x = one of the numbers, then $17 - x$ = the other number **30.** let n = the smaller odd integer, then $n + 2$ = the next consecutive odd integer **31.** let x = the smaller number, then $x + 50$ = the other number **32.** \$250 **33.** \$29 **34.** -28 and -26 **35.** 6 feet and 12 feet **36.** 15 gallons of 14% solution; 5 gallons of 22% solution

Section 2.6

37. $x < 250$ **38.** $x \geq 300$ **39.** $x \leq 200$ **40.** $x \geq 100$ **41.**

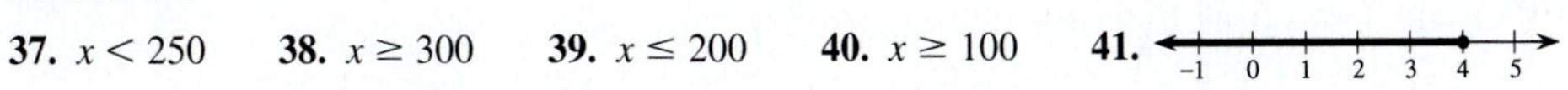

42.

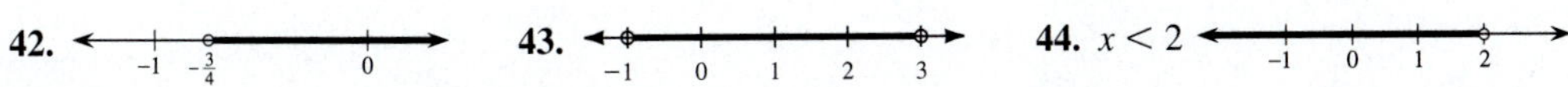

43.

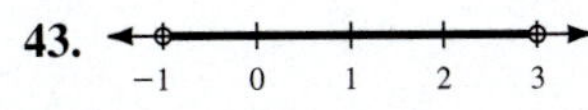

44. $x < 2$ (−1 0 1 2)

45. $2 < x \leq 5$ (0 1 2 3 4 5 6) **46.** $x \geq -1$ (−2 −1 0 1 2) **47.** $x > -2$

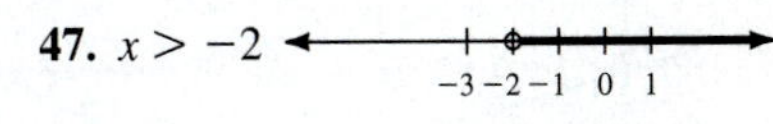

48. $0 \leq x < 1$ (0 1) **49.** $x > 1$ (−1 0 1)

50. $-2 \leq t \leq 3$ (−3 −2 −1 0 1 2 3 4) **51.** $2x + 6 \geq x + 3$; $x \geq -3$ (−3 −2 −1 0 1 2)

52. $x > -\dfrac{1}{2}$ **53.** At least \$39

Section 2.7

54. 15 minutes **55.** \$96 **56.** \$10,000 at each interest rate **57.** 3 hours **58.** 3 feet by 7 feet **59.** $b = \frac{2A}{h}$
60. $x = \frac{20 - y}{5}$

CHAPTER 2 TEST

1. $6x^2 + 8x - 8$ **2.** $2x^5 + 15$ **3.** $-2(6 - 2x)$; $-12 + 4x$ **4.** $x = -6$ **5.** $x = -19$ **6.** $2x + 5 = x - 2$; $x = -7$ **7.** $n = -3$ **8.** $x = 30$ **9.** $z = -1$ **10.** $a = 2$ **11.** $x - 5$ **12.** $.15x$ **13.** 3 and 15
14. 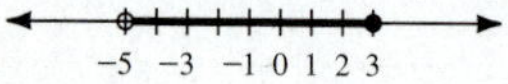
15. $x \geq 2$ 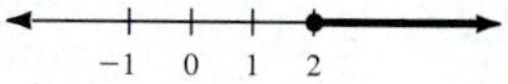
16. $-5 \leq x < 4$ **17.** $x < 18$ **18.** 8 liters
19. 3 feet by 7 feet **20.** $y = 2x - \frac{1}{2}$

CHAPTER 3

WARMING UP

1. T **2.** F **3.** T **4.** F **5.** F **6.** F **7.** T **8.** T **9.** F **10.** T

EXERCISE SET 3.1

1. $A(1, 2)$; $B(1, -4)$; $C(3, 2)$; $D(-2, 5)$; $E(-4, 0)$; $F(0, 4)$; $G(5, 0)$; $H(0, -2)$ **3.** first quadrant **5.** the origin

7. 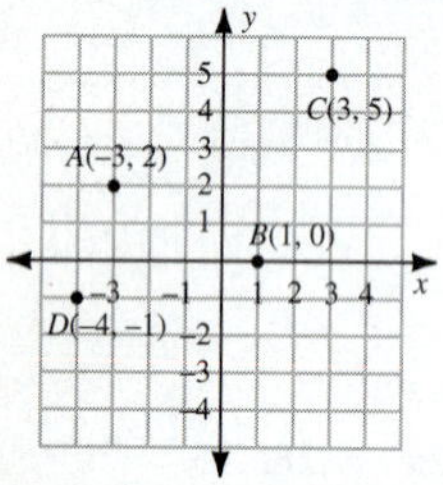

9. 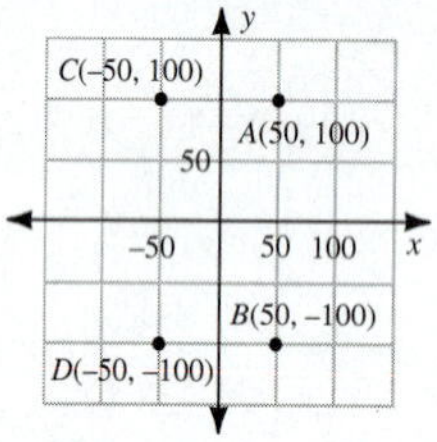

11. a solution **13.** not a solution **15.** $(0, -4)$

17.

x	y
0	-5
$-\frac{5}{3}$	0
-2	1
-3	4

19. 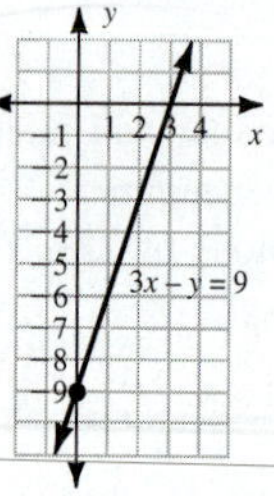

21. 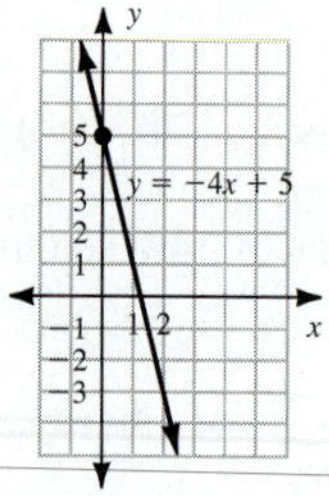

23. 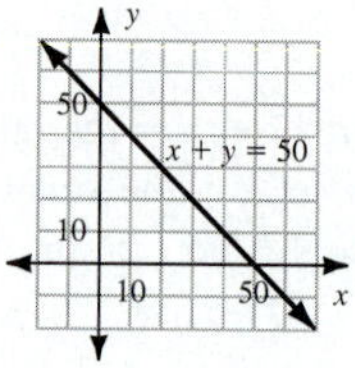

25. 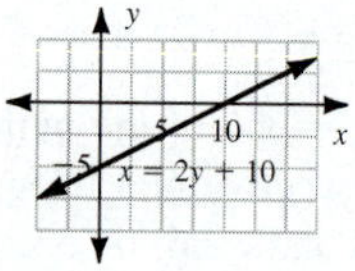

27. 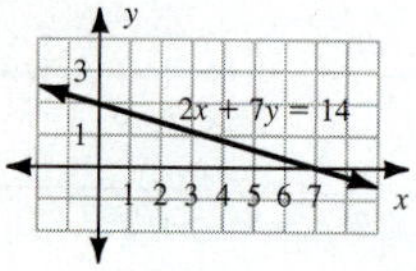

29. 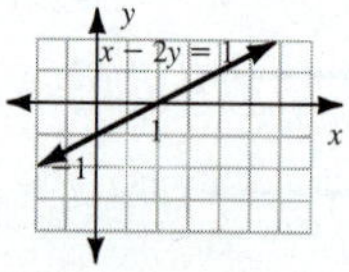

31. 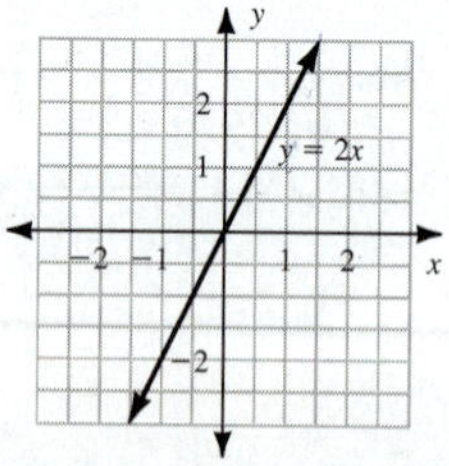

33.

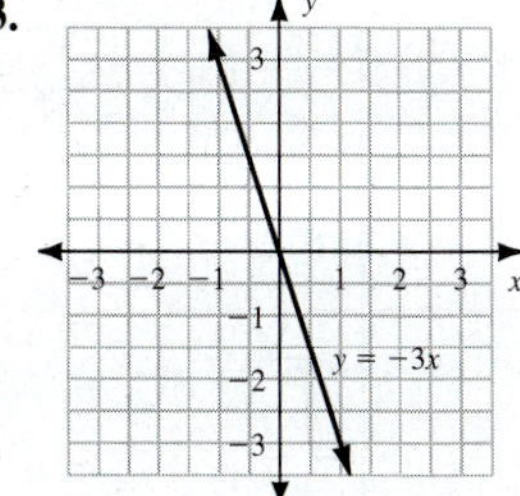

35.

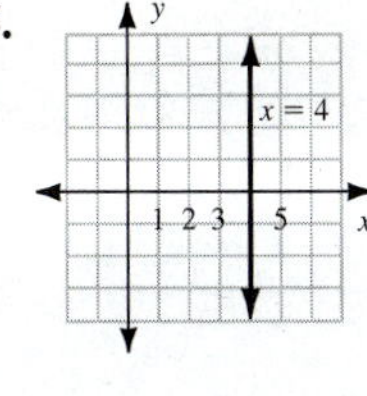

37.

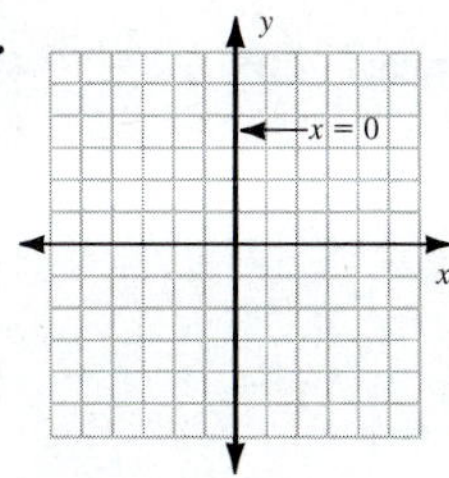

39.

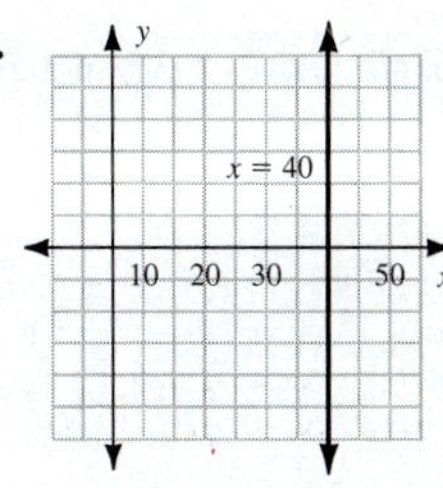

41. $y + 2x = -2$

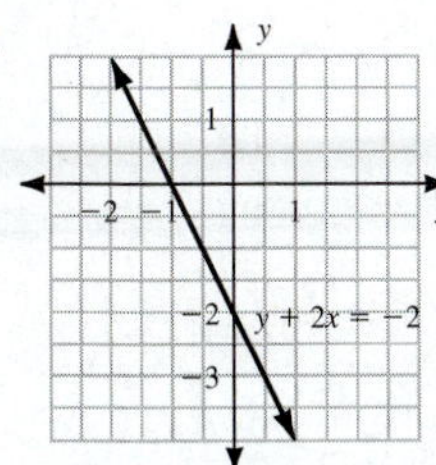

43. $y = 2(x - 5)$

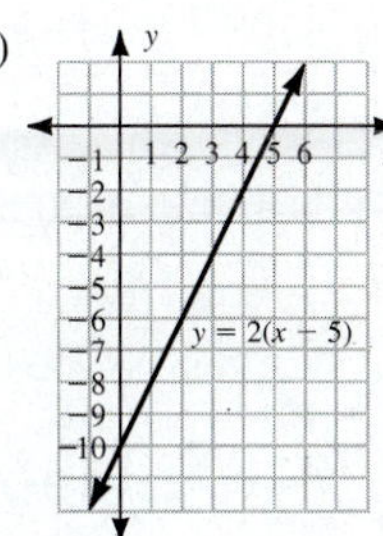

45.

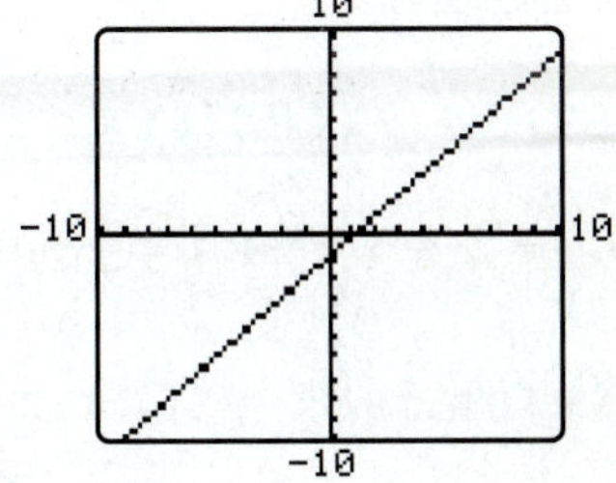

47.

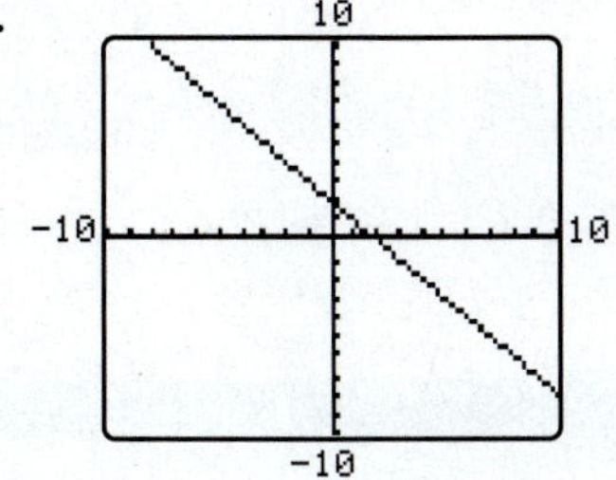

SAY IT IN WORDS

49. (x, y) is an ordered pair of real numbers, whereas $\{x, y\}$ is an unordered pair of real numbers.

REVIEW EXERCISES

51. $\frac{1}{2}$ **53.** 1

ENRICHMENT EXERCISES

1. (3, 3) **2.** (2, 7) **3.** (5, 2) **4.** $m = \frac{5}{3}$ **5.** $c = 2$

WARMING UP

1. F **2.** T **3.** F **4.** F **5.** T **6.** T **7.** F **8.** T

EXERCISE SET 3.2

1. $\frac{4}{7}$ **3.** 2 **5.** -1 **7.** undefined **9.** 9 **11.** $-\frac{3}{2}$ **13.** $-\frac{5}{8}$ **15.** $-\frac{9}{4}$ **17.** 0 **19.** undefined
21. 1.43 **23.** -0.417 **25.** parallel **27.** not parallel **29.** not parallel **31.** parallel **33.** perpendicular

35. perpendicular **37.** not perpendicular **39.** $-\frac{3}{2}$ **41.**

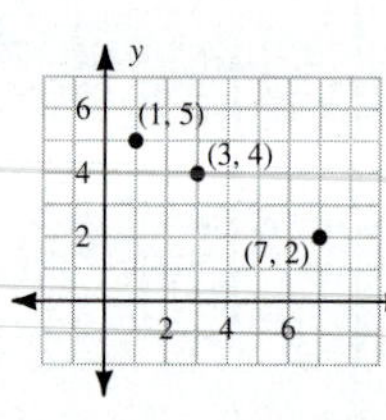

43.

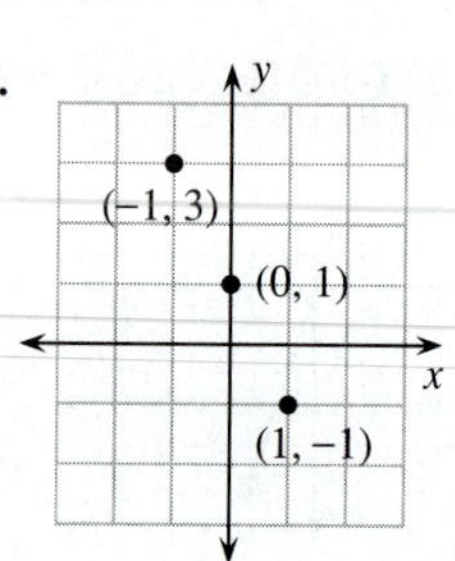

45. \$425

SAY IT IN WORDS

47. ans. vary

REVIEW EXERCISES

49. $y = 2x + 5$ **51.** $y = \frac{5}{3}x - \frac{1}{4}$

ENRICHMENT EXERCISES

1. 1 **2.** $\frac{1}{6}$ **3.** $k = 2$ **4.** $-\frac{1}{9}$ **5.** line L_1

WARMING UP

1. F **2.** T **3.** T **4.** T **5.** T **6.** F **7.** T **8.** T

EXERCISE SET 3.3

1. slope = 3; y-intercept = 2 **3.** slope = −4; y-intercept = −1 **5.** slope = 1; y-intercept = 6 **7.** slope = −1; y-intercept = −4 **9.** slope = 1; y-intercept = 0 **11.** slope = 1; y-intercept = 0 **13.** slope = −3; y-intercept = −6 **15.** slope = −5; y-intercept = 1 **17.** slope = $\frac{1}{2}$; y-intercept = −4 **19.** No, the slope is undefined. No, the line does not cross the y-axis. **21.**

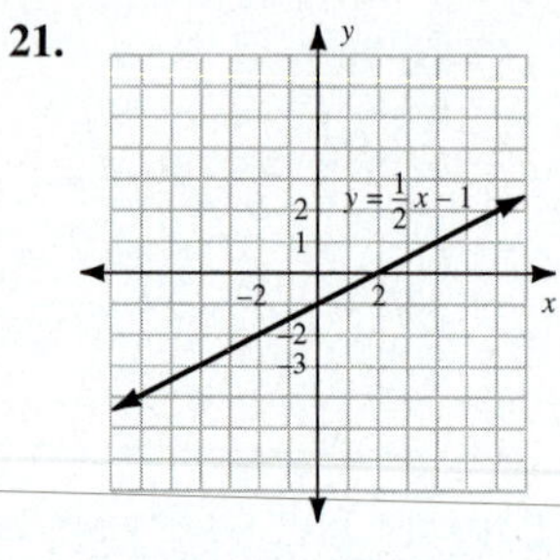

23.

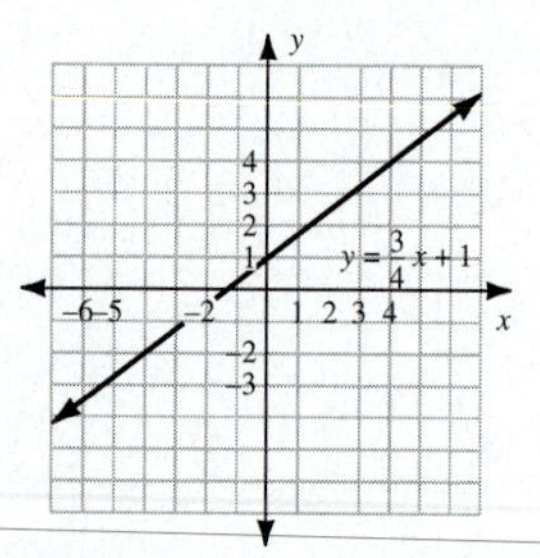

25.

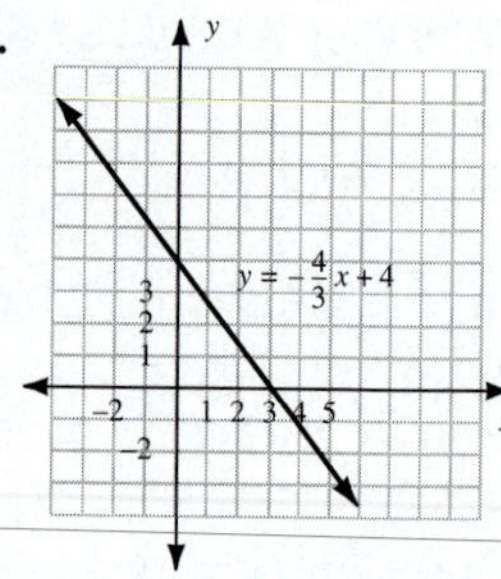

27.

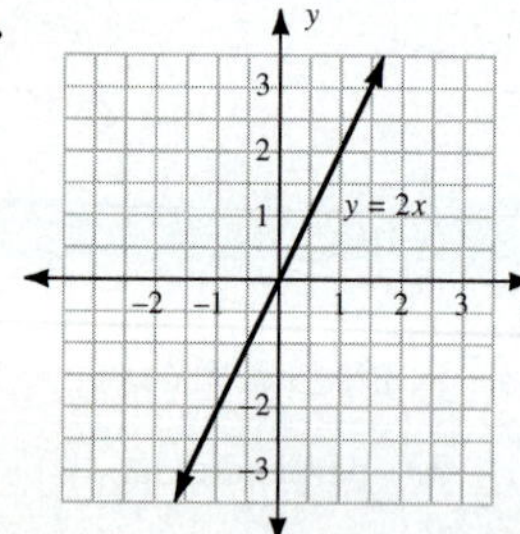

29.

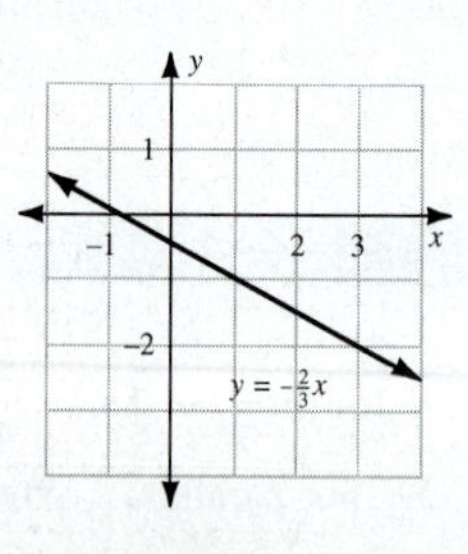

31.

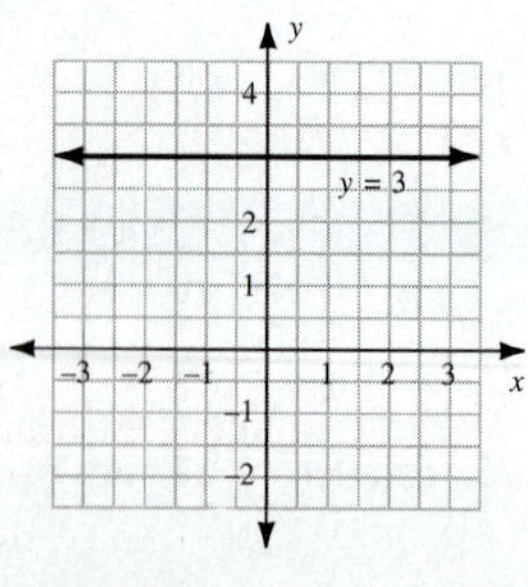

33. 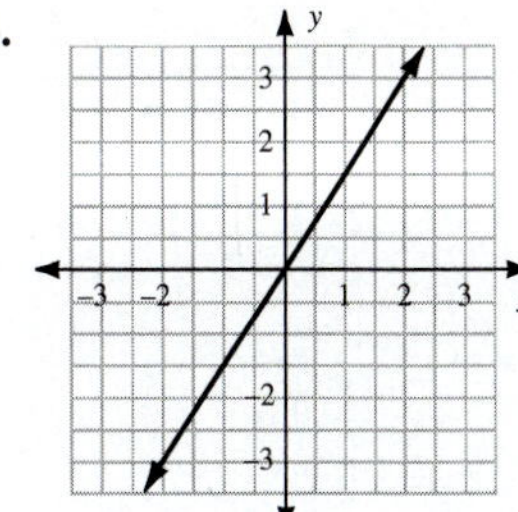

35. 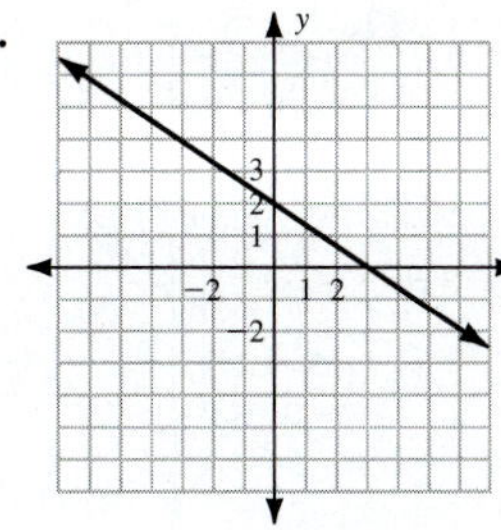

37. 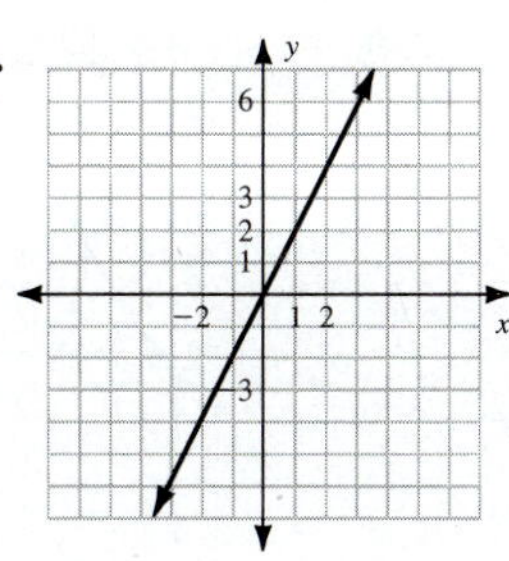

39. 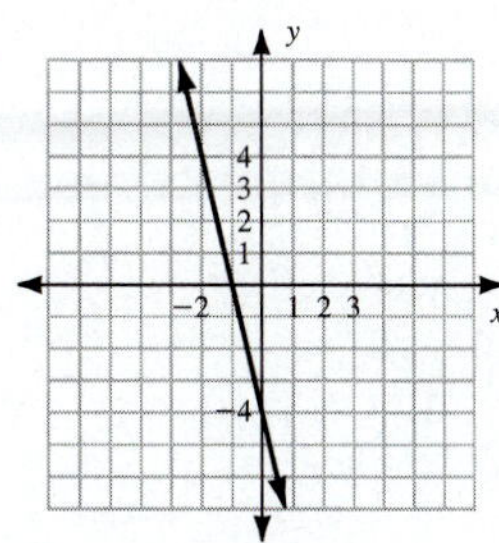

41. $-3x + y = 2$ **43.** $-x + 2y = -8$ **45.** $4x + y = 11$ **47.** $x + 2y = 11$

49. $x = 5$ **51.** $x - 2y = 2$ **53.** $y = 2x$ **55.** $y = -\frac{1}{2}x + \frac{7}{2}$ **57.** $y = -1$ **59.** $y = -7.069x - 75.2349$

61. $y = 1.384x - 4.338$ **63.** parallel **65.** not parallel

SAY IT IN WORDS

67. ans. vary

REVIEW EXERCISES

69. $x < 5$ **71.** $\frac{1}{3} \leq x < 4$

ENRICHMENT EXERCISES

1. $x = 0$ **2.** $y = 0$ **3.** $2x + y = 7$ **4.** $x + 3y = 5$ **5.** $y = -2$ **6.** $x = -5$ **7.** $y = -1$ **8.** $x = 3$

WARMING UP

1. T **2.** F **3.** T **4.** T

EXERCISE SET 3.4

1. 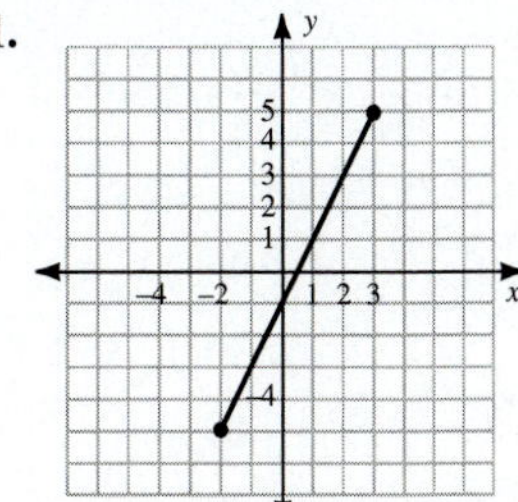

3. 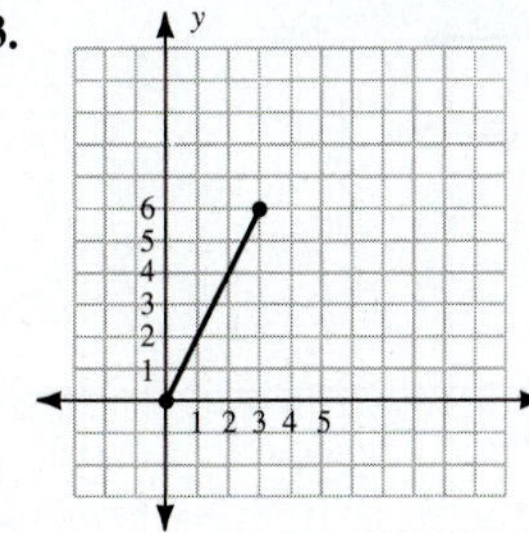

5. 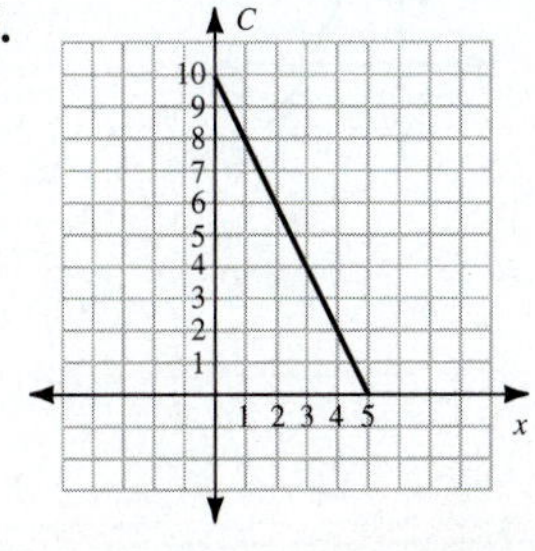

7.

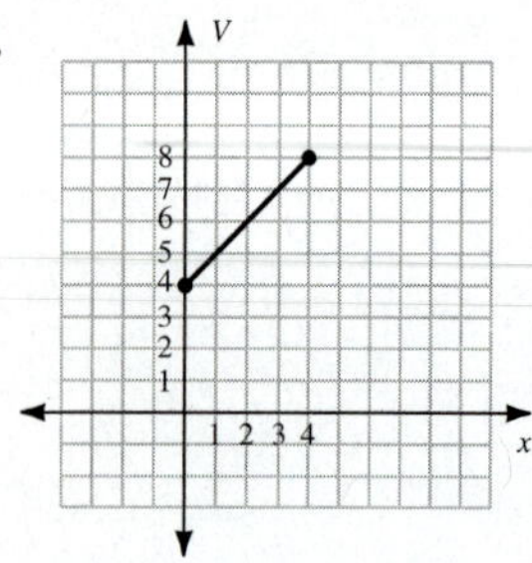

9.

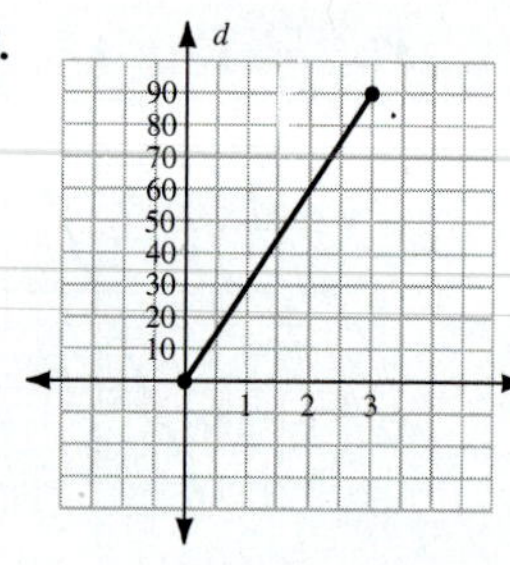

11.

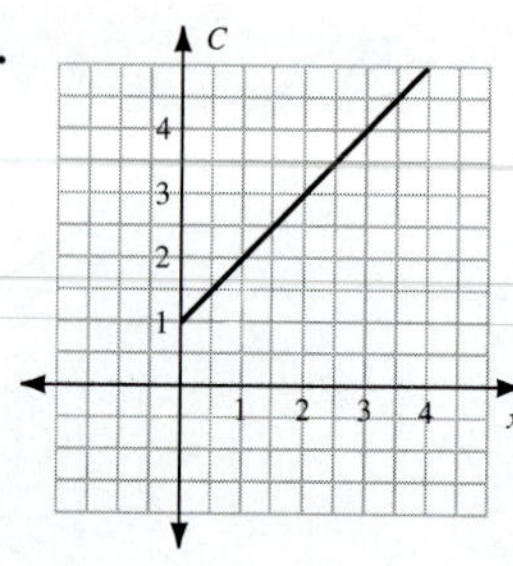

13.

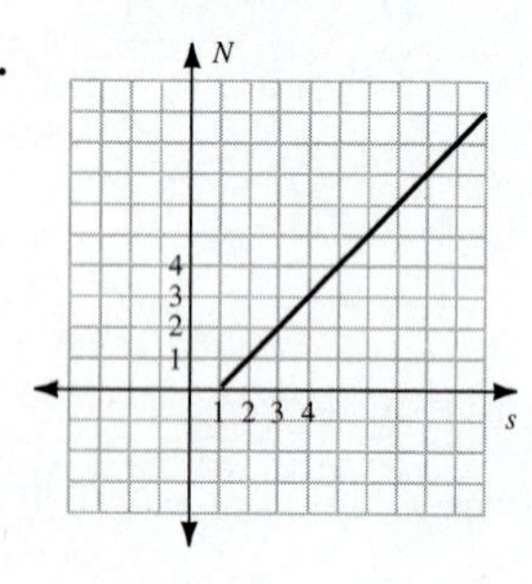

15. **(a)** $C = 10x + 20,\ x \geq 0$; **(b)**

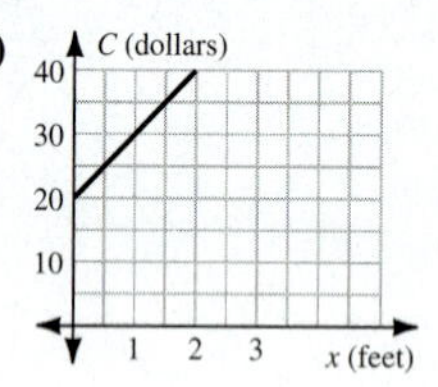

17. **(a)** $d = 20t,\ 0 \leq t \leq 3$; **(b)**

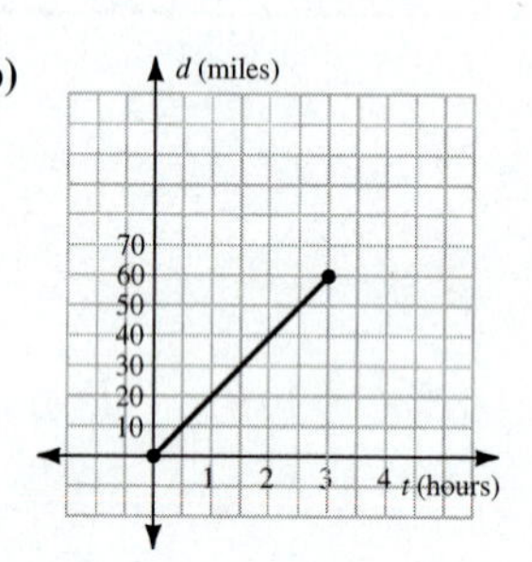

19. **(a)** $V = -2t + 50,\ 0 \leq t \leq 20$; **(b)**

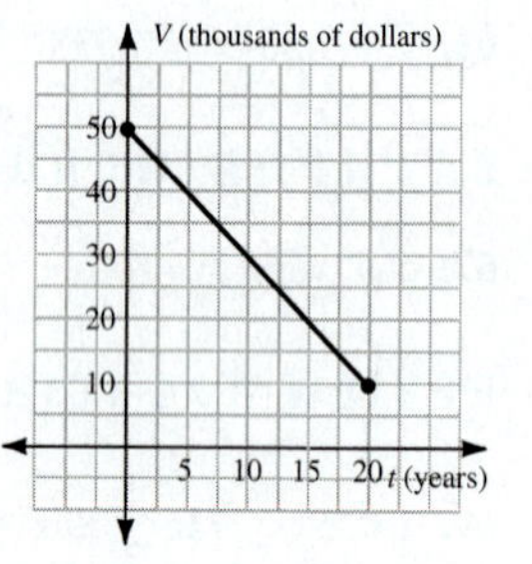

SAY IT IN WORDS

21. ans. vary

REVIEW EXERCISES

23.

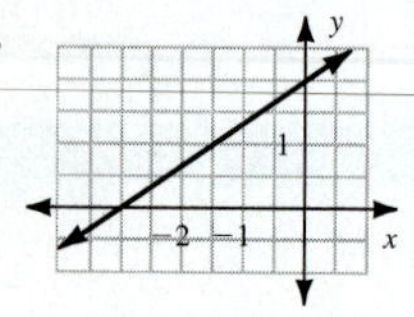

25.

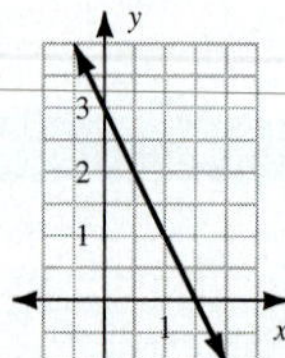

27.

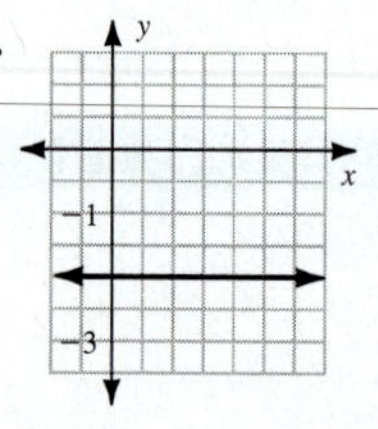

ENRICHMENT EXERCISES

1. 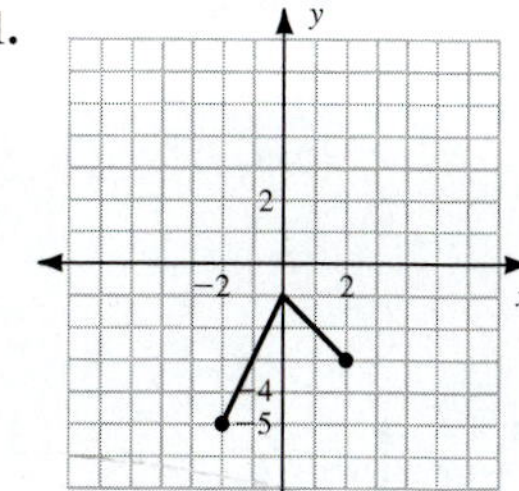

2. 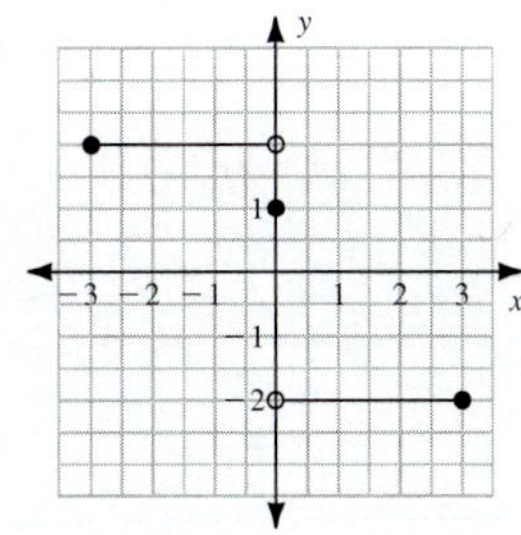

3. 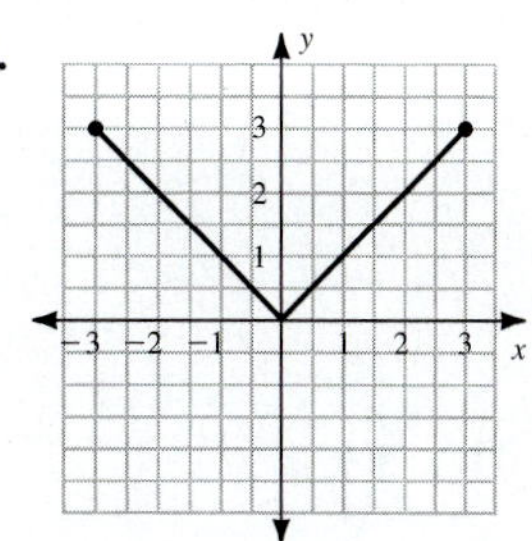

4. 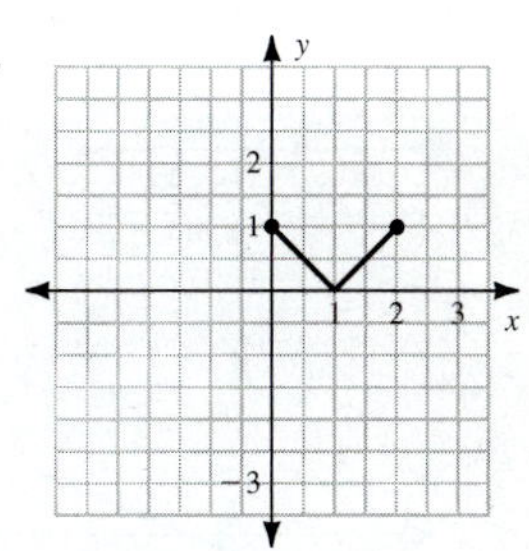

WARMING UP

1. F **2.** T **3.** F **4.** T **5.** T **6.** F

EXERCISE SET 3.5

1. domain = $\{-3, -1, 1, 2\}$; range = $\{-2, -1, 2, 3\}$; It is a function. **3.** domain = $\{2, 3, 7\}$; range = $\{-6, -5, -2, -1\}$; It is not a function. **5.** domain = $\{-5, -3, 4, 6\}$; range = $\{-3, -1, 0, 1\}$; It is a function. **7.** domain = $\{-5, -3, -2, -1\}$; range = $\{-3, -2, 2, 10\}$; It is a function. **9.** $S = \{(-3, -4), (-3, 4), (3, -4), (3, 4)\}$; domain = $\{-3, 3\}$; range = $\{-4, 4\}$; S is not a function. **11.** $S = \left\{\left(-\frac{5}{2}, 1\right), \left(-\frac{3}{2}, 0\right), (0, 1), (1, 0), (2, 1)\right\}$; domain = $\left\{-\frac{5}{2}, -\frac{3}{2}, 0, 1, 2\right\}$; range = $\{0, 1\}$; S is a function. **13.** The relation is not a function. **15.** The relation is also a function. **17.** The relation is not a function. **19.** **(a)** yes; **(b)** $\{t \mid 0 \le t \le 12\}$; **(c)** after 6 hours; **(d)** 75 mg/dl **21.** 2 **23.** 11 **25.** $3a + 2$ **27.** 5 **29.** 11 **31.** $4s^2 - 2s + 5$

SAY IT IN WORDS

33. ans. vary

REVIEW EXERCISES

35. 9 **37.** $\frac{1}{4}$ **39.** 32 **41.** 25

ENRICHMENT EXERCISES

1. 0 **2.** $\frac{1}{2}$ **3.** 2 **4.** $\frac{a}{a+1}$ **5.** $\frac{1}{3}$ **6.** $\frac{1}{4}$

CHAPTER 3 REVIEW EXERCISE SET

Section 3.1

1. $A(0, 2)$ $B(6, 0)$ $C(1, 5)$ $D(-3, -4)$ $E(-7, 4)$ $F(9, -6)$ **2.** above the x-axis **3.** to the left of the y-axis

4.

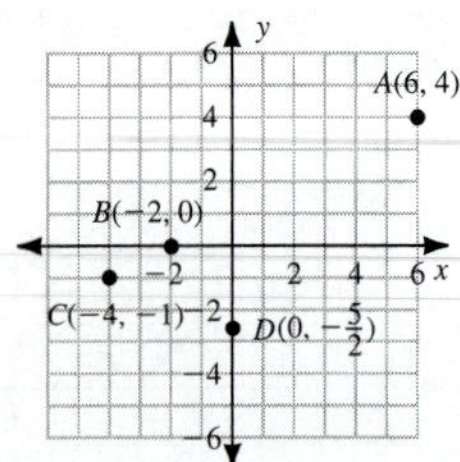

5. It is a solution. **6.** It is not a solution. **7.** (0, 5) **8.** (−2, 0) **9.** (2, 10)

10.

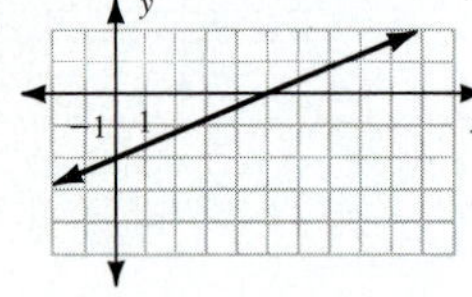

11.

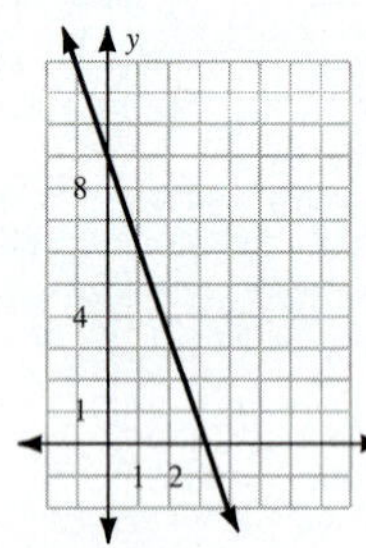

12.

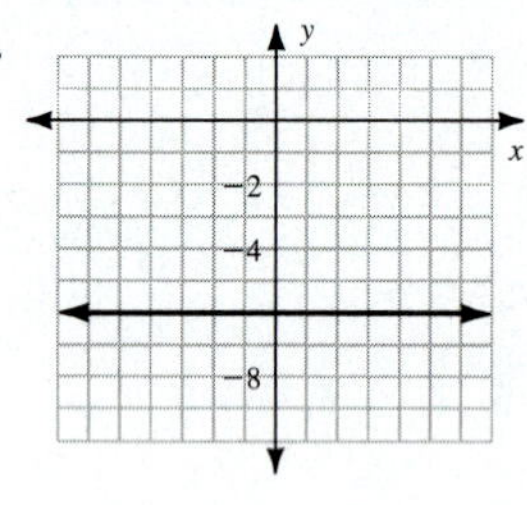

13.

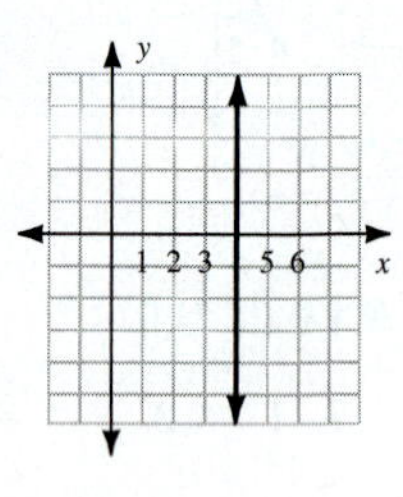

Section 3.2

14. slope $= \frac{5}{6}$ **15.** The slope is undefined. **16.** slope $= 0$ **17.** slope $= -2$ **18.** not parallel **19.** parallel

Section 3.3

20. slope $= \frac{7}{2}$; y-intercept $= -12$ **21.** slope $= \frac{5}{3}$; y-intercept $= -6$ **22. (a)**

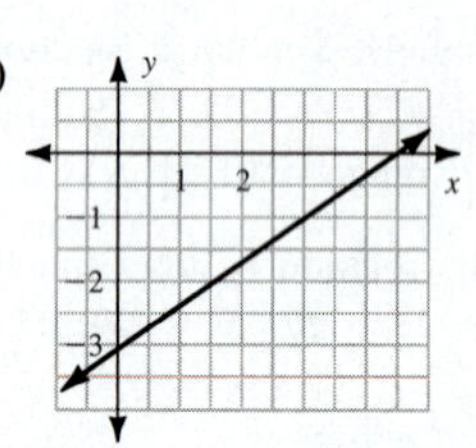

(b) $y = \frac{2}{3}x - 3$

23. (a)

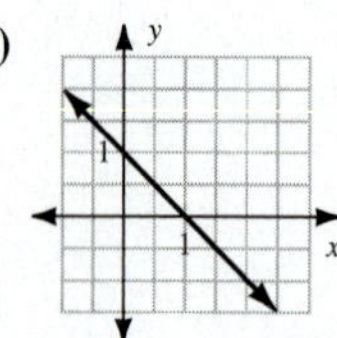

(b) $y = -x + 1$ **24.** $-x + y = 5$ **25.** $9x + 4y = -7$ **26.** $y = x + 1$

Section 3.4

27.

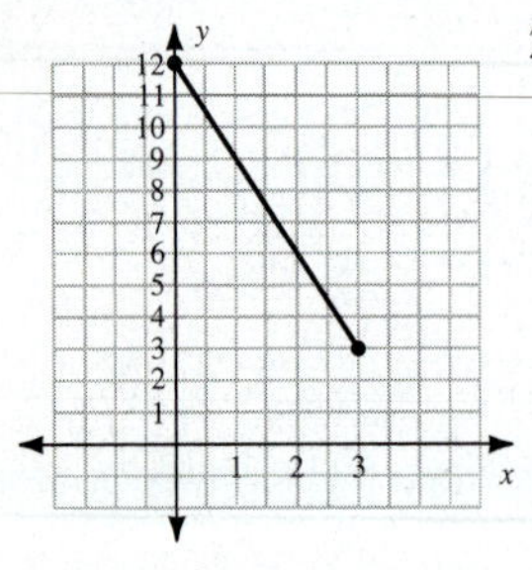

28.

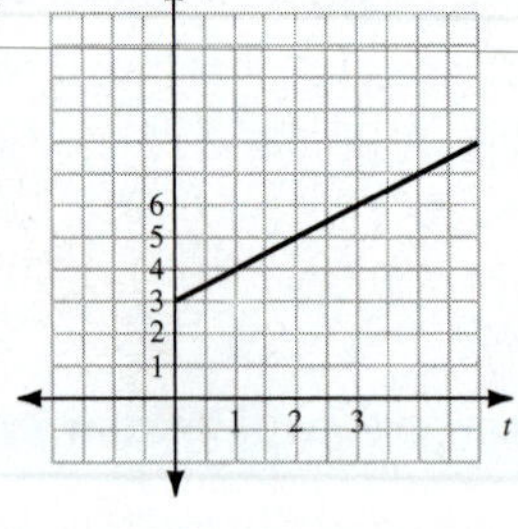

29. (a) $V = -2t + 28$; **(b)**

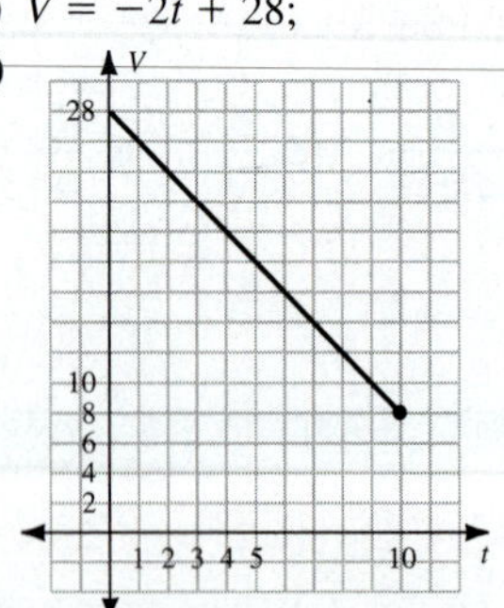

30. (a) $C = 25x + 275$; **(b)**

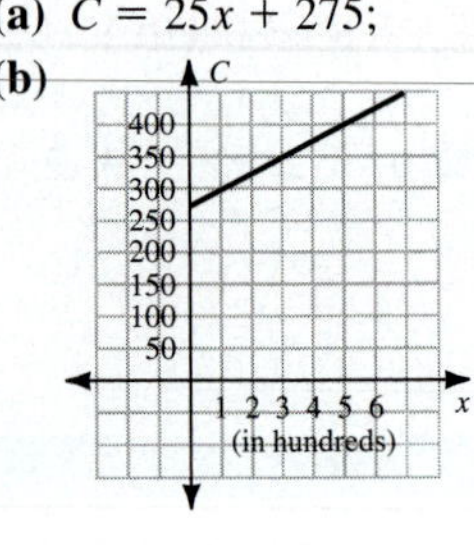

Section 3.5

31. **(a)** domain = $\{-2, -1, 0\}$; **(b)** range = $\{-1, 0, 2, 3\}$; **(c)** no; **(d)**

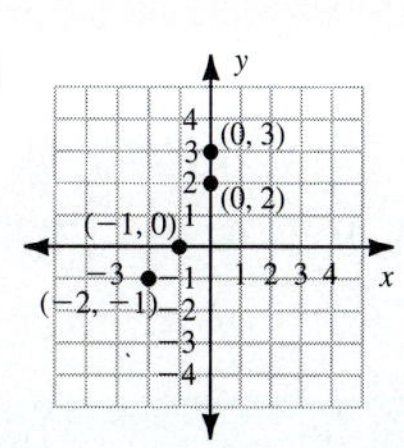

32. **(a)** 1; **(b)** −3; **(c)** $-4a^2 + 1$

CHAPTER 3 TEST

1.

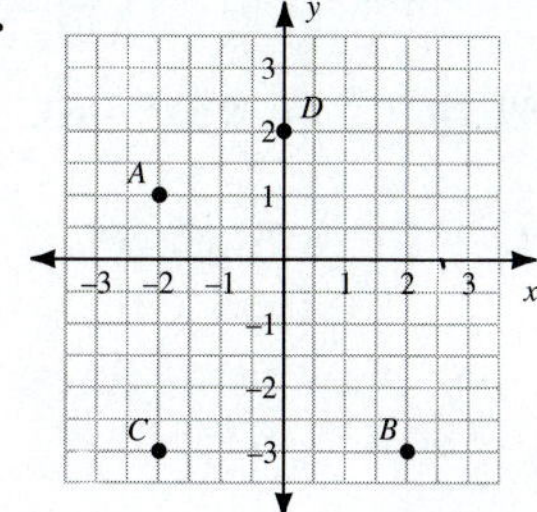

2. $A(-3, 0)$, $B(1, 2)$, $C(4, -1)$, $D(0, -1)$, $E(1, -3)$, $F(-2, 2)$

3.

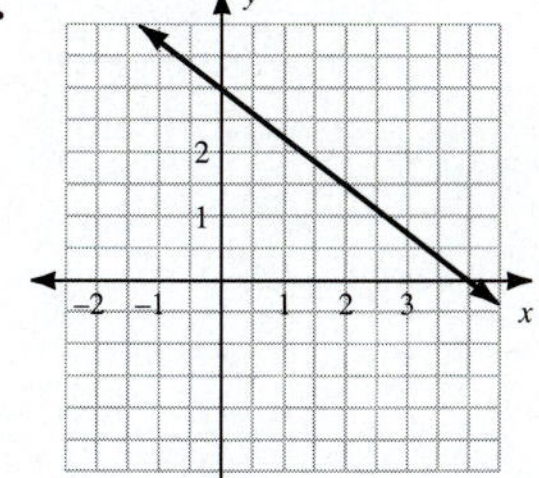

4.

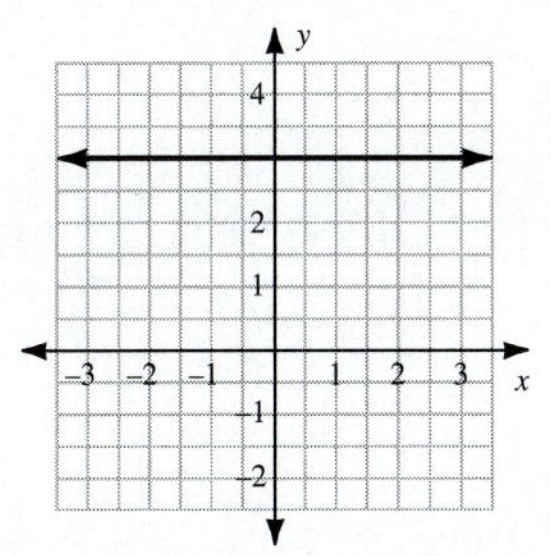

5.

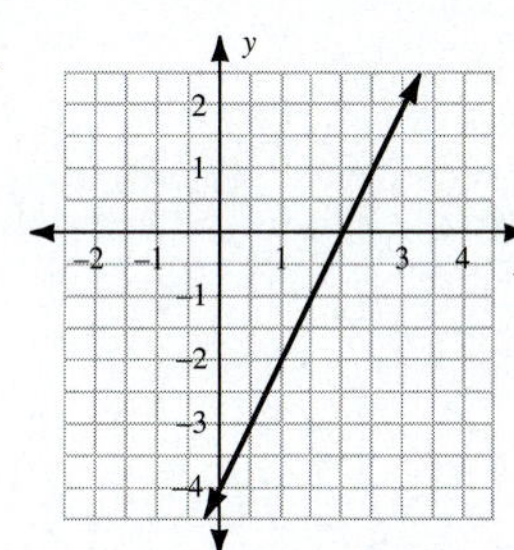

6.

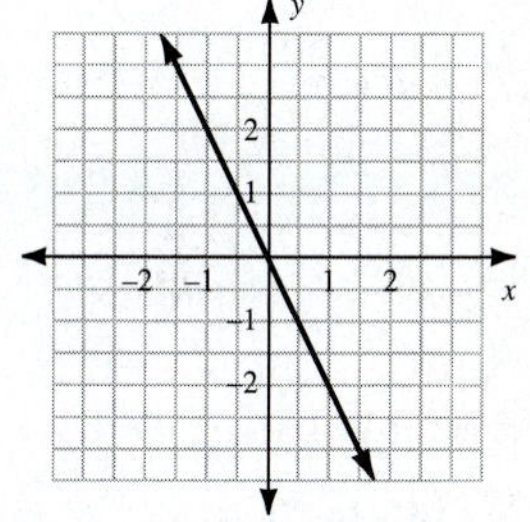

7. $m = \dfrac{3}{5}$ **8.** undefined slope **9.** $m = -3$ **10.** yes **11.** slope = −1; y-intercept = 5

12. slope = $\dfrac{2}{3}$; y-intercept = −2 **13.** $2x + y = 5$ **14.** $y = \dfrac{1}{5}x + 3$ **15.** $y = \dfrac{3}{2}x + 3$

16.

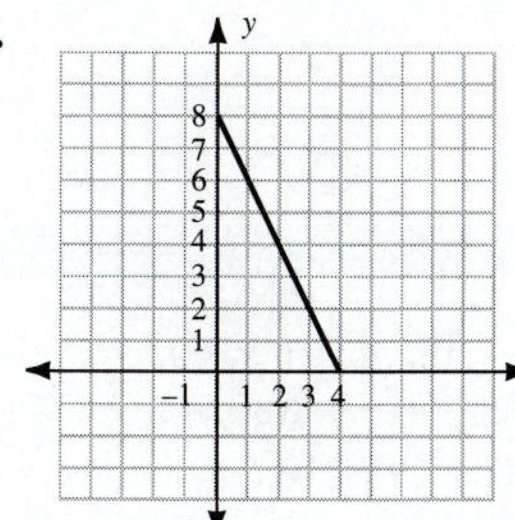

17.

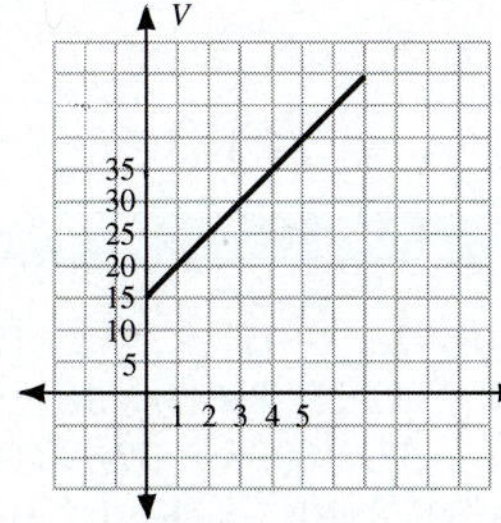

18. $C = 50x + 110$

19. **(a)** domain = $\{-4, -2, 0, 1, 2\}$;
(b) range = $\{-3, -1, 1, 3\}$;
(c) yes
(d)

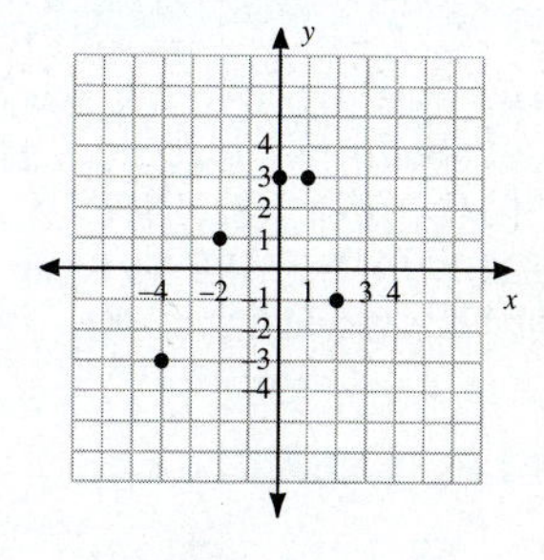

20. **(a)** 4 **(b)** −16 **(c)** $t^3 - 3t^2 + 4$

CHAPTER 4

WARMING UP

1. T **2.** F **3.** T **4.** T **5.** F **6.** F **7.** T **8.** T

EXERCISE SET 4.1

1. 2 **3.** 36 **5.** 9 **7.** -1 **9.** $\frac{8}{27}$ **11.** 36 **13.** -36 **15.** x^9 **17.** $y^{16}z^{26}$ **19.** a^7x^7 **21.** 1 **23.** x^6 **25.** b^{16} **27.** $25a^2$ **29.** x^{11} **31.** $-x^5$ **33.** $-a^4$ **35.** $27w^6$ **37.** $4m^2n^2$ **39.** $\frac{1}{16}$ **41.** $\frac{1}{10}$ **43.** $\frac{1}{a^2}$ **45.** $\frac{1}{u^8v^8}$ **47.** $\frac{1}{s^3t^6}$ **49.** 9 **51.** $\frac{1}{2}a^4$ **53.** $\frac{1}{12}$ **55.** $\frac{1}{s}$ **57.** $\frac{3}{n^2}$ **59.** m^6 **61.** $9x^3$ **63.** $\frac{1}{az}$ **65.** $\frac{49}{4}$ **67.** 4 **69.** $\frac{b^5}{a^5}$ **71.** $\frac{z^3}{8x^3}$ **73.** x^6 **75.** 125 **77.** $\frac{y^6x^4}{z^8}$ **79.** $\frac{t^5}{w^7}$ **81.** $2^5 \cdot 2^3 = 2^8$, not 4^8 **83.** $(x^3)^4 = x^{3 \cdot 4}$, not x^{3+4} **85.** $\frac{x^6}{x^3} = x^{6-3}$, not $x^{6/3}$ **87.** 4,741.632 **89.** $-7{,}464.96$

SAY IT IN WORDS

91. ans. vary

REVIEW EXERCISES

93. 100 **95.** 0.1

ENRICHMENT EXERCISES

1. x^{8n} **2.** x^{7n} **3.** $a^{3n}b^{3n}$ **4.** $\frac{v^{2n}}{u^{3n}}$

WARMING UP

1. T **2.** F **3.** T **4.** F **5.** F **6.** T **7.** F **8.** T

EXERCISE SET 4.2

1. 30,000 **3.** 0.0003 **5.** 0.0000592 **7.** 100,300 **9.** 0.6926 **11.** 0.29 **13.** 4.6 **15.** 3.591×10^3 **17.** 2.3981×10^4 **19.** 7.4 **21.** 2.1×10^{-3} **23.** 4×10^{-5} **25.** 8.103×10 **27.** 1.293002×10^6 **29.** 8.3004×10^2 **31.** 18 **33.** 30,000,000 **35.** 0.0006 **37.** 1000 **39.** 90,000 **41.** 1.9×10^{20} **43.** 2.822×10^{-4} **45.** 34 is not a number between 1 and 10, 3.4×10^3 **47.** 0.2 is not a number between 1 and 10, 2×10^{-4} **49.** 44,323 **51.** 3294

REVIEW EXERCISES

53. x^8 **55.** r^6t^6 **57.** $16x^2$ **59.** $4v^2$ **61.** $-z^3$

ENRICHMENT EXERCISE

1. 5.75064×10^{17} miles

WARMING UP

1. T **2.** T **3.** F **4.** F **5.** T **6.** T **7.** T **8.** F

EXERCISE SET 4.3

1. -3 **3.** $\frac{1}{2}$ **5.** $-\frac{2}{5}$ **7.** -1 **9.** 1 **11.** $6x^3y^5$ **13.** $-8s^3t^3$ **15.** $2sy$ **17.** $-2uv$ **19.** $-18x^3y^3$ **21.** $16v^4w^5$ **23.** $a^3b^3c^4$ **25.** $\frac{3y}{4x^3}$ **27.** $4xz^3$ **29.** $\frac{v^5w^2}{u^3}$ **31.** $\frac{5s^6}{6t}$

SAY IT IN WORDS

33. ans. vary

REVIEW EXERCISES

35. $3x$ **37.** $-2az + 6$ **39.** $3s - 2r - 2$

ENRICHMENT EXERCISES

1. a^{3n} **2.** y^{6n} **3.** $x^{3n}y^{3n}$ **4.** $\frac{s^n}{t^n}$

WARMING UP

1. F **2.** T **3.** T **4.** F **5.** T **6.** T **7.** F **8.** F

EXERCISE SET 4.4

1. $x^5 + 3x^4 - 1$; 5 **3.** $3x^4 - x^3 + x^2 + 5$; 4 **5.** $-y^2 + y + 2$; 2 **7.** z; 1 **9.** trinomial; 3 **11.** binomial; 1 **13.** monomial; 1 **15.** monomial; 0 **17.** $6x + 4$ **19.** $5x^2 + 3$ **21.** $7a^2 + a + 4$ **23.** $3z^3 - z^2 + 6z$ **25.** $x^3 + 2x^2 + 4$ **27.** $2x + 1$ **29.** $-x^2 - x + 1$ **31.** $x^3 + 2x^2 + 2x - 2$ **33.** $5.87x^2 + 1.37x - 2.68$ **35.** $-2x^2 + 3x - 2$ **37.** $y - 2$ **39.** $x^2 + x + 2$ **41.** $2s^2 - 4s - 2$ **43.** $3x^2 + 6x + 5$ **45.** $3a^3 - 7a^2 + 3$ **47.** $-2x^2 + 4x + 3$ **49.** $-m^2 + 2m + 8$ **51.** $0.86x^2 - 7.13x + 5.69$ **53.** $9a^2 - 3a + 11$ **55.** $r^5 - r^4 - r^3 + 10r^2 - 7$ **57.** $-8s^3 + 8s - 2$ **59.** $3x^2 + 8x + 2$ **61.** 9 **63.** 5 **65.** 100.768 **67.** -411 **69.** 1830

REVIEW EXERCISES

71. x^6 **73.** $6x^5$ **75.** $-18x^3$

ENRICHMENT EXERCISES

1. $-x - 4$ **2.** $4a + 1$ **3.** $6z - 4$ **4.** $-4c + 7$

WARMING UP

1. T **2.** F **3.** F **4.** F **5.** T **6.** T

EXERCISE SET 4.5

1. $2x^5 - x^3 + 2x$ **3.** $2a^4 + 6a^3$ **5.** $-6r^4 + 9r^3 - 3r^2$ **7.** $m^5 - 2m^3 + 5m$ **9.** $-12x^2 + 16x - 8$
11. $2x^3 - x^2 + 9x$ **13.** $6d^3 - 15d^2 + 9d$ **15.** $t^3 + 6t$ **17.** $u^3 - 20u$ **19.** $x^3 + xy^2$ **21.** $26.24x^3 - 6.08x^2$
23. $x^3 + 7x^2 + 13x + 4$ **25.** $6x^3 - x^2 - 10x - 4$ **27.** $3t^3 - 13t^2 - 9t + 28$ **29.** $x^2 + 4x + 3$ **31.** $y^2 + 6y + 8$
33. $a^2 + 3a - 10$ **35.** $t^2 - 6t + 8$ **37.** $x^2 + x - 6$ **39.** $a^2 - 100$ **41.** $t^4 - 12t^2 + 35$ **43.** $6v^2 + 11v + 3$
45. $21a^2 + 25a - 4$ **47.** $6s^2 - 5s + 1$ **49.** $9x^2 - 4$ **51.** $x^2 + 6x + 9$ **53.** $9z^2 - 6z + 1$
55. $A = 2x^2 + 5x + 2$ square feet **57.** $A = x^2 + 12x$ square meters **59.** $2x^2 + 80x$ square feet

SAY IT IN WORDS

61. ans. vary

REVIEW EXERCISES

63. $(x + 2)^2$ **65.** $x^2 + y^2$

ENRICHMENT EXERCISES

1. $x^3 + 3x^2 + 3x + 1$ **2.** $x^2 + 2xy + y^2$ **3.** $2xy - 2y^2$ **4.** $x^2 - \frac{1}{4}$ **5.** $x^3 - y^3$

WARMING UP

1. T **2.** F **3.** T **4.** F **5.** T **6.** F

EXERCISE SET 4.6

1. $x^2 + 4x + 4$ **3.** $z^2 - 25$ **5.** $t^2 - 4t + 4$ **7.** $4z^2 - 16$ **9.** $9a^2 - 6ab + b^2$ **11.** $25u^2 - v^2$
13. $a^6 + 2a^3c + c^2$ **15.** $9u^4 - 12u^2c + 4c^2$ **17.** $16z^2 - 8za + a^2$ **19.** $w^2 - 20wh + 100h^2$
21. $9h^2 + 6ha^2 + a^4$ **23.** $x^4 + 6x^2u^2 + 9u^4$ **25.** $4x^2 - a^2$ **27.** $z^4 - t^2$ **29.** $s^6 - 4$
31. $5.6644x^2 - 35.224x + 54.76$ **33.** $36.3609y^2 - 4251.04$ **35.** $-8x$ **37.** $3a^2 + 9$ **39.** $2x$
41. $A = 2x^2 - 8$ square feet **43.** $x^2 - 5 = (x + 2)^2$; $x = -\frac{9}{4}$ **45.** 48 and 50

SAY IT IN WORDS

47. ans. vary

REVIEW EXERCISES

49. $2x^3$ **51.** $-\frac{3}{2a^3}$

ENRICHMENT EXERCISES

1. $a^3 + 6a^2 + 12a + 8$ **2.** $y^3 - 3y^2 + 3y - 1$ **3.** $8x^3 + 36x^2 + 54x + 27$ **4.** $s^3 + 3s^2t + 3st^2 + t^3$
5. $u^3 - 3u^2w + 3uw^2 - w^3$ **6.** $\frac{1}{4}x^2 - \frac{1}{9}$

WARMING UP

1. T **2.** F **3.** T **4.** F **5.** F **6.** T

EXERCISE SET 4.7

1. $x^2 - 4x + 7$ **3.** $7z - 2 - \frac{1}{z}$ **5.** $4u^2 - 3u + 1$ **7.** $4 - \frac{2}{m} + \frac{1}{m^2}$ **9.** $-2x + 3 - \frac{1}{x}$ **11.** $-\frac{1}{2}t + 4$
13. $2r^2 - 1$ **15.** $xy^2 - 2x^2y^3$ **17.** $x - 2$ **19.** $z + 5$ **21.** $x - 6 - \frac{27}{x - 2}$ **23.** $x + 3 + \frac{4}{x + 1}$
25. $x - 7 + \frac{11}{x + 3}$ **27.** $y + 4$ **29.** $x^2 - x - 1$ **31.** $x^2 + x + 3 + \frac{-1}{x + 1}$ **33.** $2t^2 + t + 5 + \frac{3}{t - 3}$
35. $x^2 + x + 3 + \frac{2}{x - 1}$ **37.** $x^2 + 6x + 12$ **39.** $x^2 - x + 4$ **41.** $2x + 2 + \frac{1}{2x + 1}$ **43.** $x + 1$
45. $2x^2 + 2x - 3 + \frac{-2}{2x - 1}$ **47.** $3x^2 - 2x + 1$ **49.** $x^2 + x - 1 + \frac{1}{4x - 2}$

SAY IT IN WORDS

51. ans. vary

REVIEW EXERCISES

53. a^5 **55.** $2r^9$ **57.** x^3

ENRICHMENT EXERCISES

1. $x^n - 2$ **2.** $2x^n + \frac{3}{x^n}$ **3.** $y^2 - y + 1 + \frac{2y - 1}{y^2 - y}$ **4.** $4x^3 + 4x^2 - 3x + 2 + \frac{x + 1}{x^2 + x}$

CHAPTER 4 REVIEW EXERCISE SET

Section 4.1

1. x^7 **2.** x^{12} **3.** $a^4b^6c^2$ **4.** $-8z^6t^9$ **5.** $\frac{1}{4}$ **6.** $\frac{8y^6}{x^2}$ **7.** $-\frac{4a^2}{b^4}$ **8.** $\frac{3}{2u^5v^3}$ **9.** $\frac{1}{b^5}$ **10.** $\frac{1}{a^3}$

Section 4.2

11. 1.2×10^{17} **12.** 5×10^{-4} **13.** 800,000

Section 4.3

14. $15x^5$ **15.** $-8x^5y^3$ **16.** $4x$ **17.** $-\frac{3a^2c^2}{b^2}$

Section 4.4

18. $-x^2 + 4x + 2$ **19.** $-t^4 + t^3 + 5t^2 + t - 4$ **20.** 4 **21.** $8x^2 + x + 3$ **22.** $3y^3 - y^2 + y$ **23.** $2x^2 + 7x + 1$
24. $t^2 - 6t - 3$ **25.** $3t^3 - 4t^2 + t + 4$ **26.** $-3x^2 - 12x - 15$

Section 4.5

27. $-3z^6 + 6z^5$ **28.** $3s - 3s^4 + 3s^3$ **29.** $r^2 - 3r + 2$ **30.** $x^2 + 7x + 10$ **31.** $x^2 - x - 6$ **32.** $2c^2 - 8c + 8$
33. $12z^2 + 5z - 3$ **34.** $a^2 + 8a + 16$ **35.** $x^2 + 2xt - 8t^2$ **36.** $3t^3 + 3t^2 - 5t + 2$ **37.** $2w^4 + w^3 - 9w^2 - 9w$
38. $x^3 - x^2 - 3x + 3$ **39.** $-15z - 5$

Section 4.6

40. $t^2 - 9$ **41.** $a^2 + 10a + 25$ **42.** $x^2 - 12x + 36$ **43.** $9v^2 - 12vr + 4r^2$ **44.** $9t^2 - 16$ **45.** $16z^4 - 8z^2 + 1$
46. $x^4 - 6x^2y^2 + 9y^4$

Section 4.7

47. $2y^2 - 3y - 6$ **48.** $5x^2 + 8x$ **49.** $10 - \frac{1}{c}$ **50.** $x + 4$ **51.** $4x + 3 + \frac{6}{3x - 1}$ **52.** $y^2 + 3y + 9$
53. $x^2 + x - 1 + \frac{5}{x - 1}$ **54.** $2x^2 - 2x + 3$

CHAPTER 4 TEST

1. $-\frac{8}{27}$ **2.** x^{17} **3.** 1 **4.** $\frac{y^2}{16x^2}$ **5.** $\frac{b^8}{a^3c^2}$ **6.** 3.01×10^{-4} **7.** 0.00002 **8.** $-\frac{2a}{b^3}$ **9.** $-18r^4s^6t^7$
10. $\frac{x^6z^{12}}{2y^2}$ **11.** 8 **12.** $9r^5 + 8r^4 - 2r^3 + 2r^2 + r - 7$ **13.** $-6m^3 - 2$ **14.** $-4c^3 + 8c + 20$
15. $5x^8 - 10x^6 + 30x^3$ **16.** $3y^3 - 8y^2 + 9y - 10$ **17.** $a^8 - 7a^4b + 12b^2$ **18.** $9x^2 - 12x + 4$ **19.** $x^2 - 4y^2$
20. $4x^4 - 3x + 2$ **21.** $x^2 - 2x + 1$ **22.** $x^2 + x - 1 + \frac{4}{3x - 1}$

CHAPTER 5

WARMING UP

1. T **2.** T **3.** F **4.** T **5.** F **6.** T **7.** F **8.** T

EXERCISE SET 5.1

1. composite **3.** prime **5.** composite **7.** $2 \cdot 3 \cdot 7$ **9.** $2^2 \cdot 3^2$ **11.** $3 \cdot 5^2 \cdot 7$ **13.** prime **15.** $2 \cdot 5 \cdot 7$
17. $3 \cdot 7 \cdot 11$ **19.** prime **21.** 4 **23.** $3x$ **25.** 3 **27.** x^3 **29.** $4x^2y^3$ **31.** $18r^3t^3$ **33.** $3(2x + 1)$
35. $x^2(x - 1)$ **37.** $4a^2(2 + 3a^2)$ **39.** $2(x^2 + 4x + 2)$ **41.** $5(t^2 - 5t + 2)$ **43.** $4(3b^2 + 7b + 18)$ **45.** The greatest common factor is 1. **47.** $6(6p^3 - 2p^2 + 5)$ **49.** $a(2a^2 + 3a - 11)$ **51.** $y^2(14y^2 - 17y + 15)$
53. $-2(x^2 - 5)$ **55.** $-(y^2 - 1)$ **57.** $-4(x^2 - 2x + 1)$ **59.** $-x^3(x^2 + 2x - 1)$ **61.** $-2x(2x^2 + 6x + 9)$
63. $(c + d)(x + z)$ **65.** $(x^2 - 2)(a + b^2)$ **67.** $(y + x)(z + 1)$ **69.** $(x + y^2)(a + b)$ **71.** $(y + z)(y + b)$
73. $(t^2 + 3x)(c - 1)$ **75.** $(x - y)(b - c)$ **77.** $(a - b)(3 + a)$ **79.** $(a - 1)(a^2 - x)$

SAY IT IN WORDS

81. ans. vary **83.** ans. vary

REVIEW EXERCISES

85. $x^2 + x - 12$ **87.** $3x^2 - 17x + 10$ **89.** $a^2 - 4b^2$

ENRICHMENT EXERCISES

1. $9x^2(3x^3 - 2x^2 + 4x + 7)$ **2.** $12y^3(y^3 + 3y - 7y^2 - 4)$ **3.** $2(x^2 + y^2)(a - b)$ **4.** $x(x - a)(x - 2)$

WARMING UP

1. F **2.** T **3.** T **4.** F **5.** T **6.** F **7.** T **8.** T

EXERCISE SET 5.2

1. $(x+4)(x+2)$ **3.** $(t+7)(t+2)$ **5.** $(a+8)(a+2)$ **7.** $(y+4)(y+5)$ **9.** $(c+10)(c+3)$
11. $(t+2)(t-1)$ **13.** prime **15.** $(p+5)(p-3)$ **17.** $(s-3)(s-1)$ **19.** $(y-10)(y-3)$ **21.** prime
23. $(x-5)(x-7)$ **25.** $(k-4)(k+1)$ **27.** $(d+3)(d+4)$ **29.** $(y-4)(y+2)$ **31.** $(x+3a)(x+4a)$
33. $(m-6n)(m-n)$ **35.** $(x-y)(x-3y)$ **37.** $(r+8s)(r-s)$ **39.** $x(x+3)(x-2)$ **41.** $z(z+5)(z-2)$
43. $2x^2(x+2)(x+1)$ **45.** $4t^3(t+3)(t-2)$ **47.** $5a^4(a-2)(a+1)$ **49.** $2m^3(m-4)(m+1)$
51. No, since $2x+8=2(x+4)$.

SAY IT IN WORDS

53. ans. vary

REVIEW EXERCISES

55. $2a^2+5a-12$ **57.** $4y^2-5y-6$ **59.** $8x^2+6x+1$ **61.** $8x^2+30x+18$ **63.** $3x^2-16xy+5y^2$

ENRICHMENT EXERCISES

1. $(x^2+7)(x^2+3)$ **2.** $(y^2+2)(y^2+1)$ **3.** $(t^2-3)(t^2+5)$ **4.** $a(a^2+8)(a^2+3)$

WARMING UP

1. T **2.** F **3.** T **4.** T **5.** T **6.** T **7.** F **8.** T

EXERCISE SET 5.3

1. $(2x+3)(x+1)$ **3.** $(4x+1)(x+3)$ **5.** $(x-3)(2x+1)$ **7.** $(2y+1)(3y+1)$ **9.** prime
11. $(2t-1)(3t-1)$ **13.** $(4z+1)(2z-3)$ **15.** $(6x+1)(x-1)$ **17.** $(5k+2)(k+1)$ **19.** $(7v-3)(v-1)$
21. prime **23.** $(3x-1)(x+4)$ **25.** $(6n-5)(n+1)$ **27.** $(5x+3)(2x-1)$ **29.** $(3z-1)(z-3)$
31. $(2x+y)(x+y)$ **33.** $(2x-y)(x+y)$ **35.** $(2s-t)(3s+2t)$ **37.** $(2u+v)(3u-2v)$ **39.** $x^2(x+2)(2x+3)$
41. $2x(x-3)(2x-1)$ **43.** $3y^3(2y+5)(y-1)$ **45.** $t(3x-5)(2x+3)$ **47.** $3m^2(p-6)(2p+3)$
49. $(3x+2)$ and $(x+4)$ **51.** $3x+2$ inches

SAY IT IN WORDS

53. ans. vary

REVIEW EXERCISES

55. $9t^2-100$ **57.** a^2+6a+9 **59.** $9x^2-6x+1$ **61.** $x^2-4xy+4y^2$ **63.** x^3-8 **65.** $8a^3-27$

ENRICHMENT EXERCISES

1. $(2x^2+1)(3x^2-5)$ **2.** $a(x^3-2)(3x^3+1)$ **3.** $xy(y^2+2)(3y^2-1)$ **4.** $xy(x+y)(3x+2y)$

WARMING UP

1. F **2.** T **3.** F **4.** F **5.** T **6.** F **7.** T **8.** T

EXERCISE SET 5.4

1. $(x + 5)(x - 5)$ **3.** $(t + 10)(t - 10)$ **5.** $(n + p)(n - p)$ **7.** $(s + t)(s - t)$ **9.** $(2y + 1)(2y - 1)$
11. $(k + 5t)(k - 5t)$ **13.** prime **15.** $(9k + h)(9k - h)$ **17.** $(4s - 3y)(4s + 3y)$ **19.** $(z + 3)(z^2 - 3z + 9)$
21. $(4t + 1)(16t^2 - 4t + 1)$ **23.** $(x - 2)(x^2 + 2x + 4)$ **25.** $(b - 1)(b^2 + b + 1)$ **27.** $(2t - v)(4t^2 + 2tv + v^2)$
29. $(3m + 10n)(9m^2 - 30mn + 100n^2)$ **31.** $(x + 5)^2$ **33.** $(x + 2)^2$ **35.** $(t - 3)^2$ **37.** $(c - 6)^2$ **39.** $(x - y)^2$
41. $(2x + z)^2$ **43.** $7(y + 3)(y - 3)$ **45.** $-2(w + 5)(w - 5)$ **47.** $x^3(x + 9)(x - 9)$ **49.** $4y^5(y + 2)(y - 2)$
51. $6(m + 3)^2$ **53.** $10(k - 4)^2$ **55.** $3z^2(z + 2)^2$ **57.** $25x^6(x - 1)^2$ **59.** $a(a - b)^2$

SAY IT IN WORDS

61. ans. vary

REVIEW EXERCISES

63. $4(x - 3y)$ **65.** The greatest common factor is 1. **67.** $a(x - y)$

ENRICHMENT EXERCISES

1. $(z^2 + 4)(z + 2)(z - 2)$ **2.** $(a^2 + b^3)(a^4 - a^2b^3 + b^6)$ **3.** $(u^2 + v)(u^2 - v)(u^8 + u^4v^2 + v^4)$ **4.** $2s^3r^2(r - 3s)^2$
5. $224x$ **6.** $y(3x + 1)^2$

WARMING UP

1. T **2.** F **3.** T **4.** T **5.** T **6.** F **7.** T **8.** T

EXERCISE SET 5.5

1. $x^3(x^2 + 4)$ **3.** $(2t + 3)(t - 1)$ **5.** $2(y - 1)(y + 3)$ **7.** $(m + 2)(m^2 - 2m + 4)$ **9.** $(3x + 4y)(3x - 4y)$
11. $3x^2(x - 3)(x^2 + 3x + 9)$ **13.** prime **15.** $(a + 4)^2$ **17.** $(y - 9)^2$ **19.** $3(h - 2)(h - 1)$
21. $(3x + 8)(3x - 1)$ **23.** $5b^2c^2(2b + 4c - 1)$ **25.** $6(x - 1)(x + 6)$ **27.** $3t(2t + 1)(2t - 1)$
29. $2h(h + 3)(h^2 - 3h + 9)$ **31.** $m^2(m^2 + 25)$ **33.** $x(x^2 - 3x + 5)$ **35.** $(z - 3c)(z - 9c)$ **37.** $xy(3x - 2)(2x - 5)$
39. $3ab(x + 5)(x - 5)$ **41.** $bc(c - 2d)(2c - d)$ **43.** prime **45.** $(1 - 2m)(7 + 2k)$
47. $(x + y)(x - y)(a + 1)(a^2 - a + 1)$ **49.** $(y + 1)(y - 1)^2(y^2 + y + 1)$

WARMING UP

1. F **2.** F **3.** T **4.** T **5.** F **6.** T **7.** T **8.** T

EXERCISE SET 5.6

1. $\{2, 3\}$ **3.** $\left\{-3, \frac{1}{2}\right\}$ **5.** $\{0, -3\}$ **7.** $\{0, -1\}$ **9.** $\{1, 2, -1\}$ **11.** $\left\{\frac{3}{2}, -2, 2\right\}$ **13.** $\{1, 2\}$ **15.** $\{-3, 2\}$
17. $\left\{-\frac{1}{2}, -1\right\}$ **19.** $\left\{\frac{2}{3}, -3\right\}$ **21.** $\left\{-\frac{1}{2}, \frac{3}{2}\right\}$ **23.** $\left\{\frac{1}{3}, -\frac{3}{2}\right\}$ **25.** $\{\pm 2\}$ **27.** $\{\pm 4\}$ **29.** $\{1, 3\}$ **31.** $\{\pm 1\}$
33. $\{0, \pm 10\}$ **35.** $\left\{-\frac{1}{3}, 5\right\}$ **37.** $\left\{-5, \frac{10}{3}\right\}$ **39.** $\left\{-1, \frac{1}{3}\right\}$ **41.** $\{1, -3\}$ **43.** $\{1, \pm 2\}$ **45.** $\left\{\frac{1}{3}, -5\right\}$
47. $\left\{-\frac{3}{5}, 4\right\}$ **49.** $\left\{\frac{11}{2}, 2\right\}$ **51.** $\{-3, 2\}$ **53.** $\{-5, 6\}$ **55.** $\{0, \pm 3\}$ **57.** $\left\{0, \pm\frac{5}{2}\right\}$ **59.** $\left\{\frac{2}{3}, 2, -1\right\}$

SAY IT IN WORDS

61. ans. vary

REVIEW EXERCISES

63. $y = 7$ **65.** $x = -1$ **67.** $y = \frac{3}{2}x + 3$

ENRICHMENT EXERCISES

1. $\{\pm 3\}$ **2.** $\left\{\pm \frac{1}{2}\right\}$ **3.** $\left\{\pm \frac{2}{5}\right\}$ **4.** $\{2, -3\}$ **5.** $\varnothing$

WARMING UP

1. T **2.** F **3.** T **4.** T **5.** F **6.** T

EXERCISE SET 5.7

1. 6 and 7 **3.** 3 and 4 **5.** 4 and 6 **7.** -8 and -7 **9.** 2 yards by 9 yards **11.** 3 feet by 5 feet **13.** 2 feet **15.** 12 feet **17.** 50 yards by 120 yards **19.** 10 feet wide

SAY IT IN WORDS

21. ans. vary

REVIEW EXERCISES

23. $\frac{1}{3}$ **25.** $\frac{7}{3}$ **27.** 1

ENRICHMENT EXERCISES

1. 5 feet **2.** 4 inches by 6 inches

CHAPTER 5 REVIEW EXERCISE SET

Section 5.1

1. composite **2.** prime **3.** $2^3 \cdot 3^2$ **4.** $2x(x - 3)$ **5.** $y^2(y^2 - y + 1)$ **6.** $3(3t^2 - 4t + 10)$ **7.** $(x + 2)(a + b^2)$ **8.** $(y - x)(b - 4)$ **9.** $(z + 1)(1 - a)$ **10.** $(2x^2 + z^2)(2c - b)$

Section 5.2

11. $(x + 2)(x + 5)$ **12.** $(y + 3)(y - 4)$ **13.** $(m + 5)(m - 6)$ **14.** $(x - 4)(x - 5)$ **15.** prime

Section 5.3

16. $(2x + 3)(3x + 1)$ **17.** $(3y + 4)(2y - 5)$ **18.** $(2a - 5)(2a - 3)$ **19.** $(2x + 3t)(x - t)$ **20.** $2k(2k + 1)(2k - 3)$

Section 5.4

21. $(x + 10)(x - 10)$ **22.** $(3y + 2)^2$ **23.** $(2t + 3s)(2t - 3s)$ **24.** $(m + 2)(m^2 - 2m + 4)$
25. $(2a - 1)(4a^2 + 2a + 1)$ **26.** $x^3(x + 1)(x - 1)$

Section 5.5

27. $(x + y)(x - y)(t + s)$ **28.** $c^3(2c - 5)(2c - 1)$ **29.** $2t^2(t - 2)^2$ **30.** $5x^3(x + 2)(x - 2)$
31. $2(a + 4)(t + 3)(t - 3)$ **32.** $2d(2d + 3)(4d - 5)$ **33.** $2(3x - 1)(3x + 2)$ **34.** $2(y^2 + 25)$
35. $(a - 2b)(2a - b)$ **36.** $xy(2x + 3y)(3x + y)$

Section 5.6

37. $\left\{4, \frac{1}{2}\right\}$ **38.** $\{7, -3\}$ **39.** $\{0, 5\}$ **40.** $\{2, 3\}$ **41.** $\{2, -3\}$ **42.** $\{-1, 3\}$

Section 5.7

43. 4 and 5 **44.** 6 feet by 10 feet **45.** 5 inches by 7 inches **46.** 5 miles

CHAPTER 5 TEST

1. $2^2 \cdot 3 \cdot 5 \cdot 7$ **2.** $x^4(x^4 - x^2 + 1)$ **3.** $4ab^3(a^3 - 4ab^2 + 5b^5)$ **4.** $4(5x^2 - 4x + 3)$ **5.** $(a - 7)(a - 5)$
6. prime **7.** $(a - 10)(a + 2)$ **8.** $(2b + 1)(9b - 2)$ **9.** $(7x - 1)(x - 4)$ **10.** $x^2y(3x - 2y)(x + 5y)$
11. $(10x - 9)(10x + 9)$ **12.** $(4a - 5)^2$ **13.** prime **14.** $(2x - 5)(4x^2 + 10x + 25)$ **15.** $4(r^2 + 4)$
16. $(s - 8t)^2$ **17.** $(z + 4)(z^2 - 4z + 16)$ **18.** $(4m + 1)(m - 2n)$ **19.** $(a - 4)(a + 4)(a + 5)$ **20.** $x = 3$ or 4
21. $a = 0, -\frac{1}{7}$, or $\frac{3}{2}$ **22.** $x = -3$ or 1 **23.** -7 and -5 **24.** 4 feet by 6 feet

CHAPTER 6

WARMING UP

1. T **2.** T **3.** F **4.** T **5.** F **6.** T **7.** F **8.** T

EXERCISE SET 6.1

1. (a) 3 **(b)** 1 **(c)** $-\frac{3}{4}$ **3.** 5 **5.** 0 **7.** $\frac{1}{2}$ **9.** no values of x **11.** 3, -3 **13.** 3, 5 **15.** $\frac{3}{2}$, 1 **17.** $\frac{3}{x^2}$
19. $\frac{3}{x^3}$ **21.** $\frac{a}{b^3}$ **23.** $\frac{u^3}{2}$ **25.** $\frac{t}{2x^2}$ **27.** $\frac{x+1}{x+2}$ **29.** $\frac{y-1}{y+3}$ **31.** $\frac{x+y}{z-a}$ **33.** $\frac{7}{6}$ **35.** $\frac{1}{3}$ **37.** $\frac{3}{y+3}$
39. -1 **41.** $-\frac{1}{x+3}$ **43.** -2 **45.** $-\frac{2(k+3)}{3}$ **47.** $\frac{a-b}{a+b}$ **49.** $\frac{x-2y}{x+y}$ **51.** $\frac{3(x+1)}{4(2x+1)}$ **53.** Common factors, not common terms, can be divided out. **55.** 3 and 3. The rational expression $\frac{x^3+1}{x^2-x+1}$ reduces to $x + 1$.
57. An error message appears, because the denominator is 0 when $n = -1.2$.

SAY IT IN WORDS

59. ans. vary

REVIEW EXERCISES

61. $\frac{4}{3}$ **63.** $\frac{4}{3}$

ENRICHMENT EXERCISES

1. $\frac{x^2-2}{2}$ **2.** $-\frac{(y^2+1)(y-1)}{y-2}$ **3.** $\frac{x^2-x+1}{x-3}$ **4.** $\frac{y-2}{y}$

WARMING UP

1. T **2.** F **3.** F **4.** T **5.** F **6.** T

EXERCISE SET 6.2

1. $\frac{3x^2}{2}$ **3.** $5y$ **5.** $\frac{3}{uv}$ **7.** $\frac{x+3}{(x+4)^2}$ **9.** $\frac{2(t+6)^2}{3}$ **11.** $\frac{1}{x-3}$ **13.** $\frac{3(x+2)}{2(x+1)}$ **15.** $\frac{(x-5)(x+1)}{x}$
17. $-\frac{1}{3}$ **19.** $\frac{3x^2}{2y}$ **21.** $\frac{x^2(x+4)}{x-1}$ **23.** $\frac{1}{3(x-10)}$ **25.** $-\frac{1}{3}$ **27.** $\frac{z-3}{2z+1}$ **29.** $\frac{x+y}{2x-y}$

SAY IT IN WORDS

31. ans. vary

REVIEW EXERCISES

33. $\frac{1}{2(x+1)}$ **35.** $\frac{3}{x+3}$ **37.** $-\frac{2y-1}{2}$

ENRICHMENT EXERCISES

1. $\frac{1}{3x^3}$ **2.** $\frac{(x+y)^2}{x-y}$ **3.** 1

WARMING UP

1. T **2.** F **3.** T **4.** T **5.** F **6.** T **7.** F **8.** T

EXERCISE SET 6.3

1. $\frac{17}{z}$ **3.** $\frac{20}{z^2}$ **5.** $\frac{10}{b+1}$ **7.** $\frac{9}{a^2+7}$ **9.** $\frac{2}{y-3}$ **11.** $\frac{1}{t+1}$ **13.** $\frac{1}{u+3}$ **15.** 16 **17.** 36 **19.** $12y$
21. $21a^5$ **23.** $60t^4$ **25.** $3w(w+3)$ **27.** a^2-1 **29.** $(n+2)(n-1)(n-3)$ **31.** $z-1$ **33.** $c-1$
35. $\frac{13}{10}$ **37.** $\frac{13}{36}$ **39.** $\frac{6}{5y}$ **41.** $\frac{1}{6h}$ **43.** $\frac{(r-1)^2}{(r+2)(r-2)}$ **45.** $-\frac{v^2}{v^2-9}$ **47.** $\frac{5m+1}{(m-2)(m-1)(m+1)}$
49. $-\frac{z(z-3)}{(2z-3)(z+1)(z-1)}$ **51.** $\frac{2r^2+4r+3}{r(r+1)(r+2)}$ **53.** $\frac{2}{t-1}$ **55.** $-\frac{1}{y+5}$ **57.** $\frac{z^2+8z+17}{z+3}$ **59.** $\frac{5v+36}{2v+12}$
61. $\frac{2n-3}{2n}$ **63.** $x+\frac{3}{x}$; $\frac{x^2+3}{x}$ **65.** $\frac{1}{x+4}-2x$; $\frac{-2x^2-8x+1}{x+4}$

SAY IT IN WORDS

67. ans. vary

REVIEW EXERCISES

69. $\frac{5}{6}$ **71.** $\frac{2}{3}$ **73.** 4

ENRICHMENT EXERCISES

1. $\frac{3x}{10}$ **2.** $\frac{5x + 14}{10x^2}$ **3.** $\frac{17x + 10}{18x^2}$ **4.** $\frac{4x + 3}{3x^2}$

WARMING UP

1. T **2.** T **3.** F **4.** T **5.** T

EXERCISE SET 6.4

1. $\frac{2}{3}$ **3.** $\frac{b}{2a}$ **5.** $\frac{5}{4s}$ **7.** $\frac{10 + 3a}{2 + a}$ **9.** $\frac{4 - w}{5w - 3}$ **11.** $\frac{1}{z}$ **13.** $\frac{q}{3}$ **15.** c **17.** $\frac{2}{a - 3}$ **19.** $\frac{-x + 14}{7}$

SAY IT IN WORDS

21. ans. vary

REVIEW EXERCISES

23. $\{-7\}$ **25.** $\{3\}$ **27.** $\left\{2, -\frac{1}{2}\right\}$ **29.** $\left\{\frac{1}{3}, -\frac{3}{7}\right\}$

ENRICHMENT EXERCISES

1. -1 **2.** $\frac{1}{5}$ **3.** $-\frac{1}{x(x + h)}$ **4.** $-\frac{2x + h}{x^2(x + h)^2}$

WARMING UP

1. F **2.** T **3.** T **4.** F **5.** F

EXERCISE SET 6.5

1. $\{-6\}$ **3.** $\left\{-\frac{3}{2}\right\}$ **5.** $\{-6\}$ **7.** $\{4\}$ **9.** $\left\{\frac{3}{4}\right\}$ **11.** $\{1\}$ **13.** $\left\{\frac{5}{3}\right\}$ **15.** $\{-2\}$ **17.** $\varnothing$ **19.** $\{1\}$
21. $\varnothing$ **23.** $\left\{\frac{1}{2}, 3\right\}$ **25.** $\{1, 2\}$ **27.** $\{1, -4\}$ **29.** $\left\{\frac{2}{3}, -1\right\}$ **31.** $\{0, -4\}$ **33.** $\left\{-\frac{3}{2}\right\}$

SAY IT IN WORDS

35. ans. vary

REVIEW EXERCISES

37. $r = -1$ **39.** $y = \frac{1}{3}$ **41.** $x = -1$ or $x = -\frac{5}{2}$

ENRICHMENT EXERCISES

1. $\left\{\frac{AB - 4A}{2B}\right\}$ **2.** $\left\{\frac{2AB - A}{2B}\right\}$ **3.** $\{A + 2B\}$ **4.** $\left\{\frac{AB - 3A - 3B}{6B}\right\}$

WARMING UP

1. T **2.** F **3.** T **4.** T **5.** T **6.** F

EXERCISE SET 6.6

1. $\frac{15}{2}$ **3.** 14 **5.** $\frac{8}{3}$ **7.** 16 **9.** -12 or 12 **11.** -2 or 2 **13.** -4 or 4 **15.** -6 or 6 **17.** 37 1/2 miles
19. 360 **21.** $22.75 **23.** $a = 9$ cm; $b = 6$ cm **25.** $b = 10$ cm; $c = 14$ cm **27.** $x = 5$ inches

SAY IT IN WORDS

29. ans. vary

REVIEW EXERCISES

31. $x + \frac{3}{4}$ **33.** $\frac{3}{5}\left(x + \frac{10}{3}\right)$

ENRICHMENT EXERCISES

1. 8 or $\frac{2}{3}$ **2.** $-\frac{1}{2}$ or $\frac{1}{3}$ **3.** -4 or $\frac{1}{2}$ **4.** 6 feet and 9 feet

WARMING UP

1. F **2.** T **3.** T **4.** T **5.** F **6.** T

EXERCISE SET 6.7

1. $R = \frac{R_1R_2}{R_1 + R_2}$ **3.** $h = \frac{3V}{\pi r^2}$ **5.** $B = \frac{3V}{h}$ **7.** $b_1 = \frac{2A - b_2h}{h}$ **9.** $y = 3x - 5$ **11.** $y = -2x - 1$
13. $y = mx + 3m + 2$ **15.** $y = \frac{2}{5}x$ **17.** $y = -3x + 3k + h$ **19.** 6 and 9 **21.** $-\frac{3}{4}$ and $\frac{1}{2}$ **23.** 5
25. (a) $\frac{1}{9}$ (b) $\frac{4}{9}$ (c) $\frac{1}{3}$ **27.** $1\frac{1}{3}$ hrs **29.** 1.6 hrs **31.** It would take Allison 24 days. It would take John 8 days.
33. $5\frac{1}{7}$ hrs **35.** $1\frac{1}{2}$ hrs **37.** 35 mph **39.** 15.2 mph

SAY IT IN WORDS

41. yes

REVIEW EXERCISES

43. It is a solution.

ENRICHMENT EXERCISES

1. $y = \frac{2}{3}x + \frac{2}{3}$ **2.** $y = -\frac{6}{5}x - \frac{1}{5}$ **3.** $y' = \frac{2y - 3xy^2}{x}$

CHAPTER 6 REVIEW EXERCISE SET

Section 6.1

1. **(a)** $\frac{1}{2}$ **(b)** 1 **(c)** $\frac{2}{3}$ **2.** **(a)** $\frac{4}{5}$ **(b)** 0 **(c)** -1 **3.** -1 **4.** ± 2 **5.** no values **6.** 0 and 1 **7.** $\frac{2}{x^2}$
8. $\frac{y-1}{y-5}$ **9.** $\frac{2x+1}{2x-3}$ **10.** $\frac{3}{7}$ **11.** $-\frac{2x-1}{x+4}$ **12.** -1

Section 6.2

13. $\frac{2}{x}$ **14.** $\frac{1}{x+2}$ **15.** $\frac{3x^3(2x-y)}{2y(x-y)}$ **16.** $\frac{9x}{y}$ **17.** $\frac{2x-1}{x-2}$ **18.** $-\frac{3}{y+2}$ **19.** $\frac{t+1}{t(t-2)}$
20. $\frac{(x-2)(x+1)}{3x-1}$

Section 6.3

21. 84 **22.** $42a$ **23.** $5x^4$ **24.** $30b^4$ **25.** $12y^3(y+2)$ **26.** $u^2 - 4$ **27.** $(2n-1)(n+1)(n+2)$
28. $2x - 7$ **29.** $3p^2$ **30.** $\frac{14}{x}$ **31.** $\frac{12}{a}$ **32.** $\frac{7r}{r+1}$ **33.** $\frac{4}{2y+1}$ **34.** $\frac{41}{20x^2}$ **35.** $-\frac{3r}{r^2-9}$
36. $\frac{8z-11}{(z+4)(z-4)(z-1)}$ **37.** $\frac{x^2-5}{x-2}$ **38.** $\frac{x+1}{3x^2}$

Section 6.4

39. $\frac{3}{2}$ **40.** $\frac{4a^2}{5}$ **41.** $-\frac{6}{13}$ **42.** $\frac{2(3t-2)}{3(2-t)}$ **43.** $\frac{y(28+3y)}{2(9-2y)}$ **44.** $\frac{4u-7}{u+7}$

Section 6.5

45. $\{-9\}$ **46.** $\{-2, 1\}$ **47.** $r = \frac{5}{2}$ **48.** $x = 4$

Section 6.6

49. $\frac{4}{3}$ **50.** 12 **51.** -4 or 4 **52.** -6 or 6 **53.** 45 miles **54.** $x = 5$ feet and $y = 15$ feet **55.** $x = 3$ feet and $y = 3\frac{1}{2}$ feet

Section 6.7

56. $t = \frac{a - P}{Pr}$ **57.** 3 and 12 **58.** 1 week and 5 days **59.** Sue was driving $6\frac{2}{3}$ mph. Don was driving $10\frac{2}{3}$ mph.

CHAPTER 6 TEST

1. $\frac{3b^2}{7a^2}$ **2.** $\frac{3}{x+3}$ **3.** $-\frac{1}{2}$ **4.** $\frac{v}{4u^2}$ **5.** $-\frac{b(a-2)}{2}$ **6.** 1 **7.** $(3c - 1)(c - 1)(c + 1)$ **8.** $2x - 1$
9. $\frac{5}{x^3}$ **10.** $\frac{1}{5n-2}$ **11.** $\frac{16}{(y+3)(y-2)(y-5)}$ **12.** $\frac{b^2}{4a}$ **13.** $\frac{3x-4}{4x+4}$ **14.** $a = 2$ **15.** $x = 2$ or $-\frac{2}{3}$
16. $\frac{4}{3}$ **17.** -8 or 8 **18.** $2.17 **19.** $x = 8.4$ cm; $y = 11.2$ cm. **20.** $y = -2x + 6$ **21.** $\frac{3}{2}$ and $\frac{5}{2}$
22. It would take Terence 3 hours. It would take Trevor 6 hours. **23.** 23 mph

CHAPTER 7

WARMING UP

1. T **2.** F **3.** T **4.** T **5.** T **6.** F

EXERCISE SET 7.1

1. **(a)** (1, 1) is a solution. **(b)** (0, 2) is not a solution. **(c)** (5, −1) is not a solution. **3.** $x = 1$; $y = 4$ **5.** $x = 0$; $y = 4$
7. $x = 1$; $y = 2$ **9.** infinitely many solutions **11.** infinitely many solutions **13.** no solution **15.** $x = 2$; $y = -1$
17. It is a solution. **19.** (2, 3) **21.** (2, 2)

SAY IT IN WORDS

23. ans. vary

REVIEW EXERCISES

25. $x = 7$ **27.** $y = -\frac{1}{2}$ **29.** $x = \frac{5}{2}$

ENRICHMENT EXERCISES

1. $-6c_1 \neq c_2$ **2.** $-6c_1 = c_2$

WARMING UP

1. T **2.** F **3.** T **4.** T **5.** F **6.** T **7.** F **8.** T

EXERCISE SET 7.2

1. (1, 2) **3.** (5, 2) **5.** (−1, 1) **7.** (2, −2) **9.** (1, 1) **11.** (3, −1) **13.** (−2, 3) **15.** (2, −4)
17. (−4, 3) **19.** (1, −1) **21.** (3, −5) **23.** (−4, 0) **25.** no solution **27.** infinitely many solutions
29. (−6, 7) **31.** (4.899, 0.689)

SAY IT IN WORDS

33. ans. vary

REVIEW EXERCISES

35. $x = \frac{2}{5}$ **37.** $x = -\frac{1}{3}$ **39.** $z = -9$

ENRICHMENT EXERCISES

1. $a = -1, b = 1$ **2.** $s = 0, t = 2$ **3.** $u = -1; v = -2$

WARMING UP

1. T **2.** T **3.** F **4.** T **5.** F **6.** T

EXERCISE SET 7.3

1. (3, 3) **3.** (1, 2) **5.** (1, 1) **7.** (−1, 3) **9.** (1, −1) **11.** (−5, 1) **13.** (4, −3) **15.** (−3, 5) **17.** no solution **19.** infinitely many solutions **21.** (−3, −2) **23.** (−1, 4) **25.** (4, 14) **27.** (3, 3) **29.** infinitely many solutions **31.** (0, 7) **33.** 4 feet by 6 feet **35.** −1 and 9 **37.** 80 yards by 120 yards **39.** 25 hours mowing lawns and 10 hours trimming hedges **41.** $x = 0.97$, $y = -1.94$

SAY IT IN WORDS

43. ans. vary

REVIEW EXERCISES

45. $x < 4$ **47.** $x < -3$

ENRICHMENT EXERCISES

1. $\left(\frac{C_2 - C_1}{2}, \frac{C_2 + C_1}{2}\right)$ **2.** $\left(\frac{3}{1+m}, \frac{3m}{1+m}\right)$ **3.** no solution

WARMING UP

1. F **2.** T **3.** T **4.** F **5.** T **6.** T

EXERCISE SET 7.4

1. **3.**

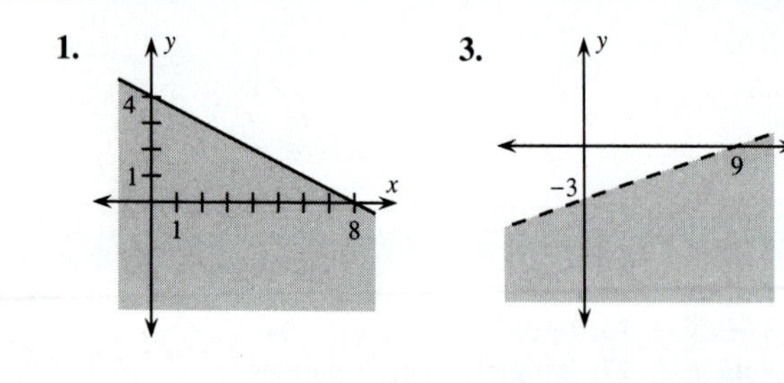

5.

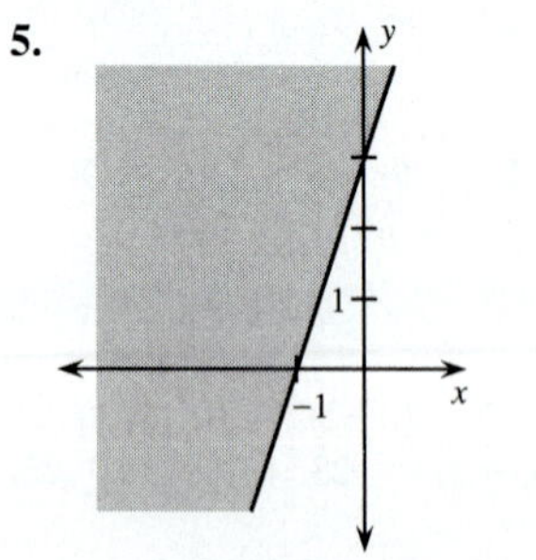

7.

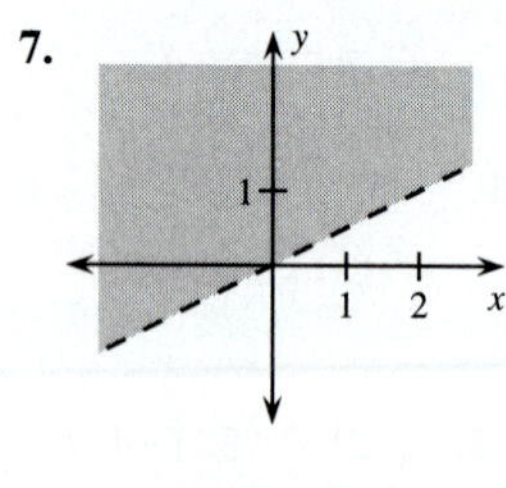

9.
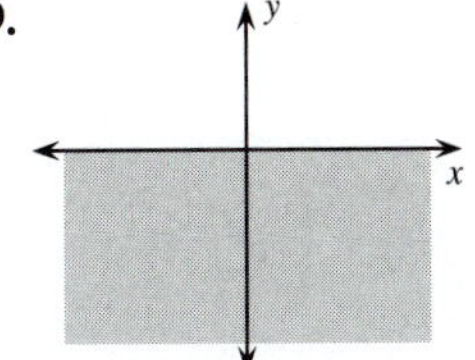

11.
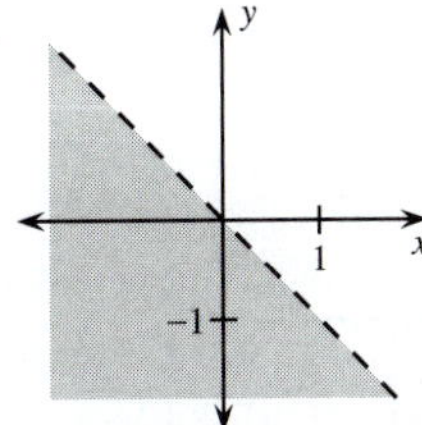

13.
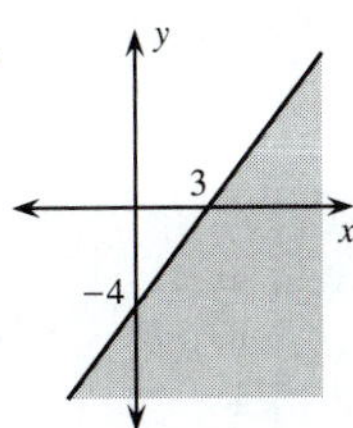

15.
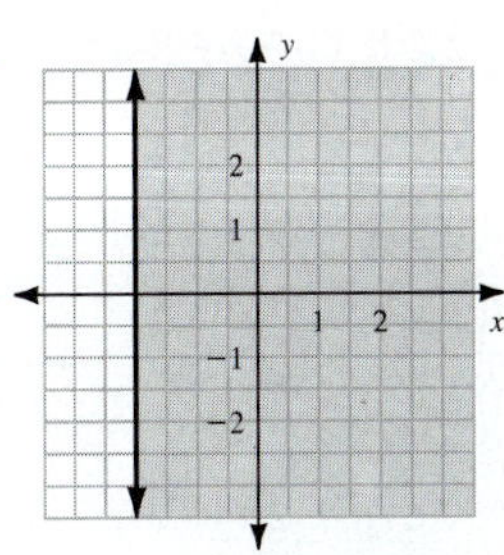

17.
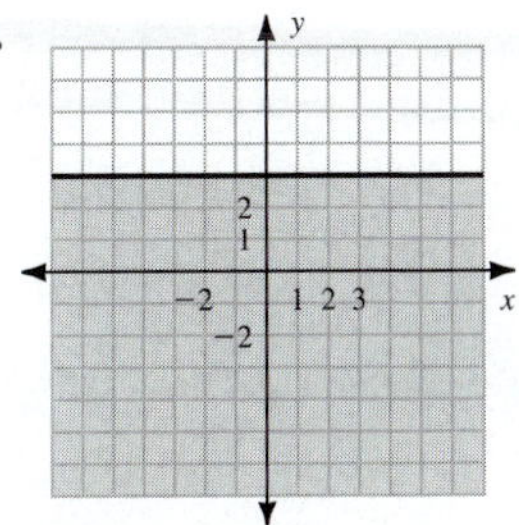

19.
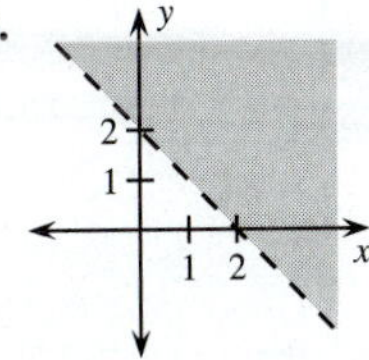

21.
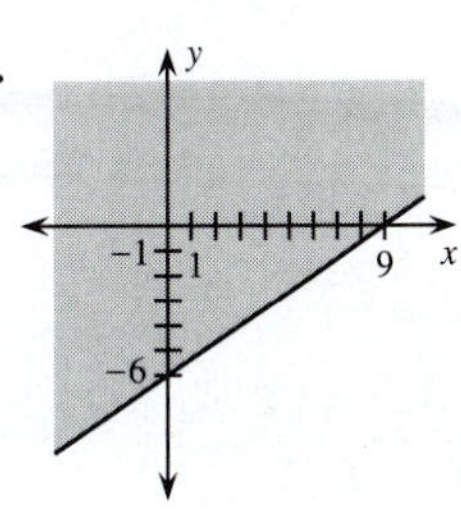

23.
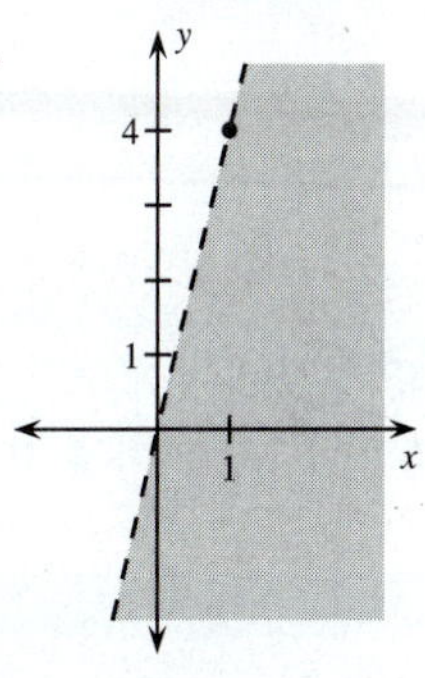

25.
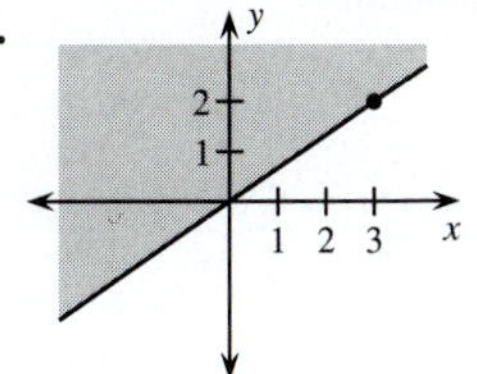

27.
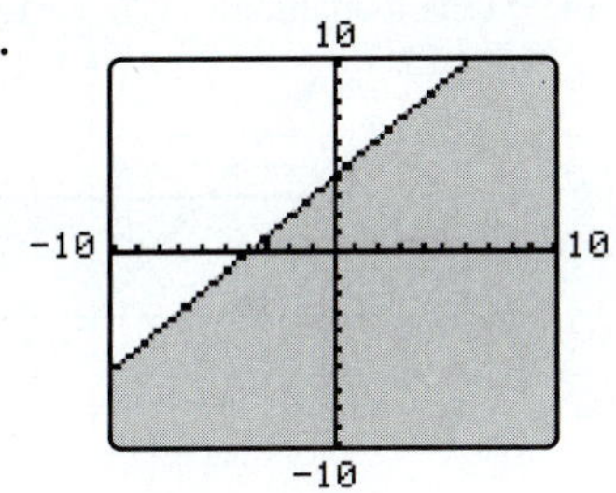

29.
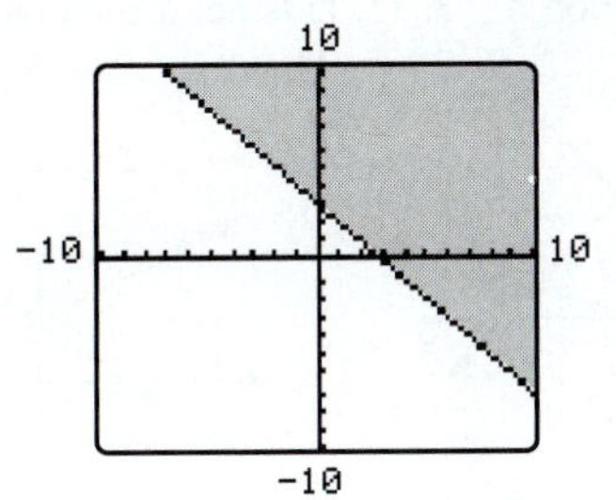

SAY IT IN WORDS

31. ans. vary

REVIEW EXERCISES

33.
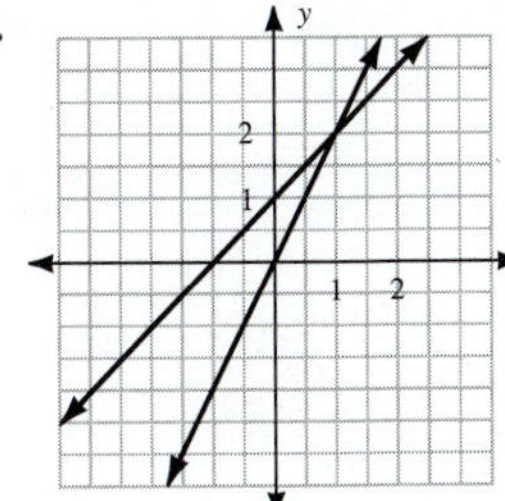

35.
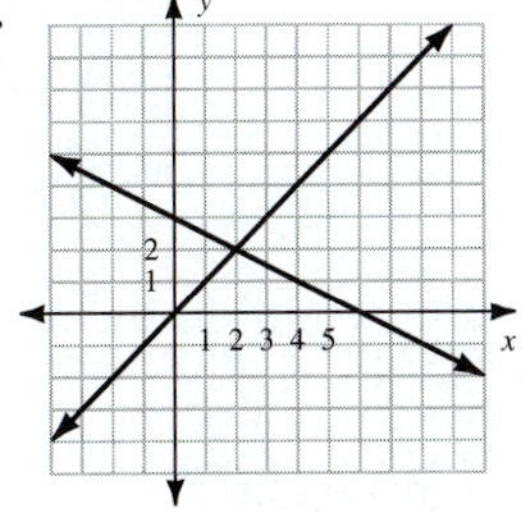

ENRICHMENT EXERCISES

1.

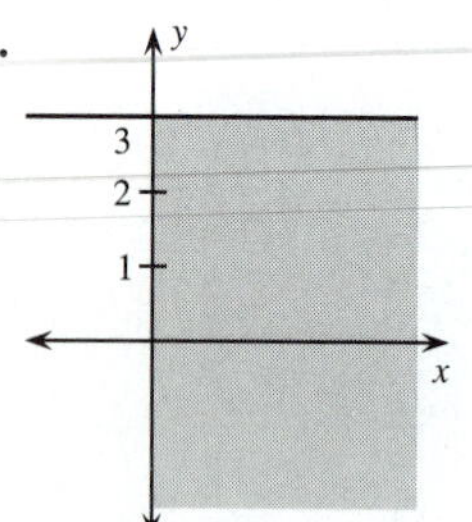

2.

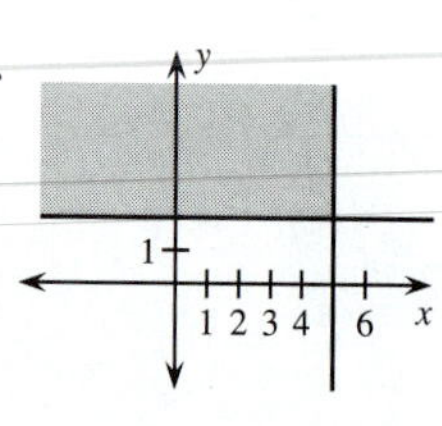

3.

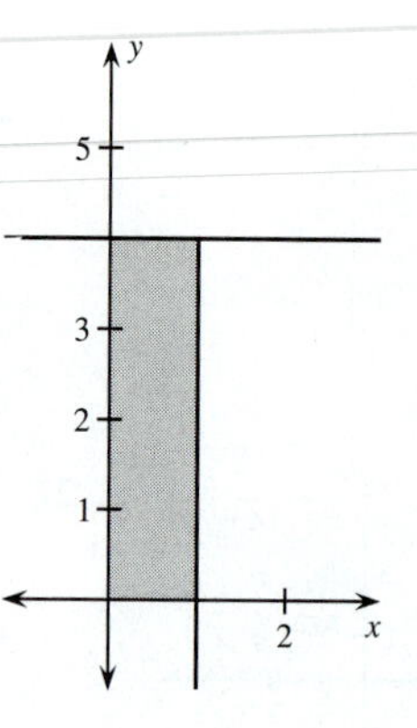

WARMING UP

1. T **2.** T **3.** T **4.** F **5.** F **6.** T

EXERCISE SET 7.5

1. (1, 1) is a solution. **3.** (5, 1) is not a solution. **5.** (4, −1) is a solution. **7.** (−1, 0) is a solution.

9.

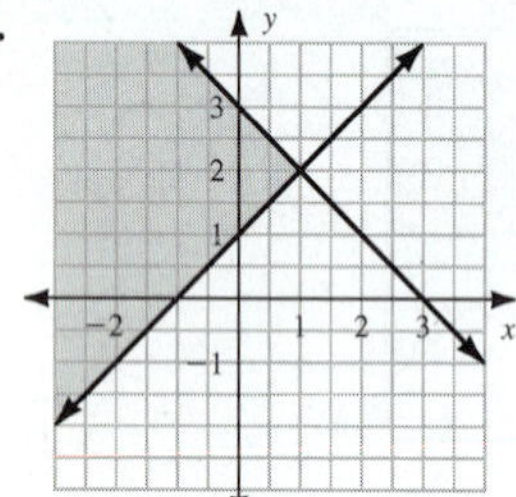

11.

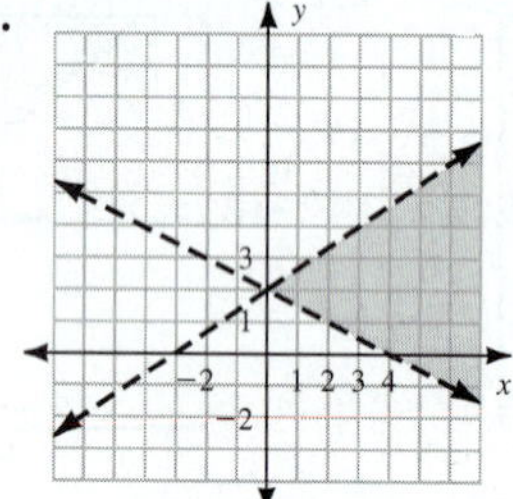

13.

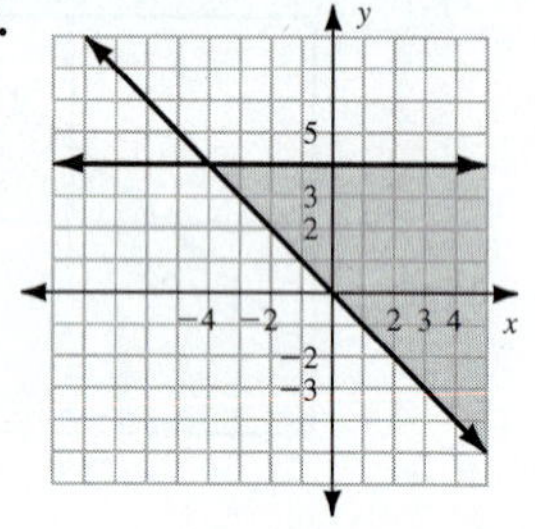

15.

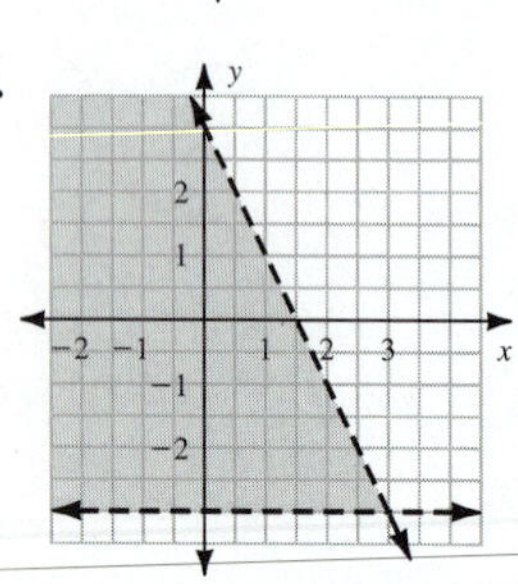

17.

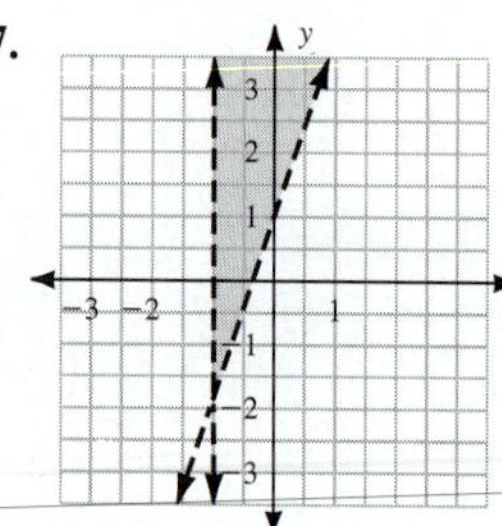

19.

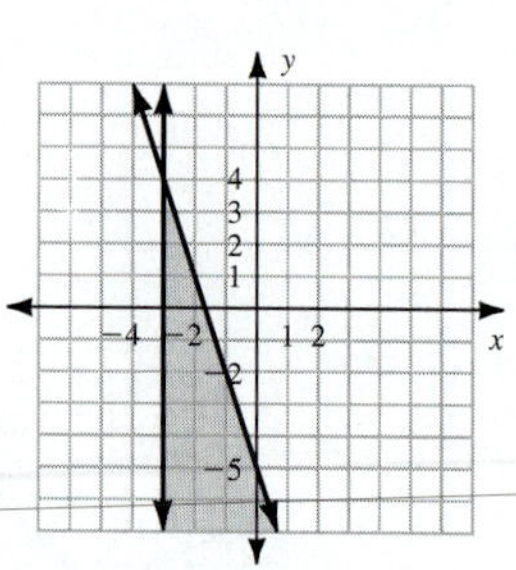

21.

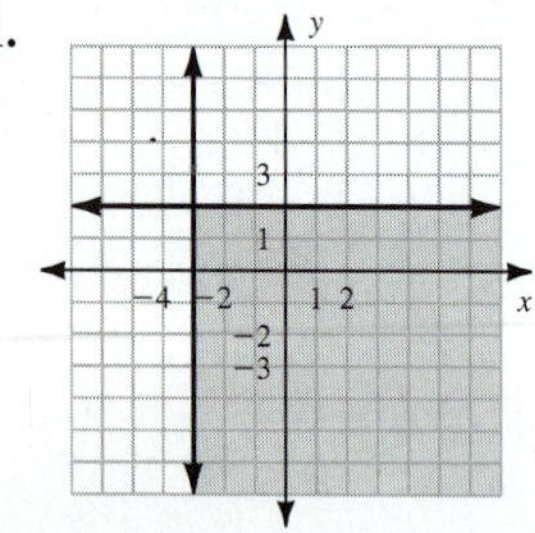

23.

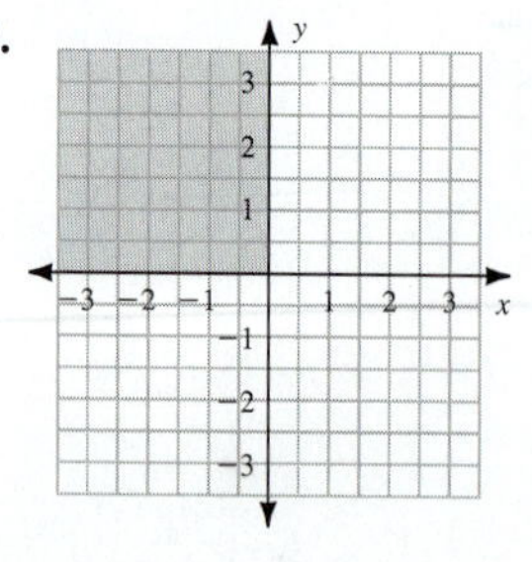

SAY IT IN WORDS

25. ans. vary

REVIEW EXERCISES

27. 8 **29.** 12 **31.** x^7 **33.** $27a^3$

ENRICHMENT EXERCISES

1.

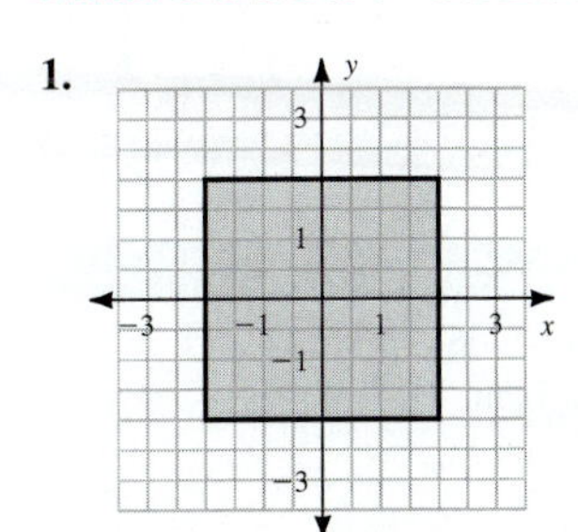

2.

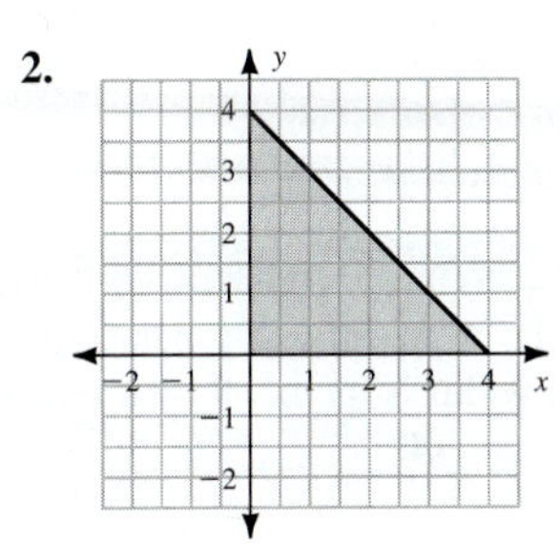

3.

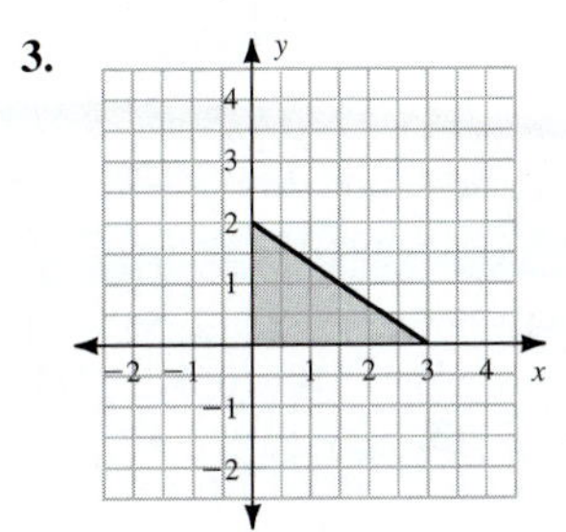

CHAPTER 7 REVIEW EXERCISE SET

Section 7.1

1. (1, −2) is a solution. **2.** (−3, −4) is not a solution. **3.** (2, 4) **4.** no solution **5.** infinitely many solutions **6.** (1, 2)

Sections 7.2 and 7.3

7. (5, −12) **8.** (1, −1) **9.** (−2, 4) **10.** (−1, 2) **11.** no solution **12.** infinitely many solutions **13.** (2, 0) **14.** (−4, 3) **15.** (3, 4) **16.** (−2, 0) **17.** (0, −1) **18.** (0, 0) **19.** 6 feet by 10 feet

Section 7.4

20.

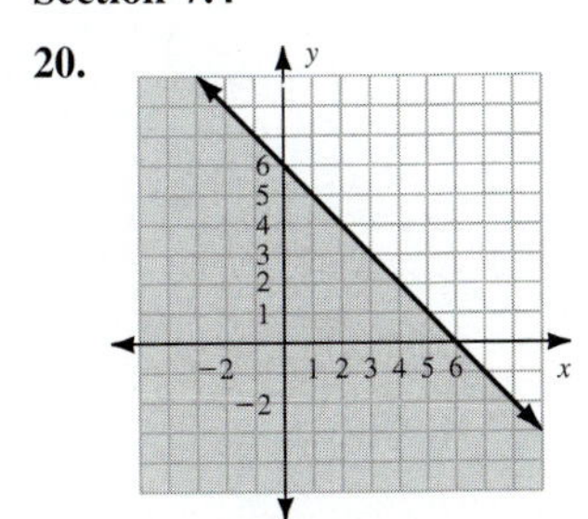

21.

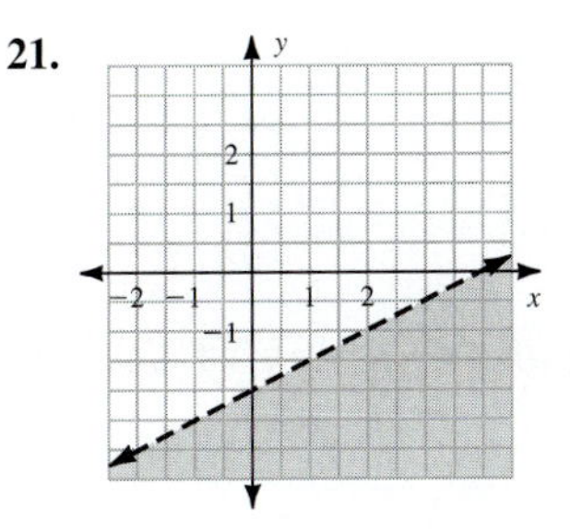

22.

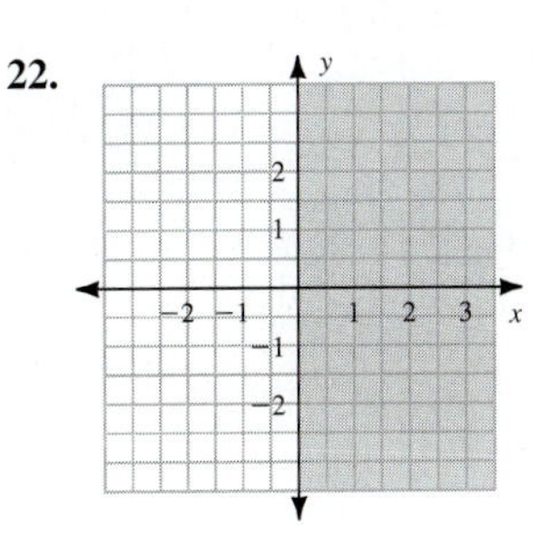

23.

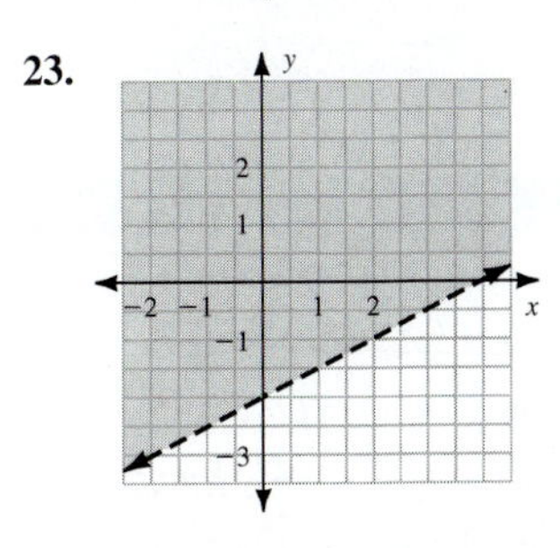

Section 7.5

24.
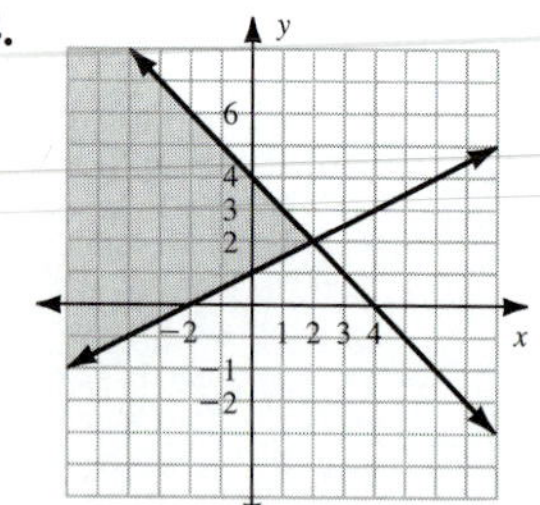

25.
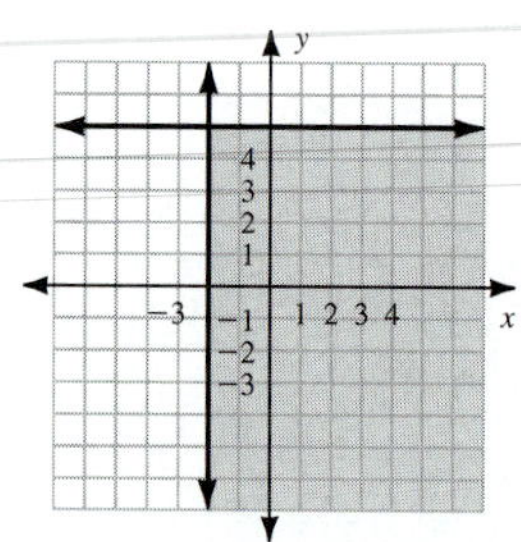

26.
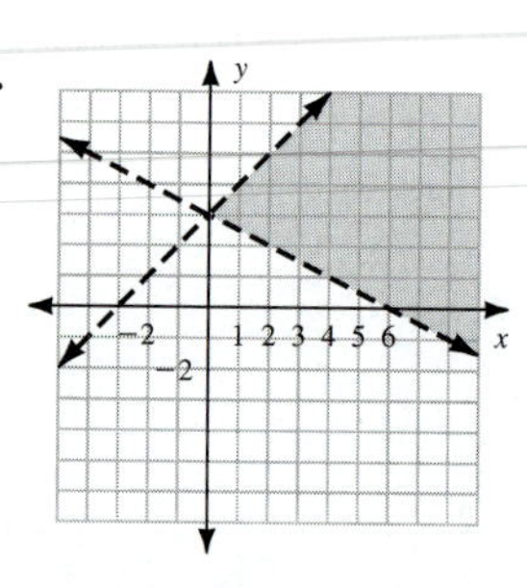

27.
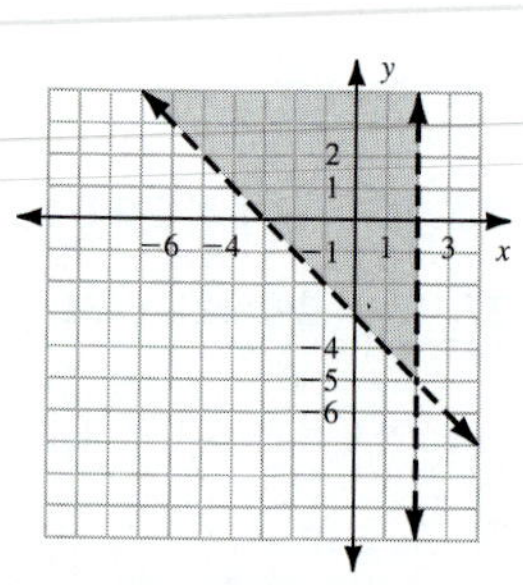

CHAPTER 7 TEST

1. (6, 1) is a solution. **2.** no solution **3.** (2, 5) **4.** (7, 1) **5.** infinitely many solutions **6.** $(-1, 2)$ **7.** $(1, -5)$ **8.** (6, 2) **9.** $(4, -3)$ **10.** (0, 1) **11.** The width is 4 cm, and the length is 5 cm.

12.
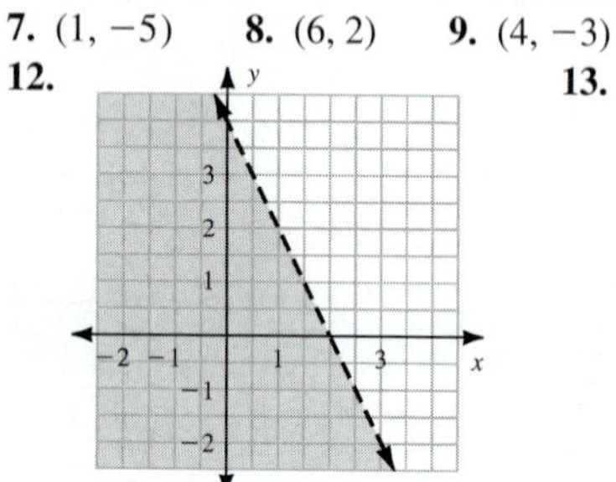

13.
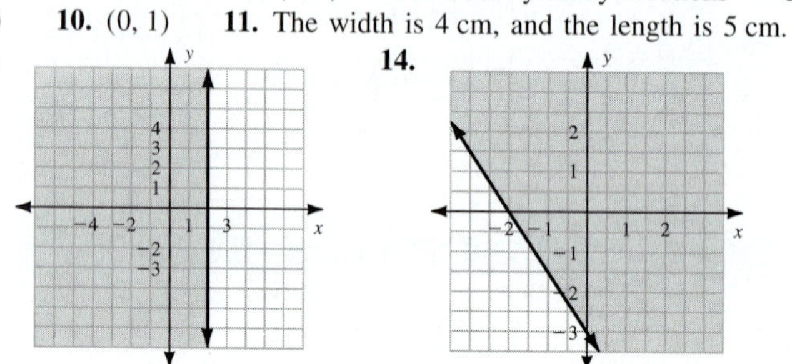

14.
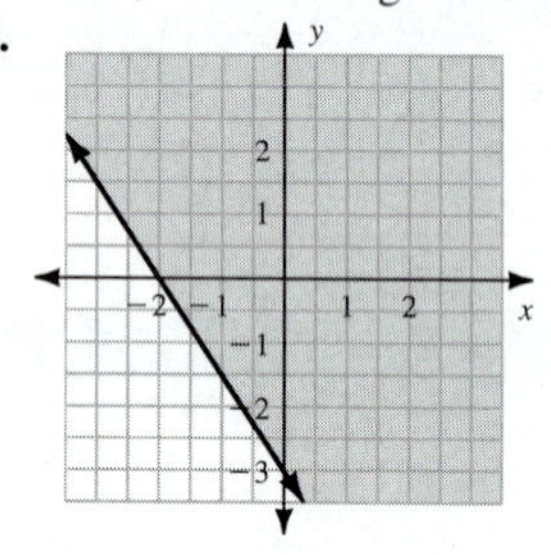

15.
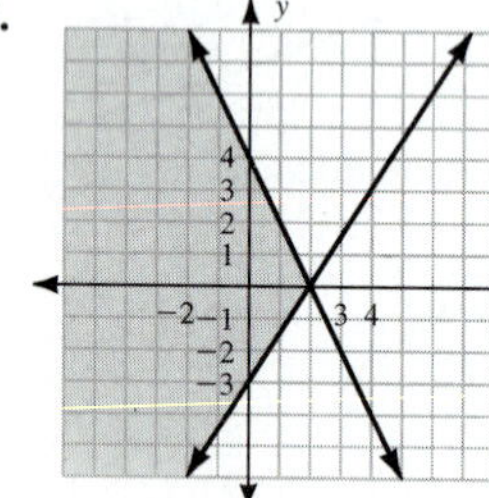

16.
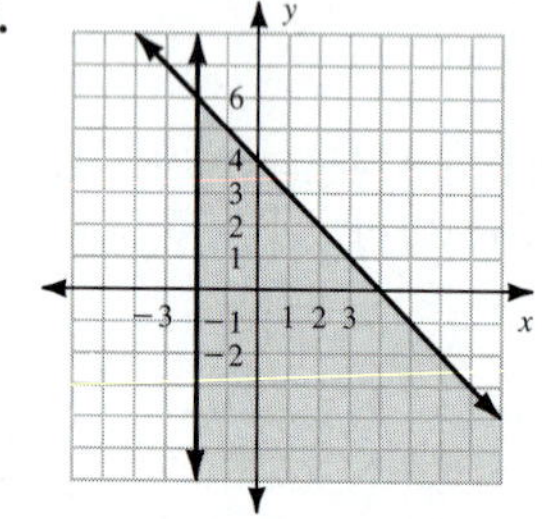

CHAPTER 8

WARMING UP

1. T **2.** T **3.** F **4.** T **5.** T **6.** F **7.** T **8.** T

EXERCISE SET 8.1

1. 7 and -7 **3.** 11 and -11 **5.** $\frac{5}{3}$ and $-\frac{5}{3}$ **7.** 0.9 and -0.9 **9.** 7 **11.** $-\frac{2}{5}$ **13.** 0.18 **15.** 3 **17.** not a real number **19.** -2 **21.** 2 **23.** -1 **25.** not a real number **27.** 2 **29.** -2 **31.** -3 **33.** not a real number **35.** -2 **37.** $4s^2t$ **39.** $3x^2y^3$ **41.** $2a^2b$ **43.** $-2a^3cz^5$ **45.** $-r^2t^5$ **47.** $8mn^2$

49. 15 cm **51.** $\frac{3}{2}$ yards **53.** $\frac{3\pi}{4}$ seconds, or approximately 2.355 seconds **55.** 80 ft/second **57.** 5.745
59. 3.593 **61.** −4.881

SAY IT IN WORDS

63. ans. vary

REVIEW EXERCISES

65. x^{10} **67.** $\frac{c^3}{d^3}$ **69.** v^8u^{12} **71.** $\frac{1}{x^3}$ **73.** $\frac{b^2}{a}$ **75.** $\frac{z^2}{x}$

ENRICHMENT EXERCISES

1. x^n **2.** a^r **3.** z^m **4.** x^n **5.** −1 **6.** 2

WARMING UP

1. T **2.** T **3.** F **4.** F **5.** T **6.** T **7.** T **8.** F

EXERCISE SET 8.2

1. $\sqrt{10}$ **3.** 2 **5.** 4 **7.** $\sqrt{3xy}$ **9.** $\sqrt[4]{6pq}$ **11.** $\frac{\sqrt{5}}{2}$ **13.** $\frac{\sqrt{2}}{x^2}$ **15.** $\frac{\sqrt[3]{11}}{4}$ **17.** $\frac{\sqrt[3]{x}}{2}$ **19.** $4\sqrt{2}$
21. $2\sqrt{3}$ **23.** $3\sqrt[3]{10}$ **25.** $2\sqrt[3]{6c}$ **27.** $-4\sqrt{2az}$ **29.** 64 **31.** x^3 **33.** $2s^3\sqrt[3]{2s}$ **35.** u^2v **37.** $2s^4\sqrt[4]{2s^2}$
39. $x\sqrt[3]{x}$ **41.** $b\sqrt{b}$ **43.** $7u^5\sqrt{u}$ **45.** $\frac{2\sqrt{3}}{3}$ **47.** $\frac{\sqrt{5x}}{5}$ **49.** $\frac{c^2\sqrt{3c}}{3}$ **51.** **(a)** 6 **(b)** 7 **(c)** 6 **(d)** 7

SAY IT IN WORDS

53. ans. vary

REVIEW EXERCISES

55. $8x^2$ **57.** $-2s^2 - 2s + 5$ **59.** $-\frac{13z}{x^2r}$ **61.** $\frac{1}{3 + 2x}$

ENRICHMENT EXERCISES

1. either $x \geq 3$ or $x \leq -1$ **2.** either $x \leq -\frac{9}{2}$ or $x \geq 2$ **3.** all real numbers **4.** $\sqrt{29}$ **5.** $2\sqrt{3}$ **6.** 3

WARMING UP

1. T **2.** F **3.** T **4.** T **5.** T **6.** F

EXERCISE SET 8.3

1. $-3\sqrt{3}$ **3.** $-4\sqrt{2}$ **5.** $13\sqrt{2}$ **7.** $\sqrt{14} + \sqrt{7}$ **9.** $-12\sqrt[3]{3}$ **11.** $8\sqrt[4]{2}$ **13.** $7\sqrt{c}$ **15.** $-3\sqrt[3]{ax}$
17. $-a^2\sqrt{a} - 2a^2$ **19.** $-\frac{4\sqrt{3}}{3}$ **21.** $\frac{11\sqrt{14}}{7}$ **23.** $\frac{\sqrt{2}}{6}$ **25.** $-\frac{5x^2\sqrt{x}}{36}$ **27.** $-2 + \sqrt{2}$ **29.** $\frac{3}{2}$

SAY IT IN WORDS

31. The difference of two radicals cannot be simplified by subtracting the radicands.

REVIEW EXERCISES

33. $6a^2bz - 8ab^2z$ **35.** $2x^2 + x - 3$ **37.** $4a^2 - 8ab + 4b^2$

ENRICHMENT EXERCISES

1. $\frac{\sqrt{7} - 1}{6}$ **2.** $2(\sqrt{3} + \sqrt{2})$ **3.** 1 **4.** -2 **5.** $-2\sqrt{2}$ **6.** $\frac{1 - 3\sqrt{2}}{2}$ **7.** $2 - 2\sqrt{3}$ **8.** $\frac{3}{2}$

WARMING UP

1. F **2.** T **3.** T **4.** F **5.** T **6.** T

EXERCISE SET 8.4

1. $12\sqrt{15}$ **3.** $6x$ **5.** $\sqrt{10} + \sqrt{22}$ **7.** $\sqrt{39} - 2\sqrt{65}$ **9.** $2 - 3\sqrt{2}$ **11.** $3\sqrt{5} - \sqrt{3}$ **13.** $6 + 6\sqrt{2}$
15. $28 - \sqrt{5}$ **17.** $2r + 5\sqrt{r} + 3$ **19.** $5 - 2\sqrt{6}$ **21.** $x + 4\sqrt{xz} + 4z$ **23.** -2 **25.** $x - 1$
27. $-\frac{2(\sqrt{3} + 4)}{13}$ **29.** $-\frac{4 - \sqrt{6}}{5}$ **31.** $-\frac{9 - 10\sqrt{2}}{7}$ **33.** $5 - \sqrt{10}$ **35.** $x + 6\sqrt{x} + 9$ **37.** $6\sqrt{5}$

SAY IT IN WORDS

39. ans. vary

REVIEW PROBLEMS

41. $x = 0$ or $x = 3$ **43.** $y = \frac{1}{3}$ or $y = 1$ **45.** $a = 1$

ENRICHMENT EXERCISES

1. $x - x^2$ **2.** $2\sqrt[3]{2} - 1$ **3.** -26 **4.** $3\sqrt[3]{3x^2} - 1$

WARMING UP

1. F **2.** T **3.** F **4.** T **5.** T **6.** F

EXERCISE SET 8.5

1. $\{16\}$ **3.** no solution **5.** $\{10\}$ **7.** $\{2\}$ **9.** $\{9\}$ **11.** $\{37\}$ **13.** $\left\{-\frac{1}{4}\right\}$ **15.** $\left\{\frac{6}{5}\right\}$ **17.** $\left\{-\frac{1}{8}\right\}$ **19.** $\{1\}$
21. $\{3\}$ **23.** no solution **25.** $\left\{\frac{1}{2}\right\}$ **27.** $\left\{\frac{1}{3}, \frac{2}{3}\right\}$ **29.** $\{0\}$ **31.** $\{2, 3\}$

SAY IT IN WORDS

33. ans. vary

REVIEW EXERCISES

35. x **37.** x^4y^4 **39.** $\frac{1}{a^5}$ **41.** $\frac{d}{c^3}$

ENRICHMENT EXERCISES

1. $\{3\}$ **2.** $\left\{\frac{16}{9}, 0\right\}$ **3.** $\{13\}$ **4.** $\{3\}$

WARMING UP

1. F **2.** T **3.** F **4.** T **5.** T **6.** T

EXERCISE SET 8.6

1. 10 **3.** -1 **5.** not a real number **7.** 64 **9.** 4 **11.** $\frac{1}{9}$ **13.** -125 **15.** 2 **17.** $4^{5/6}$ **19.** $\frac{1}{t}$
21. 1 **23.** $a^{1/2}$ **25.** $s^{2/5}$ **27.** $x^{4/3}$ **29.** a^2 **31.** w **33.** $a^{2/3}$ **35.** $\frac{1}{z^{1/4}}$ **37.** $\frac{1}{x^{9/2}}$ **39.** $2t^4r^2$
41. $9st^{5/2}$ **43.** $\frac{1}{a^4b^2}$ **45.** $x^{5/3}$ **47.** $z^{5/6}$ **49.** 9.4429 **51.** 66.5278

SAY IT IN WORDS

53. ans. vary

REVIEW EXERCISES

55. $\sqrt{6} - \sqrt{15}$ **57.** $a - 6\sqrt{a} + 9$

ENRICHMENT EXERCISES

1. $\frac{1}{x^6y^{9/2}}$ **2.** $a^{18}b^{16}$ **3.** $\frac{1}{x^{1/12}}$ **4.** $x^{11/15}$

WARMING UP

1. F **2.** T **3.** T **4.** T **5.** F **6.** T

EXERCISE SET 8.7

1. $3i$ **3.** $-i\sqrt{17}$ **5.** $2i\sqrt{7}$ **7.** $-5i\sqrt{2}$ **9.** $4 - 10i$ **11.** $2 + 2i\sqrt{2}$ **13.** $1 - 2\sqrt{5}$ **15.** $10 + i$
17. -10 **19.** $4 - 8i$ **21.** $12 - i$ **23.** $5 - 12i$ **25.** 17 **27.** $1 - i$ **29.** $-3 - i$ **31.** $-\frac{1}{3} - \frac{2}{3}i$
33. $-\frac{2}{5} + \frac{1}{5}i$

SAY IT IN WORDS

35. $a + bi$, $i = \sqrt{-1}$

REVIEW EXERCISES

37. $\left\{\frac{1}{5}, -\frac{7}{3}\right\}$ **39.** $\{\pm 7\}$

ENRICHMENT EXERCISES

1. $-2 - 2i$ **2.** $-2 + 2i$

CHAPTER 8 REVIEW EXERCISE SET

Section 8.1

1. 8 **2.** $-\frac{3}{4}$ **3.** -3 **4.** not a real number **5.** $3x^4y^2$ **6.** $-2a^2b^3c^4$

Section 8.2

7. $2\sqrt{6}$ **8.** $-2\sqrt[3]{2}$ **9.** z^2t^3 **10.** $-s^3r\sqrt{3sr}$

Sections 8.2 and 8.4

11. $\frac{\sqrt{6}}{3}$ **12.** $\frac{3\sqrt{x}}{x}$ **13.** $2\sqrt{2s}$ **14.** $3\sqrt[3]{k}$ **15.** $\frac{4\sqrt[4]{v^3}}{3}$ **16.** $-3(2 + \sqrt{5})$ **17.** $-2(1 - \sqrt{3})$
18. $\frac{(\sqrt{x} - 1)^2}{x - 1}$ **19.** $\frac{s\sqrt{t} + 2t\sqrt{s}}{2st}$ **20.** $\sqrt{c} + \sqrt{3}$

Section 8.3

21. $7\sqrt{3}$ **22.** $11\sqrt[4]{2}$ **23.** $-3a^2\sqrt{a}$ **24.** $-14\sqrt[3]{c^2}$ **25.** $-3uv\sqrt[3]{uv^2}$ **26.** $4x^2\sqrt{x}$

Section 8.4

27. $3 - 6\sqrt{2}$ **28.** $2s - 3s^2$ **29.** $x^2 - 2x^2\sqrt[3]{x^2}$ **30.** $a - 4\sqrt{ab} + 4b$ **31.** $3y\sqrt[3]{3y} - 4$

Section 8.5

32. $\{64\}$ **33.** $\{12\}$ **34.** $\{1\}$ **35.** $\left\{\frac{9}{2}\right\}$ **36.** $\left\{-\frac{3}{7}\right\}$ **37.** $\{1\}$ **38.** $\{0, 200\}$ **39.** $\{8\}$ **40.** $\left\{-\frac{20}{3}\right\}$

Section 8.6

41. $3^{3/2}$ **42.** $\frac{1}{z^{1/2}}$ **43.** $\frac{1}{x}$ **44.** s^8t^2 **45.** $w^{1/3}$ **46.** $\frac{1}{a^7}$

Section 8.7

47. $10i$ **48.** $2 - 3i$ **49.** $4 + 2i\sqrt{10}$ **50.** $1 - 2\sqrt{5}$ **51.** $3 + \frac{2}{3}i$ **52.** $10 - 6i$ **53.** $8 + 13i$ **54.** $2 + 2i$
55. $-7 - 17i$ **56.** $4 - i$ **57.** $\frac{9}{2} - 2i$ **58.** $\frac{6}{5} - \frac{8}{5}i$ **59.** $32 - 24i$

CHAPTER 8 TEST

1. 12 **2.** $-\frac{2}{3}$ **3.** $5x^2yz^5$ **4.** $-4\sqrt{5}$ **5.** $\frac{2\sqrt{7}}{7}$ **6.** $3\sqrt{5} + 6$ **7.** $\frac{2r + \sqrt{r} - 15}{4r - 25}$ **8.** $7\sqrt{3} + 2\sqrt{5}$
9. $-3a^2\sqrt{a} + a^2$ **10.** $\frac{3\sqrt{5} - 15}{5}$ **11.** $\frac{\sqrt{10}}{2}$ **12.** $-8a^2$ **13.** $4x - 12\sqrt{xy} + 9y$ **14.** $-21 - 11\sqrt{6}$

15. $w = 52$ **16.** $z = 3$ **17.** no solution **18.** x **19.** $\frac{1}{u^5}$ **20.** $3 + 2i\sqrt{6}$ **21.** $3 + 2i$ **22.** 61
23. $-\frac{3}{5} + \frac{4}{5}i$

CHAPTER 9

WARMING UP

1. F **2.** T **3.** T **4.** F **5.** T **6.** F

EXERCISE SET 9.1

1. $\{\pm 3\}$ **3.** $\{\pm 11\}$ **5.** $\left\{\pm\frac{2}{3}\right\}$ **7.** $\{\pm 3\sqrt{2}\}$ **9.** $\{\pm 5\sqrt{5}\}$ **11.** $\{2, 10\}$ **13.** $\left\{\pm\frac{4\sqrt{2}}{3}\right\}$ **15.** $\{-2 \pm 2\sqrt{7}\}$
17. $\{\pm\sqrt{6}\}$ **19.** $\{\pm 2\sqrt{5}\}$ **21.** $\{-3, 6\}$ **23.** $\{4 \pm 2\sqrt{2}\}$ **25.** $\{4 \pm 2\sqrt{14}\}$ **27.** $3\sqrt{5}$ **29.** $\frac{3 \pm 3\sqrt{2}}{2}$
31. $s = \frac{\sqrt{10}}{10}c$ **33.** $s = \frac{1}{5}c$ **35.** **(a)** $r = \sqrt{\frac{A}{\pi}}$ **(b)** $2\sqrt{5}$ inches **37.** **(a)** $r = \sqrt{\frac{3V}{\pi h}}$ **(b)** $6\sqrt{3}$ meters ≈ 10.392 meters **39.** $90\sqrt{2}$ feet **41.** $\{1.16, 5.79\}$ **43.** $r = 2\left(\sqrt{\frac{A}{P}} - 1\right)$

SAY IT IN WORDS

45. ans. vary

REVIEW EXERCISES

47. $x^2 + 6x + 9$ **49.** $x^2 - 2\sqrt{2}x + 2$

ENRICHMENT EXERCISES

1. $\left\{4, \frac{2}{3}\right\}$ **2.** $\left\{4, \frac{1}{3}\right\}$ **3.** all real numbers **4.** $\left\{-\frac{5}{2}\right\}$ **5.** length = 43.58 in., width = 24.51 in.

WARMING UP

1. T **2.** F **3.** F **4.** T **5.** T **6.** T

EXERCISE SET 9.2

1. 16 **3.** $\frac{9}{4}$ **5.** $\{-11, 3\}$ **7.** $\{1, 11\}$ **9.** $\{-8 \pm \sqrt{55}\}$ **11.** $\{-4, 18\}$ **13.** $\left\{\frac{-5 \pm \sqrt{3}}{2}\right\}$ **15.** $\left\{-\frac{3}{4}, -\frac{1}{4}\right\}$
17. $\left\{-\frac{2}{5}, -\frac{1}{5}\right\}$ **19.** $\{14, -8\}$ **21.** $\left\{-1, \frac{7}{2}\right\}$ **23.** $\{-1 \pm \sqrt{7}\}$

SAY IT IN WORDS

25. ans. vary

REVIEW EXERCISES

27. $\dfrac{2+\sqrt{2}}{5}$ **29.** $\dfrac{6-2\sqrt{3}}{3}$ **31.** $-2-\sqrt{3}$

ENRICHMENT EXERCISES

1. $\{\pm\sqrt{7}, \pm 1\}$ **2.** $\{\pm 1\}$ **3.** no real solution

WARMING UP

1. T **2.** F **3.** F **4.** T **5.** T **6.** T

EXERCISE SET 9.3

1. $a = 2$, $b = -3$, $c = 4$ **3.** $4x^2 - x + 2 = 0$, $a = 4$, $b = -1$, $c = 2$ **5.** $2x^2 + 2x - 3 = 0$, $a = 2$, $b = 2$, $c = -3$ **7.** $7x^2 + 2x - 1 = 0$, $a = 7$, $b = 2$, $c = -1$ **9.** $x^2 + x - 4 = 0$, $a = 1$, $b = 1$, $c = -4$ **11.** $\{-9, 3\}$ **13.** $\{1, 2\}$ **15.** $\left\{-4, \dfrac{1}{2}\right\}$ **17.** no real solution **19.** $\left\{\dfrac{-1\pm\sqrt{13}}{6}\right\}$ **21.** $\{-1\pm\sqrt{2}\}$ **23.** $\{2\pm\sqrt{6}\}$ **25.** $\{-3\}$ **27.** no real solution **29.** $\left\{\dfrac{1\pm 2\sqrt{2}}{2}\right\}$ **31.** $\left\{\dfrac{1\pm\sqrt{6}}{10}\right\}$ **33.** $\left\{0, -\dfrac{7}{4}\right\}$ **35.** $\left\{\dfrac{1\pm\sqrt{5}}{2}\right\}$ **37.** b is -3, so $-b$ is 3; $\left\{\dfrac{3\pm\sqrt{29}}{2}\right\}$ **39.** $-1+\sqrt{51}$ inches by $1+\sqrt{51}$ inches or approximately 6.14 inches by 8.14 inches **41.** 2.5 seconds **43.** 0.6 or -3.2

SAY IT IN WORDS

45. 1.

$$\begin{aligned} x(2x+3) &= 2 \\ 2x^2+3x-2 &= 0 \\ (2x-1)(x+2) &= 0 \end{aligned}$$

The solution set is $\left\{\dfrac{1}{2}, -2\right\}$.

2.

$$\begin{aligned} x(2x+3) &= 2 \\ 2x^2+3x &= 2 \\ x^2+\frac{3}{2}x+\frac{9}{16} &= 1+\frac{9}{16} \\ \left(x+\frac{3}{4}\right)^2 &= \frac{25}{16} \\ x+\frac{3}{4} &= \pm\frac{5}{4} \\ x &= -\frac{3}{4}\pm\frac{5}{4} \end{aligned}$$

The solution set is $\left\{\dfrac{1}{2}, -2\right\}$.

3.

$$\begin{aligned} x(2x+3) &= 2 \\ 2x^2+3x-2 &= 0 \\ x &= \frac{-3\pm\sqrt{9-4(2)(-2)}}{4} \\ &= \frac{-3\pm\sqrt{9+16}}{4} \\ &= \frac{-3\pm\sqrt{25}}{4} \\ &= \frac{-3\pm 5}{4} \end{aligned}$$

The solution set is $\left\{\dfrac{1}{2}, 2\right\}$.

REVIEW EXERCISES

47. $11i$ **49.** $-3-i\sqrt{6}$ **51.** $14+3i\sqrt{2}$

ENRICHMENT EXERCISES

1. $\{\pm 1, \pm 3\}$ **2.** $\left\{\dfrac{2\pm\sqrt{10}}{3}\right\}$ **3.** $\{\sqrt{3}, -3\sqrt{3}\}$ **4.** $\{2\sqrt{2}\pm\sqrt{10}\}$

MISCELLANEOUS QUADRATIC EQUATIONS

1. $\{2 \pm \sqrt{5}\}$ **3.** $\{0, 6\}$ **5.** no real solution **7.** $\left\{-\frac{3}{2}, \frac{1}{3}\right\}$ **9.** $\left\{\frac{1 \pm \sqrt{6}}{2}\right\}$ **11.** $\{\pm 4\}$ **13.** $\left\{\frac{5}{2}\right\}$ **15.** $\{0, 4\}$ **17.** $\{-5, 4\}$ **19.** $\left\{\frac{-11 \pm \sqrt{109}}{2}\right\}$ **21.** $\left\{-\frac{1}{2}, -\frac{3}{5}\right\}$ **23.** $\left\{-\frac{7}{2}, \frac{2}{3}\right\}$ **25.** $\{0, 9\}$ **27.** $\left\{-1 \pm \sqrt{5}\right\}$ **29.** $c = \frac{-4 \pm \sqrt{19}}{3}$ **31.** $s = 1$ or $s = \frac{5}{2}$ **33.** $\{0, 2\}$ **35.** $\{-1 \pm \sqrt{3}\}$

WARMING UP

1. F **2.** T **3.** T **4.** F **5.** T

EXERCISE SET 9.4

1. $\{\pm 7i\}$ **3.** $\{-1 \pm 2i\}$ **5.** $\left\{\frac{5 \pm 8\sqrt{2}}{3}\right\}$ **7.** $\left\{\frac{3 \pm \sqrt{13}}{2}\right\}$ **9.** $\left\{1, \frac{3}{2}\right\}$ **11.** $\left\{-\frac{3}{2} \pm \frac{\sqrt{7}}{2}i\right\}$ **13.** $\{2 \pm i\}$ **15.** $\{2 \pm \sqrt{3}\}$ **17.** $\{1 \pm \sqrt{22}\}$ **19.** $\left\{\frac{9 \pm \sqrt{77}}{2}\right\}$ **21.** 2 real solutions **23.** 1 real solution **25.** 2 complex solutions **27.** 2 real solutions **29.** 2 complex solutions

SAY IT IN WORDS

31. ans. vary

ENRICHMENT EXERCISES

1. $\left\{\frac{1}{4} \pm \frac{\sqrt{3}}{4}i\right\}$ **2.** $\left\{\frac{3 \pm 2\sqrt{3}}{6}\right\}$ **3.** $\left\{\frac{1}{4} \pm \frac{3\sqrt{2}}{8}i\right\}$ **4.** $\left\{\frac{1}{2} \pm \frac{5\sqrt{3}}{6}i\right\}$

CHAPTER 9 REVIEW EXERCISE SET

Section 9.1

1. $\{\pm 3\sqrt{3}\}$ **2.** $\{2, 8\}$ **3.** $\{3 \pm 3\sqrt{2}\}$ **4.** $\left\{\pm\frac{\sqrt{35}}{5}\right\}$ **5.** $\left\{\frac{6 \pm 4\sqrt{2}}{9}\right\}$ **6. (a)** $s = \frac{\sqrt{5}}{5}c$ **(b)** $3\sqrt{5}$ cm

Section 9.2

7. 64 **8.** $z = -1$ or $z = 11$ **9.** $x = -13$ or $x = 1$ **10.** $c = \frac{7}{2}$ or $c = \frac{1}{2}$ **11.** $t = \frac{4 \pm \sqrt{7}}{3}$ **12.** $k = -5$

Section 9.3

13. $a = 2, b = -7, c = -5$ **14.** $a = 1, b = -1, c = 12$ **15.** $a = 1, b = 4, c = -7$ **16.** $a = 21, b = -19, c = -1$ **17.** $\{-4, 7\}$ **18.** $\left\{-3, \frac{1}{3}\right\}$ **19.** $\left\{\frac{-1 \pm \sqrt{37}}{3}\right\}$ **20.** no real solution **21.** $\left\{-\frac{5}{2}, 3\right\}$ **22.** $\left\{\frac{1 \pm \sqrt{13}}{2}\right\}$ **23.** $\{\pm 5\}$ **24.** $\{-1 \pm \sqrt{5}\}$ **25.** $\left\{\frac{2 \pm 3\sqrt{2}}{3}\right\}$ **26.** $\left\{\frac{-3 \pm 2\sqrt{5}}{2}\right\}$

Section 9.4

27. height $= -2.5 + 2.5\sqrt{17} \approx 7.8$ in; base $= 2.5 + 2.5\sqrt{17} \approx 12.8$ in **28.** Yes, the car was traveling 40 mph.
29. $\left\{\frac{3}{2} \pm i\right\}$ **30.** $\{\pm 2i\sqrt{5}\}$ **31.** $\left\{\frac{1 \pm i\sqrt{11}}{3}\right\}$ **32.** $\left\{-\frac{1}{2} \pm 2i\sqrt{3}\right\}$

CHAPTER 9 TEST

1. $\left\{\pm\frac{2\sqrt{10}}{9}\right\}$ **2.** $\left\{\frac{1 \pm 2\sqrt{3}}{3}\right\}$ **3.** $\{3 \pm \sqrt{6}\}$ **4.** $\{-6, 2\}$ **5.** no real solution **6.** $\{-5 \pm \sqrt{19}\}$
7. $\left\{\frac{4 \pm \sqrt{10}}{2}\right\}$ **8.** $\left\{\frac{3}{2}, \frac{5}{2}\right\}$ **9.** $\{\pm 2\sqrt{5}\}$ **10.** $\left\{\frac{3 \pm \sqrt{13}}{2}\right\}$ **11.** $\{\pm 9i\}$ **12.** $\left\{\frac{5}{2} \pm \frac{11}{2}i\right\}$
13. $\left\{\frac{-9 \pm \sqrt{129}}{12}\right\}$ **14.** width $= -6 + \sqrt{66} \approx 2.1$ cm, length $= 6 + \sqrt{66} \approx 14.1$ cm. **15.** $\sqrt{10}$ or 3.162 seconds

APPENDIX A EXERCISE SET

1. $\frac{7}{10}$ **3.** $\frac{12}{5}$ **5.** $\frac{49}{20}$ **7.** 0.4 **9.** 3.5 **11.** 0.375 **13.** 4.93 **15.** 10.201 **17.** 4.9 **19.** 7.7
21. 9.724 **23.** 3.21 **25.** 5.32 **27.** 1.27 **29.** 3.84 **31.** 10.92 **33.** 0.3162 **35.** 1.371 **37.** 1.47
39. 3.07 **41.** 0.59 **43.** 174.6 **45.** 0.36 **47.** 0.57 **49.** 1 **51.** 0.038 **53.** 61% **55.** 2% **57.** 92.4%
59. 280% **61.** 12 **63.** 120 **65.** 32 **67.** 1.6 **69.** 4 **71.** 20% **73.** 25% **75.** $50 **77.** $4.86

ILLUSTRATION CREDITS

Chapter 1
Chapter opening photo p. 1: © Robert Gemin; unnumbered photo on p. 14: © Melissa Handy; unnumbered photo on p. 41: © Garry Gay/The Image Bank; unnumbered photo on p. 42: © Walter Iooss, Jr./The Image Bank.

Chapter 2
Chapter opening photo p. 76: © Robert Gemin; unnumbered photos pp. 113 and 123: © Melissa Handy.

Chapter 3
Chapter opening photo p. 138: © Melissa Handy.

Chapter 4
Chapter opening photo p. 206: © Melissa Handy.

Chapter 5
Chapter opening photo p. 258 and unnumbered photos pp. 295 and 297: © Melissa Handy.

Chapter 6
Chapter opening photo p. 302: © Melissa Handy.

Chapter 7
Chapter opening photo p. 362 and unnumbered photo p. 383: © Melissa Handy.

Chapter 8
Chapter opening photo p. 405: © Jerry D. Wilson.

Chapter 9
Chapter opening photo p. 459: © Robert Gemin.

Index